DICTIONARY OF
DISTANCES

DICTIONARY OF DISTANCES

ELENA DEZA
Moscow Pedagogical State University and Donghua University

MICHEL-MARIE DEZA
Ecole Normale Superieure, Paris, and Shanghai University

ELSEVIER

Amsterdam • Boston • Heidelberg • London • New York • Oxford
Paris • San Diego • San Francisco • Singapore • Sydney • Tokyo

ELSEVIER
Radarweg 29, PO Box 211, 1000 AE Amsterdam, The Netherlands
The Boulevard, Langford Lane, Kidlington, Oxford OX5 1GB, UK

First edition 2006

Library of Congress Cataloging-in-Publication Data
A catalog record for this book is available from the Library of Congress

British Library Cataloguing in Publica tion Data
A catalogue record for this book is available from the British Library

ISBN-13: 978-0-444-52087-6
ISBN-10: 0-444-52087-2

Printed and bound in The Netherlands

06 07 08 09 10   10 9 8 7 6 5 4 3 2 1

On 2 April 2006 it was 100 years from the submission by Maurice FRÉCHET of his outstanding doctoral dissertation *Sur quelques points du calcul functionnel* that introduced (within a systematic study of functional operations) the notion of metric space.

It was also 92 years from 1914, the date of publication by Felix HAUSDORFF of his famous *Grundzüge der Mengenlehre* where the theory of topological and metric spaces was created.

Let this Dictionary be our homage to the memory of these great mathematicians and their lives of dignity through the hard ways of the first half of the XX century.

Maurice FRÉCHET (1878–1973) coined in 1906 the concept of *écart* (semi-metric)

Felix HAUSDORFF (1868–1942) coined in 1914 the term *metric space*

Preface

The concept of *distance* is one of the basic ones in the whole of human experience. In everyday life it usually means some degree of closeness between two physical objects or ideas, i.e., length, time interval, gap, rank difference, coolness or remoteness, while the term *metric* is often used as a standard for a measurement. But here we consider, except in the last two chapters, the mathematical meaning of these terms. The mathematical notions of *distance metric* (i.e., a function $d(x, y)$ from $X \times X$ to the set of real numbers satisfying $d(x, y) \geqslant 0$ with equality only for $x = y$, $d(x, y) = d(y, x)$, and $d(x, y) \leqslant d(x, z) + d(z, y)$) and of *metric space* (X, d) were originated a century ago by M. Fréchet (1906) and F. Hausdorff (1914) as a special case of an infinite topological space. The *triangle inequality* above appears already in Euclid. The infinite metric spaces are seen usually as a generalization of the metric $|x - y|$ on the real numbers. Their main classes are the measurable spaces (add measure) and Banach spaces (add norm and completeness).

However, starting from K. Menger (1928) and, especially, L.M. Blumenthal (1953), an explosion of interest in finite metric spaces occurred. Another trend: many mathematical theories, in the process of their generalization, settled on the level of metric space.

Now finite distance metrics have become an essential tool in many areas of Mathematics and its applications include Geometry, Probability, Statistics, Coding/Graph Theory, Clustering, Data Analysis, Pattern Recognition, Networks, Engineering, Computer Graphics/Vision, Astronomy, Cosmology, Molecular Biology, and many other areas of science. Devising the most suitable distance metrics has become a standard task for many researchers. Especially intense ongoing searches for such distances occur, for example, in Genetics, Image Analysis, Speech Recognition, Information Retrieval. Often the same distance metric appears independently in several different areas; for example, the edit distance between words, the evolutionary distance in Biology, the Levenstein distance in Coding Theory, and the Hamming+Gap or shuffle-Hamming distance.

This body of knowledge has become too large and disparate to operate within. The number of worldwide web entries offered by Google on the topics "distance", "metric space" and "distance metric" approach 300 million (i.e., about 4% of all), 12 million and 6 million, respectively, not to mention all the printed information outside the Web, or the vast "invisible Web" of searchable databases. However, this huge amount of information on distances is too scattered: the works evaluating distance from some list usually treat very specific areas and are hardly accessible for non-experts.

Therefore, many researchers, including us, keep and cherish a collection of distances for use in their own areas of science. In view of the growing general need for an accessible interdisciplinary source for a vast multitude of researchers, we have expanded our private collection into this Dictionary. Some additional material was reworked from various encyclopedia, especially, Encyclopedia of Mathematics ([EM98]), MathWorld ([Weis99]), PlanetMath ([PM]), and Wikipedia ([WFE]). However, the majority of distances should be extracted directly from specialist literature.

The vast reservoir of concepts defined in this Dictionary, aims to be a thought-provoking archive and a valuable resource. Besides distances themselves, we have collected here many distance-related notions (especially in Chapter 1) and paradigms, enabling people from applications to get those, arcane for non-specialists, research tools, in ready-to-use fashion. This and the appearance of some distances in different contexts can be a source of new research.

At a time when over-specialization and terminology barriers isolate researchers, this Dictionary tries to be "centripetal" and "ecumenical", providing some access and altitude of vision but without taking the route of scientific vulgarization. This attempted balance defined the structure and style of the Dictionary.

The Dictionary is divided into 28 chapters grouped into 7 Parts of about the same length. The titles of parts are purposely approximative: they just allow a reader to figure out her/his area of interest and competence. For example, Parts II, III and IV, V require some culture in, respectively, pure and applied Mathematics. Part VII can be read by a layman.

The chapters are thematic lists, by areas of Mathematics or applications which can be read independently. When necessary, a chapter or a section starts with a short introduction: a field trip with the main concepts. Besides those introductions, the main properties and uses of distances are given, within items, only exceptionally. We also tried, when it was easy, to trace distances to their originator(s), but the proposed extensive bibliography has a less general ambition: it is just to provide convenient sources for a quick search.

Each chapter consists of items ordered in a way that hints of connections between them. All item titles and selected key terms can be traced in the large Subject Index (about 1400 entries); they are boldfaced unless the meaning is clear from the context. So, the definitions are easy to locate, by subject, in chapters and/or, by alphabetic order, in the index. The introductions and definitions are reader-friendly and as far as possible independent of each other; still they are interconnected, in the 3-dimensional HTML manner, by hyperlink-like boldfaced references to similar definitions.

Many nice curiosities appear in this "Who is Who" of distances. Examples of such sundry terms are: ubiquitous Euclidean distance ("as-the-crow-flies"), flower-shop metric (shortest way between two points, visiting a "flower-shop" point first), knight-move metric on a chessboard, Gordian distance of knots, Earth Mover distance, biotope distance, Procrustes distance, lift metric, post-office metric, Internet hop metric, WWW hyperlink quasi-metric, Moscow metric, dogkeeper distance. Besides abstract distances, the distances having physical meaning appear also (especially in Part VI); they range from 1.6×10^{-35} m (Planck length) to 7.4×10^{26} m (the estimated size of observable Universe, about 46×10^{60} Planck lengths).

The number of distance metrics is infinite and therefore, our Dictionary cannot enumerate all of them. But we were inspired by several successful thematic dictionaries on other infinite lists; for example, on Integer Sequences, Inequalities, Numbers, Random Processes, and by atlases of Functions, Groups, Fullerenes, etc. On the other hand, the largeness of the scope forced us often to switch into a laconic tutorial style.

The target audience consists of all researchers working on some measuring schemes and, to a certain degree, of students and a part of the general public interested in science.

We have tried to address, even if incompletely, all scientific uses of the notion of distance. However some distances did not made it into this Dictionary due to space limitations (being too specific and/or complex) or our oversight. In general, the size/interdisciplinarity cut-off, i.e., decision where to stop, was our main headache. We would be grateful to the readers who send us their favorite distances missed here. Four pages at the end are reserved for such personal additions.

We are grateful to many people for their help with this book, especially, to Jacques Beigbeder, Mathieu Dutour, Emmanuel Guerre, Jack Koolen, Jin Ho Kwak, Hiroshi Maehara, Sergey Shpectorov, Alexei Sossinsky, and Jiancang Zhuang.

Contents

Preface vii

PART I. MATHEMATICS OF DISTANCES

Chapter 1. General Definitions **2**
1.1. Basic definitions 2
1.2. Main distance-related notions 7
1.3. General distances 21

Chapter 2. Topological Spaces **31**

Chapter 3. Generalizations of Metric Spaces **36**
3.1. m-metrics 36
3.2. Indefinite metrics 37
3.3. Topological generalizations 38
3.4. Beyond numbers 40

Chapter 4. Metric Transforms **44**
4.1. Metrics on the same set 44
4.2. Metrics on set extensions 46
4.3. Metrics on other sets 48

Chapter 5. Metrics on Normed Structures **50**

PART II. GEOMETRY AND DISTANCES

Chapter 6. Distances in Geometry **62**
6.1. Geodesic Geometry 62
6.2. Projective Geometry 66

6.3. Affine Geometry 71
6.4. Non-Euclidean Geometry 73

Chapter 7. Riemannian and Hermitian Metrics **81**
7.1. Riemannian metrics and generalizations 81
7.2. Riemannian metrics in Information Theory 94
7.3. Hermitian metrics and generalizations 99

Chapter 8. Distances on Surfaces and Knots **111**
8.1. General surface metrics 111
8.2. Intrinsic metrics on surfaces 116
8.3. Distances on knots 120

Chapter 9. Distances on Convex Bodies, Cones, and Simplicial Complexes **122**
9.1. Distances on convex bodies 122
9.2. Distances on cones 127
9.3. Distances on simplicial complexes 129

PART III. DISTANCES IN CLASSICAL MATHEMATICS

Chapter 10. Distances in Algebra **134**
10.1. Group metrics 134
10.2. Metrics on binary relations 141
10.3. Lattice metrics 143

Chapter 11. Distances on Strings and Permutations **146**
11.1. Distances on general strings 147
11.2. Distances on permutations 151

Chapter 12. Distances on Numbers, Polynomials, and Matrices **154**
12.1. Metrics on numbers 154
12.2. Metrics on polynomials 158
12.3. Metrics on matrices 159

Chapter 13. Distances in Functional Analysis **165**
13.1. Metrics on function spaces 165
13.2. Metrics on linear operators 172

Chapter 14. Distances in Probability Theory **176**
14.1. Distances on random variables 177
14.2. Distances on distribution laws 178

PART IV. DISTANCES IN APPLIED MATHEMATICS

Chapter 15. Distances in Graph Theory **190**
15.1. Distances on vertices of a graph 191
15.2. Distance-defined graphs 195
15.3. Distances on graphs 197
15.4. Distances on trees 202

Chapter 16. Distances in Coding Theory **207**
16.1. Minimum distance and relatives 208
16.2. Main coding distances 211

Chapter 17. Distances and Similarities in Data Analysis **217**
17.1. Similarities and distances for numerical data 218
17.2. Relatives of Euclidean distance 221
17.3. Similarities and distances for binary data 223
17.4. Correlation similarities and distances 226

Chapter 18. Distances in Mathematical Engineering **230**
18.1. Motion planning distances 230
18.2. Cellular automata distances 234
18.3. Distances in Control Theory 234
18.4. MOEA distances 236

PART V. COMPUTER-RELATED DISTANCES

Chapter 19. Distances on Real and Digital Planes **240**
19.1. Metrics on real plane 240
19.2. Digital metrics 248

Chapter 20. Voronoi Diagram Distances **253**
20.1. Classical Voronoi generation distances 254
20.2. Plane Voronoi generation distances 256
20.3. Other Voronoi generation distances 259

Chapter 21. Image and Audio Distances **262**
21.1. Image distances 262
21.2. Audio distances 273

Chapter 22. Distances in Internet and Similar Networks **279**
22.1. Scale-free networks 279
22.2. Network-based semantic distances 281
22.3. Distances in Internet and Web 283

PART VI. DISTANCES IN NATURAL SCIENCES

Chapter 23. Distances in Biology **288**
23.1. Genetic distances for gene-frequency data 289
23.2. Distances for DNA data 292
23.3. Distances for protein data 294
23.4. Other biological distances 296

Chapter 24. Distances in Physics and Chemistry **301**
24.1. Distances in Physics 301
24.2. Distances in Chemistry 306

Chapter 25. Distances in Geography, Geophysics, and Astronomy **309**
25.1. Distances in Geography and Geophysics 309
25.2. Distances in Astronomy 312

Chapter 26. Distances in Cosmology and Theory of Relativity **318**
26.1. Distances in Cosmology 318
26.2. Distances in Theory of Relativity 324

PART VII. REAL-WORLD DISTANCES

Chapter 27. Length Measures and Scales **342**
27.1. Length scales 342
27.2. Orders of magnitude for length 346

Chapter 28. Non-Mathematical and Figurative Meaning of Distance **350**
28.1. Remoteness-related distances 350
28.2. Vision-related distances 357

28.3. Equipment distances 360
28.4. Miscellany 364

References 371

Subject Index 378

Part I

Mathematics of Distances

Chapter 1. General Definitions
Chapter 2. Topological Spaces
Chapter 3. Generalizations of Metric Spaces
Chapter 4. Metric Transforms
Chapter 5. Metrics on Normed Structures

Chapter 1

General Definitions

1.1. BASIC DEFINITIONS

- **Distance**

  Let X be a set. A function $d : X \times X \to \mathbb{R}$ is called **distance** (or **dissimilarity**) on X if, for all $x, y \in X$, it holds:

  1. $d(x, y) \geqslant 0$ (*non-negativity*);
  2. $d(x, y) = d(y, x)$ (*symmetry*);
  3. $d(x, x) = 0$.

  In Topology, it is also called **symmetric**. The vector from x to y having the length $d(x, y)$ is called *displacement*. A distance which is a squared metric, is called *quadrance*.

  For any distance d, the function, defined for $x \neq y$ by $D(x, y) = d(x, y) + c$, where $c = \max_{x,y,z \in X}(d(x, y) - d(x, z) - d(y, z))$, and $D(x, x) = 0$, is a metric.

- **Distance space**

  A **distance space** (X, d) is a set X equipped with a distance d.

- **Similarity**

  Let X be a set. A function $s : X \times X \to \mathbb{R}$ is called **similarity** (or *proximity*) on X if s is *non-negative*, *symmetric*, and if $s(x, y) \leqslant s(x, x)$ holds for all $x, y \in X$, with equality if and only if $x = y$.

  Main transforms used to obtain a distance (dissimilarity) d from a similarity s are: $d = 1 - s$, $d = \frac{1-s}{s}$, $d = \sqrt{1 - s}$, $d = \sqrt{2(1 - s^2)}$, $d = -\ln s$, $d = \arccos s$.

- **Semi-metric**

  Let X be a set. A function $d : X \times X \to \mathbb{R}$ is called **semi-metric** (or **écart**, **pseudo-metric**) on X if d is *non-negative*, *symmetric*, if $d(x, x) = 0$ holds for all $x \in X$, and if

  $$d(x, y) \leqslant d(x, z) + d(z, y)$$

  holds for all $x, y, z \in X$ (**triangle inequality**).

  For any distance d, the equality $d(x, x) = 0$ and the **strong triangle inequality** $d(x, y) \leqslant d(x, z) + d(y, z)$ imply that d is a semi-metric.

- **Metric**

Let X be a set. A function $d : X \times X \to \mathbb{R}$ is called **metric** on X if, for all $x, y, z \in X$, it holds:

1. $d(x, y) \geqslant 0$ (*non-negativity*);
2. $d(x, y) = 0$ if and only if $x = y$ (*separation* or *self-identity axiom*);
3. $d(x, y) = d(y, x)$ (*symmetry*);
4. $d(x, y) \leqslant d(x, z) + d(z, y)$ (**triangle inequality**).

- **Metric space**

A **metric space** (X, d) is a set X equipped with a metric d.

A **metric scheme** is a metric space with an integral valued metric.

- **Extended metric**

An **extended metric** is a generalization of the notion of metric: the value ∞ is allowed for a metric d.

- **Near-metric**

Let X be a set. A distance d on X is called **near-metric** if

$$0 < d(x, y) \leqslant C\big(d(x, z_1) + d(z_1, z_2) + \cdots + d(z_n, y)\big)$$

holds for all different $x, y, z_1, \ldots, z_n \in X$ and a constant $C \geqslant 1$.

- **Coarse-path metric**

Let X be a set. A metric d on X is called **coarse-path metric** if, for a fixed $C \geqslant 0$ and for every pair of points $x, y \in X$, there exists a sequence $x = x_0, x_1, \ldots, x_t = y$ for which $d(x_{i-1}, x_i) \leqslant C$ for $i = 1, \ldots, t$, and

$$d(x, y) \geqslant d(x_0, x_1) + d(x_1, x_2) + \cdots + d(x_{t-1}, x_t) - C,$$

i.e., the weakened triangle inequality $d(x, y) \leqslant \sum_{i=1}^{t} d(x_{i-1}, x_i)$ becomes an equality up to a bounded error.

- **Resemblance**

Let X be a set. A function $d : X \times X \to \mathbb{R}$ is called **resemblance** on X if d is *symmetric* and if, for all $x, y \in X$, either $d(x, x) \leqslant d(x, y)$ holds (in which case d is called *forward resemblance*), or $d(x, x) \geqslant d(x, y)$ holds (in which case d is called *backward resemblance*).

Every resemblance d induces a *strict partial order* $\prec$ on the set of all unordered pairs of elements of X by defining $\{x, y\} \prec \{u, v\}$ if and only if $d(x, y) < d(u, v)$.

For any backward resemblance d, the forward resemblance $-d$ induces the same partial order.

- **Quasi-distance**

  Let X be a set. A function $d : X \times X \to \mathbb{R}$ is called **quasi-distance** on X if d is *non-negative*, and if $d(x, x) = 0$ holds for all $x \in X$.

- **Quasi-semi-metric**

  Let X be a set. A function $d : X \times X \to \mathbb{R}$ is called **quasi-semi-metric** (or *weak metric*) on X if d is *non-negative*, if $d(x, x) = 0$ holds for all $x \in X$, and if

$$d(x, y) \leqslant d(x, z) + d(z, y)$$

  holds for all $x, y, z \in X$ (**oriented triangle inequality**).

- **Albert quasi-metric**

  An **Albert quasi-metric** d is a quasi-semi-metric on X with *weak definiteness*, i.e., for all $x, y \in X$ the equality $d(x, y) = d(y, x)$ implies $x = y$.

- **Weak quasi-metric**

  A **weak quasi-metric** d is a quasi-semi-metric on X with *weak symmetry*, i.e., for all $x, y \in X$, $d(x, y) = 0$ if and only if $d(y, x) = 0$.

- **Quasi-metric**

  Let X be a set. A function $d : X \times X \to \mathbb{R}$ is called **quasi-metric** on X if $d(x, y) \geqslant 0$ holds for all $x, y \in X$ with equality if and only if $x = y$, and if

$$d(x, y) \leqslant d(x, z) + d(z, y)$$

  holds for all $x, y, z \in X$ (**oriented triangle inequality**). A *quasi-metric space* (X, d) is a set X equipped with a quasi-metric d.

  For any quasi-metric d, the function $D(x, y) = d(x, y) + d(y, x)$ is a metric.

- **$2k$-gonal distance**

  An $2k$-**gonal distance** d is a distance on X which satisfies the $2k$-**gonal inequality**

$$\sum_{1 \leqslant i < j \leqslant n} b_i b_j d(x_i, x_j) \leqslant 0$$

  for all $b \in \mathbb{Z}^n$ with $\sum_{i=1}^n b_i = 0$ and $\sum_{i=1}^n |b_i| = 2k$, and for all distinct elements $x_1, \dots, x_n \in X$.

- **Distance of negative type**

  A **distance of negative type** d is a distance on X which is $2k$-**gonal** for any $k \geqslant 1$, i.e., satisfies the **negative type inequality**

$$\sum_{1 \leqslant i < j \leqslant n} b_i b_j d(x_i, x_j) \leqslant 0$$

for all $b \in \mathbb{Z}^n$ with $\sum_{i=1}^{n} b_i = 0$, and for all distinct elements $x_1, \ldots, x_n \in X$.

A distance can be of negative type without being a semi-metric. Cayley proved that a metric d is an L_2-**metric** if and only if d^2 is a distance of negative type.

- **$(2k + 1)$-gonal distance**

An $(2k + 1)$-**gonal distance** d is a distance on X which satisfies the $(2k + 1)$-**gonal inequality**

$$\sum_{1 \leqslant i < j \leqslant n} b_i b_j d(x_i, x_j) \leqslant 0$$

for all $b \in \mathbb{Z}^n$ with $\sum_{i=1}^{n} b_i = 1$ and $\sum_{i=1}^{n} |b_i| = 2k + 1$, and for all distinct elements $x_1, \ldots, x_n \in X$.

The $(2k + 1)$-gonal inequality with $k = 1$ is the usual triangle inequality. The $(2k + 1)$-gonal inequality implies the $2k$-**gonal inequality**.

- **Hypermetric**

A **hypermetric** d is a distance on X which is $(2k + 1)$-**gonal** for any $k \geqslant 1$, i.e., satisfies the **hypermetric inequality**

$$\sum_{1 \leqslant i < j \leqslant n} b_i b_j d(x_i, x_j) \leqslant 0$$

for all $b \in \mathbb{Z}^n$ with $\sum_{i=1}^{n} b_i = 1$, and for all distinct elements $x_1, \ldots, x_n \in X$. Any hypermetric is a semi-metric and a **distance of negative type**. Any L_1-**metric** is a hypermetric.

- **Ptolemaic metric**

A **Ptolemaic metric** d is a metric on X which satisfies the **Ptolemaic inequality**

$$d(x, y)d(u, z) \leqslant d(x, u)d(y, z) + d(x, z)d(y, u)$$

(shown by Ptolemy to hold in the Euclidean space) for all $x, y, u, z \in X$.

A metric space $(V, \|x - y\|)$ (where $(V, \|.\|)$ is a *normed vector space*) is *Ptolemaic* if and only if it is an **inner product space**.

- **Assouad pseudo-distance**

An **Assouad pseudo-distance** (or **weak ultrametric**) d is a distance on X such that for a constant $C \geqslant 1$ the inequality

$$0 < d(x, y) \leqslant C \max\{d(x, z), d(z, y)\}$$

holds for all $x, y, z \in X$, $x \neq y$.

The term **pseudo-distance** is also used, in some applications, for a **pseudo-metric** (i.e., a semi-metric), for a **quasi-distance**, for a **near-metric**, for a distance which can be infinite, for a distance with an error, etc.

- **Ultrametric**

An **ultrametric** (or **non-Archimedean metric**) d is a metric on X which satisfies the following strengthened version of the triangle inequality:

$$d(x, y) \leqslant \max\{d(x, z), d(z, y)\}$$

for all $x, y, z \in X$. So, at least two of $d(x, y)$, $d(z, y)$ and $d(x, z)$ are the same.

A metric d is ultrametric if and only if d^{α} is a metric for any real positive number α. Any ultrametric satisfies the **four-point inequality**.

- **Four-point inequality metric**

A metric d on X satisfies the **four-point inequality** if the following strengthened version of the triangle inequality holds: for all $x, y, z, u \in X$

$$d(x, y) + d(z, u) \leqslant \max\{d(x, z) + d(y, u), d(x, u) + d(y, z)\}.$$

Equivalently, among the three sums $d(x, y)+d(z, u)$, $d(x, z)+d(y, u)$, $d(x, u)+d(y, z)$ the two largest sums are equal.

A metric satisfies the four-point inequality if and only if it is a **tree-like metric**.

Any metric, satisfying the four-point inequality, is a **Ptolemaic metric**.

A **bush metric** is a metric for which all four-point inequalities are equalities, i.e., $d(x, y) + d(u, z) = d(x, u) + d(y, z)$ holds for any $u, x, y, z \in X$.

- **Relaxed four-point inequality metric**

A metric d on X satisfies the **relaxed four-point inequality** if, for all $x, y, z, u \in X$, among the three sums

$$d(x, y) + d(z, u), \quad d(x, z) + d(y, u), \quad d(x, u) + d(y, z)$$

at least two (not necessarily two largest) are equal.

A metric satisfies the relaxed four-point inequality if and only if it is a **relaxed tree-like metric**.

- **δ-hyperbolic metric**

Given a number $\delta \geqslant 0$, a metric d on a set X is called **δ-hyperbolic** if it satisfies the **Gromov δ-hyperbolic inequality** (another weakening of the **four-point inequality**): for all $x, y, z, u \in X$ it holds

$$d(x, y) + d(z, u) \leqslant 2\delta + \max\{d(x, z) + d(y, u), d(x, u) + d(y, z)\}.$$

A metric space (X, d) is δ-hyperbolic if and only if it holds

$$(x.y)_{x_0} \geqslant \min\{(x.z)_{x_0}, (y.z)_{x_0}\} - \delta$$

for all $x, y, z \in X$ and for any $x_0 \in X$, where $(x.y)_{x_0} = \frac{1}{2}(d(x_0, x)+d(x_0, y)-d(x, y))$ is the **Gromov product** of the points x and y of X with respect of base-point $x_0 \in X$.

A metric space (X, d) is 0-hyperbolic exactly when d satisfies the **four-point inequality**. Every bounded metric space of diameter D is D-hyperbolic. The n-dimensional *hyperbolic space* is ln 3-hyperbolic.

- **Gromov product similarity**

  Given a metric space (X, d) with a fixed point $x_0 \in X$, the **Gromov product similarity** (or *Gromov product, covariance*) $(.)_{x_0}$ is a similarity on X, defined by

$$(x.y)_{x_0} = \frac{1}{2}\big(d(x, x_0) + d(y, x_0) - d(x, y)\big).$$

  If X is a *measure space* with $d(x, y) = \mu(x \triangle y)$, then $(x.y)_{\emptyset} = \mu(x \cap y)$. If d is a **distance of negative type**, i.e., $d(x, y) = d_E^2(x, y)$ for a subset X of an Euclidean space $\mathbb{E}^n$, then $(x.y)_0$ is the usual *inner product* on $\mathbb{E}^n$. The function $d_{x_0}(x, y) = C - (x.y)_{x_0}$ (called *Farris transform* in Phylogenetics) with $C \geqslant \max_{u,v \in X} d(u, v)$ is a metric. It is an **ultrametric** if and only if d satisfies the **four-point inequality**.

1.2. MAIN DISTANCE-RELATED NOTIONS

- **Metric ball**

  Given a metric space (X, d), the **metric ball** (or *closed metric ball*) with center $x_0 \in X$ and radius $r > 0$ is defined by $\overline{B}(x_0, r) = \{x \in X: d(x_0, x) \leqslant r\}$, and the *open metric ball* with center $x_0 \in X$ and radius $r > 0$ is defined by $B(x_0, r) = \{x \in X: d(x_0, x) < r\}$.

  The **metric sphere** with center $x_0 \in X$ and radius $r > 0$ is defined by $S(x_0, r) = \{x \in X: d(x_0, x) = r\}$.

  For the **norm metric** on an n-dimensional *normed vector space* $(V, \|.\|)$, the closed metric ball $\overline{B}^n = \{x \in V: \|x\| \leqslant 1\}$ is called *unit ball*, and the set $S^{n-1} = \{x \in V: \|x\| = 1\}$ is called *unit sphere* (or *unit hypersphere*). In a two-dimensional vector space, a metric ball (closed or open) is called *metric disk* (closed or open, respectively).

- **Metric topology**

  A **metric topology** is a *topology* on X induced by a metric d on X.

  More exactly, given a metric space (X, d), define the *open set* in X as an arbitrary union of (finitely or infinitely many) open **metric balls** $B(x, r) = \{y \in X: d(x, y) < r\}$, $x \in X, r \in \mathbb{R}, r > 0$. A *closed set* is defined now as the complement of an open set. The **metric topology** on (X, d) is defined as the set of all open sets of X. A topological space which can arise in this way from a metric space is called **metrizable space**.

- **Metrization theorems**

  Metrization theorems are theorems which give sufficient conditions for a topological space to be **metrizable**, i.e., with topology which is a **metric topology**.

- **Metric interval**

Given two different points $x, y \in X$ of a metric space (X, d), the **metric interval** between x and y is the set

$$I(x, y) = \{z \in X: \ d(x, y) = d(x, z) + d(z, y)\}.$$

A metric space (X, d) is called **antipodal metric space** (or *diametrical metric space*) if, for any $x \in X$, there exists the *antipode*, i.e., an unique $x' \in X$ such that $I(x, x') = X$.

A metric space (X, d) is called **distance monotone metric space** if, for any interval $I(x, x')$ and $y \in X \backslash I(x, x')$, there exists $x'' \in I(x, x')$ with $d(y, x'') > d(x, x')$.

- **Metric triangle**

Three different points $x, y, z \in X$ of a metric space (X, d) form a **metric triangle** if the **metric intervals** $I(x, y)$, $I(y, z)$ and $I(z, x)$ intersect only in the common end points.

A **metric tree** is a metric space all of whose metric triangles are degenerated.

- **Modular metric space**

A metric space (X, d) is called **modular** if for any three different points $x, y, z \in X$ there exist $u \in I(x, y) \cap I(y, z) \cap I(z, x)$.

This should not be confused with **modular distance** and **modulus metric**.

- **Metric quadrangle**

Four different points $x, y, z, u \in X$ of a metric space (X, d) form a **metric quadrangle** if $x, z \in I(y, u)$ and $y, u \in I(x, z)$. It holds $d(x, y) = d(z, u)$ and $d(x, u) = d(y, z)$ in such a metric quadrangle.

A metric space (X, d) is called *weakly spherical* if, for any three different points $x, y, z \in X$ with $y \in I(x, z)$, there exists $u \in X$ such that x, y, z, u form a metric quadrangle.

- **Metric curve**

A **metric curve** (or, simply, *curve*) γ in a metric space (X, d) is a continuous mapping $\gamma : I \to X$ from an interval I of $\mathbb{R}$ into X. A curve is called *simple* if it is injective. A curve $\gamma : [a, b] \to X$ is called *Jordan curve* (or *simple closed curve*) if it does not cross itself, and $\gamma(a) = \gamma(b)$. The *length* $l(\gamma)$ of a curve $\gamma : [a, b] \to X$ is defined by $l(\gamma) = \sup\{\sum_{1 \leqslant i \leqslant n} d(\gamma(t_i), \gamma(t_{i-1})): \ n \in \mathbb{N}, a = t_0 < t_1 < \cdots < t_n = b\}$. A *rectifiable curve* is a curve with the finite length.

- **Geodesic**

A *geodesic segment* (or *shortest path*) in a metric space is a locally shortest *curve* between two points. In other words, a geodesic segment $[x, y]$ from x to y is an *isometric embedding* $\gamma : [a, b] \to X$ with $\gamma(a) = x$ and $\gamma(b) = y$. A **geodesic** is a locally isometric embedding of the whole $\mathbb{R}$ in X. A *metric straight line* (or *minimizing geodesic*) is a geodesic which is minimal between any two of its points.

A **geodesic metric space** is a metric space in which any two points are joined by a geodesic segment. Given a metric space (X, d), the **total convexity** of a set $M \subset X$ means that for any two points of M any geodesic segment connecting them lies entirely in M.

- **Metric convexity**

Given a metric space (X, d), the **metric convexity** of a set $M \subset X$ means that, for any different points $x, y \in M$ and any $0 < \lambda < 1$, there exists $z \in M$ such that $d(x, z) = \lambda d(x, y)$, and $d(z, y) = (1 - \lambda)d(x, y)$.

- **Proximity graph of metric space**

The **proximity graph** (or *underlying graph*) of a metric space (X, d) is a graph with the vertex-set X and xy being an edge if $I(x, y) = \{x, y\}$, i.e., no third point $z \in X$, for which $d(x, y) = d(x, z) + d(z, y)$, exists.

- **Menger convexity**

A metric space (X, d) is called **Menger-convex** (or *M-convex*) if, for any different $x, y \in X$, we have $|I(x, y)| > 2$, i.e., there exists a third point $z \in X$ for which $d(x, y) = d(x, z) + d(z, y)$. The **Menger-convexity** of a set $M \subset X$ means that, moreover, $z \in M$ if $x, y \in M$. There exist discrete Menger-convex metric spaces.

- **Hyperconvexity**

A metric space (X, d) is called **hyperconvex** (or **injective**) if it is **Menger-convex** and its closed **metric balls** have the *infinite Helly property*, i.e., any family of closed balls $\overline{B}(x_i, r_i)$ with centers x_i and radii r_i, $i \in I$, satisfying $d(x_i, x_j) \leqslant r_i + r_j$ for all $i, j \in I$, has non-empty intersection (cf. **injective metric space**).

- **Metric entropy**

Given $\varepsilon > 0$, the **metric entropy** (or *ε-entropy*, *relative ε-entropy*) $H_\varepsilon(M, X)$ of a set M, lying in a metric space (X, d), is defined by

$$H_\varepsilon(M, X) = \log_2 N_\varepsilon(M, X),$$

where $N_\varepsilon(M, X)$ is the smallest number of points in an *ε-net* for the metric space (M, d), i.e., a set of points such that the union of ε-balls, centered at those points, covers M.

The *absolute ε-entropy* of a set M is the number $H_\varepsilon(M) = \inf H_\varepsilon(M, X)$, where the infimum is taken over all metric spaces (X, d) such that $M \subset X$.

- **Metric dimension**

For a metric space (X, d) and any real number $q > 0$, let $N_X(q)$ be the minimal number of sets with diameter at most q which are needed in order to cover X (cf. **metric entropy**). The number $\lim_{q \to 0} \frac{\ln N(q)}{\ln(1/q)}$ (if it exists) is called **metric dimension** (or **Minkowski–Bouligand dimension**, *Minkowski dimension*, *packing dimension*, *box-counting dimension*) of X.

If the limit above does not exist, then the following notions of dimension are considered:

1. The number $\underline{\lim}_{q \to 0} \frac{\ln N(q)}{\ln(1/q)}$ is called **lower metric dimension** (or *lower box dimension, Pontryagin–Schnirelmann dimension*, **lower Minkowski dimension**);
2. The number $\overline{\lim}_{q \to 0} \frac{\ln N(q)}{\ln(1/q)}$ is called **upper metric dimension** (or *entropy dimension*, **Kolmogorov–Tihomirov dimension**, *upper box dimension*).

In the mathematical literature other, less prominent, notions of metric dimension also occur. For example,

1. For any $c > 1$, the *metric dimension* $\dim_c(X)$ of a finite metric space (X, d) is the least dimension of a real *normed space* $(V, \|.\|)$ such that there is an embedding $f : X \to V$ with $d(x, y) \geqslant \|f(x) - f(y)\| \geqslant \frac{1}{c} d(x, y)$;
2. The *dimension* of a finite metric space (X, d) is the least dimension n of an Euclidean space $\mathbb{E}^n$ such that $(X, f(d))$ is its metric subspace, where the minimum is taken over all continuous monotone increasing functions $f(t)$ of $t \geqslant 0$;
3. The *metric dimension* of a metric space is the minimum size of its **metric basis**, i.e., of its smallest subset S such that no two points have the same distances to all points of S.
4. The *equilateral dimension* of a metric space is the maximum cardinality of its *equilateral* (or *equidistant*) subset, i.e., such that any two its distinct points are at the same distance. For a normed space, this dimension is equal to the maximum number of translates of its unit ball that pairwise touch.

- **Hausdorff dimension**

For a metric space (X, d) and any real $p, q > 0$, let $M_p^q(X) = \inf \sum_{i=1}^{+\infty} (diam(A_i))^p$, where the infimum is taken over all countable coverings $\{A_i\}_i$ of X with the diameter of A_i less than q. The **Hausdorff dimension** (or *Hausdorff–Besicovitch dimension, capacity dimension, fractal dimension*) $\dim_{Haus}(X, d)$ of X is defined by

$$\inf\Big\{p: \lim_{q \to 0} M_p^q(X) = 0\Big\}.$$

Any countable metric space have Hausdorff dimension 0; Hausdorff dimension of the Euclidean space $\mathbb{E}^n$ is equal to n.

For each **totally bounded** metric space, its Hausdorff dimension is bounded from above by its **metric dimension** and from below by its **topological dimension**.

- **Topological dimension**

For any compact metric space (X, d) its **topological dimension** (or **Lebesgue covering dimension**) is defined by

$$\inf_{d'}\big\{\dim_{Haus}(X, d')\big\},$$

where d' is any metric on X topologically equivalent to d, and $\dim_{Haus}$ is the **Hausdorff dimension**.

In general, the **topological dimension** of a topological space X is the smallest integer n such that, for any finite open covering of X, there exists a (finite open) sub-covering (i.e., a refinement of it) with no point of X belonging to more than $n + 1$ elements.

- **Assouad–Nagata dimension**

The **Assouad–Nagata dimension** of a metric space (X, d) is the smallest integer n for which there exists a constant $C > 0$ such that, for all $s > 0$, there exists a covering of X by its subsets of diameter at most Cs with no point of X belonging to more than $n + 1$ elements.

For a metric space, its **topological dimension** does not exceed its Assouad–Nagata dimension. A metric space (X, d) has finite Assouad–Nagata dimension if and only if it has finite *doubling dimension*, i.e., the smallest integer N such that every **metric ball** can be covered by a family of at most N metric balls of half the radius.

The **asymptotic dimension** of a metric space (X, d) was introduced by Gromov; it is the smallest integer n such that, for all $s > 0$, there exist a constant $D = D(s)$ and a covering of X by its subsets of diameter at most D with no point of X belonging to more than $n + 1$ elements.

We say that a metric space (X, d) has **Godsil–McKay dimension** $n \geqslant 0$ if there exist an element $x_0 \in X$ and two positive constants c and C such that $ck^n \leqslant |\{x \in X: d(x, x_0) \leqslant k\}| \leqslant Ck^n$ holds for every integer $k \geqslant 0$. This notion was introduced in [GoMc80] for the **path metric** of a countable locally finite graph. It was proved there that if the group $\mathbb{Z}^n$ acts faithfully and with a finite number of orbits on the vertices of the graph, then this dimension is equal to n.

- **Fractal**

For a metric space, its **topological dimension** does not exceed its **Hausdorff dimension**. A **fractal** is a metric space for which this inequality is strict. (Originally, Mandelbrot defined a fractal as a point set with non-integer Hausdorff dimension.) For example, the *Cantor set*, which is an 0-dimensional topological space, has the Hausdorff dimension $\frac{\ln 2}{\ln 3}$.

The term **fractal** is used also, more generally, for *self-similar* (i.e., roughly, looking similar at any scale) object (usually, a subset of $\mathbb{R}^n$).

- **Length of metric space**

Fremlin's length of metric space (X, d) is one-dimensional Hausdorff outer measure on X.

Hejcman's length $lng(Y)$ of a subset Y of metric space (X, d) is

$$\sup\left\{lng(A): A \subset Y, |A| < \infty\right\}.$$

Here $lng(\emptyset) = 0$ and, for a finite subset A of X, $lng(A) = \min \sum_{i=1}^{n} d(x_{i-1}, x_i)$ over all sequences $x_0, \dots, x_n$ such that $\{x_i : i = 0, 1, \dots, n\} = A$.

Schechtman's length of finite metric space (X, d) is $\inf \sqrt{\sum_{i=1}^{n} a_i^2}$ over all sequences $a_1, \dots, a_n$ of positive numbers such that there exists a sequence $X_0, \dots, X_n$ of partitions of X with following properties:

1. $X_0 = \{X\}$ and $X_n = \{\{x\}: x \in X\}$,

2. X_i refines X_{i-1} for $i = 1, \dots, n$,
3. For $i = 1, \dots, n$ and $B, C \subset A \in X_{i-1}$ with $B, C \in X_i$, there exists a one-to-one map f from B onto C such that $d(x, f(x)) \leqslant a_i$ for all $x \in B$.

- ***D*-chromatic number**

Given a metric space (X, d) and a set D of positive real numbers, the ***D*-chromatic number** of (X, d) is the standard *chromatic number* of the ***D*-distance graph** of (X, d), i.e., the graph with the vertex-set X and the edge-set $\{xy: \ d(x, y) \in D\}$. Usually, (X, d) is an *l_p-space* and $D = \{1\}$ (*Benda–Perles chromatic number*) or $D = [1 - \varepsilon, 1 + \varepsilon]$ (the chromatic number of the *ε-unit distance graph*).

- **Polychromatic number**

For a metric space (X, d), it is the minimum number of colors needed to color all the points $x \in X$ so that for each color class C_i, there is a distance d_i such that no two points of C_i are at distance d_i.

For any integer $t > 0$, the ***t*-distance chromatic number** of (X, d) is the minimum number of colors needed to color all the points $x \in X$ so that any two points whose distance apart is $\leqslant t$ have distinct colors.

For any integer $t > 0$, the ***t*-th Babai number** of (X, d) is the minimum number of colors needed to color all the points $x \in X$ so that, for any set D of positive distances with $|D| \leqslant t$, any two points whose distance belongs to D have distinct colors.

- **Rendez-vous number**

Given a metric space (X, d), its **rendez-vous number** (or *Gross number, magic number*) is a positive real number $r(X, d)$ (if it exists), defined by the property that for each integer n and all (not necessarily distinct) $x_1, \dots, x_n \in X$ there exists $x \in X$ such that

$$r(X, d) = \frac{1}{n}\sum_{i=1}^{n} d(x_i, x).$$

If for a metric space (X, d) the rendez-vous number $r(X, d)$ exists, then it is said that (X, d) has the **average distance property** and its **magic constant** is defined by $\frac{r(X,d)}{diam(X,d)}$, where $diam(X, d) = \max_{x,y \in X} d(x, y)$ is the **diameter** of (X, d).

Every compact connected metric space has the average distance property. The *unit ball* $\{x \in V: \ \|x\| \leqslant 1\}$ of a **Banach space** $(V, \|.\|)$ has the average distance property with the rendez-vous number 1.

- **Metric radius**

The **metric radius** of a set $M \subset X$ in a metric space (X, d) is the infimum of radii of **metric balls** which contain M.

The **covering radius** of a set $M \subset X$ is $\max_{x \in X} \min_{y \in M} d(x, y)$ (**directed Hausdorff distance** from X to M), i.e., the smallest number R such that the balls of radius R with centers at the elements of M cover X. The **packing radius** of M is the largest r such that the balls of radius r with centers at the elements of M are pairwise disjoint.

A m-subset M of a metric space (X, d) is called **minimax distance design of size** m if $\max_{x\in X} \min_{y\in M} d(x, y) = \min\{\max_{x\in X} \min_{y\in S} d(x, y) : S \subset X, |S| = m\}$ holds, and it is called **maximum distance design of size** m if $\min_{x\in M} \min_{y\in M\setminus\{y\}} d(x, y) = \max\{\min_{x\in S} \min_{y\in S\setminus\{y\}} d(x, y) : S \subset X, |S| = m\}$ holds.

- **Metric diameter**

The **diameter** $diam(M)$ of a set $M \subset X$ in a metric space (X, d) is defined by

$$\sup_{x,y\in M} d(x, y).$$

The **diameter graph** of M has, as vertices, all points $x \in M$ with $d(x, y) = diam(M)$ for some $y \in M$; it has, as edges, all pairs of its vertices at distance $diam(M)$ in (X, d).

The value

$$diam(X, d) = \sup_{x,y\in X} d(x, y)$$

is called **diameter** of the metric space (X, d). The numbers

$$\sum_{x,y\in M,\ x\neq y} \frac{1}{d^2(x, y)} \quad \text{and} \quad \frac{1}{|M|(|M|-1)} \sum_{x,y\in M,\ x\neq y} d(x, y)$$

are called, respectively, **energy** and **average distance** of the set M.

In Chemistry, the number $\sum_{x,y\in M, x\leqslant y} d(x, y)$ is called *Wiener number* of M.

- **Eccentricity**

Given a finite metric space (X, d), the **eccentricity** of a point $x \in X$ is the number $e(x) = \max_{y\in X} d(x, y)$. The numbers $\max_{x\in X} e(x)$ and $\min_{x\in X} e(x)$ are the **diameter** and the **radius** of (X, d), respectively. Some authors call *radius* the half of diameter.

The sets $\{x \in X : \max_{y\in X} d(x, y) \leqslant \max_{y\in X} d(z, y)$ for any $z \in X\}$ and $\{x \in X : \sum_{y\in X} d(x, y) \leqslant \sum_{y\in X} d(z, y)$ for any $z \in X\}$ are, respectively, the **metric center** (or *eccentricity center*) and the **metric median** (or *distance center*) of (X, d).

- **Steiner ratio**

Given a metric space (X, d) and a finite subset V of X, consider the complete weighted graph $G = (V, E)$ with the vertex-set V and edge-weights $d(x, y)$ for all $x, y \in V$.

A *spanning tree* T in G is a subset of $|V| - 1$ edges forming a tree on V with the *weight* $d(T)$ equal to the sum of weights of its edges. Let MST_V be a *minimal spanning tree* in G, i.e., a spanning tree in G with the minimal weight $d(MST_V)$.

A *minimal Steiner tree* on V is a tree SMT_V such that its vertex-set is a subset of X containing V, and $d(SMT_V) = \inf_{M\subset X:\, V\subset M} d(MST_M)$.

The **Steiner ratio** $St(X, d)$ of the metric space (X, d) is defined by

$$\inf_{V\subset X} \frac{d(SMT_V)}{d(MST_V)}.$$

For any metric space (X, d) we have $\frac{1}{2} \leqslant St(X, d) \leqslant 1$. For the l_2**-metric** (i.e., the Euclidean metric) on $\mathbb{R}^2$, it is equal to $\frac{\sqrt{3}}{2}$, while for the l_1**-metric** on $\mathbb{R}^2$ it is equal to $\frac{2}{3}$.

- **Order of congruence**

  A metric space (X, d) has the **order of congruence** n if every finite metric space which is not *isometrically embeddable* in (X, d) has a subspace with at most n points which is not isometrically embeddable in (X, d).

- **Midset**

  Given a metric space (X, d) and distinct points $y, z \in X$, the **midset** (or *bisector*) of X is the set $\{x \in X : d(x, y) = d(x, z)\}$ of *midpoints* x.

  A metric space is said to have *n-points midset property* if, for every pair of its points, the midset has exactly n points.

- **Metric basis**

  Given a metric space (X, d), a set $M \subset X$ is called **metric basis** of X if the following condition holds: $d(x, s) = d(y, s)$ for all $s \in M$ implies $x = y$. For $x \in X$, the numbers $d(x, s)$, $s \in M$, are called **metric coordinates** of x.

  Every largest affine independent subset of an *affine space* (i.e., a translation of a vector space), taken with the Euclidean metric, is a minimal metric basis.

- **Element of best approximation**

  Given a metric space (X, d) and a subset $M \subset X$, an element $u_0 \in M$ is called **element of best approximation** to a given element $x \in X$ if $d(x, u_0) = \inf_{u \in M} d(x, u)$, i.e., if $d(x, u_0)$ is the **point-set distance** $d(x, M)$.

  A **Chebyshev set** (or *gated set*) in a metric space (X, d) is a subset of X containing an unique element of best approximation for every $x \in X$.

- **Metric projection**

  Given a metric space (X, d) and a subset $M \subset X$, the **metric projection** is a multi-valued mapping associating to each element $x \in X$ the set of **elements of best approximation** from the set M (cf. **distance map**).

  The set M is a **Chebyshev set** if and only if the corresponding metric projection is a single-valued mapping.

- **Chebyshev center**

  Given a metric space (X, d) and a bounded subset $M \subset X$, the *Chebyshev radius* of the set M is $\inf_{x \in X} \sup_{y \in M} d(x, y)$, and a **Chebyshev center** of M is an element $x_0 \in X$ realizing this infimum.

- **Distance map**

  Given a metric space (X, d) and a subset $M \subset X$, the **distance map** is a function $f_M : X \to \mathbb{R}_{\geqslant 0}$, where $f_M(x) = \inf_{u \in M} d(x, u)$ is the **point-set distance** $d(x, M)$ (cf. **metric projection**).

If the boundary $B(M)$ of the set M is defined, then the **signed distance function** g_M is defined on X by $g_M(x) = -\inf_{u \in B(M)} d(x, u)$ for $x \in M$ and $g(x) = \inf_{u \in B(M)} d(x, u)$, otherwise. If M is a (closed and orientable) manifold in $\mathbb{R}^n$, then g_M is the solution of the *eikonal equation* $|\nabla g| = 1$ for its *gradient* ∇.

Distance maps are used in Robot Motion (M being the set of obstacle points) and, especially, in Image Processing (M being the set of all or only boundary pixels of the image). For $X = \mathbb{R}^2$, the graph $\{(x, f_M(x)): x \in X\}$ of $d(x, M)$ is called *Voronoi surface* of M.

- **Metric transform**

A **metric transform** is a distance obtained as a function of a given metric (cf. Chapter 4).

- **Discrete dynamic system**

A **discrete dynamic system** is a pair consisting of a non-empty metric space (X, d), called *phase space*, and a continuous mapping $f : X \to X$, called *evolution law*. For any $x \in X$, its *orbit* is the sequence $\{f^n(x)\}_n$; here $f^n(x) = f(f^{n-1}(x))$ with $f^0(x) = x$. The orbit of $x \in X$ is called *periodic* if $f^n(x) = x$ for some $n > 0$.

Usually, the discrete dynamic systems are studied (for example, in Control Theory) in the context of stability of systems; Chaos Theory concerns itself with the systems with maximal possible instability.

An *attractor* is a closed subset A of X such that there exists an *open neighborhood* U of A with the property that $\lim_{n \to \infty} d(f^n(b), A) = 0$ for every $b \in U$. Here $d(x, A) = \inf_{y \in A} d(x, y)$ is the **point-set distance**.

A dynamic system is called (topologically or Devaney) *chaotic* if it is *regular* (i.e., X has a dense subset of elements having periodic orbits) and *transitive* (i.e., for any two non-empty open subsets A, B of X, there exists a number n such that $f^n(A) \cap B \neq \emptyset$).

- **Metric cone**

The **metric cone** is a collection of all semi-metrics on the set $V_n = \{1, \dots, n\}$.

- **Distance matrix**

Given a finite metric space $(X = \{x_1, \dots, x_n\}, d)$, its **distance matrix** is the symmetric $n \times n$ matrix $((d_{ij}))$, where $d_{ij} = d(x_i, x_j)$ for any $1 \leqslant i, j \leqslant n$.

Let s denote the number of different non-zero values of d_{ij}. The metric space (X, d) is said to have **strength** t if, for any integers $p, q \geqslant 0$ with $p+q \leqslant t$, there is a polynomial $f_{pq}(s)$ of degree at most $\min\{p, q\}$ such that $((d_{ij}^{2p}))((d_{ij}^{2q})) = ((f_{pq}(d_{ij}^2)))$.

- **Cayley–Menger matrix**

Given a finite metric space $(X = \{x_1, \dots, x_n\}, d)$, its **Cayley–Menger matrix** is the symmetric $(n+1) \times (n+1)$ matrix

$$CM(X, d) = \begin{pmatrix} 0 & e \\ e^T & D \end{pmatrix},$$

where $D = ((d_{ij}))$ is the **distance matrix** of (X, d), and e is the n-vector all components of which are 1. The determinant of $CM(X, d)$ is called *Cayley–Menger determinant.*

- **Gram matrix**

Given elements $v_1, \dots, v_k$ of an Euclidean space, their **Gram matrix** is the symmetric $k \times k$ matrix

$$G(v_1, \dots v_k) = \big(\big(\langle v_i, v_j \rangle\big)\big)$$

of pairwise *inner products* of $v_1, \dots, v_k$.

An $k \times k$ matrix is positive-semi-definite if and only if it is a Gram matrix. An $k \times k$ matrix is positive-definite if and only if it is a Gram matrix with linearly independent defining vectors.

We have $G(v_1, \dots v_k) = \frac{1}{2}((d_E^2(v_0, v_i) + d_E^2(v_0, v_j) - d_E^2(v_i, v_j)))$, i.e., the inner product $\langle , \rangle$ is the **Gromov product similarity** of the **squared Euclidean distance** d_E^2. A $k \times k$ matrix $((d_E^2(v_i, v_j)))$ is a **distance of negative type**; all such $k \times k$ matrices form the (non-polyhedral) closed convex cone of all such distances on a k-set.

The determinant of a Gram matrix is called *Gram determinant*; it is equal to the square of the *k-dimensional volume* of the *parallelotope* constructed on $v_1, \dots v_k$.

- **Isometry**

Given metric spaces (X, d_X) and (Y, d_Y), a function $f : X \to Y$ is called an *isometric embedding* of X into Y if it is injective, and, for all $x, y \in X$, the equality $d_Y(f(x), f(y)) = d_X(x, y)$ holds.

An **isometry** is a bijective isometric embedding. Two metric spaces are called **isometric** (or *isometrically isomorphic*) if there exists an isometry between them. An isometry of a metric space (X, d) onto itself is called **motion**.

A property of metric spaces which is invariant with respect to isometries (completeness, boundedness, etc.) is called **metric property** (or *metric invariant*).

A *path isometry* (or *arcwise isometry*) is a mapping from X into Y (not necessarily bijective) preserving the lengths of curves.

- **Symmetric metric space**

A metric space (X, d) is called **symmetric** if, for any point $p \in X$, there exists a *symmetry* relative to that point, i.e., a **motion** f_p of this metric space such that $f_p(f_p(x)) = x$ for all $x \in X$, and p is an isolated fixed point of f_p.

- **Homogeneous metric space**

A metric space (X, d) is called **homogeneous** (or *highly transitive*) if, for each two finite isometric subsets $Y = \{y_1, \dots, y_m\}$ and $Z = \{z_1, \dots, z_m\}$ of X, there exists a **motion** of X mapping Y to Z. A metric space is called *point-homogeneous* if, for any two points of it, there exists a motion mapping one of the points to the other. In general, a *homogeneous space* is a set together with a given transitive group of *symmetries*.

A metric space (X, d) is called (Grünbaum–Kelly) **metrically homogeneous metric space** if $\{d(x, z) : z \in X\} = \{d(y, z) : z \in X\}$ for any $x, y \in X$.

- ***c*-uniformly perfect metric space**

Every proper closed **metric ball** of radius r in a metric space has diameter at most $2r$. A metric space is called c**-uniformly perfect**, $0 < c \leqslant 1$, if this diameter is at least $2cr$.

- **Homeomorphic metric spaces**

Two metric spaces (X, d_X) and (Y, d_Y) are called **homeomorphic** (or *topologically isomorphic*) if there exists a *homeomorphism* from X to Y, i.e., a bijective function $f : X \to Y$ such that f and f^{-1} are *continuous* (the preimage of every open set in Y is open in X).

Two metric spaces (X, d_X) and (Y, d_Y) are called *uniformly isomorphic* if there exists a bijective function $f : X \to Y$ such that f and f^{-1} are *uniformly continuous* functions. The function f is *uniformly continuous* if, for any $\varepsilon > 0$, there exists $\delta > 0$ such that, for any $x, y \in X$, inequality $d_X(x, y) < \delta$ implies inequality $d_Y(f(x), f(y)) < \varepsilon$. A continuous function f is uniformly continuous if X is compact.

- ***C*-quasi-conformal metrical mapping**

Given metric spaces (X, d_X) and (Y, d_Y), a homeomorphism $f : X' \to Y'$ (where $X' \subset X$ and $Y' \subset Y$ are open sets) is called C**-quasi-conformal metrical mapping** if there exists a constant $C \geqslant 1$ such that

$$\limsup_{r\to 0} \frac{\max\{d_Y(f(x), f(y)):\ d(x, y) \leqslant r\}}{\min\{d_Y(f(x), f(y)):\ d(x, y) \geqslant r\}} \leqslant C$$

holds for each $x \in X'$.

- **Lipschitz mapping**

Let c be a positive constant. Given metric spaces (X, d_X) and (Y, d_Y), a function $f : X \to Y$ is called **Lipschitz mapping** (or, more exactly, *c-Lipschitz mapping*) if the inequality

$$d_Y\big(f(x), f(y)\big) \leqslant c d_X(x, y)$$

holds for all $x, y \in X$. The minimal such c, i.e.,

$$\sup_{x,y\in X,\ x\neq y} \frac{d_Y(f(x), f(y))}{d_X(x, y)},$$

is called *dilatation* of f.

An c-Lipschitz mapping is called **short mapping** if $c = 1$, and is called **contraction** if $c < 1$. Every contraction from a **complete** metric space into itself has an unique fixed point.

- **Bi-Lipschitz mapping**

Let $c > 1$ be a positive constant. Given metric spaces (X, d_X) and (Y, d_Y), a function $f : X \to Y$ is called **bi-Lipschitz mapping** (or *c-bi-Lipschitz mapping*, *c-embedding*) if there exists a positive real number r such that, for any $x, y \in X$, we have the following inequalities:

$$rd_X(x, y) \leqslant d_Y\big(f(x), f(y)\big) \leqslant crd_X(x, y).$$

The smallest c for which f is an c-bi-Lipschitz mapping is called **distortion** of f. Bourgain proved that every k-point metric space c-embeds into an Euclidean space with distortion $O(\ln k)$.

Two metrics d_1 and d_2 on X are called **bi-Lipschitz equivalent metrics** if there are positive constants c and C such that $cd_1(x, y) \leqslant d_2(x, y) \leqslant Cd_1(x, y)$ holds for all $x, y \in X$. In other words, the identity mapping is a bi-Lipschitz mapping from (X, d_1) into (X, d_2).

- **Dilation**

  Given a metric space (X, d) and a positive real number r, a function $f : X \to X$ is called **dilation** if $d(f(x), f(y)) = rd(x, y)$ holds for any $x, y \in X$.

- **Metric Ramsey number**

  For a given class $\mathcal{M}$ of metric spaces (usually, l_p-*spaces*), an integer $n \geqslant 1$, and a real number $c \geqslant 1$, the **metric Ramsey number** (or c-*metric Ramsey number*) $R_{\mathcal{M}}(c, n)$ is the largest integer m such that every n-point metric space has a subspace of size m that c-embeds into a member of $\mathcal{M}$ (see [BLMN05]).

- **c-isomorphism of metric spaces**

  Given two metric spaces (X, d_X) and (Y, d_Y), the *Lipschitz norm* $\|.\|_{Lip}$ on the set of all injective mappings $f : X \to Y$ is defined by

  $$\|f\|_{Lip} = \sup_{x,y\in X,\ x\neq y} \frac{d_Y(f(x), f(y))}{d_X(x, y)}.$$

  Two metric spaces X and Y are called c-**isomorphic** if there exists an injective mapping $f : X \to Y$ such that $\|f\|_{Lip}\|f^{-1}\|_{Lip} \leqslant c$.

- **Quasi-isometry**

  Given metric spaces (X, d_X) and (Y, d_Y), a function $f : X \to Y$ is called **quasi-isometry** if there exist real numbers $C > 0$ and c such that

  $$C^{-1}d_X(x, y) - c \leqslant d_Y\big(f(x), f(y)\big) \leqslant Cd(x, y) + c,$$

  and $Y = \bigcup_{x\in X} B_{d_Y}(f(x), c)$, i.e., for every point $y \in Y$, there exists a point $x \in X$ such that $d_Y(y, f(x)) \leqslant c$.

  A quasi-isometry with $C = 1$ is called **coarse isometry**.

- **Coarse embedding**

Given metric spaces (X, d_X) and (Y, d_Y), a function $f : X \to Y$ is called **coarse embedding** if there exist non-decreasing functions $\rho_1, \rho_2 : [0, \infty) \to [0, \infty)$ such that $\rho_1(d_X(x, y)) \leqslant d_Y(f(x), f(y)) \leqslant \rho_2(d_X(x, y))$ for all $x, y \in X$, and $\lim_{t\to\infty} \rho_1(t) = +\infty$.

Metrics d_1 and d_2 on X are called **coarsely equivalent metrics** if there exist non-decreasing functions $f, g : [0, \infty) \to [0, \infty)$ such that $d_1 \leqslant f(d_2)$ and $d_2 \leqslant g(d_1)$.

- **Short mapping**

Given metric spaces (X, d_X) and (Y, d_Y), a function $f : X \to Y$ is called **short mapping** (or *non-expansive mapping*, *semi-contraction*) if

$$d_Y\big(f(x), f(y)\big) \leqslant d_X(x, y)$$

holds for all $x, y \in X$. The function f is called *strictly short* if the inequality is strict for all $x \neq y$. A **submetry** is a short mapping such that image of any **metric ball** is a metric ball of the same radius. Any surjective short mapping $f : X \to X$ is an **isometry** if and only if (X, d_X) is a compact metric space.

Two subsets A and B of a metric space (X, d) are called (W.T. Gowers) **similar** if there exist short mappings $f : A \to X$, $g : B \to X$ and a small $\varepsilon > 0$ such that every point of A is within ε of some point of B, every point of B is within ε of some point of A, and $|d(x, g(f(x))) - d(y, f(g(y)))| \leqslant \varepsilon$ for every $x \in A$ and $y \in B$.

- **Category of metric spaces**

A *category* Ψ consists of a class $Ob\,\Psi$, whose elements are called *objects of the category*, and a class $Mor\,\Psi$, elements of which are called *morphisms of the category*. These classes have to satisfy the following conditions:

1. To each ordered pair of objects A, B is associated a set $H(A, B)$ of morphisms;
2. Each morphism belongs to only one set $H(A, B)$;
3. The composition $f \cdot g$ of two morphisms $f : A \to B$, $g : C \to D$ is defined if $B = C$ in which case it belongs to $H(A, D)$;
4. The composition of morphisms is associative;
5. Each set $H(A, A)$ contains, as an *identity*, a morphism id_A such that $f \cdot id_A = \alpha$ and $id_A \cdot g = g$ for any morphisms $f : X \to A$ and $g : A \to Y$.

The **category of metric spaces**, denoted by *Met* (see [Isbe64]), is a category which has metric spaces as objects and **short mappings** as morphisms. An unique **injective hull** exists in this category for every one of its objects; it can be identified with its **tight span**. In *Met*, the *monomorphisms* are injective short mappings, and *isomorphisms* are **isometries**.

- **Injective metric space**

A metric space (X, d) is called **injective** (or **hyperconvex**) if, for every *isometric embedding* $f : X \to X'$ of (X, d) into another metric space (X', d'), there exists a **short**

mapping f' from X' into X with $f' \cdot f = id_X$, i.e., X is a *retract* of X'. Equivalently, X is an *absolute retract*, i.e., a retract of every metric space into which it embeds isometrically.

- **Injective hull**

  The notion of **injective hull** is a generalization of the notion of **Cauchy completion**. Given a metric space (X, d), it can be embedded isometrically into an **injective** metric space $(\hat{X}, \hat{d})$; given any such *isometric embedding* $f : X \to \hat{X}$, there exists an unique smallest injective subspace $(\overline{X}, \overline{d})$ of $(\hat{X}, \hat{d})$ containing $f(X)$ which is called **injective hull** of X. It is isometrically identified with the **tight span** of (X, d).

  The metric space coincides with its injective hull if and only if it is an injective metric space.

- **Tight extension**

  An extension (X', d') of a metric space (X, d) is called **tight extension** if, for every semi-metric d'' on X' satisfying the conditions $d''(x_1, x_2) = d(x_1, x_2)$ for all $x_1, x_2 \in X$, and $d''(y_1, y_2) \leqslant d'(y_1, y_2)$ for any $y_1, y_2 \in X'$, one has $d''(y_1, y_2) = d'(y_1, y_2)$ for all $y_1, y_2 \in X'$.

  The **tight span** is the *universal tight extension* of X, i.e., it contains, up to canonical isometries, every tight extension of X, and it has no proper tight extension itself.

- **Tight span**

  Given a metric space (X, d) of finite diameter, consider the set $\mathbb{R}^X = \{f : X \to \mathbb{R}\}$. The **tight span** $T(X, d)$ of (X, d) is defined as the set $T(X, d) = \{f \in \mathbb{R}^X : \ f(x) = \sup_{y \in X}(d(x, y) - f(y))$ for all $x \in X\}$, endowed with the metric induced on $T(X, d)$ by the *sup norm* $\|f\| = \sup_{x \in X} |f(x)|$.

  The set X can be identified with the set $\{h_x \in T(X, d) : \ h_x(y) = d(y, x)\}$ or, equivalently, with the set $T^0(X, d) = \{f \in T(X, d) : \ 0 \in f(X)\}$. The **injective hull** $(\overline{X}, \overline{d})$ of X can be isometrically identified with the tight span $T(X, d)$ by

  $$\overline{X} \to T(X, d), \quad \overline{x} \to h_{\overline{x}} \in T(X, d) : \ h_{\overline{x}}(y) = \overline{d}\big(f(y), \overline{x}\big).$$

  For example, if $X = \{x_1, x_2\}$, then $T(X, d)$ is the interval of length $d(x_1, x_2)$. A metric space coincides with its tight span if and only if it is an **injective** metric space.

  The tight span of a metric space (X, d) of finite diameter can be considered as a polytopal complex. The dimension of this complex is called **Dress dimension** (or *combinatorial dimension*) of (X, d).

- **Real tree**

  A **complete** metric space (X, d) is called **real tree** (or $\mathbb{R}$-*tree*) if, for all $x, y \in X$, there exists an unique **metric curve** from x to y, and this curve is a *geodesic segment*. The real trees are exactly **tree-like** metric spaces which are **geodesic**.

  If X is the set of all bounded subsets of $\mathbb{R}$ containing their infima with the metric on X defined by $d(A, B) = 2 \max\{\sup x \Delta y, \inf x, \inf y\} - (\inf x + \inf y)$, then the metric space (X, d) is a real tree.

The **tree-like** metric spaces are by definition the metric subspaces of the real trees; real trees are exactly the **injective** metric spaces among tree-like spaces. If (X, d) is a finite metric space, then the **tight span** $T(X, d)$ is a real tree and can be viewed as an edge-weighted graph-theoretical tree. A metric space (X, d) is a real tree if and only if it is complete, arc-wise connected, and satisfies the **four-point inequality**.

1.3. GENERAL DISTANCES

- **Discrete metric**

  Given a set X, the **discrete metric** (or **trivial metric**) d is a metric on X, defined by $d(x, y) = 1$ for all distinct $x, y \in X$ (and $d(x, x) = 0$). The metric space (X, d) is called *discrete metric space*.

- **Indiscrete semi-metric**

  Given a set X, the **indiscrete semi-metric** d is a semi-metric on X, defined by $d(x, y) = 0$ for all $x, y \in X$.

- **Equidistant metric**

  Given a set X and a positive real number t, the **equidistant metric** d is a metric on X, defined by $d(x, y) = t$ for all distinct $x, y \in X$ (and $d(x, x) = 0$).

- **(1, 2)-B-metric**

  Given a set X, the $(1, 2)$-B**-metric** d is a metric on X such that, for any $x \in X$, the number of points $y \in X$ with $d(x, y) = 1$ is at most B, and all other distances are equal to 2. The $(1, 2)$-B**-metric** is the **truncated metric** of a graph with maximal vertex degree B.

- **Induced metric**

  An **induced metric** (or *relative metric*) is a restriction d' of a metric d on a set X to a subset X' of X.

  A metric space (X', d') is called **metric subspace** of the metric space (X, d), and the metric space (X, d) is called **metric extension** of (X', d').

- **Dominating metric**

  Given metrics d and d_1 on a set X, d_1 **dominates** d if $d_1(x, y) \geqslant d(x, y)$ for all $x, y \in X$.

- **Equivalent metrics**

  Two metrics d_1 and d_2 on a set X are called **equivalent** if they define the same *topology* on X, i.e., if, for every point $x_0 \in X$, every open **metric ball** with center at x_0 defined with respect to d_1, contains an open metric ball with the same center but defined with respect to d_2, and conversely.

Two metrics d_1 and d_2 are equivalent if and only if, for every $\varepsilon > 0$ and every $x \in X$, there exists $\delta > 0$ such that $d_1(x, y) \leqslant \delta$ implies $d_2(x, y) \leqslant \varepsilon$ and, conversely, $d_2(x, y) \leqslant \delta$ implies $d_1(x, y) \leqslant \varepsilon$.

- **Complete metric**

Given a metric space (X, d), a sequence $\{x_n\}_n$, $x_n \in X$, is said to have **convergence to** $x^* \in X$ if $\lim_{n\to\infty} d(x_n, x^*) = 0$, i.e., for any $\varepsilon > 0$, there exists $n_0 \in \mathbb{N}$ such that $d(x_n, x^*) < \varepsilon$ for any $n > n_0$.

A sequence $\{x_n\}_n$, $x_n \in X$, is called *Cauchy sequence* if, for any $\varepsilon > 0$, there exists $n_0 \in \mathbb{N}$ such that $d(x_n, x_m) < \varepsilon$ for any $m, n > n_0$.

A metric space (X, d) is called **complete metric space** if every its *Cauchy sequence* converges. In this case the metric d is called **complete metric**.

- **Cauchy completion**

Given a metric space (X, d), its **Cauchy completion** is a metric space (X^*, d^*) on the set X^* of all equivalence classes of *Cauchy sequences*, where the sequence $\{x_n\}_n$ is called *equivalent to* $\{y_n\}_n$ if $\lim_{n\to\infty} d(x_n, y_n) = 0$. The metric d^* on X^* is defined by

$$d^*(x^*, y^*) = \lim_{n\to\infty} d(x_n, y_n)$$

for any $x^*, y^* \in X^*$, where $\{x_n\}_n$ (respectively, $\{y_n\}_n$) is any element in the equivalence class x^* (respectively, y^*).

The Cauchy completion (X^*, d^*) is unique, up to isometry, **complete** metric space, into which the metric space (X, d) embeds as a *dense* metric subspace.

The Cauchy completion of the metric space $(\mathbb{Q}, |x - y|)$ of rational numbers is the *real line* $(\mathbb{R}, |x - y|)$. A **Banach space** is the Cauchy completion of a *normed vector space* $(V, \|.\|)$ with the **norm metric** $\|x - y\|$. A **Hilbert space** correspond to the case an *inner product norm* $\|x\| = \sqrt{\langle x, x\rangle}$.

- **Bounded metric**

A metric (distance) d on a set X is called **bounded** if there exists a constant $C > 0$ such that $d(x, y) \leqslant C$ for any $x, y \in X$.

For example, given a metric d on X, the metric D on X, defined by $D(x, y) = \frac{d(x,y)}{1+d(x,y)}$, is bounded with $C = 1$.

A metric space (X, d) with a bounded metric d is called **bounded metric space**.

- **Totally bounded metric space**

A metric space (X, d) is called **totally bounded** if, for every positive real number r, there exist finitely many open **metric balls** of radius r whose union is equal to X. Every totally bounded metric space is **bounded** and **separable**.

- **Separable metric space**

A metric space is called **separable** if it contains a countable *dense* subset, i.e., some countable subset with which all its elements can be approached.

A metric space is separable if and only if it is **second-countable**, and if and only if it is **Lindelöf**. Every **totally bounded** metric space is separable.

- **Metric compactum**

A **metric compactum** (or *compact metric space*) is a metric space in which every sequence has *Cauchy subsequence*, and those subsequences are convergent. A metric space is compact if and only if it is **totally bounded** and **complete**. A subset of the Euclidean space $\mathbb{E}^n$ is compact if and only if it is bounded and closed.

- **Proper metric space**

A metric space is called **proper** if every closed **metric ball** in this space is compact. Every proper metric space is **complete**.

- **UC metric space**

A metric space is called **UC metric space** (or *Atsuji space*) if any continuous function from it into an arbitrary metric space is *uniformly continuous*.

Every **metric compactum** is an UC metric space. Every UC metric space is **complete**.

- **Polish space**

A **Polish space** is a **complete separable** metric space. A metric space is called **Souslin space** if it is a continuous image of a Polish space.

A **metric triple** (or *mm-space*) is a Polish space (X, d) with a *Borel probability measure* μ, i.e., a non-negative real function μ on the *Borel sigma-algebra* $\mathcal{F}$ of X with the following properties: $\mu(\emptyset) = 0$, $\mu(X) = 1$, and $\mu(\bigcup_n A_n) = \sum_n \mu(A_n)$ for any finite or countable collection of pairwise disjoint sets $A_n \in \mathcal{F}$.

Given a topological space (X, τ), a *sigma-algebra* on X is a collection $\mathcal{F}$ of subsets of X with the following properties: $\emptyset \in \mathcal{F}$, $X \backslash U \in \mathcal{F}$ for $U \in \mathcal{F}$, and $\bigcup_n A_n \in \mathcal{F}$ for a finite or countable collection $\{A_n\}_n$, $A_n \in \mathcal{F}$. The sigma-algebra on X which is related to the topology of X, i.e., consists of all open and closed sets of X, is called *Borel sigma-algebra* of X. Any metric space is a *Borel space*, i.e., a set, equipped with a Borel sigma-algebra.

- **Norm metric**

Given a *normed vector space* $(V, \|.\|)$, the **norm metric** on V is defined by

$$\|x - y\|.$$

The metric space $(V, \|x - y\|)$ is called **Banach space** if it is **complete**. Examples of norm metrics are l_p- and L_p**-metrics**, in particular, the **Euclidean metric**. On $\mathbb{R}$ all l_p-metrics coincide with the **natural metric** $|x - y|$ (cf. Chapter 5).

- **Path metric**

Given a *connected graph* $G = (V, E)$, its **path metric** d_{path} is a metric on V, defined as the length (i.e., the number of edges) of a shortest path connecting two given vertices x and y from V (cf. Chapter 15).

- **Editing metric**

  Given a finite set X and a finite set $\mathcal{O}$ of (unary) *editing operations* on X, the **editing metric** on X is the **path metric** of the graph with the vertex-set X and xy being an edge if y can be obtained from x by one of the operations from $\mathcal{O}$.

- **Gallery metric**

  A *chamber system* is a set X (whose elements are referred to as *chambers*) equipped with n equivalence relations $\sim_i$, $1 \leqslant i \leqslant n$. A *gallery* is a sequence of chambers $x_1, \dots, x_m$ such that $x_i \sim_j x_{i+1}$ for every i and some j depending on i. The **gallery metric** is an extended metric on X which is the length of the shortest gallery connecting x and $y \in X$ (and is equal to ∞ if there is no connecting gallery). The gallery metric is the **path metric** of the graph with the vertex-set X and xy being an edge if $x \sim_i y$ for some $1 \leqslant i \leqslant n$.

- **Riemannian metric**

  Given a connected n-dimensional smooth *manifold* M^n, its **Riemannian metric** is a collection of positive-definite symmetric bilinear forms $((g_{ij}))$ on the tangent spaces of M^n which varies smoothly from point to point. The length of a curve γ on M^n is expressed as $\int_\gamma \sqrt{\sum_{i,j} g_{ij} dx_i dx_j}$, and the **intrinsic metric** on M^n, sometimes called also **Riemannian distance** (between points of M^n), is defined as the infimum of lengths of curves, connecting any two given points $x, y \in M^n$ (cf. Chapter 7).

- **Projective metric**

  A **projective metric** d is a continuous metric on $\mathbb{R}^n$ which satisfies the condition

  $$d(x, z) = d(x, y) + d(y, z)$$

  for any collinear points x, y, z lying in that order on a common line. The Hilbert 4th problem asked in 1900 to classify such metrics; it is done only for dimension $n = 2$ ([Amba76]); cf. Chapter 6.

  Every **norm metric** on $\mathbb{R}^n$ is projective. Every projective metric on $\mathbb{R}^2$ is a **hypermetric**.

- **Convex distance function**

  Given a convex region $B \subset \mathbb{R}^n$ that contains the origin in its interior, the **convex distance function** $d_B(x, y)$ is defined by

  $$\min\{\lambda : \ x - y \in \lambda B\}.$$

  If B is centrally-symmetric with respect to the origin, then d_B is a **Minkowskian metric** whose unit ball is B.

- **Product metric**

  Given n metric spaces (X_1, d_1), (X_2, d_2), ..., (X_n, d_n), the **product metric** is a metric on the *Cartesian product* $X_1 \times X_2 \times \cdots \times X_n = \{x = (x_1, x_2, \dots, x_n) : \ x_1 \in X_1, \dots, x_n \in X_n\}$, defined as a function of $d_1, \dots, d_n$ (cf. Chapter 4).

- **Hamming metric**

The **Hamming metric** d_H is a metric on $\mathbb{R}^n$, defined by

$$\left|\{i:\ 1 \leqslant i \leqslant n,\ x_i \neq y_i\}\right|.$$

On binary vectors $x, y \in \{0, 1\}^n$ the Hamming metric and the l_1**-metric** coincide.

- **Lee metric**

Given $m, n \in \mathbb{N}$, $m \geqslant 2$, the **Lee metric** d_{Lee} is a metric on $\mathbb{Z}_m^n = \{0, 1, \dots, m-1\}^n$, defined by

$$\sum_{1 \leqslant i \leqslant n} \min\{|x_i - y_i|, m - |x_i - y_i|\}.$$

The metric space (Z_m^n, d_{Lee}) is a discrete analog of the *elliptic space*.

- **Symmetric difference metric**

Given a *measure space* $(\Omega, \mathcal{A}, \mu)$, the **symmetric difference semi-metric** (or *measure semi-metric*) $d_\triangle$ is a semi-metric on the set $\mathcal{A}_\mu = \{A \in \mathcal{A}:\ \mu(A) < \infty\}$, defined by $\mu(A \triangle B)$, were $A \triangle B = (A \cup B) \backslash (A \cap B)$ is the *symmetric difference* of the sets A and $B \in \mathcal{A}_\mu$.

The value $d_\triangle(A, B) = 0$ if and only if $\mu(A \triangle B) = 0$, i.e., if A and B are equal *almost everywhere*. Identifying two sets $A, B \in \mathcal{A}_\mu$ if $\mu(A \triangle B) = 0$, we obtain the **symmetric difference metric** (or **Fréchet–Nikodym–Aronszayn distance**, **measure metric**).

If μ is the *counting measure*, i.e., $\mu(A) = |A|$ is the number of elements in A, then $d_\triangle(A, B) = |A \triangle B|$. In this case $|A \triangle B| = 0$ if and only if $A = B$. The **Johnson distance** between k-sets A and B is $\frac{|A \triangle B|}{2} = k - |A \cap B|$.

- **Enomoto–Katona metric**

Given a finite set X and an integer k, $2k \leqslant |X|$, the **Enomoto–Katona metric** is the distance between unordered pairs (X_1, X_2) and Y_1, Y_2 of disjoint k-subsets of X, defined by

$$\min\{|X_1 \setminus Y_1| + |X_2 \setminus Y_2|, |X_1 \setminus Y_2| + |X_2 \setminus Y_1|\}.$$

- **Steinhaus distance**

Given a *measure space* $(\Omega, \mathcal{A}, \mu)$, the **Steinhaus distance** d_{St} is a semi-metric on the set $\mathcal{A}_\mu = \{A \in \mathcal{A}:\ \mu(A) < \infty\}$, defined by

$$\frac{\mu(A \triangle B)}{\mu(A \cup B)} = 1 - \frac{\mu(A \cap B)}{\mu(A \cup B)}$$

if $\mu(A \cup B) > 0$ (and is equal to 0 if $\mu(A) = \mu(B) = 0$). It becomes a metric on the set of equivalence classes of elements from $\mathcal{A}_\mu$; here $A, B \in \mathcal{A}_\mu$ are called *equivalent* if $\mu(A \triangle B) = 0$.

If $d_\triangle$ is the **symmetric difference metric**, then $d_{St} = 2d_\triangle^\emptyset$, where, for a given metric d on a set X and a given point $p \in X$, the **transform metric** d^p on X is defined by

$$d^p(x,y) = \frac{d(x,y)}{d(x,p) + d(y,p) + d(x,y)}.$$

The **biotope distance** (or **Tanimoto distance**) $\frac{|A \triangle B|}{|A \cup B|}$ is the special case of Steinhaus distance, obtained for the *counting measure* $\mu(A) = |A|$.

- **Point-set distance**

Given a metric space (X,d), the **point-set distance** $d(x,A)$ between a point $x \in X$ and a subset A of X is defined as

$$\inf_{y \in A} d(x,y).$$

For any $x, y \in X$ and for any non-empty subset A of X, we have the following version of the triangle inequality: $d(x,A) \leqslant d(x,y) + d(y,A)$ (cf. **distance map**).

- **Set-set distance**

Given a metric space (X,d), the **set-set distance** between two subsets A and B of X is defined by

$$\inf_{x \in A,\, y \in B} d(x,y).$$

In Data Analysis, the set-set distance between clusters is called **single linkage**, while $\sup_{x \in A,\, y \in B} d(x,y)$ is called **complete linkage**.

- **Hausdorff metric**

Given a metric space (X,d), the **Hausdorff metric** (or *two-sided Hausdorff distance*) d_{Haus} is a metric on the family $\mathcal{F}$ of all compact subsets of X, defined by

$$\max\{d_{dHaus}(A,B), d_{dHaus}(B,A)\},$$

where $d_{dHaus}(A,B) = \max_{x \in A} \min_{y \in B} d(x,y)$ is the **directed Hausdorff distance** (or *one-sided Hausdorff distance*) from A to B. In other words, $d_{Haus}(A,B)$ is the minimal number ε (called also **Blaschke distance**) such that closed ε-neighborhood of A contains B and closed ε-neighborhood of B contains A. It holds also that $d_{Haus}(A,B)$ is equal to

$$\sup_{x \in X} \big| d(x,A) - d(x,B) \big|,$$

where $d(x,A) = \min_{y \in A} d(x,y)$ is the **point-set distance**. The Hausdorff metric is not a **norm metric**.

If the above definition is extended for non-compact closed subsets A and B of X, then $d_{Haus}(A,B)$ can be infinite, i.e., it becomes an extended metric. For not necessarily closed subsets A and B of X, the **Hausdorff semi-metric** between them is defined as the Hausdorff metric between their closures. If X is finite, d_{Haus} is a metric on the class of all subsets of X.

M.D. Wills proved that the Hausdorff distance between two non-empty bounded closed *convex* subsets of a metric space with a **norm metric** is equal to the Hausdorff distance between their boundaries.

- **L_p-Hausdorff distance**

Given a finite metric space (X, d), the **L_p-Hausdorff distance** ([Badd92]) between two subsets A and B of X is defined by

$$\left(\sum_{x \in X} \left|d(x, A) - d(x, B)\right|^p\right)^{\frac{1}{p}},$$

where $d(x, A)$ is the **point-set distance**. The usual **Hausdorff metric** corresponds to the case $p = \infty$.

- **Generalized G-Hausdorff metric**

Given a group $(G, \cdot, e)$ acting on a metric space (X, d), the **generalized G-Hausdorff metric** between two closed bounded subsets A and B of X is defined by

$$\min_{g_1, g_2 \in G} d_{Haus}\big(g_1(A), g_2(B)\big),$$

where d_{Haus} is the **Hausdorff metric**. If $d(g(x), g(y)) = d(x, y)$ for any $g \in G$ (i.e., if the metric d is *left-invariant* with respect of G), then above metric is equal to $\min_{g \in G} d_{Haus}(A, g(B))$.

- **Gromov–Hausdorff metric**

The **Gromov–Hausdorff metric** is a metric on the set of all *isometry classes* of compact metric spaces, defined by

$$\inf d_{Haus}\big(f(X), g(Y)\big)$$

for any two classes X^* and Y^* with the representatives X and Y, respectively, where d_{Haus} is the **Hausdorff metric**, and the minimum is taken over all metric spaces M and all *isometric embeddings* $f : X \to M$, $g : Y \to M$. The corresponding metric space is called *Gromov–Hausdorff space*.

- **Fréchet metric**

Let (X, d) be a metric space. Consider a set $\mathcal{F}$ of all continuous mappings $f : A \to X$, $g : B \to X, \dots,$ where $A, B, \dots$ are subsets of $\mathbb{R}^n$, homeomorphic to $[0, 1]^n$ for a fixed dimension $n \in \mathbb{N}$.

The **Fréchet semi-metric** d_F is a semi-metric on $\mathcal{F}$, defined by

$$\inf_{\sigma} \sup_{x \in A} d\big(f(x), g\big(\sigma(x)\big)\big),$$

where the infimum is taken over all orientation preserving homeomorphisms $\sigma : A \to B$. It becomes the **Fréchet metric** on the set of equivalence classes $f^* = \{g\colon d_F(g, f) = 0\}$.

- **Banach–Mazur distance**

The **Banach–Mazur distance** d_{BM} between two n-dimensional *normed spaces* V and W is defined by

$$\ln \inf_T \|T\| \cdot \|T^{-1}\|,$$

where the infimum is taken over all isomorphisms $T : V \to W$. It can be written also as $\ln d(V, W)$, where the number $d(V, W)$ is the smallest positive $d \geqslant 1$ such that $\overline{B}^n_W \subset T(\overline{B}^n_V) \subset d\,\overline{B}^n_W$ for some linear invertible transformation $T : V \to W$. Here $\overline{B}^n_V = \{x \in V\colon \|x\|_V \leqslant 1\}$ and $\overline{B}^n_W = \{x \in W\colon \|x\|_W \leqslant 1\}$ are the *unit balls* of the normed spaces $(V, \|.\|_V)$ and $(W, \|.\|_W)$, respectively.

$d_{BM}(V, W) = 0$ if and only if V and W are *isometric*, and it becomes a metric on the set X^n of all equivalence classes of n-dimensional normed spaces, where $V \sim W$ if they are isometric. The pair (X^n, d_{BM}) is a compact metric space which is called **Banach–Mazur compactum**.

Gluskin–Khrabrov distance (or *modified Banach–Mazur distance*) is defined by

$$\inf\left\{||T||_{X\to Y}\colon\ |det\,T| = 1\right\} \cdot \left\{||T||_{Y\to X}\colon\ |det\,T| = 1\right\}.$$

- **Lipschitz distance**

Given two metric spaces (X, d_X) and (Y, d_Y), the *Lipschitz norm* $\|.\|_{Lip}$ on the set of all injective functions $f : X \to Y$ is defined by

$$\|f\|_{Lip} = \sup_{x,y\in X,\, x\neq y} \frac{d_Y(f(x), f(y))}{d_X(x, y)}.$$

The **Lipschitz distance** between metric spaces (X, d_X) and (Y, d_Y) is defined by

$$\ln \inf_f \|f\|_{Lip} \cdot \|f^{-1}\|_{Lip},$$

where the infimum is taken over all bijective functions $f : X \to Y$. Equivalently, it is infimum of numbers $\ln \alpha$ such that there exists a bijective **bi-Lipschitz mapping** between (X, d_X) and (Y, d_Y) with constants $\exp(-\alpha)$, $\exp(\alpha)$. It becomes a metric on the set of all isometry classes of compact metric spaces.

This distance is an analog to the **Banach–Mazur distance** and, in the case of finite-dimensional real Banach spaces, coincides with it. It coincides also with the **Hilbert projective metric** on non-negative projective spaces, obtained by starting with $\mathbb{R}^n_+$ and identifying any point x with cx, $c > 0$.

- **Lipschitz distance between measures**

Given a compact metric space (X, d), the *Lipschitz semi-norm* $\|.\|_{Lip}$ on the set of all functions $f : X \to \mathbb{R}$ is defined by

$$\|f\|_{Lip} = \sup_{x,y \in X,\, x \neq y} \frac{|f(x) - f(y)|}{d(x, y)}.$$

The **Lipschitz distance between measures** μ and ν on X is defined by

$$\sup_{\|f\|_{Lip} \leqslant 1} \int f d(\mu - \nu).$$

If μ and ν are probability measures, then it is **Kantorovich–Mallows–Monge–Wasserstein metric**.

An analog of the Lipschitz distance between measures for the *state space* of *unital C^*-algebra* is the **Connes metric**.

- **Compact quantum metric space**

Let V be a *normed space* (or, more generally, a locally convex *topological vector space*), and let V' be its *continuous dual*, i.e., the set of all continuous linear functionals f on V. The *weak* topology* (or *Gelfand topology*) on V' is defined as the weakest (i.e., with the fewest open sets) topology on V' such that, for every $x \in V$, the map $F_x : V' \to \mathbb{R}$ defined by $F_x(f) = f(x)$ for all $f \in V'$, remains continuous.

An *order-unit space* is a *partially ordered* real (complex) vector space $(A, \preceq)$ with a distinguished element e, called *order unit*, which satisfies the following properties:

1. For any $a \in A$, there exists $r \in \mathbb{R}$ with $a \preceq re$;
2. If $a \in A$ and $a \preceq re$ for all positive $r \in \mathbb{R}$, then $a \preceq 0$ (*Archimedean property*).

The main example of an order-unit space is the vector space of all self-adjoint elements in an *unital C^*-algebra* with the identity element being order unit. Here an *C^*-algebra* is a *Banach algebra* over $\mathbb{C}$ equipped with a special *involution*. It is called *unital* if it has an *unit* (multiplicative identity element); such C^*-algebras are also called, roughly, *compact non-commutative topological spaces*. The typical example of an unital C^*-algebra is a complex algebra of linear operators on a complex **Hilbert space** which is topologically closed in the norm topology of operators, and is closed under the operation of taking adjoints of operators.

The *state space* of an order-unit space $(A, \preceq, e)$ is the set $S(A) = \{f \in A'_+ : \ \|f\| = 1\}$ of *states*, i.e., continuous linear functionals f with $\|f\| = f(e) = 1$.

Rieffel's **compact quantum metric space** is a pair $(A, \|.\|_{Lip})$, where $(A, \preceq, e)$ is an order-unit space, and $\|.\|_{Lip}$ is a semi-norm on A (with values in $[0, +\infty]$), called *Lipschitz semi-norm*, which satisfies the following conditions:

1. For $a \in A$, it holds $\|a\|_{Lip} = 0$ if and only if $a \in \mathbb{R}e$;
2. the metric $d_{Lip}(f, g) = \sup_{a \in A:\, \|a\|_{Lip} \leqslant 1} |f(a) - g(a)|$ generates on the state space $S(A)$ its weak* topology.

So, one has an usual metric space $(S(A), d_{Lip})$. If the order-unit space $(A, \preceq, e)$ is an C^*-algebra, then d_{Lip} is the **Connes metric**, and if, moreover, the C^*-algebra is non-commutative, the metric space $(S(A), d_{Lip})$ is called **non-commutative metric space**.

The expression *quantum metric space* comes from the belief, by many experts in Quantum Gravity and String Theory, that the Planck-scale geometry of *space-time* is similar to one coming from such non-commutative C^*-algebras. For example, Non-commutative Field Theory supposes that on sufficiently small (quantum) distances, the spatial coordinates do not commute, i.e., it is impossible to measure exactly the position of a particle with respect to more than one axis.

- **Universal metric space**

  A metric space (U, d) is called **universal** for a collection $\mathcal{M}$ of metric spaces if any metric space (M, d_M) from $\mathcal{M}$ is *isometrically embeddable* in (U, d), i.e., there exists a mapping $f : M \to U$ which satisfies to $d_M(x, y) = d(f(x), f(y))$ for any $x, y \in M$.

  The **Urysohn space** is a **homogeneous complete separable** metric space which is the universal metric space for all **Polish** (i.e., complete separable) metric spaces.

  The **Hilbert cube** is the universal metric space for the class of metric spaces with a countable base.

  The *graphic metric space* of the *random graph* (which can be defined as the set of all prime numbers $p \equiv 1 \pmod 4$ with pq being an edge if p is a quadratic residue modulo q) is the universal metric space for any finite or countable metric space with distances 0, 1 and 2 only. It is a discrete analog of the Urysohn space.

- **Constructive metric space**

  A **constructive metric space** is a pair (X, d), where X is some set of constructive objects (usually, words over an alphabet), and d is an algorithm converting any pair of elements of X into a constructive real number $d(x, y)$ such that d becomes a metric on X.

- **Effective metric space**

  Let $\{x_n\}_{n\in\mathbb{N}}$ be a sequence of elements from a given **complete** metric space (X, d) such that the set $\{x_n : n \in \mathbb{N}\}$ is *dense* in (X, d). Let $\mathcal{N}(m, n, k)$ be the *Cantor number* of a triple $(n, m, k) \in \mathbb{N}^3$, and let $\{q_k\}_{k\in\mathbb{N}}$ be a fixed total standard numbering of the set $\mathbb{Q}$ of rational numbers.

  The triple $(X, d, \{x_n\}_{n\in\mathbb{N}})$ is called **effective metric space** ([Hemm02]) if the set $\{\mathcal{N}(n, m, k) : d(x_m, x_n) < q_k\}$ is recursively enumerable.

Chapter 2

Topological Spaces

A *topological space* (X, τ) is a set X with a *topology* τ, i.e., a collection of subsets of X with the following properties:

1. $X \in \tau, \emptyset \in \tau$;
2. If $A, B \in \tau$, then $A \cap B \in \tau$;
3. For any collection $\{A_\alpha\}_\alpha$, if all $A_\alpha \in \tau$, then $\bigcup_\alpha A_\alpha \in \tau$.

The sets in τ are called *open sets*, and their complements are called *closed sets*. A *base* of the topology τ is a collection of open sets such that every open set is an union of sets in the base. The coarsest topology has two open sets, the empty set and X, and is called *trivial topology* (or *indiscrete topology*). The finest topology contains all subsets as open sets, and is called *discrete topology*.

In a metric space (X, d) define the *open ball* as the set $B(x, r) = \{y \in X : d(x, y) < r\}$, where $x \in X$ (the *center* of the ball), and $r \in \mathbb{R}, r > 0$ (the *radius* of the ball). A subset of X which is the union of (finitely or infinitely many) open balls, is called *open set*. Equivalently, a subset U of X is called *open* if, given any point $x \in U$, there exists a real number $\varepsilon > 0$ such that, for any point $y \in X$ with $d(x, y) < \varepsilon$, $y \in U$. Any metric space is a topological space, the topology (**metric topology**, *topology induced by the metric* d) being the set of all open sets. The metric topology is always T_4 (see below a list of topological spaces). A topological space which can arise in this way from a metric space, is called **metrizable space**. A *semi-metric topology* is a topology on X induced similarly by a semi-metric d on X. In general, this topology is not even T_0.

Given a topological space (X, τ), a *neighborhood* of a point $x \in X$ is a set containing an open set which in turn contains x. The *closure* of a subset of a topological space is the smallest closed set, which contains it. An *open cover* of X is a collection $\mathcal{L}$ of open sets, the union of which is X; its *subcover* is a cover $\mathcal{K}$ such that every member of $\mathcal{K}$ is a member of $\mathcal{L}$; its *refinement* is a cover $\mathcal{K}$, where every member of $\mathcal{K}$ is a subset of some member of $\mathcal{L}$. A collection of subsets of X is called *locally finite* if every point of X has a neighborhood which meets only finitely many of these subsets. A subset $A \subset X$ is called *dense* if it has non-empty intersection with every non-empty open set, or, equivalently, if the only closed set containing it is X. In a metric space (X, d), a *dense set* is a subset $A \subset X$ such that, for any $x \in X$ and any $\varepsilon > 0$, there exists $y \in A$, satisfies to $d(x, y) < \varepsilon$. A *local base* of a point $x \in X$ is a collection $\mathcal{U}$ of neighborhoods of x such that every neighborhood of x contains some member of $\mathcal{U}$.

A function from one topological space to another is called *continuous* if the preimage of every open set is open. Roughly, given $x \in X$, all points close to x map to points close to $f(x)$. A function f from one metric space (X, d_X) to another metric space (Y, d_Y) is *continuous* at the point $c \in X$ if, for any positive real number ε, there exists a positive real number δ such that all $x \in X$ satisfying $d_X(x, c) < \delta$ will also satisfy $d_Y(f(x), f(y)) < \varepsilon$; the function is continuous on an interval I if it is continuous at any point of I.

The following classes of topological spaces (up to T_4) include any metric space.

- ***T_0*-space**

  An T_0**-space** (or *Kolmogorov space*) is a topological space (X, τ) fulfilling the *T_0-separation axiom*: for every two points $x, y \in X$ there exists an open set U such that $x \in U$ and $y \notin U$, or $y \in U$ and $x \notin U$ (every two points are *topologically distinguishable*).

- ***T_1*-space**

  An T_1**-space** (or *Fréchet space*) is a topological space (X, τ) fulfilling the *T_1-separation axiom*: for every two points $x, y \in X$ there exist two open sets U and V such that $x \in U$, $y \notin U$, and $y \in V$, $x \notin V$ (every two points are *separated*). T_1-spaces are always T_0.

- ***T_2*-space**

  An T_2**-space** (or **Hausdorff space**, *separated space*) is a topological space (X, τ) fulfilling the *T_2-axiom*: every two points $x, y \in X$ have disjoint neighborhoods. T_2-spaces are always T_1.

- **Regular space**

  A **regular space** is a topological space in which every neighborhood of a point contains a closed neighborhood of the same point.

- ***T_3*-space**

  An T_3**-space** (or *Vietoris space, regular Hausdorff space*) is a topological space which is T_1 and **regular**.

- **Completely regular space**

  A **completely regular space** (or *Tychonoff space*) is a **Hausdorff space** (X, τ) in which any closed set A and any $x \notin A$ are *functionally separated*.

  Two subsets A and B of X are *functionally separated* if there exists a continuous function $f : X \to [0, 1]$ such that $f(x) = 0$ for any $x \in A$, and $f(y) = 1$ for any $y \in B$.

- **Moore space**

  A **Moore space** is a **regular space** with a *development*.

  A *development* is a sequence $\{\mathcal{U}_n\}_n$ of open covers such that, for every $x \in X$ and every open set A containing x, there exists n such that $St(x, \mathcal{U}_n) = \bigcup\{U \in \mathcal{U}_n : x \in U\} \subset A$, i.e., $\{St(x, \mathcal{U}_n)\}_n$ is a *neighborhood base* at x.

- **Normal space**

  A **normal space** is a topological space in which, for any two disjoint closed sets A and B, there exist two disjoint open sets U and V such that $A \subset U$, and $B \subset V$.

- **T_4-space**

An **T_4-space** (or *Tietze space, normal Hausdorff space*) is a topological space which is T_1 and **normal**. Any metric space (X, d) is an T_4-space.

- **Completely normal space**

A **completely normal space** is a topological space in which any two separated sets have disjoint neighborhoods.

Sets A and B are *separated* in X if each is disjoint from the other's closure.

- **T_5-space**

An **T_5-space** (or *completely normal Hausdorff space*) is a topological space which is **completely normal** and T_1. T_5-spaces are always T_4.

- **Separable space**

A **separable space** is a topological space which has a countable dense subset.

- **Lindelöf space**

A **Lindelöf space** is a topological space in which every open cover has a countable subcover.

- **First-countable space**

A topological space is called **first-countable** if every point has a countable local base. Any metric space is first-countable.

- **Second-countable space**

A topological space is called **second-countable** if its topology has a countable base. Second-countable spaces are always **separable**, **first-countable**, and **Lindelöf**.

For metric spaces the properties of being second-countable, **separable**, and **Lindelöf** are all equivalent.

The Euclidean space $\mathbb{E}^n$ with its usual topology is second-countable.

- **Baire space**

A **Baire space** is a topological space in which every intersection of countably many dense open sets is dense.

- **Connected space**

A topological space (X, τ) is called **connected** if it is not the union of a pair of disjoint non-empty open sets. In this case the set X is called *connected set.*

A topological space (X, τ) is called **locally connected** if every point $x \in X$ has a local base consisting of connected sets.

A topological space (X, τ) is called **path-connected** (or 0-*connected*) if for every points $x, y \in X$ there is a *path* γ from x to y, i.e., a continuous function $\gamma : [0, 1] \to X$ with $\gamma(x) = 0, \gamma(y) = 1$.

A topological space (X, τ) is called **simply connected** (or 1-*connected*) if it consists of one piece, and has no circle-shaped "holes" or "handles" or, equivalently, if every continuous curve of X is *contractible*, i.e., can be reduced to one of its points by a *continuous deformation*.

- **Paracompact space**

A topological space is called **paracompact** if every its open cover has an open locally finite refinement. Any metric space (X, d) is paracompact.

- **Locally compact space**

A topological space is called **locally compact** if every point has a local base consisting of compact neighborhoods. Roughly speaking, every small portion of the space looks like a small portion of a **compact space**. The Euclidean spaces $\mathbb{E}^n$ are locally compact. The spaces $\mathbb{Q}_p$ of *p-adic numbers* are locally compact.

- **Totally bounded space**

A topological space is called **totally bounded** if it can be covered by finitely many subsets of any fixed size. A metric space is totally bounded if for every positive real number r there exist finitely many *open balls* of radius r, whose union is equal to X. Every totally bounded metric space is **bounded**.

- **Compact space**

A topological space (X, τ) is called **compact** if every open cover of X has a finite subcover. In this case the set X is called *compact set*.

Compact spaces are always **Lindelöf**, **totally bounded**, and **paracompact**. A metric space is compact if and only if it is **complete** and **totally bounded**. A subset of an Euclidean space $\mathbb{E}^n$ is compact if and only if it is closed and bounded.

There exists a number of topological properties which are equivalent to compactness in metric spaces, but are inequivalent in general topological spaces. Thus, a metric space is compact if and only if it is a *sequentially compact space* (every sequence has a convergent subsequence), or a *countably compact space* (every countable open cover has a finite subcover), or a *pseudo-compact space* (every real-valued continuous function on the space is bounded), or a *weakly countably compact space* (every infinite subset has an accumulation point).

- **Locally convex space**

A *topological vector space* is a real (complex) vector space V which is a **Hausdorff space** with continuous vector addition and scalar multiplication. It is called **locally convex** if its topology has a base, where each member is a *convex* set.

A subset A of V is called *convex* if, for all $x, y \in A$ and all $t \in [0, 1]$, the point $tx + (1 - t)y \in A$, i.e., every point on the *line segment* connecting x and y belongs to A.

Any metric space $(V, \|x - y\|)$ on a real (complex) vector space V with a **norm metric** $\|x - y\|$ is a locally convex space; each point of V has a local base consisting of *convex* sets.

- **Countably-normed space**

A **countably-normed space** is a **locally convex** space (V, τ) whose topology is defined using a countable set of *compatible norms* $\|.\|_1, \dots, \|.\|_n, \dots$. It means, that if a sequence $\{x_n\}_n$ of elements of V that is fundamental in the norms $\|.\|_i$ and $\|.\|_j$ converges to zero in one of these norms, then it also converges in the other. A countably-normed space is a **metrizable space**, and its metric can be defined by

$$\sum_{n=1}^{\infty} \frac{1}{2^n} \frac{\|x - y\|_n}{1 + \|x - y\|_n}.$$

- **Hyperspace**

A **hyperspace** of a topological space (X, τ) is a topological space on the set $CL(X)$ of all non-empty closed (or, moreover, compact) subsets of X. The topology of a hyperspace of X is called *hypertopology*. Examples of such *hit-and-miss topology* are the *Vietoris topology*, and the *Fell topology*. Examples of such *weak hyperspace topology* are the *Hausdorff metric topology*, and the *Wijsman topology*.

- **Discrete space**

A **discrete space** is a topological space (X, τ) with the *discrete topology*. It can be considered as the metric space (X, d) with the **discrete metric**: $d(x, x) = 0$, and $d(x, y) = 1$ for $x \neq y$.

- **Indiscrete space**

An **indiscrete space** is a topological space (X, τ) with the *indiscrete topology*. It can be considered as the semi-metric space (X, d) with the **indiscrete semi-metric**: $d(x, y) = 0$ for any $x, y \in X$.

- **Metrizable space**

A topological space is called **metrizable** if it is homeomorphic to a metric space. Metrizable spaces are always T_2 and **paracompact** (and, hence, **normal** and **completely regular**), and **first-countable**.

A topological space is called **locally metrizable** if every its point has a metrizable neighborhood.

Chapter 3

Generalizations of Metric Spaces

Some immediate generalizations of the notion of metric, namely **quasi-metric**, **near-metric**, **extended metric**, were defined in Chapter 1. Here were give some generalizations in the direction of Topology, Probability, Algebra, etc.

3.1. m-METRICS

- **m-hemi-metric**

  Let X be a set. A function $d : X^{m+1} \to \mathbb{R}$ is called **m-hemi-metric** if d is *non-negative*, i.e., $d(x_1, \dots, x_{n+1}) \geqslant 0$ for all $x_1, \dots, x_{n+1} \in X$, if d is *totally symmetric*, i.e., satisfies $d(x_1, \dots, x_{m+1}) = d(x_{\pi(1)}, \dots, x_{\pi(m+1)})$ for all $x_1, \dots, x_{m+1} \in X$ and for any permutation π of $\{1, \dots, m+1\}$, if d is *zero conditioned*, i.e., $d(x_1, \dots, x_{m+1}) = 0$ if and only if $x_1, \dots, x_{m+1}$ are not pairwise distinct, and if, for all $x_1, \dots, x_{m+2} \in X$, d satisfies to the **m-simplex inequality**:

$$d(x_1, \dots, x_{m+1}) \leqslant \sum_{i=1}^{m+1} d(x_1, \dots, x_{i-1}, x_{i+1}, \dots, x_{m+2}).$$

- **2-metric**

  Let X be a set. A function $d : X^3 \to \mathbb{R}$ is called **2-metric** if d is *non-negative, totally symmetric, zero conditioned*, and satisfies the **tetrahedron inequality**

$$d(x_1, x_2, x_3) \leqslant d(x_4, x_2, x_3) + d(x_1, x_4, x_3) + d(x_1, x_2, x_4).$$

  It is the most important case $m = 2$ of the **m-hemi-metric**.

- **(m, s)-super-metric**

  Let X be a set, and let s be a positive real number. A function $d : X^{m+1} \to \mathbb{R}$ is called **(m, s)-super-metric** ([DeDu03]) if d is *non-negative, totally symmetric, zero conditioned*, and satisfies to the **(m, s)-simplex inequality**:

$$sd(x_1, \dots, x_{m+1}) \leqslant \sum_{i=1}^{m+1} d(x_1, \dots, x_{i-1}, x_{i+1}, \dots, x_{m+2}).$$

  An (m, s)-super-metric is an **m-hemi-metric** if $s \geqslant 1$.

3.2. INDEFINITE METRICS

- **Indefinite metric**

An **indefinite metric** (or *G-metric*) on a real (complex) vector space V is a *bilinear* (in complex case, *sesquilinear*) *form* G on V, i.e., a function $G : V \times V \to \mathbb{R}$ ($\mathbb{C}$), such that, for any $x, y, z \in V$ and for any scalars α, β, we have the following properties: $G(\alpha x + \beta y, z) = \alpha G(x, z) + \beta G(y, z)$, and $G(x, \alpha y + \beta z) = \overline{\alpha} G(x, z) + \overline{\beta} G(y, z)$, where $\overline{\alpha} = \overline{a + bi} = a - bi$ denotes the *complex conjugation*.

If G is a positive-definite symmetric form, then it is an *inner product* on V, and one can use it to canonically introduce a *norm* and the corresponding **norm metric** on V. In the case of a general form G, there is neither a norm, nor a metric canonically related to G, and the term **indefinite metric** only recalls the close relation of positive-definite bilinear forms with certain metrics in vector spaces (cf. Chapters 7 and 26).

The pair (V, G) is called *space with an indefinite metric*. A finite-dimensional space with an indefinite metric is called *bilinear metric space*. A **Hilbert space** H, endowed with a continuous G-metric, is called *Hilbert space with an indefinite metric*. The most important example of such space is an *J-space*.

A subspace L in a space (V, G) with an indefinite metric is called *positive subspace*, *negative subspace*, or *neutral subspace*, depending on whether $G(x, x) > 0$, $G(x, x) < 0$, or $G(x, x) = 0$ for all $x \in L$.

- **Hermitian G-metric**

A **Hermitian G-metric** is an **indefinite metric** G^H on a complex vector space V such that, for all $x, y \in V$, we have the equality

$$G^H(x, y) = \overline{G^H(y, x)},$$

where $\overline{\alpha} = \overline{a + bi} = a - bi$ denotes the *complex conjugation*.

- **Regular G-metric**

A **regular G-metric** is a continuous **indefinite metric** G on a **Hilbert space** H over $\mathbb{C}$, generated by an invertible *Hermitian operator* T by the formula

$$G(x, y) = \langle T(x), y \rangle,$$

where $\langle , \rangle$ is the *inner product* on H.

A *Hermitian operator* on a Hilbert space H is a *linear operator* T on H, defined on a *dense domain* $D(T)$ of H such that $\langle T(x), y \rangle = \langle x, T(y) \rangle$ for any $x, y \in D(T)$. A bounded Hermitian operator is either defined on the whole of H, or can be so extended by continuity, and then $T = T^*$. On a finite-dimensional space a Hermitian operator can be described by a *Hermitian matrix* $((a_{ij})) = ((\overline{a}_{ji}))$.

- **J-metric**

An J**-metric** is a continuous **indefinite metric** G on a **Hilbert space** H over $\mathbb{C}$, defined by a certain *Hermitian involution* J on H by the formula

$$G(x, y) = \langle J(x), y \rangle,$$

where $\langle , \rangle$ is the *inner product* on H.

An *involution* is a mapping H onto H whose square is the *identity mapping*. The involution J may be represented as $J = P_+ - P_-$, where P_+ and P_- are orthogonal projections in H, and $P_+ + P_- = H$. The *rank of indefiniteness* of the J-metric is defined as $\min\{\dim P_+, \dim P_-\}$.

The space (H, G) is called J*-space*. An J-space with the finite rank of indefiniteness is called *Pontryagin space*.

3.3. TOPOLOGICAL GENERALIZATIONS

- **Partial metric space**

A **partial metric space** is a pair (X, d), where X is a set, and d is a non-negative symmetric function $d : X \times X \to \mathbb{R}$ such that $d(x, x) \leqslant d(x, y)$ for all $x, y \in X$ (*axiom of small self-distances*), and

$$d(x, y) \leqslant d(x, z) + d(z, y) - d(z, z)$$

for all $x, y, z \in X$ (**sharp triangle inequality**). Any partial metric d satisfies the following *local triangle axiom*: $\lim_{n\to\infty}(d(x, y_n) - d(x, x)) = 0$, $\lim_{n\to\infty}(d(y_n, z_n) - d(y_n, y_n)) = 0$ imply $\lim_{n\to\infty}(d(x, z_n) - d(x, x)) = 0$ for any $x \in X$ and any two sequences $\{y_n\}_n$ and $\{z_n\}_n$ of elements of X.

- **τ-distance space**

An τ**-distance space** is a pair (X, f), where X is a topological space, and f is an Aamri-Moutawakil's τ*-distance* on X, i.e., a non-negative function $f : X \times X \to \mathbb{R}$ such that, for any $x \in X$ and any neighborhood U of x, there exists $\varepsilon > 0$ with $\{y \in X : f(x, y) < \varepsilon\} \subset U$.

Any distance space (X, d) is an τ-distance space for the topology τ_f defined as follows: $A \in \tau_f$ if, for any $x \in X$, there exists $\varepsilon > 0$ with $\{y \in X : f(x, y) < \varepsilon\} \subset A$. However, there exist non-metrizable τ-distance spaces. An τ-distance $f(x, y)$ neither need be symmetric, nor vanish for $x = y$; for example, $e^{|x-y|}$ is an τ-distance on $X = \mathbb{R}$ with usual topology.

- **Proximity space**

A **proximity space** is a set X with a binary relation δ on the *power set* $P(X)$ of all its subsets which satisfies the following axioms:

1. $A\delta B$ if and only if $B\delta A$ (*symmetry*);
2. $A\delta(B \cup C)$ if and only if $A\delta B$ or $A\delta C$ (*additivity*);
3. $A\delta A$ if and only if $A \neq \emptyset$ (*reflexivity*).

The relation δ defines a *proximity structure* (a *proximity*) on X. If $A\delta B$ fails, the sets A and B are called *remote sets*.

Every metric space (X, d) is a proximity space: define $A\delta B$ if and only if $d(A, B) = \inf_{x\in A, y\in B} d(x, y) = 0$.

- **Uniform space**

Those topological spaces provide a generalization of metric spaces, based on **set-set distances** instead of point-point distances.

An **uniform space** is a set X with an *uniform structure* $\mathcal{U}$ – a non-empty collection of subsets of $X \times X$, called *entourages*, with the following properties:

1. Every subset of $X \times X$ which contains a set of $\mathcal{U}$, belongs to $\mathcal{U}$;
2. Every finite intersection of sets of $\mathcal{U}$ belongs to $\mathcal{U}$;
3. Every set of $\mathcal{U}$ contains the set $\{(x, x): x \in X\} \subset X \times X$;
4. If V belongs to $\mathcal{U}$, then the set $\{(y, x): (x, y) \in V\}$ belongs to $\mathcal{U}$;
5. If V belongs to $\mathcal{U}$, then there exists $V' \in \mathcal{U}$ such that $(x, z) \in V$, whenever $(x, y), (y, z) \in V'$.

Every metric space (X, d) is an uniform space. An entourage in (X, d) is a subset of $X \times X$ which contains the set $V_\varepsilon = \{(x, y) \in X \times X: d(x, y) < \varepsilon\}$ for some positive real number ε.

Every uniform space is a **proximity space**: define that set A is *near* to the set B if $A \times B$ has non-empty intersection with any entourage.

Every uniform space is a **completely regular** topological space, and, conversely, on every completely regular space can be defined an uniform structure.

- **Approach space**

Those topological spaces provide a generalization of metric spaces, based on **point-set distances** instead of point-point distances.

An **approach space** is a pair (X, D), where X is a set, and D is a **point-set distance**, i.e., a function $X \times P(X) \to [0, \infty]$ (here $P(X)$ is the set of all subsets of X) satisfying, for all $x \in X$ and all $A, B \in P(X)$, to the following conditions:

1. $D(x, \{x\}) = 0$;
2. $D(x, \{\emptyset\}) = \infty$;
3. $D(x, A \cup B) = \min\{D(x, A), D(x, B)\}$;
4. $D(x, A) \leqslant D(x, A^\varepsilon) + \varepsilon$ for any $\varepsilon \in [0, \infty]$, where $A^\varepsilon = \{x: D(x, A) \leqslant \varepsilon\}$ is the "ε-ball" with the center x.

Every metric space (X, d) (moreover, any extended quasi-semi-metric space) is an approach space: define $D(x, A) = d(x, A) = \inf_{y\in A} d(x, y)$.

Given a **locally compact separable** metric space (X, d) and the family $\mathcal{F}$ of its non-empty closed subsets, the **Baddeley–Molchanov distance function** gives a tool for another generalization. It is a function $D : X \times \mathcal{F} \to \mathbb{R}$ which is lower semi-continuous with respect to its first argument, measurable with respect to the second, and satisfies the following two conditions: $F = \{x \in X : D(x, F) \leqslant 0\}$ for $F \in \mathcal{F}$, and $D(x, F_1) \geqslant D(x, F_2)$ for $x \in X$ whenever $F_1, F_2 \in \mathcal{F}$ and $F_1 \subset F_2$.

Additional conditions $D(x, \{y\}) = D(y, \{x\})$, and $D(x, F) \leqslant D(x, \{y\}) + D(y, F)$ for all $x, y \in X$ and for every $F \in \mathcal{F}$, provide analogs of symmetry and triangle inequality. The case $D(x, F) = \inf_{y \in F} d(x, y)$ corresponds to the usual point-set distance.

- **Metric bornology**

  Given a topological space X, a *bornology* of X is any family $\mathcal{A}$ of proper subsets A of X such that the following conditions hold:

  1. $\bigcup_{A \in \mathcal{A}} A = X$;
  2. $\mathcal{A}$ is an *ideal*, i.e., contains all subsets and finite unions of its members.

     The family $\mathcal{A}$ is a **metric bornology** ([Beer99]) if, moreover, it holds:
  3. $\mathcal{A}$ contains a countable base;
  4. For any $A \in \mathcal{A}$ there exists $A' \in \mathcal{A}$ such that the closure of A coincides with the interior of A'.

  The metric bornology is called *trivial* if $\mathcal{A}$ is the *power set* (i.e., the set of all subsets) $P(X)$ of X; such metric bornology corresponds to the family of bounded sets of some bounded metric. For any non-compact **metrizable** topological space X, there exists an unbounded metric compatible with this topology. A non-trivial metric bornology on such space X corresponds to the family of bounded subsets with respect to some such unbounded metric. A non-compact metrizable topological space X admits uncountably many distinct such non-trivial metric bornologies.

3.4. BEYOND NUMBERS

- **Probabilistic metric space**

  A notion of **probabilistic metric space** is a generalization of the notion of metric space (see, for example, [ScSk83]) in two ways: distances become a probability distributions, and the sum in the triangle inequality becomes a **triangle operation**.

  Formally, let A be the set of all *probability distribution functions*, whose support lies in $[0, \infty]$. For any $a \in [0, \infty]$ define $\varepsilon_a \in A$ by $\varepsilon_a(x) = 1$ if $x > a$ or $x = \infty$, and $\varepsilon_a = 0$, otherwise. Functions in A are ordered by defining $F \leqslant G$ to mean $F(x) \leqslant G(x)$ for all $x \geqslant 0$. A commutative and associative operation τ on A is called **triangle operation** if it satisfy to $\tau(F, \varepsilon_0) = F$ for any $F \in A$ and $\tau(E, F) \leqslant \tau(G, H)$ whenever $E \leqslant G$, $F \leqslant H$.

  A **probabilistic metric space** is a triple (X, d, τ), where X is a set, d is a function $X \times X \to A$, and τ is a triangle operation such that, for any $p, q, r \in X$, it holds:

  1. $d(p, q) = \varepsilon_0$ if and only if $p = q$;

2. $d(p,q) = d(q,p)$;
3. $d(p,r) \leqslant \tau(d(p,q), d(q,r))$.

The inequality 3. becomes the triangle inequality if τ is the usual addition on $\mathbb{R}$.

For any $x \geqslant 0$, the value $d(p,q)$ at x can be interpreted as "the probability that the distance between p and q is less than x"; it was approach of K. Menger, who proposed in 1942 the original version, *statistical metric space*, of this notion. Several notions of *fuzzy metric space* were proposed within this framework.

- **Generalized metric**

Let X be a set. Let $(G, +, \leqslant)$ be an *ordered semi-group* (not necessarily commutative) having a least element 0. A function $d : X \times X \to G$ is called **generalized metric** if the following conditions hold:

1. $d(x,y) = 0$ if and only if $x = y$;
2. $d(x,y) \leqslant d(x,z) + d(z,y)$ for all $x, y \in X$;
3. $d(x,y) = \overline{d(y,x)}$, where $\overline{\alpha}$ is a fixed order-preserving *involution* of G.

The pair (X,d) is called **generalized metric space**.

If the condition 2. and "only if" in 1. above are dropped, we obtain a **generalized distance** d, and a **generalized distance space** (X,d).

- **Distance on building**

A *Coxeter group* is a group $(W, \cdot, 1)$ generated by the elements $\{w_1, \dots, w_n : (w_i w_j)^{m_{ij}} = 1, 1 \leqslant i, j \leqslant n\}$. Here $M = ((m_{ij}))$ is a *Coxeter matrix*, i.e., an arbitrary symmetric $n \times n$ matrix with $m_{ii} = 1$, and other values are positive integers or ∞. The *length* $l(x)$ of $x \in W$ is the smallest number of generators $w_1, \dots, w_n$ needed to represent x.

Let X be a set, and let $(W, \cdot, 1)$ be a Coxeter group. The pair (X,d) is called *building* over $(W, \cdot, 1)$ if the function $d : X \times X \to W$, called **distance on building**, has the following properties:

1. $d(x,y) = 1$ if and only if $x = y$;
2. $d(y,x) = (d(x,y))^{-1}$;
3. the relation $\sim_i$, defined by $x \sim_i y$ if $d(x,y) = 1$ or w_i, is an equivalence relation;
4. given $x \in X$ and an equivalence class C of $\sim_i$, there exists an unique $y \in C$ such that $d(x,y)$ is *shortest* (i.e., of smallest length), and $d(x,y') = d(x,y)w_i$ for any $y' \in C$, $y' \neq y$.

The **gallery distance on building** d' is an usual metric on X, defined by $l(d(x,y))$. The distance d' is the **path metric** in the graph with the vertex-set X and xy being an edge if $d(x,y) = w_i$ for some $1 \leqslant i \leqslant n$. The gallery distance on building is a special case of **gallery metric** (of *chamber system* X).

- **Boolean metric space**

A *Boolean algebra* (or *Boolean lattice*) is a *distributive lattice* $(B, \vee, \wedge)$ admitting least element 0 and greatest element 1 such that every $x \in B$ has a *complement* $\overline{x}$ with $x \vee \overline{x} = 1$ and $x \wedge \overline{x} = 0$.

Let X be a set, and let $(B, \vee, \wedge)$ be a Boolean algebra. The pair (X, d) is called **Boolean metric space** over B if the function $d : X \times X \to B$ has the following properties:

1. $d(x, y) = 0$ if and only if $x = y$;
2. $d(x, y) \leqslant d(x, z) \vee d(z, y)$ for all $x, y, z \in X$.

- **Space over algebra**

A **space over algebra** is a metric space with a differential-geometric structure, whose points can be provided with coordinates from some *algebra*, as the rule, associative with identity.

A *module* over an algebra is a generalization of a vector space over a field, its definition can be obtained from the definition of a vector space by replacing the field by the associative algebra with identity. An *affine space over an algebra* is a similar generalization of an *affine space* over a field. In affine spaces over algebras one can specify a Hermitian metric, while in the case of commutative algebras even a quadratic metric can be given. To do this one defines in an unital module a *scalar product* $\langle x, y\rangle$, in the first case with the property $\langle x, y\rangle = J(\langle y, x\rangle)$, where J is an *involution* of the algebra, and in the second case with the property $\langle y, x\rangle = \langle x, y\rangle$.

The n-dimensional *projective space over an algebra* is defined as the variety of one-dimensional submodules of an $(n+1)$-dimensional unital module over this algebra. The introduction of a *scalar product* $\langle x, y\rangle$ in an unital module makes it possible to define in a projective space constructed by means of this module Hermitian, or, in the case of commutative algebra, quadratic elliptic and hyperbolic metrics. The metric invariant of the points of these spaces is the *cross-ratio* $W = \langle x, x\rangle^{-1}\langle x, y\rangle\langle y, y\rangle^{-1}\langle y, x\rangle$. If W is a real number, then the invariant w, for which $W = \cos^2 w$, is called **distance** between x and y **in the space over algebra**.

- **Partially ordered distance**

Let X be a set. Let $(G, \leqslant)$ be a *partially ordered set* with a least element g_0 such that $G' = G\backslash\{g_0\}$ is non-empty and, for any $g_1, g_2 \in G'$, there exist $g_3 \in G'$ such that $g_3 \leqslant g_1$ and $g_3 \leqslant g_2$.

A **partially ordered distance** is a function $d : X \times X \to G$ such that, for any $x, y \in X$, $d(x, y) = g_0$ if and only if $x = y$.

Consider the following possible properties:

1. For any $g_1 \in G'$, there exists $g_2 \in G'$ such that, for any $x, y \in X$, from $d(x, y) \leqslant g_2$ it follows $d(y, x) \leqslant g_1$;
2. For any $g_1 \in G'$, there exist $g_2, g_3 \in G'$ such that, for any $x, y, z \in X$, from $d(x, y) \leqslant g_2$ and $d(y, z) \leqslant g_3$ it follows $d(x, z) \leqslant g_1$;
3. For any $g_1 \in G'$, there exists $g_2 \in G'$ such that, for any $x, y, z \in X$, from $d(x, y) \leqslant g_2$ and $d(y, z) \leqslant g_2$ it follows $d(y, x) \leqslant g_1$;
4. G' has no first element;
5. $d(x, y) = d(y, x)$ for any $x, y \in X$;

6. For any $g_1 \in G'$, there exists $g_2 \in G'$ such that, for any $x, y, z \in X$, from $d(x, y) <^* g_2$ and $d(y, z) <^* g_2$ it follows $d(x, z) <^* g_1$; here $p <^* q$ means that either $p < q$, or p is not comparable to q;
7. The order relation $<$ is a total ordering of G.

In terms of above properties, d is called: the **Appert partially ordered distance** if 1. and 2. hold; the **Golmez partially ordered distance of first type** if 4., 5., and 6. hold; the **Golmez partially ordered distance of second type** if 3., 4., and 5. hold; the **Kurepa–Fréchet distance** if 3., 4., 5., and 7. hold.

Chapter 4

Metric Transforms

There are many ways to obtain new distances (metrics) from given distances (metrics). Metric transforms give new distances as a functions of given metrics (or given distances) on the same set X. If one obtains the metric, it is called **transform metric**. We give some important examples of transform metrics in the first section.

Given a metric on a set X, one can construct a new metric on an extension of X; similarly, given a collection of metrics on sets $X_1, \dots, X_n$, one can obtain a new metric on an extension of $X_1, \dots, X_n$. Examples of such operations are given in the second section.

Given a metric on X, there are many distances on other structures, connected with X, for example, on the set of all subsets of X. Main distances of such kind are considered in the third section.

4.1. METRICS ON THE SAME SET

- **Metric transform**

  A **metric transform** is a distance on a set X, obtained as a function of given metrics (or given distances) on X.

  In particular, given a continuous monotone increasing function $f(x)$ of $x \geqslant 0$, called *scale*, and a distance space (X, d), one obtains other distance space (X, d_f), called *scale metric transform* of X, defining $d_f(x, y) = f(d(x, y))$. For every finite distance space (X, d), there exists a scale f, such that (X, d_f) is a metric subspace of an Euclidean space $\mathbb{E}^n$.

  If (X, d) is a metric space and f is a continuous differentiable strictly increasing scale with $f(0) = 0$ and non-increasing f', then (X, d_f) is a metric space (cf. **functional transform metric**).

- **Transform metric**

  A **transform metric** is a metric on a set X which is a **metric transform**, i.e., is obtained as a function of a given metric (or given metrics) on X. In particular, transform metrics can be obtained from a given metric d (or given metrics d_1 and d_2) on X by any of the following operations:

  1. $\alpha d(x, y), \alpha > 0$ (**α-scaled metric**, or **dilated metric**);
  2. $\min\{t, d(x, y)\}$ (**t-truncated metric**);
  3. $d(x, y) + \alpha, \alpha \geqslant 0$, for $x \neq y$;

4. $\frac{d(x,y)}{1+d(x,y)}$;
5. $d^p(x,y) = \frac{d(x,y)}{d(x,p)+d(y,p)+d(x,y)}$, where p is an fixed element of X;
6. $\max\{d_1(x,y), d_2(x,y)\}$;
7. $\alpha d_1(x,y) + \beta d_2(x,y)$, where $\alpha, \beta > 0$ (cf. **metric cone**).

- **Functional transform metric**

Let $f : \mathbb{R} \to \mathbb{R}$ be a twice differentiable real function, defined for $x \geqslant 0$ such that $f(0) = 0$, $f'(x) > 0$ for all $x \geqslant 0$, and $f''(x) \leqslant 0$ for all $x \geqslant 0$. In fact, f is *concave* on $[0, \infty)$; in particular, $f(x+y) \leqslant f(x) + f(y)$.

Given a metric space (X, d), the **functional transform metric** d_f is a **transform metric** on X, defined by

$$f\big(d(x,y)\big).$$

Metrics d_f and d are equivalent. If d is a metric on X, then, for example, αd ($\alpha > 0$), d^α ($0 < \alpha < 1$), $\ln(1+d)$, $\operatorname{arcsinh} d$, $\operatorname{arccosh} d$, and $\frac{d}{1+d}$ are functional transform metrics on X.

- **Power transform metric**

Let $0 < \alpha \leqslant 1$. Given a metric space (X, d), the **power transform metric** is a **functional transform metric** on X, defined by

$$\big(d(x,y)\big)^\alpha.$$

For a given metric d on X and any $\alpha > 1$, the function d^α is a distance on X. It is a metric if and only if d is an **ultrametric**.

- **Schoenberg transform metric**

Let $\lambda > 0$. Given a metric space (X, d), the **Schoenberg transform metric** is a **functional transform metric** on X, defined by

$$1 - e^{-\lambda d(x,y)}.$$

- **g-transform metric**

Given a metric space (X, d), let $g : X \to X$ be an injective function on X. The **g-transform metric** is a **transform metric** on X, defined by

$$d\big(g(x), g(y)\big).$$

- **Internal metric**

Given a metric space (X, d) in which every pair of points x, y is joined by a *rectifiable curve*, the **internal metric** (or **interior metric, induced intrinsic metric**) D is a **transform metric** on X, obtained from d as the infimum of the lengths of all rectifiable curves connecting two given points x and $y \in X$.

The metric d on X is called **intrinsic metric** (or *length metric*) if it coincides with its internal metric. In this case, the metric space (X, d) is called **length space**.

4.2. METRICS ON SET EXTENSIONS

- **Extension distances**

If d is a distance on $V_n = \{1, \dots, n\}$, and $\alpha \in \mathbb{R}, \alpha > 0$, then the following extension distances (see, for example, [DeLa97]) are used.

The **gate extension distance** $gat = gat_\alpha^d$ is a distance on $V_{n+1} = \{1, \dots, n+1\}$, defined by the following conditions:

1. $gat(1, n+1) = \alpha$;
2. $gat(i, n+1) = \alpha + d(1, i)$ if $2 \leqslant i \leqslant n$;
3. $gat(i, j) = d(i, j)$ if $1 \leqslant i < j \leqslant n$.

The distance gat_0^d is called **gate** 0**-extension** or, simply, 0**-extension** of d.

If $\alpha \geqslant \max_{2 \leqslant i \leqslant n} d(1, i)$, then the **antipodal extension distance** $ant = ant_\alpha^d$ is a distance on V_{n+1}, defined by the following conditions:

1. $ant(1, n+1) = \alpha$;
2. $ant(i, n+1) = \alpha - d(1, i)$ if $2 \leqslant i \leqslant n$;
3. $ant(i, j) = d(i, j)$ if $1 \leqslant i < j \leqslant n$.

If $\alpha \geqslant \max_{1 \leqslant i, j \leqslant n} d(i, j)$, then the **full antipodal extension distance** $Ant = Ant_\alpha^d$ is a distance on $V_{2n} = \{1, \dots, 2n\}$, defined by the following conditions:

1. $Ant(i, n+i) = \alpha$ if $1 \leqslant i \leqslant n$;
2. $Ant(i, n+j) = \alpha - d(i, j)$ if $1 \leqslant i \neq j \leqslant n$;
3. $Ant(i, j) = d(i, j)$ if $1 \leqslant i \neq j \leqslant n$;
4. $Ant(n+i, n+j) = d(i, j)$ if $1 \leqslant i \neq j \leqslant n$.

It is obtained by apply the antipodal extension operation iteratively n times, starting from d.

The **spherical extension distance** $sph = sph_\alpha^d$ is a distance on V_{n+1}, defined by the following conditions:

1. $sph(i, n+1) = \alpha$ if $1 \leqslant i \leqslant n$;
2. $sph(i, j) = d(i, j)$ if $1 \leqslant i < j \leqslant n$.

- **1-sum distance**

Let d_1 be a distance on a set X_1, let d_2 be a distance on a set X_2, and suppose that $X_1 \cap X_2 = \{x_0\}$. The 1**-sum distance** of d_1 and d_2 is the distance d on $X_1 \cup X_2$, defined by the following conditions:

$$d(x, y) = \begin{cases} d_1(x, y), & \text{if } x, y \in X_1, \\ d_2(x, y), & \text{if } x, y \in X_2, \\ d(x, x_0) + d(x_0, y), & \text{if } x \in X_1,\ y \in X_2. \end{cases}$$

In Graph Theory, the 1-sum distance is a **path metric**, corresponding to the clique 1-sum operation for graphs.

- **Product metric**

Given n metric spaces $(X_1, d_1), (X_2, d_2), \dots, (X_n, d_n)$, the **product metric** is a metric on the *Cartesian product*

$$X_1 \times X_2 \times \cdots \times X_n = \{x = (x_1, x_2, \dots, x_n): \; x_1 \in X_1, \dots, x_n \in X_n\},$$

defined as a function of $d_1, \dots, d_n$. The simplest product metrics are defined by

1. $\sum_{i=1}^{n} d_i(x_i, y_i)$;
2. $(\sum_{i=1}^{n} d_i^p(x_i, y_i))^{\frac{1}{p}}$, $1 < p < \infty$;
3. $\max_{1 \leqslant i \leqslant n} d_i(x_i, y_i)$;
4. $\sum_{i=1}^{n} \frac{1}{2^i} \frac{d_i(x_i, y_i)}{1+d_i(x_i, y_i)}$;
5. $\min_{1 \leqslant i \leqslant n} \{d_i(x_i, y_i), 1\}$.

Last two metrics are **bounded** and can be extended to the product of countably many metric spaces.

If $X_1 = \cdots = X_n = \mathbb{R}$, and $d_1 = \cdots = d_n = d$, where $d(x, y) = |x - y|$ is the **natural metric** on $\mathbb{R}$, all product metrics above induce the Euclidean topology on the n-dimensional space $\mathbb{R}^n$. They do not coincide with the Euclidean metric on $\mathbb{R}^n$, but they are equivalent to it. In particular, the set $\mathbb{R}^n$ with the Euclidean metric can be considered as the Cartesian product $\mathbb{R} \times \cdots \times \mathbb{R}$ of n copies of the *real line* $(\mathbb{R}, d)$ with the product metric, defined by $\sqrt{\sum_{i=1}^{n} d^2(x_i, y_i)}$.

- **Fréchet product metric**

Let (X, d) be a metric space with a **bounded** metric d. Let $X^\infty = X \times \cdots \times X \cdots = \{x = (x_1, \dots, x_n, \dots): \; x_1 \in X_1, \dots, x_n \in X_n, \dots\}$ be the *product space* of X.

The **Fréchet product metric** is a **product metric** on X^∞, defined by

$$\sum_{n=1}^{\infty} A_n d(x_n, y_n),$$

where $\sum_{n=1}^{\infty} A_n$ is any convergent series of positive terms. Usually, $A_n = \frac{1}{2^n}$ is used.

A metric (sometimes called *Fréchet metric*) on the set of all sequences $\{x_n\}_n$ of real (complex) numbers, defined by

$$\sum_{n=1}^{\infty} A_n \frac{|x_n - y_n|}{1 + |x_n - y_n|},$$

where $\sum_{n=1}^{\infty} A_n$ is any convergent series of positive terms, is a Fréchet product metric of countably many copies of $\mathbb{R}$ ($\mathbb{C}$). Usually, $A_n = \frac{1}{n!}$ or $A_n = \frac{1}{2^n}$ are used.

- **Hilbert cube metric**

The *Hilbert cube* $I^{\aleph_0}$ is the *Cartesian product* of countable many copies of the interval [0, 1], equipped with the metric

$$\sum_{i=1}^{\infty} 2^{-i}|x_i - y_i|$$

(cf. **Fréchet product metric**). It also can be identified up to homeomorphisms with compact metric space formed by all sequences $\{x_n\}_n$ of real numbers such that $0 \leqslant x_n \leqslant \frac{1}{n}$, where the metric is defined as $\sqrt{\sum_{n=1}^{\infty}(x_n - y_n)^2}$.

- **Warped product metric**

Let (X, d_X) and (Y, d_Y) be two **complete length spaces**, and let $f : X \to \mathbb{R}$ be a positive continuous function. Given a curve $\gamma : [a, b] \to X \times Y$, consider its projections $\gamma_1 : [a, b] \to X$ and $\gamma_2 : [a, b] \to Y$ to X and Y, and define the length of γ by the formula $\int_a^b \sqrt{|\gamma_1'|^2(t) + f^2(\gamma_1(t))|\gamma_2'|^2(t)dt}$.

The **warped product metric** is a metric on $X \times Y$, defined as the infimum of lengths of all rectifiable curves, connected two given points in $X \times Y$ (see [BuIv01]).

4.3. METRICS ON OTHER SETS

Given a metric space (X, d), one can construct several distances between some subsets of X. The main such distances are: the **point-set distance** $d(x, A) = \inf_{y \in A} d(x, y)$ between a point $x \in X$ and a subset $A \subset X$, the **set-set distance** $\inf_{x \in A,\, y \in B} d(x, y)$ between two subsets A and B of X, and the **Hausdorff metric** between compact subsets of X, which are considered in chapter 1. In this section we list some other distances of such kind.

- **Line-line distance**

The **line-line distance** is the **set-set distance** in $\mathbb{E}^3$ between two *skew* lines, i.e., two straight lines that do not lie in a plane. It is the length of the segment of their common perpendicular whose end points lie on the lines. For l_1 and l_2 with equations l_1: $x = p + qt$, $t \in \mathbb{R}$, and l_2: $x = r + st$, $t \in \mathbb{R}$, the distance is given by

$$\frac{|\langle r - p, q \times s\rangle|}{\|q \times s\|_2},$$

where $\times$ is the *cross product* on $\mathbb{E}^3$, $\langle , \rangle$ is the *inner product* on $\mathbb{E}^3$, and $\|.\|_2$ is the Euclidean norm. For $x = (x_1, x_2, x_3)$, $y = (y_1, y_2, y_3)$, one has $x \times y = (x_2y_3 - x_3y_2, x_3y_1 - x_1y_3, x_1y_2 - x_2y_1)$.

Vertical distance between lines l_1 and l_2 is the length of the vertical segment with one endpoint on l_1 and one endpoint on l_2, provided a unique such segment exists.

- **Point-line distance**

The **point-line distance** is the **point-set distance** between a point and a line.

In $\mathbb{E}^2$, the distance between a point $z = (z_1, z_2)$ and a line $l: ax_1 + bx_2 + c = 0$ is given by

$$\frac{|az_1 + bz_2 + c|}{\sqrt{a^2 + b^2}}.$$

In $\mathbb{E}^3$, the distance between a point z and a line $l: x = p + qt, t \in \mathbb{R}$, is given by

$$\frac{\|q \times (p - z)\|_2}{\|q\|_2},$$

where $\times$ is the *cross product* on $\mathbb{E}^3$, and $\|.\|_2$ is the Euclidean norm.

- **Point-plane distance**

The **point-line distance** is the **point-set distance** in $\mathbb{E}^3$ between a point and a plane. The distance between a point $z = (z_1, z_2, z_3)$ and a plane $\alpha: ax_1 + bx_2 + cx_3 + d = 0$ is given by

$$\frac{|az_1 + bz_2 + cz_3 + d|}{\sqrt{a^2 + b^2 + c^2}}.$$

- **Prime number distance**

The **prime number distance** is the **point-set distance** in $(\mathbb{N}, |n - m|)$ between a number $n \in \mathbb{N}$ and the set of prime numbers $P \subset \mathbb{N}$. It is the absolute difference between n and the nearest prime number.

- **Distance up to nearest integer**

The **distance up to nearest integer** is the **point-set distance** in $(\mathbb{R}, |x - y|)$ between a number $x \in \mathbb{R}$ and the set of integers $\mathbb{Z} \subset \mathbb{R}$, i.e., $\min_{n \in \mathbb{Z}} |x - n|$.

- **Busemann metric of sets**

Given a metric space (X, d), the **Busemann metric of sets** (see [Buse55]) is a metric on the set of all non-empty closed subsets of X, defined by

$$\sup_{x \in X} \big| d(x, A) - d(x, B) \big| e^{-d(p,x)},$$

where p is a fixed point of X, and $d(x, A) = \min_{y \in A} d(x, y)$ is the **point-set distance**. Instead of weighting factor $e^{-d(p,x)}$, one can take any distance transform function which decrease fast enough (cf. also L_p-**Hausdorff distance**, and the list of variations of the **Hausdorff metric** in Chapter 21).

Chapter 5

Metrics on Normed Structures

In this chapter we consider a special class of metrics, defined on some *normed structures*, as the norm of difference between two given elements. This structure can be a group (with a *group norm*), a vector space (with a *vector norm* or, simply, a *norm*), a vector lattice (with a *Riesz norm*), a field (with a *valuation*), etc.

- **Group norm metric**

  A **group norm metric** is a metric on a *group* $(G, +, 0)$, defined by

$$\|x + (-y)\| = \|x - y\|,$$

  where $\|.\|$ is a *group norm* on G, i.e., a function $\|.\| : G \to \mathbb{R}$ such that, for all $x, y \in G$, we have the following properties:

  1. $\|x\| \geqslant 0$, with $\|x\| = 0$ if and only if $x = 0$;
  2. $\|x\| = \| - x\|$;
  3. $\|x + y\| \leqslant \|x\| + \|y\|$ (triangle inequality).

  Any group norm metric d is **right-invariant**, i.e., $d(x, y) = d(x + z, y + z)$ for any $x, y, z \in G$. On the other hand, any right-invariant (as well as any **left-invariant**, and, in particular, any **bi-invariant**) metric d on G is a group norm metric, since one can define a group norm on G by $\|x\| = d(x, 0)$.

- ***F*-norm metric**

  A *vector space* (or *linear space*) over a *field* $\mathbb{F}$ is a set V equipped with operations of *vector addition* $+ : V \times V \to V$ and *scalar multiplication* $\cdot : F \times V \to V$ such that $(V, +, 0)$ forms an *Abelian group* (where $0 \in V$ is the *zero vector*), and, for all *vectors* $x, y \in V$ and any *scalars* $a, b \in \mathbb{F}$, we have the following properties: $1 \cdot x = x$ (where 1 is the multiplicative unit of $\mathbb{F}$), $(ab) \cdot x = a \cdot (b \cdot x)$, $(a + b) \cdot x = a \cdot x + b \cdot x$, and $a \cdot (x + y) = a \cdot x + a \cdot y$. A vector space over the field $\mathbb{R}$ of real numbers is called *real vector space*. A vector space over the field $\mathbb{C}$ of complex numbers is called *complex vector space*.

  An ***F*-norm metric** is a metric on a real (complex) vector space V, defined by

$$\|x - y\|_F,$$

  where $\|.\|_F$ is an *F-norm* on V, i.e., a function $\|.\|_F : V \to \mathbb{R}$ such that, for all $x, y \in V$ and for any scalar a with $|a| = 1$, we have the following properties:

1. $\|x\|_F \geqslant 0$, with $\|x\|_F = 0$ if and only if $x = 0$;
2. $\|ax\|_F = \|x\|_F$;
3. $\|x + y\|_F \leqslant \|x\|_F + \|y\|_F$ (triangle inequality).

An F-norm is called *p-homogeneous* if $\|ax\|_F = |a|^p \|x\|_F$.

Any F-norm metric d is a **translation invariant metric**, i.e., $d(x, y) = d(x + z, y + z)$ for all $x, y, z \in V$. Conversely, if d is a translation invariant metric on V, then $\|x\|_F = d(x, 0)$ is an F-norm on V.

- **F^*-metric**

An F^***-metric** is an F**-norm metric** $\|x - y\|_F$ on a real (complex) vector space V such that the operations of scalar multiplication and vector addition are *continuous* with respect to $\|.\|_F$. It means, that $\|.\|_F$ is a function $\|.\|_F : V \to \mathbb{R}$ such that, for all $x, y, x_n \in V$ and for all scalars a, a_n, we have the following properties:

1. $\|x\|_F \geqslant 0$, with $\|x\|_F = 0$ if and only if $x = 0$;
2. $\|ax\|_F = \|x\|_F$ for all a with $|a| = 1$;
3. $\|x + y\|_F \leqslant \|x\|_F + \|y\|_F$;
4. $\|a_n x\|_F \to 0$ if $a_n \to 0$;
5. $\|ax_n\|_F \to 0$ if $x_n \to 0$;
6. $\|a_n x_n\|_F \to 0$ if $a_n \to 0$, $x_n \to 0$.

The metric space $(V, \|x - y\|_F)$ with an F^*-metric is called *F^*-space*. Equivalently, an F^*-space is a metric space (V, d) with a **translation invariant metric** d such that the operation of scalar multiplication and vector addition are continuous with respect to this metric.

A *modular space* is an F^*-space $(V, \|.\|_F)$ in which the F-norm $\|.\|_F$ is defined by

$$\|x\|_F = \inf\left\{\lambda > 0\colon\ \rho\left(\frac{x}{\lambda}\right) < \lambda\right\},$$

and ρ is a *metrizing modular* on V, i.e., a function $\rho : V \to [0, \infty]$ such that, for all $x, y, x_n \in V$ and for all scalars a, a_n, we have the following properties:

1. $\rho(x) = 0$ if and only if $x = 0$;
2. $\rho(ax) = \rho(x)$ implies $|a| = 1$;
3. $\rho(ax + by) \leqslant \rho(x) + \rho(y)$ implies $a, b \geqslant 0$, $a + b = 1$;
4. $\rho(a_n x) \to 0$ if $a_n \to 0$ and $\rho(x) < \infty$;
5. $\rho(ax_n) \to 0$ if $\rho(x_n) \to 0$ (*metrizing property*);
6. For any $x \in V$, there exists $k > 0$ such that $\rho(kx) < \infty$.

A **complete** F^*-space is called *F-space*. A **locally convex** F-space is known as *Fréchet space* in Functional Analysis.

- **Norm metric**

A **norm metric** is a metric on a real (complex) vector space V, defined by

$$\|x - y\|,$$

where $\|.\|$ is a *norm* on V, i.e., a function $\|.\| : V \to \mathbb{R}$ such that, for all $x, y \in V$ and for any scalar a, we have the following properties:

1. $\|x\| \geqslant 0$, with $\|x\| = 0$ if and only if $x = 0$;
2. $\|ax\| = |a|\|x\|$;
3. $\|x + y\| \leqslant \|x\| + \|y\|$ (triangle inequality).

Therefore, a norm $\|.\|$ is an 1-*homogeneous* F-*norm*. The vector space $(V, \|.\|)$ is called *normed vector space* or, simply, *normed space*.

On any given finite-dimensional vector space all norms are equivalent. Every finite-dimensional normed space is **complete**. Any metric space can be embedded isometrically in some normed vector space as a closed linearly independent subset.

The *norm-angular distance* between x and y is defined by

$$d(x, y) = \left\| \frac{x}{\|x\|} - \frac{y}{\|y\|} \right\|.$$

L. Maligranda remarked the following sharpening of the triangle inequality in normed spaces: for any $x, y \in V$, it holds

$$\big(2-d(x,-y)\big)\min\big(\|x\|.\|y\|\big) \leqslant \|x\|+\|y\|-\|x+y\| \leqslant \big(2-d(x,-y)\big)\max\big(\|x\|.\|y\|\big).$$

- **Semi-norm semi-metric**

A **semi-norm semi-metric** is a semi-metric on a real (complex) vector space V, defined by

$$\|x - y\|,$$

where $\|.\|$ is a *semi-norm* (or *pre-norm*) on V, i.e., a function $\|.\| : V \to \mathbb{R}$ such that, for all $x, y \in V$ and for any scalar a, we have the following properties:

1. $\|x\| \geqslant 0$, with $\|0\| = 0$;
2. $\|ax\| = |a|\|x\|$;
3. $\|x + y\| \leqslant \|x\| + \|y\|$ (triangle inequality).

The vector space $(V, \|.\|)$ is called *semi-normed vector space*. Many *normed vector spaces*, in particular, **Banach spaces**, are defined as the quotient space by the subspace of elements of semi-norm zero.

A *quasi-normed space* is a vector space V, on which a *quasi-norm* is given. A *quasi-norm* on V is a non-negative function $\|.\| : V \to \mathbb{R}$ which satisfies the same axioms as a norm, except for the triangle inequality which is replaced by the weaker requirement: there exists a constant $C > 0$ such that, for all $x, y \in V$, we have the following inequality:

$$\|x + y\| \leqslant C\big(\|x\| + \|y\|\big)$$

(cf. **near-metric**). An example of a quasi-normed space, that is not normed, is the *Lebesgue space* $L_p(\Omega)$ with $0 < p < 1$ in which a quasi-norm is defined by

$$\|f\| = \left(\int_{\Omega} |f(x)|^p \, dx\right)^{1/p}, \quad f \in L_p(\Omega).$$

- **Banach space**

A **Banach space** (or *B-space*) is a **complete** metric space $(V, \|x-y\|)$ on a vector space V with a norm metric $\|x - y\|$. Equivalently, it is the complete *normed space* $(V, \|.\|)$. In this case, the norm $\|.\|$ on V is called *Banach norm*. Some examples of Banach spaces are:

1. l_p^n-*spaces*, l_p^∞-*spaces*, $1 \leqslant p \leqslant \infty$, $n \in \mathbb{N}$;
2. The space C of convergent numerical sequences with the norm $\|x\| = \sup_n |x_n|$;
3. The space C_0 of numerical sequences which converge to zero with the norm $\|x\| = \max_n |x_n|$;
4. The space $C_{[a,b]}^p$, $1 \leqslant p \leqslant \infty$, of continuous functions on $[a, b]$ with the L_p-*norm* $\|f\|_p = (\int_a^b |f(t)|^p \, dt)^{\frac{1}{p}}$;
5. The space C_K of continuous functions on a compactum K with the norm $\|f\| = \max_{t \in K} |f(t)|$;
6. The space $(C_{[a,b]})^n$ of functions on $[a, b]$ with continuous derivatives up to and including the order n with the norm $\|f\|_n = \sum_{k=0}^n \max_{a \leqslant t \leqslant b} |f^{(k)}(t)|$;
7. The space $C^n[I^m]$ of all functions defined in an m-dimensional cube that are continuously differentiable up to and including the order n with the norm of uniform boundedness in all derivatives of order at most n;
8. The space $M_{[a,b]}$ of bounded measurable functions on $[a, b]$ with the norm

$$\|f\| = \operatorname*{ess\,sup}_{a \leqslant t \leqslant b} |f(t)| = \inf_{e, \mu(e)=0} \sup_{t \in [a,b] \setminus e} |f(t)|;$$

9. The space $A(\Delta)$ of functions analytic in the open *unit disk* $\Delta = \{z \in \mathbb{C}: |z| < 1\}$ and continuous in the closed disk $\overline{\Delta}$ with the norm $\|f\| = \max_{z \in \overline{\Delta}} |f(z)|$;
10. The **Lebesgue spaces** $L_p(\Omega)$, $1 \leqslant p \leqslant \infty$;
11. The *Sobolev spaces* $W^{k,p}(\Omega)$, $\Omega \subset \mathbb{R}^n$, $1 \leqslant p \leqslant \infty$, of functions f on Ω such that f and its derivatives, up to some order k, have a finite L_p-*norm*, with the norm $\|f\|_{k,p} = \sum_{i=0}^k \|f^{(i)}\|_p$;
12. The *Bohr space* AP of almost-periodic functions with the norm

$$\|f\| = \sup_{-\infty < t < +\infty} |f(t)|.$$

A finite-dimensional real Banach space is called *Minkowskian space*. A norm metric of a Minkowskian space is called **Minkowskian metric**. In particular, any l_p**-metric** is a Minkowskian metric.

All n-dimensional Banach spaces are pairwise isomorphic; their set becomes compact if one introduces the **Banach–Mazur distance** by $d_{BM}(V, W) = \ln \inf_T \|T\| \cdot \|T^{-1}\|$, where the infimum is taken over all operators which realize an isomorphism $T : V \to W$.

- **l_p-metric**

The **l_p-metric** d_{l_p}, $1 \leqslant p \leqslant \infty$, is a norm metric on $\mathbb{R}^n$ (or on $\mathbb{C}^n$), defined by

$$\|x - y\|_p,$$

where the l_p*-norm* $\|.\|_p$ is defined by

$$\|x\|_p = \left(\sum_{i=1}^{n} |x_i|^p\right)^{\frac{1}{p}}.$$

For $p = \infty$, we obtain $\|x\|_\infty = \lim_{p\to\infty} \sqrt[p]{\sum_{i=1}^{n} |x_i|^p} = \max_{1\leqslant i\leqslant n} |x_i|$. The metric space $(\mathbb{R}^n, d_{l_p})$ is abbreviated as l_p^n and is called l_p^n*-space*.

The **l_p-metric**, $1 \leqslant p \leqslant \infty$, on the set of all sequences $x = \{x_n\}_{n=1}^{\infty}$ of real (complex) numbers, for which the sum $\sum_{i=1}^{\infty} |x_i|^p$ (for $p = \infty$, the sum $\sum_{i=1}^{\infty} |x_i|$) is finite, is defined by

$$\left(\sum_{i=1}^{\infty} |x_i - y_i|^p\right)^{\frac{1}{p}}.$$

For $p = \infty$, we obtain $\max_{i\geqslant 1} |x_i - y_i|$. This metric space is abbreviated as l_p^∞ and is called l_p^∞*-space*.

Most important are l_1-, l_2- and l_∞-metrics; the l_2-metric on $\mathbb{R}^n$ is also called **Euclidean metric**. The l_2-metric on the set of all sequences $\{x_n\}_n$ of real (complex) numbers, for which $\sum_{i=1}^{\infty} |x_i|^2 < \infty$, is also known as **Hilbert metric**.

- **Euclidean metric**

The **Euclidean metric** (or **Pythagorean distance**, **as-crow-flies distance**) d_E is a metric on $\mathbb{R}^n$, defined by

$$\|x - y\|_2 = \sqrt{(x_1 - y_1)^2 + \cdots + (x_n - y_n)^2}.$$

It is the ordinary **l_2-metric** on $\mathbb{R}^n$. The metric space $(\mathbb{R}^n, d_E)$ is abbreviated as $\mathbb{E}^n$ and is called **Euclidean space** (or *real Euclidean space*). Sometimes, the expression "Euclidean space" stands for the case $n = 3$, as opposed to the *Euclidean plane* for the case $n = 2$. The *Euclidean line* (or *real line*) is obtained for $n = 1$, i.e., it is the metric space $(\mathbb{R}, |x - y|)$ with the **natural metric** $|x - y|$.

In fact, $\mathbb{E}^n$ is an **inner product space** (and even **Hilbert space**), i.e., $d_E(x, y) = \|x - y\|_2 = \sqrt{\langle x - y, x - y\rangle}$, where $\langle x, y\rangle$ is an *inner product* on $\mathbb{R}^n$ which is given in a suitably chosen (Cartesian) coordinate system by the formula $\langle x, y\rangle = \sum_{i=1}^{n} x_i y_i$. In standard coordinate system one has $\langle x, y\rangle = \sum_{i,j} g_{ij} x_i y_j$, where $g_{ij} = \langle e_i, e_j\rangle$, and the **metric tensor** $((g_{ij}))$ is a positive-definite symmetric $n \times n$ matrix.

In general, an Euclidean space is defined as a space, the properties of which are described by the axioms of *Euclidean Geometry*.

- **Unitary metric**

The **unitary metric** (or *complex Euclidean metric*) is the l_2-**metric** on $\mathbb{C}^n$, defined by

$$\|x - y\|_2 = \sqrt{|x_1 - y_1|^2 + \cdots + |x_n - y_n|^2}.$$

The metric space $(\mathbb{C}^n, \|x - y\|_2)$ is called *unitary space* (or *complex Euclidean space*). For $n = 1$, we obtain the *complex plane* (or *Argand plane*), i.e., the metric space $(\mathbb{C}, |z - u|)$ with the **complex modulus metric** $|z - u|$; here $|z| = |z_1 + iz_2| = \sqrt{z_1^2 + z_2^2}$ is the *complex modulus* (cf. also **quaternion metric**).

- L_p**-metric**

An L_p**-metric** d_{L_p}, $1 \leqslant p \leqslant \infty$, is a norm metric on $L_p(\Omega, \mathcal{A}, \mu)$, defined by

$$\|f - g\|_p$$

for any $f, g \in L_p(\Omega, \mathcal{A}, \mu)$. The metric space $(L_p(\Omega, \mathcal{A}, \mu), d_{L_p})$ is called L_p**-space** (or **Lebesgue space**).

Here Ω is a set, and $\mathcal{A}$ is an σ*-algebra* of subsets of Ω, i.e., a collection of subsets of Ω satisfying the following properties:

1. $\Omega \in \mathcal{A}$;
2. If $A \in \mathcal{A}$, then $\Omega \backslash A \in \mathcal{A}$;
3. If $A = \bigcup_{i=1}^{\infty} A_i$ with $A_i \in \mathcal{A}$, then $A \in \mathcal{A}$.

A function $\mu : \mathcal{A} \to \mathbb{R}_{\geqslant 0}$ is called *measure* on $\mathcal{A}$ if it is *additive*, i.e., $\mu(\bigcup_{i\geqslant 1} A_i) = \sum_{i\geqslant 1} \mu(A_i)$ for all pairwise disjoint sets $A_i \in \mathcal{A}$, and satisfies $\mu(\emptyset) = 0$. A *measure space* is a triple $(\Omega, \mathcal{A}, \mu)$.

Given a function $f : \Omega \to \mathbb{R}(\mathbb{C})$, its L_p*-norm* is defined by

$$\|f\|_p = \left(\int_{\Omega} |f(\omega)|^p \mu(d\omega)\right)^{\frac{1}{p}}.$$

Let $L_p(\Omega, \mathcal{A}, \mu) = L_p(\Omega)$ denotes the set of all functions $f : \Omega \to \mathbb{R}\ (\mathbb{C})$ which satisfy the condition $\|f\|_p < \infty$. Strictly speaking, $L_p(\Omega, \mathcal{A}, \mu)$ consists of equivalence classes of functions, where two functions are *equivalent* if they are equal *almost everywhere*, i.e., the set on which they differ has measure zero. The set $L_\infty(\Omega, \mathcal{A}, \mu)$ is the set of equivalence classes of measurable functions $f : \Omega \to \mathbb{R}\ (\mathbb{C})$ whose absolute values are bounded almost everywhere.

The most classical example of an L_p-metric is d_{L_p} on the set $L_p(\Omega, \mathcal{A}, \mu)$, where Ω is the open interval (0, 1), $\mathcal{A}$ is the *Borel sigma-algebra* on (0, 1), and μ is the *Lebesgue measure*. This metric space is abbreviated by $L_p(0, 1)$ and is called $L_p(0, 1)$*-space*.

In the same way, one can define the L_p-metric on the set $C_{[a,b]}$ of all real (complex) continuous functions on $[a, b]$: $d_{L_p}(f, g) = (\int_a^b |f(x) - g(x)|^p dx)^{\frac{1}{p}}$. For $p = \infty$, $d_{L_\infty}(f, g) = \max_{a \leqslant x \leqslant b} |f(x) - g(x)|$. This metric space is abbreviated by $C^p_{[a,b]}$ and is called $C^p_{[a,b]}$*-space*.

If $\Omega = \mathbb{N}$, $\mathcal{A} = 2^\Omega$ is the collection of all subsets of Ω, and μ is the *cardinality measure* (i.e., $\mu(A) = |A|$ if A is a finite subset of Ω, and $\mu(A) = \infty$, otherwise), then the metric space $(L_p(\Omega, 2^\Omega, |.|), d_{L_p})$ coincides with the space l_p^∞.

If $\Omega = V_n$ is a set of cardinality n, $\mathcal{A} = 2^{V_n}$, and μ is the cardinality measure, then the metric space $(L_p(V_n, 2^{V_n}, |.|), d_{L_p})$ coincides with the space l_p^n.

- **Dual metrics**

The l_p**-metric** and the l_q**-metric**, $1 < p, q < \infty$, are called **dual** if $1/p + 1/q = 1$.

In general, when dealing with a *normed vector space* $(V, \|.\|_V)$, one is interested in the *continuous* linear functionals from V into the base field ($\mathbb{R}$ or $\mathbb{C}$). These functionals form a **Banach space** $(V', \|.\|_{V'})$, called *continuous dual* of V. The norm $\|.\|_{V'}$ on V' is defined by $\|T\|_{V'} = \sup_{\|x\|_V \leqslant 1} |T(x)|$.

The continuous dual for the metric space l_p^n (l_p^∞) is l_q^n (l_q^∞, respectively). The continuous dual of l_1^n (l_1^∞) is l_∞^n (l_∞^∞, respectively). The continuous duals of the Banach spaces C (consisting of all convergent sequences, with l_∞**-metric**) and C_0 (consisting of the sequences converging to zero, with l_∞**-metric**) are both naturally identified with l_1^∞.

- **Inner product space**

An **inner product space** (or *pre-Hilbert space*) is a metric space $(V, \|x - y\|)$ on a real (complex) vector space V with an *inner product* $\langle x, y\rangle$ such that the norm metric $\|x - y\|$ is constructed using the *inner product norm* $\|x\| = \sqrt{\langle x, x\rangle}$.

An *inner product* $\langle , \rangle$ on a real (complex) vector space V is a *symmetric bilinear* (in complex case, *sesquilinear*) form on V, i.e., a function $\langle , \rangle : V \times V \longrightarrow \mathbb{R}$ $(\mathbb{C})$ such that, for all $x, y, z \in V$ and for all scalars α, β, we have the following properties:

1. $\langle x, x\rangle \geqslant 0$, with $\langle x, x\rangle = 0$ if and only if $x = 0$;
2. $\langle x, y\rangle = \overline{\langle y, x\rangle}$, where $\overline{\alpha} = \overline{a + bi} = a - bi$ denotes the *complex conjugation*;
3. $\langle \alpha x + \beta y, z\rangle = \alpha\langle x, z\rangle + \beta\langle y, z\rangle$.

For a complex vector space, an inner product is called also *Hermitian inner product*, and the corresponding metric space is called *Hermitian inner product space*.

A norm $\|.\|$ in a *normed space* $(V, \|.\|)$ is generated by an inner product if and only if, for all $x, y \in V$, we have: $\|x + y\|^2 + \|x - y\|^2 = 2(\|x\|^2 + \|y\|^2)$.

- **Hilbert space**

A **Hilbert space** is an **inner product space** which, as a metric space, is **complete**. More precisely, a Hilbert space is a complete metric space $(H, \|x - y\|)$ on a real (complex) vector space H with an *inner product* $\langle,\rangle$ such that the norm metric $\|x - y\|$ is constructed using the *inner product norm* $\|x\| = \sqrt{\langle x, x\rangle}$. Any Hilbert space is a **Banach space**.

An example of a Hilbert space is the set of all sequences $x = \{x_n\}_n$ of real (complex) numbers such that $\sum_{i=1}^{\infty} |x_i|^2$ converges, with the **Hilbert metric**, defined by

$$\left(\sum_{i=1}^{\infty} |x_i - y_i|^2\right)^{\frac{1}{2}}.$$

Other examples of Hilbert spaces are any L_2**-space**, and any finite-dimensional inner product space. In particular, any Euclidean space is a Hilbert space.

A direct product of two **Hilbert spaces** is called *Liouville space* (or *line space*, *extended Hilbert space*).

- **Riesz norm metric**

A *Riesz space* (or *vector lattice*) is a partially ordered vector space $(V_{Ri}, \preceq)$ in which the following conditions hold:

1. The vector space structure and the partial order structure are compatible, i.e., from $x \preceq y$ follows that $x+z \preceq y+z$, and from $x \succ 0$, $a \in \mathbb{R}$, $a > 0$ follows that $ax \succ 0$;
2. For any two elements $x, y \in V_{Ri}$, there exist *join* $x \vee y \in V_{Ri}$ and *meet* $x \wedge y \in V_{Ri}$ (cf. Chapter 10).

The **Riesz norm metric** is a norm metric on V_{Ri}, defined by

$$\|x - y\|_{Ri},$$

where $\|.\|_{Ri}$ is a *Riesz norm* on V_{Ri}, i.e., a norm such that, for any $x, y \in V_{Ri}$, the inequality $|x| \preceq |y|$, where $|x| = (-x) \vee (x)$, implies $\|x\|_{Ri} \leqslant \|y\|_{Ri}$.

The space $(V_{Ri}, \|.\|_{Ri})$ is called *normed Riesz space*. In the case of completeness, it is called *Banach lattice*.

- **Banach–Mazur compactum**

The **Banach–Mazur distance** d_{BM} between two n-dimensional *normed spaces* $(V, \|.\|_V)$ and $(W, \|.\|_W)$ is defined by

$$\ln \inf_T \|T\| \cdot \|T^{-1}\|,$$

where the infimum is taken over all isomorphisms $T : V \rightarrow W$. It is a metric on the set X^n of all equivalence classes of n-dimensional normed spaces, where $V \sim W$ if and only if they are *isometric*. Then pair (X^n, d_{BM}) is a compact metric space which is called **Banach–Mazur compactum**.

- **Quotient metric**

Given a *normed space* $(V, \|.\|_V)$ with a norm $\|.\|_V$ and a closed subspace W of V, let $(V/W, \|.\|_{V/W})$ be the normed space of cosets $x + W = \{x + w : w \in W\}$, $x \in V$, with the *quotient norm* $\|x + W\|_{V/W} = \inf_{w \in W} \|x + w\|_V$.

The **quotient metric** is a norm metric on V/W, defined by

$$\|(x + W) - (y + W)\|_{V/W}.$$

- **Tensor norm metric**

Given *normed spaces* $(V, \|.\|_V)$ and $(W, \|.\|_W)$, a norm $\|.\|_{\otimes}$ on the *tensor product* $V \otimes W$ is called *tensor norm* (or *cross norm*) if $\|x \otimes y\|_{\otimes} = \|x\|_V \|y\|_W$ for all *decomposable* tensors $x \otimes y$.

The **tensor product metric** is a norm metric on $V \otimes W$, defined by

$$\|z - t\|_{\otimes}.$$

For any $z \in V \otimes W$, $z = \sum_j x_j \otimes y_j$, $x_j \in V$, $y_j \in W$, its *projective norm* (or π-*norm*) is defined by $\|z\|_{pr} = \inf \sum_j \|x_j\|_V \|y_j\|_W$, where the infimum is taken over all representation of z as a sum of decomposable vectors. It is the largest tensor norm on $V \otimes W$.

- **Valuation metric**

A **valuation metric** is a metric on a *field* $\mathbb{F}$, defined by

$$\|x - y\|,$$

where $\|.\|$ is a *valuation* on $\mathbb{F}$, i.e., a function $\|.\| : \mathbb{F} \to \mathbb{R}$ such that, for all $x, y \in \mathbb{F}$, we have the following properties:

1. $\|x\| \geqslant 0$, with $\|x\| = 0$ if and only if $x = 0$;
2. $\|xy\| = \|x\| \, \|y\|$,
3. $\|x + y\| \leqslant \|x\| + \|y\|$ (triangle inequality).

If $\|x + y\| \leqslant \max\{\|x\|, \|y\|\}$, the valuation $\|.\|$ is called *non-Archimedean*. In this case, the valuation metric is an **ultrametric**. The simplest example of a valuation is the *trivial valuation* $\|.\|_{tr}$: $\|0\|_{tr} = 0$, and $\|x\|_{tr} = 1$ for $x \in \mathbb{F}\backslash\{0\}$. It is non-Archimedean.

There are different definitions of valuation in Mathematics. Thus, the function $\nu : \mathbb{F} \to \mathbb{R} \cup \{\infty\}$ is called *valuation* if $\nu(x) \geqslant 0$, $\nu(0) = \infty$, $\nu(xy) = \nu(x) + \nu(y)$, and $\nu(x + y) \geqslant \min\{\nu(x), \nu(y)\}$ for all $x, y \in \mathbb{F}$. The valuation $\|.\|$ can be obtained from the function ν by the formula $\|x\| = \alpha^{\nu(x)}$ for some fixed $0 < \alpha < 1$ (cf. p-**adic metric**). The *Kürschäk valuation* $|.|_{Krs}$ is defined as a function $|.|_{Krs} : \mathbb{F} \to \mathbb{R}$ such that $|x|_{Krs} \geqslant 0$, $|x|_{Krs} = 0$ if and only if $x = 0$, $|xy|_{Krs} = |x|_{Krs}|y|_{Krs}$, and $|x + y|_{Krs} \leqslant C \max\{|x|_{Krs}, |y|_{Krs}\}$ for all $x, y \in \mathbb{F}$ and for some positive constant C, called *constant of valuation*. If $C \leqslant 2$, one obtains the ordinary definition of the valuation $\|.\|$ which is non-Archimedean if $C \leqslant 1$. In general, any $|.|_{Krs}$ is *equivalent* to some $\|.\|$, i.e., $|.|^p_{Krs} = \|.\|$ for some $p > 0$. At last, given an *ordered group* $(G, \cdot, e, \leqslant)$ equipped with zero, the *Krull valuation* is defined as a function $|.| : \mathbb{F} \to G$ such that $|x| = 0$ if and only if $x = 0$, $|xy| = |x||y|$, and $|x + y| \leqslant \max\{|x|, |y|\}$ for any $x, y \in \mathbb{F}$. It is a generalization of the definition of non-Archimedean valuation $\|.\|$ (cf. **generalized metric**).

- **Power series metric**

Let $\mathbb{F}$ be an arbitrary algebraic field, and let $\mathbb{F}\langle x^{-1}\rangle$ be the field of power series of the form $w = \alpha_{-m}x^m + \cdots + \alpha_0 + \alpha_1 x^{-1} + \cdots$, $\alpha_i \in \mathbb{F}$. Given $l > 1$, a *non-Archimedean valuation* $\|.\|$ on $\mathbb{F}\langle x^{-1}\rangle$ is defined by

$$\|w\| = \begin{cases} l^m, & \text{if } w \neq 0, \\ 0, & \text{if } w = 0. \end{cases}$$

The **power series metric** is the **valuation metric** $\|w - v\|$ on $\mathbb{F}\langle x^{-1}\rangle$.

Part II

Geometry and Distances

Chapter 6. Distances in Geometry
Chapter 7. Riemannian and Hermitian Metrics
Chapter 8. Distances on Surfaces and Knots
Chapter 9. Distances on Convex Bodies, Cones, and Simplicial Complexes

Chapter 6

Distances in Geometry

Geometry arose as the field of knowledge dealing with spatial relationships. It was one of the two fields of pre-modern Mathematics, the other being the study of numbers. In modern times, geometric concepts have been generalized to a high level of abstraction and complexity.

6.1. GEODESIC GEOMETRY

In Mathematics, the notion of "geodesic" is a generalization of the notion of "straight line" to curved spaces. This term is taken from *Geodesy*, the science of measuring the size and the shape of the Earth.

Given a metric space (X, d), a **metric curve** γ is a continuous function $\gamma : I \to X$, where I is an *interval* (i.e., non-empty connected subset) of $\mathbb{R}$. If γ is r times continuously differentiable, it is called *regular curve* of class C^r; if $r = \infty$, γ is called *smooth curve*.

In general, a curve may cross itself. A curve is called *simple curve* (or *arc*, *path*) if it does not cross itself, i.e., if it is injective. A curve $\gamma : [a, b] \to X$ is called *Jordan curve* (or *simple closed curve*) if it does not cross itself, and $\gamma(a) = \gamma(b)$.

The *length* (which may be equal to ∞) $l(\gamma)$ of a curve $\gamma : [a, b] \to X$ is defined by $\sup \sum_{i=1}^{n} d(\gamma(t_{i-1}), \gamma(t_i))$, where the supremum is taken over all finite decompositions $a = t_0 < t_1 < \cdots < t_n = b$, $n \in \mathbb{N}$, of $[a, b]$. A curve with finite length is called *rectifiable*. For each regular curve $\gamma : [a, b] \to X$ define the *natural parameter* s of γ by $s = s(t) = l(\gamma|_{[a,t]})$, where $l(\gamma|_{[a,t]})$ is the length of the part of γ, corresponding to the interval $[a, t]$. The parametrization $\gamma = \gamma(s)$ is called *natural*. In this parametrization, for any $t_1, t_2 \in I$, one has $l(\gamma|_{[t_1,t_2]}) = |t_2 - t_1|$, and $l(\gamma) = |b - a|$.

The length of any curve $\gamma : [a, b] \to X$ is at least the distance between its end points: $l(\gamma) \geqslant d(\gamma(a), \gamma(b))$. The curve γ, for which $l(\gamma) = d(\gamma(a), \gamma(b))$, is called *geodesic segment* (or *shortest path*) from $x = \gamma(a)$ to $y = \gamma(b)$, and denoted by $[x, y]$. Thus, a geodesic segment is a shortest join of its endpoints; it is an *isometric embedding* of $[a, b]$ in X. In general, geodesic segments need not exist, except for a trivial case when segment consists of one point only. Moreover, a geodesic segment joining two points need not be unique.

A *geodesic* is a curve which extends indefinitely in both directions and behaves locally like a segment, i.e., is everywhere locally a distance minimizer. More exactly, a curve $\gamma : \mathbb{R} \to X$, given in the *natural parametrization*, is called *geodesic* if, for any $t \in \mathbb{R}$, there exists a *neighborhood* U of t such that, for any $t_1, t_2 \in U$, we have $d(\gamma(t_1), \gamma(t_2)) = |t_1 -$

$t_2|$. Thus, any geodesic is a *locally isometric embedding* of the whole $\mathbb{R}$ in X. A geodesic is called *metric straight line* (or *minimizing geodesic*) if the equality $d(\gamma(t_1), \gamma(t_2)) = |t_1 - t_2|$ holds for all $t_1, t_2 \in \mathbb{R}$. Such geodesic is an isometric embedding of the whole real line $\mathbb{R}$ in X. A geodesic is called *metric great circle* if it is an isometric embedding of a circle $S^1(0, r)$ in X. In general, geodesics need not exists.

- **Geodesic metric space**

  A metric space (X, d) is called **geodesic** if any two points in X can be joined by a *geodesic segment*, i.e., for any two points $x, y \in X$, there is an **isometry** from the segment $[0, d(x, y)]$ into X.

  Any *complete Riemannian space* is a geodesic metric space.

- **Geodesic distance**

  The **geodesic distance** (or **shortest path distance**) is the length of a *geodesic segment* (i.e., a *shortest path*) between two points.

- **Intrinsic metric**

  The metric d on X is called **intrinsic metric** (or *length metric*) if the distance $d(x, y)$ between any pair x, y of points in X is equal to the infimum of lengths of curves connecting these points. A metric space (X, d) with the intrinsic metric d is called **length space** (or *path metric space*, *inner metric space*).

  If, moreover, any pair x, y of points can be joined by a curve of length $d(x, y)$, then the metric d is called *strictly intrinsic*, and the length space (X, d) is a **geodesic** metric space.

  Given a metric space (X, d) in which every pair of points is joined by a rectifiable curve, the **induced intrinsic metric** (or **internal metric**, **interior metric**) D on X is defined as the infimum of the lengths of all rectifiable curves, connecting two given points $x, y \in X$.

- **Space of geodesics**

  A **space of geodesics** (or *G-space*) is a metric space (X, d) with the geometry characterized by the fact that extensions of geodesics, defined as locally shortest lines, are unique. Such geometry is a generalization of *Hilbert Geometry* (see [Buse55]).

  More exactly, an G-space (X, d) is defined by the following conditions:

  1. It is *finitely compact*, i.e., a bounded infinite set in X has at least one *accumulation point*;
  2. It is **Menger-convex**, i.e., for any different $x, y \in X$, there exists a third point $z \in X$, $z \neq x, y$, such that $d(x, z) + d(z, y) = d(x, y)$;
  3. It is *locally extendable*, i.e., for any $a \in X$, there exists $r > 0$ such that, for any distinct points x, y in the ball $B(a, r)$, there exists z distinct from x and y such that $d(x, y) + d(y, z) = d(x, z)$;
  4. It is *uniquely extendable*, i.e., if in 3. above two points z_1 and z_2 were found, so that $d(y, z_1) = d(y, z_2)$, then $z_1 = z_2$.

The existence of geodesic segments is ensured by finite compactness and Menger-convexity: any two points of a finitely compact Menger-convex set X can be joined by a geodesic segment in X. The existence of geodesics is ensured by the axiom of local prolongation: if a finitely compact Menger-convex set X is locally extendable, then there exists a geodesic contains a given segment. Finally, the uniqueness of prolongation ensures the assumption of Differential Geometry that a *line element* determines a geodesic in one way only.

All *Riemannian* and *Finsler spaces* are G-spaces. An one-dimensional G-space is a metric straight line or a metric great circle. Any two-dimensional G-space is a topological *manifold*.

Every G-space is a *chord space*, i.e., a metric space with distinguished geodesics (see [BuPh87]).

- **Desarguesian space**

A **Desarguesian space** is a **space of geodesics** (X, d) in which the role of geodesics is played by ordinary straight lines. Thus, X may be topologically mapped into a *projective space* $\mathbb{R}P^n$ so that each geodesic of X is mapped into a straight line of $\mathbb{R}P^n$. Any X mapped into $\mathbb{R}P^n$ must either cover all of $\mathbb{R}P^n$, and, in such a case, the geodesics of X are all metric great circles of the same length, or X may be considered as an open *convex* subset of an affine space A^n.

A space (X, d) of geodesics is a Desarguesian space if and only if the following conditions hold:

1. The geodesic passing through two different points is unique;
2. For dimension $n = 2$, both the direct and the converse *Desargues theorems* are valid, and, for dimension $n > 2$, any three points in X lie in one plane.

Among *Riemannian spaces*, the only Desarguesian spaces are Euclidean, *hyperbolic*, and *elliptic spaces*. An example of the non-Riemannian Desarguesian space is the *Minkowskian space* which can be regarded as the prototype of all non-Riemannian spaces, including *Finsler spaces*.

- **Space of elliptic type**

A **space of elliptic type** is a **space of geodesics** in which the geodesic through two points is unique, and all geodesics are the metric great circles of the same length.

- **Straight space**

A **straight space** is a **space of geodesics** in which extension of a geodesic is possible in the large. Any *geodesic* in a straight space is a metric straight line, and is uniquely determined by any two of its points. Straight spaces are simply connected spaces without conjugate points. Any two-dimensional straight space is homeomorphic to the plane.

Minkowskian spaces and all simply-connected *Riemannian spaces* of non-positive curvature (including Euclidean and *hyperbolic spaces*) are straight spaces.

- **Gromov-hyperbolic metric space**

A metric space (X, d) is called **Gromov-hyperbolic** if it is **geodesic** and **δ-hyperbolic** for some $\delta \geqslant 0$.

Any complete simply connected *Riemannian space* of *sectional curvature* $k \leqslant -a^2$ is Gromov-hyperbolic metric space with $\delta = \frac{\ln 3}{a}$. Any Euclidean space $\mathbb{E}^n$ with $n > 1$ is not Gromov-hyperbolic. An important class of Gromov-hyperbolic metric spaces are *hyperbolic groups*, i.e., finitely generated groups whose **word metric** is δ-hyperbolic for some $\delta \geqslant 0$. A metric space is a **real tree** exactly when it is Gromov-hyperbolic metric space with $\delta = 0$.

A geodesic metric space (X, d) is δ-hyperbolic if and only if it is *Rips* 4δ*-hyperbolic*, i.e., each of its *geodesic triangles* (the union of three *geodesic segments* $[x, y]$, $[x, z]$, $[y, z]$) is 4δ*-thin* (or 4δ*-slim*): every side of the triangle is contained in the 4δ*-neighborhood* of the other two sides (a 4δ*-neighborhood* of a subset $A \subset X$ is the set $\{b \in X\colon \inf_{a \in A} d(b, a) < 4\delta\}$).

- **CAT(k) space**

Let (X, d) be a **complete geodesic** metric space. Let M^2 be a simply connected two-dimensional *Riemannian manifold* of *constant curvature* $k \leqslant 0$ (for $k = 0$ and -1, it is the Euclidean plane $\mathbb{E}^2$ and the real *hyperbolic plane* H^2, respectively).

A *triangle* T in X consists of three points in X together with three *geodesic segments* joining them pairwise; the segments are called the *sides of the triangle*. For a triangle $T \subset X$, a *comparison triangle* for T in M^2 is a triangle $T' \subset M^2$ together with a map f_T which sends each side of T isometrically onto a side of T'. A triangle T is said to satisfy the Gromov's *CAT(k) inequality* (for Cartan, Alexandrov and Toponogov) if, for every $x, y \in T$, we have

$$d(x, y) \leqslant d_{M^2}\big(f_T(x), f_T(y)\big),$$

where f_T is the map associated to a comparison triangle for T in M^2. So, the geodesic triangle T is at least as "thin" as its comparison triangle in M^2.

A **CAT(k) space** is a **proper** (i.e., all closed metric balls are compact) **geodesic** metric space in which every triangle satisfies the CAT(k) inequality.

Gromov hyperbolic metric spaces are CAT(0) spaces and a generalization of CAT(-1) spaces.

CAT(0) spaces are called also *Hadamard spaces*, because they are generalizations of *Hadamard manifolds* which are simply connected, complete Riemannian manifolds such that the sectional curvature is non-positive. A CAT(0) space is not a manifold, in general; it can be a tree, for example.

An *Alexandrov space with non-positive curvature* (or *non-positively curved space*) is a metric space (X, d) in which, for any $x \in X$, there exists $r > 0$ such that the closed metric ball $\overline{B}(x, r) = \{y \in X\colon\ d(x, y) \leqslant r\}$ is a CAT(0) space with respect to d.

- **Tits metric**

  Let (X, d) be a **CAT(0) space**. An *unit speed geodesic ray* in X is a curve $\alpha : [0, +\infty) \to X$ which realizes the distance between any two of its points, i.e., α is an *isometric embedding* of $[0, \infty) \subset \mathbb{R}$ into X. Two unit speed geodesic rays α^1, α^2 in X are called *asymptotic* if there is a constant $C \geqslant 0$ such that $\lim_{t\to\infty} d(\alpha^1(t), \alpha^2(t)) \leqslant C$; the corresponding equivalence class is denoted by $\alpha^1_\infty (= \alpha^2_\infty)$. The set $\partial_\infty X$ of all equivalence classes of asymptotic geodesic rays of X is called *boundary of X at infinity*.

  The **Tits metric** (or *asymptotic angle of divergence*) is a metric on $\partial_\infty X$, defined by

$$2 \arcsin\left(\frac{\rho}{2}\right)$$

  for all $\alpha^1_\infty, \alpha^2_\infty \in \partial_\infty X$, where $\rho = \lim_{t\to+\infty} \frac{1}{t} d(\alpha^1(t), \alpha^2(t))$. The set $\partial_\infty X$ equipped with the Tits metric is called *Tits boundary* of X.

- **Projectively flat space**

  A metric space is called **projectively flat** if it locally admits a *geodesic mapping* (i.e., a mapping preserving geodesics) into an Euclidean space.

6.2. PROJECTIVE GEOMETRY

Projective Geometry is a branch of Geometry dealing with the properties and invariants of geometric figures under *projection*. Affine Geometry and Euclidean Geometry are subsets of Projective Geometry.

An n-dimensional *projective space* $\mathbb{F}P^n$ is the space of one-dimensional vector subspaces of a given $(n+1)$-dimensional vector space V over a field $\mathbb{F}$. The basic construction is to form the set of equivalence classes of non-zero vectors in V under the relation of scalar proportionality. This idea goes back to mathematical descriptions of *perspective*. The use of a basis of V allows the introduction of *homogeneous coordinates* of a point in $\mathbb{K}P^n$ which are usually written as $(x_1 : x_2 : \ldots : x_n : x_{n+1})$ – a vector of length $n+1$, other than $(0 : 0 : 0 : \ldots : 0)$. Two sets of coordinates that are proportional denote the same point of the projective space. Any point of projective space which can be represented as $(x_1 : x_2 : \ldots : x_n : 0)$ is called *point at infinity*. The part of a projective space $\mathbb{K}P^n$ not "at infinity", i.e., the set of points of the projective space which can be represented as $(x_1 : x_2 : \ldots : x_n : 1)$, is an n-dimensional *affine space* A^n.

The notation $\mathbb{R}P^n$ denotes the *real projective space* of dimension n, i.e., the space of one-dimensional vector subspaces of $\mathbb{R}^{n+1}$. The notation $\mathbb{C}P^n$ denotes the *complex projective space* of complex dimension n. The projective space $\mathbb{R}P^n$ carries a natural structure of a compact smooth n-dimensional *manifold*. It can be viewed as the space of lines through the zero element 0 of $\mathbb{R}^{n+1}$ (i.e., as a *ray space*). It can be viewed as the set $\mathbb{R}^n$, considered as an *affine space*, together with its points at infinity. It can be viewed also as the set of points of an n-dimensional sphere in $\mathbb{R}^{n+1}$ with identified diametrically-opposite points.

The projective points, projective straight lines, projective planes, ..., projective hyperplanes of $\mathbb{K}P^n$ are one-dimensional, two-dimensional, three-dimensional, ...,

n-dimensional subspaces of V, respectively. Any two projective straight lines in a projective plane have one and only one common point. A *projective transformation* (or *collineation, projectivity*) is a bijection of a projective space onto itself, preserving collinearity (the property of points to be on one line) in both directions. Any projective transformation is a composition of a pair of *perspective projections*. Projective transformations do not preserve sizes or angles but do preserve *type* (that is, points remain points, and lines remain lines), *incidence* (that is, whether a point lies on a line), and *cross-ratio*. Here, given four collinear points $x, y, z, t \in \mathbb{F}P^n$, their *cross-ratio* is defined by $(x, y, z, t) = \frac{(x-z)(y-t)}{(y-z)(x-t)}$, where $\frac{x-z}{x-t}$ denotes the ratio $\frac{f(x)-f(z)}{f(x)-f(t)}$ for some affine bijection f from the straight line $l_{x,y}$ through the points x and y onto $\mathbb{K}$. Given four projective straight lines l_x, l_y, l_z, l_t, containing points x, y, z, t, respectively, and passing through a given point, their *cross-ratio*, defined by $(l_x, l_y, l_t, l_t) = \frac{\sin(l_x,l_z)\sin(l_y,l_t)}{\sin(l_y,l_z)\sin(l_x,l_t)}$, coincides with (x, y, z, t). The cross-ratio of four complex numbers x, y, z, t is given by $(x, y, z, t) = \frac{(x-z)(y-t)}{(y-z)(x-t)}$. It is real if and only if the four numbers are either collinear or cocyclic.

- **Projective metric**

  Given a subset D of a projective space $\mathbb{R}P^n$, the **projective metric** d is a metric on D such that shortest paths with respect to this metric are parts of or entire projective straight lines. It is assumed that the following conditions hold:

  1. D does not belong to a hyperplane;
  2. For any three non-collinear points $x, y, z \in D$, the triangle inequality holds in the strict sense: $d(x, y) < d(x, z) + d(z, y)$;
  3. If x, y are different points in D, then the intersection of the straight line $l_{x,y}$ through x and y with D is either all of $l_{x,y}$, and forms a *metric great circle* (i.e., is isometric to a circle), or is obtained from $l_{x,y}$ by discarding some segment (which can be reduced to a point), and forms a *metric straight line* (i.e., is isometric to the whole $\mathbb{R}$).

  The metric space (D, d) is called *projective metric space*. The problem to determine all projective metrics is the so-called *fourth problem of Hilbert*; it is decided only for dimension $n = 2$. In fact, given a smooth measure on the space of hyperplanes in $\mathbb{R}P^n$, define the distance between any two points $x, y \in \mathbb{R}P^n$ as one-half the measure of all hyperplanes intersecting the line segment joining x and y. The obtained metric is projective. It is the *Busemann's construction* of projective metrics. For $n = 2$, Ambartzumian ([Amba76]) proved that all projective metrics can be obtained from the Busemann's construction.

  In a projective metric space there are no, simultaneously, both types of straight lines: they are either all metric straight lines, or they are all metric great circles of the same length (*Hamel's theorem*). Spaces of the first kind are called *open*. They coincide with subspaces of an affine space; the geometry of open projective metric spaces is a *Hilbert Geometry*. *Hyperbolic Geometry* is a Hilbert Geometry in which there exist reflections at all straight lines. Thus, the set D has Hyperbolic Geometry if and only if it is the interior of an ellipsoid. The geometry of open projective metric spaces whose subsets coincide with all of affine space, is a *Minkowski Geometry*. *Euclidean Geometry* is a

Hilbert Geometry and a Minkowski Geometry, simultaneously. Spaces of the second kind are called *closed*; they coincide with the whole of $\mathbb{R}P^n$. *Elliptic Geometry* is the geometry of a projective metric space of the second kind.

- **Strip projective metric**

The **strip projective metric** ([BuKe53]) is a **projective metric** on the strip $St = \{x \in \mathbb{R}^2 : -\pi/2 < x_2 < \pi/2\}$, defined by

$$\sqrt{(x_1 - y_1)^2 + (x_2 - y_2)^2} + |\tan x_2 - \tan y_2|.$$

Note, that St with the ordinary Euclidean metric $\sqrt{(x_1 - y_1)^2 + (x_2 - y_2)^2}$ is not a *projective metric space* .

- **Half-plane projective metric**

The **half-plane projective metric** ([BuKe53]) is a **projective metric** on $\mathbb{R}^2_+ = \{x \in \mathbb{R}^2 : x_2 > 0\}$, defined by

$$\sqrt{(x_1 - y_1)^2 + (x_2 - y_2)^2} + \left|\frac{1}{x_2} - \frac{1}{y_2}\right|.$$

- **Hilbert projective metric**

Given a set H, the **Hilbert projective metric** h is a **complete projective metric** on H. It means, that H contains, together with two arbitrary distinct points x and y also the points z and t for which $h(x, z) + h(z, y) = h(x, y)$, $h(x, y) + h(y, t) = h(x, t)$, and is homeomorphic to a *convex* set in an n-dimensional affine space A^n, the geodesics in H being mapped to straight lines of A^n. The metric space (H, h) is called *Hilbert projective space*, and the geometry of a Hilbert projective space is called *Hilbert Geometry*.

Formally, let D be a non-empty *convex* open set in A^n with the boundary ∂D not containing two proper coplanar but non-collinear segments (ordinary the boundary of D is a strictly convex closed curve, and D is its interior). Let $x, y \in D$ be located on a straight line which intersects ∂D at z and t, z is on the side of y, and t is on the side of x. Then the Hilbert metric h on D is defined by

$$\frac{r}{2} \ln(x, y, z, t),$$

where (x, y, z, t) is the *cross-ratio* of x, y, z, t, and r is a fixed positive constant.

The metric space (D, h) is a **straight space**. If D is an ellipsoid, then h is the **hyperbolic metric**, and defines *Hyperbolic Geometry* on D. On the *unit disk* $\Delta = \{z \in \mathbb{C} : |z| < 1\}$ the metric h coincides with the **Cayley–Klein–Hilbert metric**. If $n = 1$, the metric h makes D isometric to the Euclidean line.

If ∂D contains coplanar but non-collinear segments, a projective metric on D can be given by $h(x, y) + d(x, y)$, where d is any **Minkowskian metric** (usually, the Euclidean metric).

• **Minkowskian metric**

The **Minkowskian metric** (or **Minkowski–Hölder distance**) is the **norm metric** of a finite-dimensional real **Banach space**.

Formally, let $\mathbb{R}^n$ be an n-dimensional real vector space, let K be a *symmetric convex body* in $\mathbb{R}^n$, i.e., an open neighborhood of the origin which is bounded, convex, and *symmetric* ($x \in K$ if and only if $-x \in K$). Then the *Minkowski functional* $\|.\|_K : \mathbb{R}^n \to [0, \infty)$, defined by

$$\|x\|_K = \inf\left\{\alpha > 0 \colon \frac{x}{\alpha} \in \partial K\right\},$$

is a *norm* on $\mathbb{R}^n$, and the Minkowskian metric m on $\mathbb{R}^n$ is defined by

$$\|x - y\|_K.$$

The metric space $(\mathbb{R}^n, m)$ is called *Minkowskian space*. It can be considered as an n-dimensional affine space A^n with a metric m in which the role of the *unit ball* is played by a given centrally-symmetric convex body. The geometry of a Minkowskian space is called *Minkowski Geometry*. For a strictly convex symmetric body Minkowskian metric is a **projective metric**, and $(\mathbb{R}^n, m)$ is a **straight space**. A Minkowski Geometry is Euclidean if and only if its *unit sphere* is an ellipsoid.

The Minkowskian metric m is proportional to the Euclidean metric d_E on every given line l, i.e., $m(x, y) = \phi(l) d_E(x, y)$. Thus, the Minkowskian metric can be considered as a metric which is defined in the whole affine space A^n and has the property that the *affine ratio* $\frac{ac}{ab}$ of any three collinear points a, b, c (cf. Section 3) is equal to their *distance ratio* $\frac{m(a,c)}{m(a,b)}$.

• **Busemann metric**

The **Busemann metric** ([Buse55]) is a metric on the real n-dimensional projective space $\mathbb{R}P^n$, defined by

$$\min\left\{\sum_{i=1}^{n+1}\left|\frac{x_i}{\|x\|} - \frac{y_i}{\|y\|}\right|, \sum_{i=1}^{n+1}\left|\frac{x_i}{\|x\|} + \frac{y_i}{\|y\|}\right|\right\}$$

for any $x = (x_1 : \ldots : x_{n+1})$, $y = (y_1 : \ldots : y_{n+1}) \in \mathbb{R}P^n$, where $\|x\| = \sqrt{\sum_{i=1}^{n+1} x_1^2}$.

• **Flag metric**

Given an n-dimensional *projective space* $\mathbb{F}P^n$, the **flag metric** d is a metric on $\mathbb{F}P^n$, defined by an *flag*, i.e., an *absolute* consisting of a collection of m-planes α_m, $m = 0, \ldots, n-1$, with α_{i-1} belonging to α_i for all $i \in \{1, \ldots, n-1\}$. The metric space $(\mathbb{F}P^n, d)$ is abbreviated by F^n and is called *flag space*.

If one chooses an affine coordinate system $(x_i)_i$ in a space F^n, so that the vectors of the lines passing through the $(n - m - 1)$-plane α_{n-m-1} are defined by the condition $x_1 = \cdots = x_m = 0$, then the flag metric $d(x, y)$ between the points $x = (x_1, \ldots, x_n)$

and $y = (y_1, \dots, y_n)$ is defined by

$$d(x, y) = |x_1 - y_1|, \quad \text{if } x_1 \neq y_1,$$
$$d(x, y) = |x_2 - y_2|, \quad \text{if } x_1 = y_1, x_2 \neq y_2, \dots,$$
$$d(x, y) = |x_k - y_k|, \quad \text{if } x_1 = y_1, \dots, x_{k-1} = y_{k-1}, x_k \neq y_k, \dots.$$

- **Projective determination of a metric**

The **projective determination of a metric** is an introduction, in subsets of a projective space, of a metric such that these subsets become isomorphic to an Euclidean, *hyperbolic*, or *elliptic space*.

To obtain an *Euclidean determination of a metric* in $\mathbb{R}P^n$, one should distinguish in this space an $(n-1)$-dimensional hyperplane π, called *hyperplane at infinity*, and define $\mathbb{E}^n$ as the subset of the projective space obtained by removing from it this hyperplane π. In terms of homogeneous coordinates, π consists of all points $(x_1 : \dots : x_n : 0)$, and $\mathbb{E}^n$ consists of all points $(x_1 : \dots : x_n : x_n)$ with $x_n \neq 0$. Hence, it can be written as $\mathbb{E}^n = \{x \in \mathbb{R}P^n : x = (x_1 : \dots : x_n : 1)\}$. The Euclidean metric d_E on $\mathbb{E}^n$ is defined by

$$\sqrt{\langle x - y, x - y \rangle},$$

where, for any $x = (x_1 : \dots : x_n : 1), y = (y_1 : \dots : y_n : 1) \in \mathbb{E}^n$, one has $\langle x, y \rangle = \sum_{i=1}^{n} x_i y_i$.

To obtain a *hyperbolic determination of a metric* in $\mathbb{R}P^n$, a set D of interior points of a real oval hypersurface Ω of order two in $\mathbb{R}P^n$ is considered. The **hyperbolic metric** d_{hyp} on D is defined by

$$\frac{r}{2} \left| \ln(x, y, z, t) \right|,$$

where z and t are the points of intersection of the straight line $l_{x,y}$ through the points x and y with Ω, (x, y, z, t) is the *cross-ratio* of the points x, y, z, t, and r is a fixed positive constant. If, for any $x = (x_1 : \dots : x_{n+1}), y = (y_1 : \dots : y_{n+1}) \in \mathbb{R}P^n$, the *scalar product* $\langle x, y \rangle = -x_1 y_1 + \sum_{i=2}^{i+1} x_i y_i$ is defined, the hyperbolic metric on the set $D = \{x \in \mathbb{R}P^n : \langle x, x \rangle < 0\}$ can be written as

$$r \operatorname{arccosh} \frac{|\langle x, y \rangle|}{\sqrt{\langle x, x \rangle \langle y, y \rangle}},$$

where r is a fixed positive constant, and arccosh denotes the non-negative values of the inverse hyperbolic cosine.

To obtain an *elliptic determination of a metric* in $\mathbb{R}P^n$, one should consider, for any $x = (x_1 : \dots : x_{n+1}), y = (y_1 : \dots : y_{n+1}) \in \mathbb{R}P^n$, the *inner product* $\langle x, y \rangle = \sum_{i=1}^{n+1} x_i y_i$. The **elliptic metric** d_{ell} on $\mathbb{R}P^n$ is defined now by

$$r \arccos \frac{|\langle x, y \rangle|}{\sqrt{\langle x, x \rangle \langle y, y \rangle}},$$

where r is a fixed positive constant, and arccos is the inverse cosine in $[0, \pi]$.

In all the considered cases, some hypersurfaces of the second order remain invariant under given **motions**, i.e., projective transformations preserving a given metric. These hypersurfaces are called *absolutes*. In the case of an Euclidean determination of a metric, the absolute is an imaginary $(n-2)$-dimensional oval surface of order two, in fact, the degenerate absolute $x_1^2 + \cdots + x_n^2 = 0$, $x_{n+1} = 0$. In the case of a hyperbolic determination of a metric, the absolute is a real $(n-1)$-dimensional oval hypersurface of order two, in the simplest case, the absolute $-x_1^2 + x_2^2 + \cdots + x_{n+1}^2 = 0$. In the case of an elliptic determination of a metric, the absolute is an imaginary $(n-1)$-dimensional oval hypersurface of order two, in fact, the absolute $x_1^2 + \cdots + x_{n+1}^2 = 0$.

6.3. AFFINE GEOMETRY

An n-dimensional *affine space* over a field $\mathbb{F}$ is a set A^n (the elements of which are called *points* of the affine space) to which corresponds an n-dimensional vector space V over $\mathbb{F}$ (called *space associated to* A^n) such that, for any $a \in A^n$, $A = a + V = \{a + v : v \in V\}$. In the other words, if $a = (a_1, \dots, a_n)$, $b = (b_1, \dots, b_n) \in A^n$, then the vector $\overrightarrow{ab} = (b_1 - a_1, \dots, b_n - a_n)$ belongs to V. In an affine space, one can add a vector to a point to get another point, and subtract points to get vectors, but one cannot add points, since there is no origin. Given points $a, b, c, d \in A^n$ such that $c \neq d$, and the vectors $\overrightarrow{ab}$ and $\overrightarrow{cd}$ are collinear, the scalar λ, defined by $\overrightarrow{ab} = \lambda \overrightarrow{cd}$, is called *affine ratio* of ab and cd, and is denoted by $\frac{ab}{cd}$.

An *affine transformation* (or *affinity*) is a bijection of A^n onto itself which preserves *collinearity* (i.e., all points lying on a line initially, still lie on a line after transformation) and *ratios of distances* (for example, the midpoint of a line segment remains the midpoint after transformation). In this sense, affine indicates a special class of *projective transformations* that do not move any objects from the affine space to the plane at infinity or conversely. Any affine transformation is a composition of *rotations*, *translations*, *dilations*, and *shears*. The set of all affine transformations of A^n forms a group $Aff(A^n)$, called *general affine group* of A^n. Each element $f \in Aff(A)$ can be given by a formula $f(a) = b$, $b_i = \sum_{j=1}^n p_{ij} a_j + c_j$, where $((p_{ij}))$ is an invertible matrix.

The subgroup of $Aff(A^n)$, consisting of affine transformations with $det((p_{ij})) = 1$, is called *equi-affine group* of A^n. An *equi-affine space* is an affine space with the equi-affine group of transformations. The fundamental invariants of an equi-affine space are volumes of parallelepipeds. In an *equi-affine plane* A^2, any two vectors v_1, v_2 have an invariant $|v_1 \times v_2|$ (the modulus of their cross product) – the surface area of the parallelogram constructed on v_1 and v_2. Given a non-rectilinear curve $\gamma = \gamma(t)$, its *affine parameter* (or *equi-affine arc length*) is an invariant parameter, defined by $s = \int_{t_0}^t |\gamma' \times \gamma''|^{1/3} dt$. The invariant $k = \frac{d^2\gamma}{ds^2} \times \frac{d^3\gamma}{ds^3}$ is called *equi-affine curvature* of γ. Passing to the general affine group, two more invariants of the curve are considered: the *affine arc length* $\sigma = \int k^{1/2} ds$, and the *affine curvature* $\mathrm{k} = \frac{1}{k^{3/2}} \frac{dk}{ds}$.

For A^n, $n > 2$, the *affine parameter* (or *equi-affine arch length*) of a curve $\gamma = \gamma(t)$ is defined by $s = \int_{t_0}^t |(\gamma', \gamma'', \dots, \gamma^{(n)})|^{\frac{2}{n(n+1)}} dt$, where the invariant $(v_1, \dots, v_n)$ is the (oriented) volume spanned by the vectors $v_1, \dots, v_n$, which is equal to the determinant of the $n \times n$ matrix whose i-th column is the vector v_i.

- **Affine distance**

  Given an *affine plane* A^2, a *line element* (a, l_a) of A^2 consists of a point $a \in A^2$ together with a straight line $l_a \subset A^2$ passing through a.

  The **affine distance** is a distance on the set of all line elements of A^2, defined by

$$2f^{1/3},$$

  where, for a given line elements (a, l_a) and (b, l_b), f is the surface area of the triangle abc if c is the point of intersection of the straight lines l_a and l_b. The affine distance between (a, l_a) and (b, l_b) can be interpreted as the affine length of the arc ab of a parabola such that l_a and l_b are tangent to the parabola at the point a and b, respectively.

- **Affine pseudo-distance**

  Let A^2 be an *equi-affine plane*, and let $\gamma = \gamma(s)$ be a curve in A^2, defined as a function of the *affine parameter* s. The **affine pseudo-distance** dp_{aff} for A^2 is defined by

$$dp_{aff}(a,b) = \left| \overrightarrow{ab} \times \frac{d\gamma}{ds} \right|,$$

  i.e., is equal to the surface area of the parallelogram constructed on the vectors $\overrightarrow{ab}$ and $\frac{d\gamma}{ds}$, where b is an arbitrary point in A^2, a is a point on γ, and $\frac{d\gamma}{ds}$ is the tangent vector to the curve γ at the point a.

  The **affine pseudo-distance** for an *equi-affine space* A^3 can be defined in a similar manner as

$$\left| \left(\overrightarrow{ab}, \frac{d\gamma}{ds}, \frac{d^2\gamma}{ds^2} \right) \right|,$$

  where $\gamma = \gamma(s)$ is a curve in A^3, defined as a function of the *affine parameter* s, $b \in A^3$, a is a point of γ, and the vectors $\frac{d\gamma}{ds}, \frac{d^2\gamma}{ds^2}$ are obtained at the point a.

  For A^n, $n > 3$, we have $dp_{aff}(a,b) = |(\overrightarrow{ab}, \frac{d\gamma}{ds}, \dots, \frac{d^{n-1}\gamma}{ds^{n-1}})|$. For an arbitrary parametrization $\gamma = \gamma(t)$, one obtains $dp_{aff}(a,b) = |(\overrightarrow{ab}, \gamma', \dots, \gamma^{(n-1)})||(\gamma', \dots, \gamma^{(n-1)})|^{\frac{1-n}{1+n}}$.

- **Affine metric**

  The **affine metric** is a metric on a *non-developable surface* $r = r(u_1, u_2)$ in an *equi-affine space* A^3, given by its **metric tensor** $((g_{ij}))$:

$$g_{ij} = \frac{a_{ij}}{|det((a_{ij}))|^{1/4}},$$

  where $a_{ij} = (\partial_1 r, \partial_2 r, \partial_{ij} r)$, $i, j \in \{1, 2\}$.

6.4. NON-EUCLIDEAN GEOMETRY

The term **non-Euclidean Geometry** describes both *Hyperbolic Geometry* (or *Lobachevsky Geometry*, *Lobachevsky–Bolyai–Gauss Geometry*) and *Elliptic Geometry* (sometimes called also *Riemannian Geometry*) which are contrasted with *Euclidean Geometry* (or *Parabolic Geometry*). The essential difference between Euclidean and non-Euclidean Geometry is the nature of parallel lines. In Euclidean Geometry, if we start with a line l and a point a, which is not on l, then we can only draw one line through a that is parallel to l. In Hyperbolic Geometry there are infinitely many lines through a parallel to l. In Elliptic Geometry, parallel lines do not exist.

The *Spherical Geometry* is also "non-Euclidean", but it fails the axiom that any two points determine exactly one line.

- **Spherical metric**

  Let $S^n(0, r) = \{x \in \mathbb{R}^{n+1} : \sum_{i=1}^{n+1} x_i^2 = r^2\}$ be a sphere in $\mathbb{R}^{n+1}$ with the center 0 and the radius $r > 0$.

  The **spherical metric** (or **great circle metric**) d_{sph} is a metric on $S^n(0, r)$, defined by

$$r \arccos\left(\frac{|\sum_{i=1}^{n+1} x_i y_i|}{r^2}\right),$$

  where arccos is the inverse cosine in $[0, \pi]$. It is the length of the *great circle* arc, passing through x and y. In terms of the standard *inner product* $\langle x, y\rangle = \sum_{i=1}^{n+1} x_i y_i$ on $\mathbb{R}^{n+1}$, the spherical metric can be written as $r \arccos \frac{|\langle x,y\rangle|}{\sqrt{\langle x,x\rangle\langle y,y\rangle}}$.

  The metric space $(S^n(0, r), d_{sph})$ is called *n-dimensional spherical space*. It is a space of curvature $1/r^2$, and r is the radius of curvature. It is a model of n-dimensional *Spherical Geometry*. The great circles of the sphere are its geodesics, all geodesics are closed and of the same length. (See, for example, [Blum70].)

- **Elliptic metric**

  Let $\mathbb{R}P^n$ be the real n-dimensional projective space. The **elliptic metric** d_{ell} is a metric on $\mathbb{R}P^n$, defined by

$$r \arccos \frac{|\langle x, y\rangle|}{\sqrt{\langle x, x\rangle\langle y, y\rangle}},$$

  where, for any $x = (x_1 : \ldots : x_{n+1})$, $y = (y_1 : \ldots : y_{n+1}) \in \mathbb{R}P^n$, one has $\langle x, y\rangle = \sum_{i=1}^{n+1} x_i y_i$, r is a fixed positive constant, and arccos is the inverse cosine in $[0, \pi]$.

  The metric space $(\mathbb{R}P^n, d_{ell})$ is called *n-dimensional elliptic space*. It is a model of n-dimensional *Elliptic Geometry*. It is the space of curvature $1/r^2$, and r is the radius of curvature. As $r \to \infty$, the metric formulas of Elliptic Geometry yield formulas of Euclidean Geometry (or become meaningless).

  If $\mathbb{R}P^n$ is viewed as the set $E^n(0, r)$, obtained from the sphere $S^n(0, r) = \{x \in \mathbb{R}^{n+1} : \sum_{i=1}^{n+1} x_i^2 = r^2\}$ in $\mathbb{R}^{n+1}$ with the center 0 and the radius r by identifying diametrically-opposite points, then the elliptic metric on $E^n(0, r)$ can be written as $d_{sph}(x, y)$ if

$d_{sph}(x, y) \leqslant \frac{\pi}{2}r$, and as $\pi r - d_{sph}(x, y)$ if $d_{sph}(x, y) > \frac{\pi}{2}r$, where d_{sph} is the **spherical metric** on $S^n(0, r)$. Thus, no two points of $E^n(0, r)$ have distance exceeding $\frac{\pi}{2}r$. The elliptic space $(E^2(0, r), d_{ell})$ is called *Poincaré sphere*.

If $\mathbb{R}P^n$ is viewed as the set E^n of lines through the zero element 0 in $\mathbb{R}^{n+1}$, then the elliptic metric on E^n is defined as the angle between the corresponding subspaces.

An n-dimensional elliptic space is a *Riemannian space* of constant positive curvature. It is the only such space which is topologically equivalent to a projective space. (See, for example, [Blum70], [Buse55].)

- **Hermitian elliptic metric**

Let $\mathbb{C}P^n$ be the n-dimensional complex projective space. The **Hermitian elliptic metric** d^H_{ell} (see, for example, [Buse55]) is a metric on $\mathbb{C}P^n$, defined by

$$r \arccos \frac{|\langle x, y\rangle|}{\sqrt{\langle x, x\rangle\langle y, y\rangle}},$$

where, for any $x = (x_1 : \ldots : x_{n+1})$, $y = (y_1 : \ldots : y_{n+1}) \in \mathbb{C}P^n$, one has $\langle x, y\rangle = \sum_{i=1}^{n+1} \overline{x}_i y_i$, r is a fixed positive constant, and arccos is the inverse cosine in $[0, \pi]$.

The metric space $(\mathbb{C}P^n, d^H_{ell})$ is called n-dimensional *Hermitian elliptic space* (cf. **Fubini–Study metric**).

- **Elliptic plane metric**

The **elliptic plane metric** is the **elliptic metric** on the *elliptic plane* $\mathbb{R}P^2$. If $\mathbb{R}P^2$ is viewed as the *Poincaré sphere* (i.e., a sphere in $\mathbb{R}^3$ with identified diametrically-opposite points) of diameter 1 tangent to the extended complex plane $\overline{\mathbb{C}} = \mathbb{C} \cup \{\infty\}$ at the point $z = 0$, then, under the stereographic projection from the "north pole" (0, 0, 1), $\overline{\mathbb{C}}$ with identified points z and $-\frac{1}{\overline{z}}$ is a model of the elliptic plane, and the elliptic plane metric d_{ell} on it is defined by its *line element* $ds^2 = \frac{|dz|^2}{(1+|z|^2)^2}$.

- **Pseudo-elliptic distance**

The **pseudo-elliptic distance** (or *elliptic pseudo-distance*) dp_{ell} is a distance on the extended complex plane $\overline{\mathbb{C}} = \mathbb{C} \cup \{\infty\}$ with identified points z and $-\frac{1}{\overline{z}}$, defined by

$$\left|\frac{z - u}{1 + \overline{z}u}\right|.$$

In fact, $dp_{ell}(z, u) = \tan d_{ell}(z, u)$, where d_{ell} is the **elliptic plane metric**.

- **Hyperbolic metric**

Let $\mathbb{R}P^n$ be the n-dimensional real projective space. Let, for any $x = (x_1 : \ldots : x_{n+1})$, $y = (y_1 : \ldots : y_{n+1}) \in \mathbb{R}P^n$, the *scalar product* $\langle x, y\rangle = -x_1 y_1 + \sum_{i=2}^{n+1} x_i y_i$ be considered.

The **hyperbolic metric** d_{hyp} is a metric on the set $H^n = \{x \in \mathbb{R}P^n : \langle x, x\rangle < 0\}$, defined by

$$r \operatorname{arccosh} \frac{|\langle x, y\rangle|}{\sqrt{\langle x, x\rangle\langle y, y\rangle}},$$

where r is a fixed positive constant, and arccosh denotes the non-negative values of the inverse hyperbolic cosine. In this construction, the points of H^n can be viewed as the one-spaces of the *pseudo-Euclidean space* $\mathbb{R}^{n,1}$ inside the cone $C = \{x \in \mathbb{R}^{n,1} : \langle x, x\rangle = 0\}$.

The metric space (H^n, d_{hyp}) is called *n-dimensional hyperbolic space*. It is a model of n-dimensional *Hyperbolic Geometry*. It is the space of curvature $-1/r^2$, and r is the radius of curvature. Replacement of r by ir transforms all metric formulas of Hyperbolic Geometry into the corresponding formulas of Elliptic Geometry. As $r \to \infty$, both systems yield formulas of Euclidean Geometry (or become meaningless).

If H^n is viewed as the set $\{x \in \mathbb{R}^n : \sum_{i=1}^n x_i^2 < K\}$, where $K > 1$ is an arbitrary fixed constant, the hyperbolic metric can be written as

$$\frac{r}{2} \ln \frac{1 + \sqrt{1 - \gamma(x, y)}}{1 - \sqrt{1 - \gamma(x, y)}}, \quad \text{where } \gamma(x, y) = \frac{(K - \sum_{i=1}^n x_i^2)(K - \sum_{i=1}^n y_i^2)}{(K - \sum_{i=1}^n x_i y_i)^2},$$

and r is a positive number with $\tanh \frac{1}{r} = \frac{1}{\sqrt{K}}$.

If H^n is viewed as a submanifold of the $(n + 1)$-dimensional *pseudo-Euclidean space* $\mathbb{R}^{n,1}$ with the scalar product $\langle x, y\rangle = -x_1 y_1 + \sum_{i=2}^{n+1} x_i y_i$ (in fact, as the top sheet $\{x \in \mathbb{R}^{n,1} : \langle x, x\rangle = -1, x_1 > 0\}$ of the two-sheeted *hyperboloid of revolution*), then the hyperbolic metric on H^n is induced from the **pseudo-Riemannian metric** on $R^{n,1}$ (cf. **Lorentz metric**).

An n-dimensional hyperbolic space is a *Riemannian space* of constant negative curvature. It is the only such space which is **complete** and topologically equivalent to an Euclidean space. (See, for example, [Blum70], [Buse55].)

- **Hermitian hyperbolic metric**

Let $\mathbb{C}P^n$ be the n-dimensional complex projective space. Let, for any $x = (x_1 : \ldots : x_{n+1})$, $y = (y_1 : \ldots : y_{n+1}) \in \mathbb{C}P^n$, the *scalar product* $\langle x, y\rangle = -\overline{x}_1 y_1 + \sum_{i=2}^{n+1} \overline{x}_i y_i$ be considered.

The **Hermitian hyperbolic metric** d_{hyp}^H (see, for example, [Buse55]) is a metric on the set $\mathbb{C}H^n = \{x \in \mathbb{C}P^n : \langle x, x\rangle < 0\}$, defined by

$$r \operatorname{arccosh} \frac{|\langle x, y\rangle|}{\sqrt{\langle x, x\rangle\langle y, y\rangle}},$$

where r is a fixed positive constant, and arccosh denotes the non-negative values of the inverse hyperbolic cosine.

The metric space $(\mathbb{C}H^n, d_{hyp}^H)$ is called n-dimensional *Hermitian hyperbolic space*.

• **Poincaré metric**

The **Poincaré metric** d_P is the **hyperbolic metric** for the *Poincaré disk model* (or *conformal disk model*) of Hyperbolic Geometry. In this model every point of the *unit disk* $\Delta = \{z \in \mathbb{C}: \ |z| < 1\}$ is called *hyperbolic point*, the disk Δ itself is called *hyperbolic plane*, circular arcs (and diameters) in Δ which are orthogonal to the *absolute* $\Omega = \{z \in \mathbb{C}: \ |z| = 1\}$ are called *hyperbolic straight lines*. Every point of Ω is called *ideal point*. The angular measurements in this model are the same as in Hyperbolic Geometry. The Poincaré metric on Δ is defined by its *line element*

$$ds^2 = \frac{|dz|^2}{(1-|z|^2)^2} = \frac{dz_1^2 + dz_2^2}{(1-z_1^2-z_2^2)^2}.$$

The distance between two points z and u of Δ can be written as

$$\frac{1}{2}\ln\frac{|1-z\overline{u}|+|z-u|}{|1-z\overline{u}|-|z-u|} = \operatorname{arctanh}\frac{|z-u|}{|1-z\overline{u}|}.$$

In terms of *cross-ratio*, it is equal to

$$\frac{1}{2}\ln(z,u,z^*,u^*) = \frac{1}{2}\ln\frac{(z^*-z)(u^*-u)}{(z^*-u)(u^*-z)},$$

where z^* and u^* are the points of intersection of the hyperbolic straight line passing through z and u with Ω, z^* on the side of u, and u^* on the side of z.

In the *Poincaré half-plane model* of Hyperbolic Geometry the *hyperbolic plane* is the upper half-plane $H^2 = \{z \in \mathbb{C}: \ z_2 > 0\}$, and the *hyperbolic lines* are semi-circles and half-lines which are orthogonal to the real axis. The *absolute* (i.e., the set of *ideal points*) is the real axis together with the point at infinity. The angular measurements in the model are the same as in Hyperbolic Geometry. The *line element* of the **Poincaré metric** on H^2 is given by

$$ds^2 = \frac{|dz|^2}{(\Im z)^2} = \frac{dz_1^2 + dz_2^2}{z_2^2}.$$

The distance between two points z, u can be written as

$$\frac{1}{2}\ln\frac{|z-\overline{u}|+|z-u|}{|z-\overline{u}|-|z-u|} = \operatorname{arctanh}\frac{|z-u|}{|z-\overline{u}|}.$$

In terms of cross-ratio, it is equal to

$$\frac{1}{2}\ln(z,u,z^*,u^*) = \frac{1}{2}\ln\frac{(z^*-z)(u^*-u)}{(z^*-u)(u^*-z)},$$

where z^* is the ideal point of the half-line emanating from z and passing through u, and u^* is the ideal point of the half-line emanating from u and passing through z.

In general, the **hyperbolic metric** in any *domain* $D \subset \mathbb{C}$ with at least three boundary points is defined as the preimage of the Poincaré metric in Δ under a *conformal mapping* $f : D \to \Delta$. Its *line element* has the form

$$ds^2 = \frac{|f'(z)|^2|dz|^2}{(1-|f(z)|^2)^2}.$$

The distance between two points z and u in D can be written as

$$\frac{1}{2}\ln\frac{|1-f(z)\overline{f(u)}|+|f(z)-f(u)|}{|1-f(z)\overline{f(u)}|-|f(z)-f(u)|}.$$

- **Pseudo-hyperbolic distance**

The **pseudo-hyperbolic distance** (or **Gleason distance**, *hyperbolic pseudo-distance*) dp_{hyp} is a metric on the *unit disk* $\Delta = \{z \in \mathbb{C}\colon\ |z| < 1\}$, defined by

$$\left|\frac{z-u}{1-\overline{z}u}\right|.$$

In fact, $dp_{hyp}(z,u) = \tanh d_P(z,u)$, where d_P is the **Poincaré metric** on Δ.

- **Cayley–Klein–Hilbert metric**

The **Cayley–Klein–Hilbert metric** d_{CKH} is the **hyperbolic metric** for the *Klein model* (or *projective disk model*, *Beltrami–Klein model*) for Hyperbolic Geometry. In this model the *hyperbolic plane* is realized as the *unit disk* $\Delta = \{z \in \mathbb{C}\colon\ |z| < 1\}$, and the *hyperbolic straight lines* are realized as the chords of Δ. Every point of the *absolute* $\Omega = \{z \in \mathbb{C}\colon\ |z| = 1\}$ is called *ideal point*. The angular measurements in this model are distorted. The **Cayley–Klein–Hilbert metric** on Δ is given by its **metric tensor** $((g_{ij}))$, $i, j = 1, 2$:

$$g_{11} = \frac{r^2(1-z_2^2)}{(1-z_1^2-z_2^2)^2}, \quad g_{12} = \frac{r^2 z_1 z_2}{(1-z_1^2-z_2^2)^2}, \quad g_{22} = \frac{r^2(1-z_1^2)}{(1-z_1^2-z_2^2)^2},$$

where r is an arbitrary positive constant. The distance between points z and u in Δ can be written as

$$r \operatorname{arccosh}\left(\frac{1-z_1u_1-z_2u_2}{\sqrt{1-z_1^2-z_2^2}\sqrt{1-u_1^2-u_2^2}}\right),$$

where arccosh denotes the non-negative values of the inverse hyperbolic cosine.

- **Weierstrass metric**

Given a real n-dimensional **inner product space** $(\mathbb{R}^n, \langle,\rangle)$, $n \geqslant 2$, the **Weierstrass metric** d_W is a metric on $\mathbb{R}^n$, defined by

$$\operatorname{arccosh}\left(\sqrt{1+\langle x,x\rangle}\sqrt{1+\langle y,y\rangle}-\langle x,y\rangle\right),$$

where arccosh denotes the non-negative values of the inverse hyperbolic cosine.

Here, $(x, \sqrt{1+\langle x, x\rangle}) \in \mathbb{R}^n \oplus \mathbb{R}$ are the *Weierstrass coordinates* of $x \in \mathbb{R}^n$, and the metric space $(\mathbb{R}^n, d_W)$ can be identified with the *Weierstrass model* of Hyperbolic Geometry.

The **Cayley–Klein–Hilbert metric**

$$d_{CKH}(x, y) = \operatorname{arccosh} \frac{1-\langle x, y\rangle}{\sqrt{1-\langle x, x\rangle}\sqrt{1-\langle y, y\rangle}}$$

on the set $B^n = \{x \in \mathbb{R}^n\colon\ \langle x, x\rangle < 1\}$ can be obtained from d_W by $d_{GKH}(x, y) = d_W(\mu(x), \mu(y))$, where $\mu : \mathbb{R}^n \to B^n$ is the *Weierstrass mapping*: $\mu(x) = \frac{x}{\sqrt{1-\langle x,x\rangle}}$.

- **Quasi-hyperbolic metric**

Given a *domain* $D \subset \mathbb{R}^n$, $n \geqslant 2$, the **quasi-hyperbolic metric** is a metric on D, defined by

$$\inf_{\gamma\in\Gamma} \int_\gamma \frac{|dz|}{\rho(z)},$$

where the infimum is taken over the set Γ of all rectifiable curves connecting x and y in D, $\rho(z) = \inf_{u\in\partial D} \|z-u\|_2$ is the distance between z and the boundary ∂D of D, and $\|.\|_2$ is the Euclidean norm on $\mathbb{R}^n$.

For $n = 2$, one can define the **hyperbolic metric** on D by

$$\inf_{\gamma\in\Gamma} \int_\gamma \frac{2|f'(z)|}{1-|f(z)|^2}|dz|,$$

where $f : D \to \Delta$ is any conformal mapping of D onto the *unit disk* $\Delta = \{z \in \mathbb{C}\colon |z| < 1\}$. For $n \geqslant 3$, this metric is defined only for the half-hyperplane H^n and for the *open unit ball* B^n as the infimum over all $\gamma \in \Gamma$ of the integrals $\int_\gamma \frac{|dz|}{z_n}$ and $\int_\gamma \frac{2|dz|}{1-\|z\|_2^2}$, respectively.

- **Apollonian metric**

Let $D \subset \mathbb{R}^n$, $D \neq \mathbb{R}^n$, be a *domain* such that the complement of D is not contained in a hyperplane or a sphere.

The **Apollonian metric** (or **Barbilian metric**, [Barb35]) is a metric on D, defined by

$$\sup_{a,b\in\partial D} \ln \frac{\|a-x\|_2\|b-y\|_2}{\|a-y\|_2\|b-x\|_2},$$

where ∂D is the boundary of D, and $\|.\|_2$ is the Euclidean norm on $\mathbb{R}^n$.

- **Half-Apollonian metric**

Given a *domain* $D \subset \mathbb{R}^n$, $D \neq \mathbb{R}^n$, the **half-Apollonian metric** is a metric on D, defined by

$$\sup_{a \in \partial D} \left| \ln \frac{\|a - y\|_2}{\|a - x\|_2} \right|,$$

where ∂D is the boundary of D, and $\|.\|_2$ is the Euclidean norm on $\mathbb{R}^n$.

- **Gehring metric**

  Given a *domain* $D \subset \mathbb{R}^n$, $D \neq \mathbb{R}^n$, the **Gehring metric** (or $\tilde{j}_D$-*metric*) is a metric on D, defined by

$$\frac{1}{2} \ln\left(\left(1 + \frac{\|x - y\|_2}{\rho(x)}\right)\left(1 + \frac{\|x - y\|_2}{\rho(y)}\right)\right),$$

  where $\|.\|_2$ is the Euclidean norm on $\mathbb{R}^n$, and $\rho(x) = \inf_{u \in \partial D} \|x - u\|_2$ is the distance between x and the boundary ∂D of D.

- **Vuorinen metric**

  Given a *domain* $D \subset \mathbb{R}^n$, $D \neq \mathbb{R}^n$, the **Vuorinen metric** (or j_D-*metric*) is a metric on D, defined by

$$\ln\left(1 + \frac{\|x - y\|_2}{\min\{\rho(x), \rho(y)\}}\right),$$

  where $\|.\|_2$ is the Euclidean norm on $\mathbb{R}^n$, and $\rho(x) = \inf_{u \in \partial D} \|x - u\|_2$ is the distance between x and the boundary ∂D of D.

- **Ferrand metric**

  Given a *domain* $D \subset \mathbb{R}^n$, $D \neq \mathbb{R}^n$, the **Ferrand metric** is a metric on D, defined by

$$\inf_{\gamma \in \Gamma} \int_\gamma \sup_{a,b \in \partial D} \frac{\|a - b\|_2}{\|z - a\|_2 \|z - b\|_2} |dz|,$$

  where the infimum is taken over the set Γ of all rectifiable curves connecting x and y in D, ∂D is the boundary of D, and $\|.\|_2$ is the Euclidean norm on $\mathbb{R}^n$.

- **Seittenranta metric**

  Given a *domain* $D \subset \mathbb{R}^n$, $D \neq \mathbb{R}^n$, the **Seittenranta metric** (or **distance ratio metric**, *cross-ratio metric*) is a metric on D, defined by

$$\sup_{a,b \in \partial D} \ln\left(1 + \frac{\|a - x\|_2 \|b - y\|_2}{\|a - b\|_2 \|x - y\|_2}\right),$$

  where ∂D is the boundary of D, and $\|.\|_2$ is the Euclidean norm on $\mathbb{R}^n$.

- **Modulus metric**

  Let $D \subset \mathbb{R}^n$, $D \neq \mathbb{R}^n$, be a *domain*, whose boundary ∂D has positive capacity.

  The **modulus metric** (or **conformal metric**) is a metric on D, defined by

$$\inf_{C_{xy}} M\big(\Delta(C_{xy}, \partial D, D)\big),$$

where $M(\Gamma)$ is the *conformal modulus* of the curve family Γ, and C_{xy} is a continuum such that for some $\gamma : [0, 1] \to D$ we have the following properties: $C_{xy} = \gamma([0, 1])$, $\gamma(0) = x$, and $\gamma(1) = y$ (cf. **extremal metric**).

- **Ferrand second metric**

Let $D \subset \mathbb{R}^n$, $D \neq \mathbb{R}^n$, be a *domain* such that $|\mathbb{R}^n \backslash \{D\}| \geqslant 2$. The **Ferrand second metric** is a metric on D, defined by

$$\left(\inf_{C_x, C_y} M(\Delta(C_x, C_y, D))\right)^{\frac{1}{1-n}},$$

where $M(\Gamma)$ is the *conformal modulus* of the curve family Γ, and C_z, $z = x, y$, is a continuum such that for some $\gamma_z : [0, 1] \to D$ we have the following properties: $C_z = \gamma([0, 1))$, $z \in |\gamma_z|$, and $\gamma_z(t) \to \partial D$ as $t \to 1$ (cf. **extremal metric**).

- **Parabolic distance**

The **parabolic distance** is a metric on $\mathbb{R}^{n+1}$, considered as $\mathbb{R}^n \times \mathbb{R}$, defined by

$$\sqrt{(x_1 - y_1)^2 + \cdots + (x_n - y_n)^2 + |t_x - t_y|^{1/m}}, \quad m \in \mathbb{N},$$

for any $x = (x_1, \ldots, x_n, t_x)$, $y = (y_1, \ldots, y_n, t_y) \in \mathbb{R}^n \times \mathbb{R}$.

The space $\mathbb{R}^n \times \mathbb{R}$ can be interpreted as multidimensional *space-time*.

Usually, the value $m = 2$ is applied. There exist some variants of the parabolic distance, for example, the parabolic distance

$$\sup\{|x_1 - y_1|, |x_2 - y_2|^{1/2}\}$$

on $\mathbb{R}^2$, or the **half-space parabolic distance** on $\mathbb{R}^3_+ = \{x \in \mathbb{R}^3 : x_1 \geqslant 0\}$, defined by

$$\frac{|x_1 - y_1| + |x_2 - y_2|}{\sqrt{x_1} + \sqrt{x_2} + \sqrt{|x_2 - y_2|}} + \sqrt{|x_3 - y_3|}.$$

Chapter 7

Riemannian and Hermitian Metrics

Riemannian Geometry is a multi-dimensional generalization of the intrinsic geometry of two-dimensional surfaces in the Euclidean space $\mathbb{E}^3$. It studies *real smooth manifolds* equipped with **Riemannian metrics**, i.e., collections of positive-definite symmetric bilinear forms $((g_{ij}))$ on their tangent spaces which varies smoothly from point to point. The geometry of such (*Riemannian*) manifolds is based on the *line element* $ds^2 = \sum_{i,j} g_{ij} dx_i dx_j$. This gives in particular local notions of angle, length of curves, and volume. From those some other global quantities can be derived, by integrating local contributions. Thus, the value ds is interpreted as the length of the vector $(dx_1, \dots, dx_n)$; the arc length of a curve γ is expressed by $\int_\gamma \sqrt{\sum_{i,j} g_{ij} dx_i dx_j}$; then the **intrinsic metric** on a Riemannian manifold is defined as the infimum of lengths of curves joining two given points of the manifold. Therefore, a Riemannian metric is not an ordinary metric, but it induced an ordinary metric, in fact, the intrinsic metric, sometimes called **Riemannian distance**, on any connected Riemannian manifold; a Riemannian metric is an infinitesimal form of the corresponded Riemannian distance.

As particular special cases of Riemannian Geometry, there occur two standard types, *Elliptic Geometry* and *Hyperbolic Geometry*, of *Non-Euclidean Geometry*, as well as *Euclidean Geometry* itself.

If the bilinear forms $((g_{ij}))$ are non-degenerate but indefinite, one obtains the *Pseudo-Riemannian Geometry*. In the case of dimension four (and *signature* $(1, 3)$) it is the main object of the General Theory of Relativity. If $ds = F(x_1, \dots, x_n, dx_1, \dots, dx_n)$, where F is a real positive-definite convex function which can not be given as the square root of a symmetric bilinear form (as in the Riemannian case), one obtains the *Finsler Geometry* generalizing Riemannian Geometry.

Hermitian Geometry studies *complex manifolds* equipped with **Hermitian metrics**, i.e., collections of positive-definite symmetric sesquilinear forms on their tangent spaces, which varies smoothly from point to point. It is a complex analog of Riemannian Geometry. A special class of Hermitian metrics form **Kähler metrics** which have closed fundamental form w. A generalization of Hermitian metrics give **complex Finsler metrics** which can not be written in terms of a bilinear symmetric positive-definite sesqulinear form.

7.1. RIEMANNIAN METRICS AND GENERALIZATIONS

A *real n-dimensional manifold with boundary M^n* is a **Hausdorff space** in which every point has an open neighborhood homeomorphic to either an open subset of $\mathbb{E}^n$, or an open subset of the closed half of $\mathbb{E}^n$. The set of points which have an open neighborhood homeomorphic to $\mathbb{E}^n$ is called *interior* (of the manifold); it is always non-empty. The complement of the interior is called *boundary* (of the manifold); it is an $(n-1)$-dimensional manifold. If the boundary of M^n is empty, one obtains a *real n-dimensional manifold without boundary*.

A manifold without boundary is called *closed* if it is compact, and *open*, otherwise.

An open set of M^n together with a homeomorphism between the open set and an open set of $\mathbb{E}^n$ is called *coordinate chart*. A collection of charts which cover M^n is called *atlas* on M^n. The homeomorphisms of two overlapping charts provide a transition mapping from a subset of $\mathbb{E}^n$ to some other subset of $\mathbb{E}^n$. If all these mappings are continuously differentiable, then M^n is called *differentiable manifold*. If all the connecting mappings are k times continuously differentiable, then the manifold is called C^k *manifold*; if they are infinitely often differentiable, then the manifold is called *smooth manifold* (or C^∞ *manifold*).

An atlas of a manifold is called *oriented* if the coordinate transformations between charts are all positive, i.e., the Jacobians of the coordinate transformations between any two charts are positive at every point. An *orientable manifold* is a manifold admitting an oriented atlas.

Manifolds inherit many local properties of the Euclidean space. In particular, they are locally path-connected, locally compact, and locally metrizable.

Associated with every point on a differentiable manifold is a *tangent space* and its dual, a *cotangent space*. Formally, let M^n be an C^k manifold, $k \geqslant 1$, and p is a point of M^n. Fix a chart $\varphi : U \to \mathbb{E}^n$, where U is an open subset of M^n containing p. Suppose two curves $\gamma^1 : (-1, 1) \to M^n$ and $\gamma^2 : (-1, 1) \to M^n$ with $\gamma^1(0) = \gamma^2(0) = p$ are given such that $\varphi \cdot \gamma^1$ and $\varphi \cdot \gamma^2$ are both differentiable at 0. Then γ^1 and γ^2 are called *tangent* at 0 if the ordinary derivatives of $\varphi \cdot \gamma^1$ and $\varphi \cdot \gamma^2$ coincide at 0: $(\varphi \cdot \gamma^1)'(0) = (\varphi \cdot \gamma^2)'(0)$. If the functions $\varphi \cdot \gamma^i : (-1, 1) \to \mathbb{E}^n$, $i = 1, 2$, are given by n real-valued component functions $(\varphi \cdot \gamma^i)_1(t), \dots, (\varphi \cdot \gamma^i)_n(t)$, the condition above means, that their Jacobians $(\frac{d(\varphi \cdot \gamma^i)_1(t)}{dt}, \dots, \frac{d(\varphi \cdot \gamma^i)_n(t)}{dt})$ coincide at 0. This is an equivalence relation, and the equivalence class $\gamma'(0)$ of the curve γ is called *tangent vector* of M^n at p. The *tangent space* $T_p(M^n)$ of M^n at p is defined as the set of all tangent vectors at p. The function $(d\varphi)_p : T_p(M^n) \to \mathbb{E}^n$, defined by $(d\varphi)_p(\gamma'(0)) = (\varphi \cdot \gamma)'(0)$ is bijective, and can be used to transfer the vector space operations from $\mathbb{E}^n$ over to $T_p(M^n)$.

All the tangent spaces $T_p(M^n)$, $p \in M^n$, "glued together", form the *tangent bundle* $T(M^n)$ of M^n. Any element of $T(M^n)$ is a pair (p, v), where $v \in T_p(M^n)$. If for an open neighborhood U of p the function $\varphi : U \to \mathbb{R}^n$ is a coordinate chart, then preimage V of U in $T(M^n)$ admits a mapping $\psi : V \to \mathbb{R}^n \times \mathbb{R}^n$, defined by $\psi(p, v) = (\varphi(p), d\varphi(p))$. It defines structure of smooth $2n$-dimensional manifold on $T(M^n)$. The *cotangent bundle* $T^*(M^n)$ of M^n is obtained in similar manner using cotangent spaces $T_p^*(M^n)$, $p \in M^n$.

A *vector field* on a manifold M^n is a *section* of its tangent bundle $T(M^n)$, i.e., a smooth function $f : M^n \to T(M^n)$ which assigns to every point $p \in M^n$ a vector $v \in T_p(M^n)$.

A *connection* (or *covariant derivative*) is a way of specifying a derivative of a vector field along another vector field on a manifold. Formally, the covariant derivative ∇ of a vector u (defined at a point $p \in M^n$) in the direction of the vector v (defined at the same point p) is a rule that defines a third vector at p, called $\nabla_v u$, which has the properties of a derivative. A Riemannian metric uniquely defines a special covariant derivative called

Levi-Civita connection. It is the torsion-free connection ∇ of the tangent bundle, preserving the given Riemannian metric.

The *Riemann curvature tensor* R is the standard way to express the *curvature* of *Riemannian manifolds*. The Riemann curvature tensor can be given in terms of the Levi-Civita connection ∇ by the following formula:

$$R(u, v)w = \nabla_u \nabla_v w - \nabla_v \nabla_u w - \nabla_{[u,v]} w,$$

where $R(u, v)$ is a linear transformation of the tangent space of the manifold M^n; it is linear in each argument. If $u = \frac{\partial}{\partial x_i}$, $v = \frac{\partial}{\partial x_j}$ are coordinate vector fields, then $[u, v] = 0$, and the formula simplifies to $R(u, v)w = \nabla_u \nabla_v w - \nabla_v \nabla_u w$, i.e., the curvature tensor measures anti-commutativity of the covariant derivative. The linear transformation $w \to R(u, v)w$ is also called *curvature transformation*.

The *Ricci curvature tensor* (or *Ricci curvature*) *Ric* is obtained as the trace of the full curvature tensor R. It can be thought of as a Laplacian of the Riemannian metric tensor in the case of Riemannian manifolds. Ricci curvature tensor is a linear operator on the tangent space at a point. Given an orthonormal basis $(e_i)_i$ in the tangent space $T_p(M^n)$, we have

$$Ric(u) = \sum_i R(u, e_i)e_i.$$

The result does not depend on the choice of an orthonormal basis. Starting with dimension four, the Ricci curvature does not describe the curvature tensor completely.

Ricci scalar (or *scalar curvature*) *Sc* of a Riemannian manifold M^n is the full trace of the curvature tensor; given an orthonormal basis $(e_i)_i$ at $p \in M^n$, we have

$$Sc = \sum_{i,j} \langle R(e_i, e_j)e_j, e_i \rangle = \sum_i \langle Ric(e_i), e_i \rangle.$$

Sectional curvature $K(\sigma)$ of a Riemannian manifold M^n is defined as the *Gauss curvature* of an σ*-section* at a point $p \in M^n$. Here, given an 2-plane σ in the tangent space $T_p(M^n)$, an σ*-section* is a locally-defined piece of surface which has the plane σ as a tangent plane at p, obtained from geodesics which start at p in the directions of the image of σ under the exponential mapping.

- **Metric tensor**

The **metric tensor** (or *basic tensor*, *fundamental tensor*) is a symmetric tensor of rank 2, that is used to measure distances and angles in a real n-dimensional differentiable manifold M^n. Once a local coordinate system $(x_i)_i$ is chosen, the metric tensor appears as a real symmetric $n \times n$ matrix $((g_{ij}))$.

The assignment of a metric tensor on an n-dimensional differentiable manifold M^n introduces a *scalar product* (i.e., symmetric bilinear but, in general, not positive-definite form) $\langle , \rangle_p$ on the tangent space $T_p(M^n)$ at any point $p \in M^n$, defined by

$$\langle x, y \rangle_p = g_p(x, y) = \sum_{i,j} g_{ij}(p) x_i y_j,$$

where $g_{ij}(p)$ is a value of the metric tensor at the point $p \in M^n$, and $x = (x_1, \dots, x_n)$, $y = (y_1, \dots, y_n) \in T_p(M^n)$. The collection of all these scalar products is called **metric** g with the metric tensor $((g_{ij}))$. The length ds of the vector $(dx_1, \dots, dx_n)$ is expressed by the quadratic differential form

$$ds^2 = \sum_{i,j} g_{ij}\, dx_i\, dx_j,$$

which is called *line element* (or *first fundamental form*) of the metric g. The *length* of a curve γ is expressed by the formula $\int_\gamma \sqrt{\sum_{i,j} g_{ij} dx_i dx_j}$. In general case it may be real, purely imaginary, or zero (an *isotropic curve*).

The *signature* of a metric tensor is the pair (p, q) of positive (p) and negative (q) *eigenvalues* of the matrix $((g_{ij}))$. The signature is said to be *indefinite* if both p and q are non-zero, and *positive-definite* if $q = 0$. A Riemannian metric is a metric g with a positive-definite signature $(p, 0)$, and a pseudo-Riemannian metric is a metric g with an indefinite signature (p, q).

- **Non-degenerate metric**

A **non-degenerate metric** is a metric g with the metric tensor $((g_{ij}))$, for which the metric discriminant $\det((g_{ij})) \neq 0$. All Riemannian and pseudo-Riemannian metrics are non-degenerate.

A **degenerate metric** is a metric g with the metric tensor $((g_{ij}))$ for which the metric discriminant $\det((g_{ij})) = 0$ (cf. **semi-Riemannian metric** and **semi-pseudo-Riemannian metric**). A manifold with a degenerate metric is called *isotropic manifold*.

- **Diagonal metric**

A **diagonal metric** is a metric g with a metric tensor $((g_{ij}))$ which is zero for $i \neq j$. The Euclidean metric is a diagonal metric, as its metric tensor has the form $g_{ii} = 1$, $g_{ij} = 0$ for $i \neq j$.

- **Riemannian metric**

Consider a real n-dimensional differentiable manifold M^n in which each tangent space is equipped with an *inner product* (i.e., a symmetric positive-definite bilinear form) which varies smoothly from point to point.

A **Riemannian metric** on M^n is a collection of inner products $\langle , \rangle_p$ on the tangent spaces $T_p(M^n)$, one for each $p \in M^n$.

Every inner product $\langle , \rangle_p$ is completely defined by inner products $\langle e_i, e_j \rangle_p = g_{ij}(p)$ of elements $e_1, \dots, e_n$ of standard basis in $\mathbb{E}^n$, i.e., by real symmetric and positive-definite $n \times n$ matrix $((g_{ij})) = ((g_{ij}(p)))$, called **metric tensor**. In fact, $\langle x, y \rangle_p = \sum_{i,j} g_{ij}(p) x_i y_j$, where $x = (x_1, \dots, x_n)$, $y = (y_1, \dots, y_n) \in T_p(M^n)$. The smooth function g completely determines the Riemannian metric.

A Riemannian metric on M^n is not an ordinary metric on M^n. However, for a connected manifold M^n, every Riemannian metric on M^n induces an ordinary metric on M^n, in fact, the **intrinsic metric** of M^n; for any points $p, q \in M^n$ the **Riemannian distance** between them is defined as

$$\inf_{\gamma} \int_0^1 \left\langle \frac{d\gamma}{dt}, \frac{d\gamma}{dt} \right\rangle^{\frac{1}{2}} dt = \inf_{\gamma} \int_0^1 \sqrt{\sum_{i,j} g_{ij} \frac{dx_i}{dt} \frac{dx_j}{dt}}\, dt,$$

where the infimum is taken over all rectifiable curves $\gamma : [0, 1] \to M^n$, connecting p and q.

A *Riemannian manifold* (or *Riemannian space*) is a real n-dimensional differentiable manifold M^n equipped with a Riemannian metric. The theory of Riemannian spaces is called *Riemannian Geometry*. The simplest examples of Riemannian spaces are Euclidean spaces, *hyperbolic spaces*, and *elliptic spaces*. A Riemannian space is called *complete* if it is a **complete** metric space.

- **Conformal structure**

A **conformal structure** on a vector space V is a class of pairwise-homothetic Euclidean metrics on V. Any Euclidean metric d_E on V defines a conformal structure $\{\lambda d_E : \lambda > 0\}$.

A **conformal structure** on a manifold is a field of conformal structures on the tangent spaces or, equivalently, a class of *conformally equivalent Riemannian metrics*. Two Riemannian metrics g and h on a smooth manifold M^n are called *conformally equivalent* if $g = f \cdot h$ for some positive function f on M^n, called *conformal factor*.

- **Conformal space**

The **conformal space** (or *inversive space*) is the Euclidean space $\mathbb{E}^n$ extended by an ideal point (at infinity). Under *conformal* transformations, i.e., continuous transformations preserving local angles, the ideal point can be taken to an ordinary point. Therefore, in a conformal space a sphere is indistinguishable from a plane: a plane is a sphere passing through the ideal point.

Conformal spaces are considered in *Conformal Geometry* (or *Angle-Preserving Geometry*, *Möbius geometry*, *Inversive Geometry*) in which properties of figures are studied that are invariant under conformal transformations. It is the set of transformations that map spheres into spheres, i.e., generated by the Euclidean transformations together with *inversions* which in coordinate form are conjugate to $x_i \to \frac{r^2 x_i}{\sum_j x_j^2}$, where r is the radius of the inversion. An inversion in a sphere becomes an everywhere well-defined automorphism of period two. Any angle inverts into an equal angle.

The two-dimensional conformal space is the *Riemann sphere*, on which the conformal transformations are given by the *Möbius transformations* $z \to \frac{az+b}{cz+d}$, $ad - bc \neq 0$.

In general, a *conformal mapping* between two Riemannian manifolds is a diffeomorphism between them such that the pulled back metric is *conformally equivalent* to the original one. A **conformal Euclidean space** is a *Riemannian space* admitting a conformal mapping onto an Euclidean space.

In the General Theory of Relativity, conformal transformations are considered on the *Minkowski space* $\mathbb{R}^{1,3}$ extended by two ideal points.

- **Space of constant curvature**

A **space of constant curvature** is a *Riemannian space* M^n for which the sectional curvature $K(\sigma)$ is constant in all two-dimensional directions σ.

A **space form** is a connected complete space of constant curvature. A **flat space** is a space form of zero curvature.

The Euclidean space and the flat torus are space forms of zero curvature (i.e., flat spaces), the sphere is a space form of positive curvature, the *hyperbolic space* is a space form of negative curvature.

- **Generalized Riemannian spaces**

A **generalized Riemannian space** is a metric space with the **intrinsic metric**, subject to certain restrictions on the curvature. Such spaces include *spaces of bounded curvature*, *Riemannian spaces*, etc. Generalized Riemannian spaces differ from Riemannian spaces not only by greater generality, but also by the fact that they are defined and investigated on the basis of their metric alone, without coordinates.

A *space of bounded curvature* ($\leqslant k$ and $\geqslant k'$) is a generalized Riemannian space, defined by the condition: for any sequence of *geodesic triangles* T_n contracting to a point we have

$$k \geqslant \overline{\lim}\frac{\overline{\delta}(T_n)}{\sigma(T_n^0)} \geqslant \underline{\lim}\frac{\overline{\delta}(T_n)}{\sigma(T_n^0)} \geqslant k',$$

where a *geodesic triangle* $T = xyz$ is the triplet of geodesic segments $[x, y]$, $[y, z]$, $[z, x]$ (the sides of T) connecting in pairs three different points x, y, z, $\overline{\delta}(T) = \alpha + \beta + \gamma - \pi$ is the *excess* of the geodesic triangle T, and $\sigma(T^0)$ is the area of an Euclidean triangle T^0 with the sides of the same lengths. The **intrinsic metric** on the space of bounded curvature is called **metric of bounded curvature**. Such a space turns out to be Riemannian under two additional conditions: local compactness of the space (this ensures the condition of local existence of geodesics), and local extendibility of geodesics. If in this case $k = k'$, it is a Riemannian space of constant curvature k (cf. **space of geodesics**).

A space of curvature $\leqslant k$ is defined by the condition $\overline{\lim}\frac{\overline{\delta}(T_n)}{\sigma(T_n^0)} \leqslant k$. In such space any point has a neighborhood in which the sum $\alpha+\beta+\gamma$ of the angles of a geodesic triangle T does not exceed the sum $\alpha_k + \beta_k + \gamma_k$ of the angles of a triangle T^k with sides of the same lengths in a space of constant curvature k. The intrinsic metric of such space is called k**-concave metric**.

A space of curvature $\geqslant k$ is defined by the condition $\underline{\lim}\frac{\overline{\delta}(T_n)}{\sigma(T_n^0)} \geqslant k$. In such space any point has a neighborhood in which $\alpha + \beta + \gamma \geqslant \alpha_k + \beta_k + \gamma_k$ for triangles T and T^k. The intrinsic metric of such space is called K**-concave metric**.

An *Alexandrov space* is a generalized Riemannian space with upper, lower or integral curvature bounds.

- **Complete Riemannian metric**

A Riemannian metric g on a manifold M^n is called **complete** if M^n forms a **complete** metric space with respect to g. Any Riemannian metric on a compact manifold is complete.

- **Ricci-flat metric**

A **Ricci-flat metric** is a Riemannian metric with vanished Ricci curvature tensor.

A *Ricci-flat manifold* is a Riemannian manifold equipped with a Ricci-flat metric. Ricci-flat manifolds represent vacuum solutions to the *Einstein field equation*, and are special cases of *Kähler–Einstein manifolds*. Important Ricci-flat manifolds are *Calabi–Yau manifolds*, and *hyper-Kähler manifolds*.

- **Osserman metric**

An **Osserman metric** is a Riemannian metric for which the Riemannian curvature tensor R is *Osserman*. It means, that the eigenvalues of the *Jacobi operator* $\mathcal{J}(x) : y \rightarrow R(y, x)x$ are constant on the *unit sphere* S^{n-1} in $\mathbb{E}^n$, i.e., they are independent of the unit vectors x.

- **G-invariant metric**

An G**-invariant metric** is a Riemannian metric g on a differentiable manifold M^n, that does not change under any of the transformations of a given *Lie group* $(G, \cdot, id)$ of transformations. The group $(G, \cdot, id)$ is called *group of motions* (or *group of isometries*) of the Riemannian space (M^n, g).

- **Ivanov–Petrova metric**

Let R be the Riemannian curvature tensor of a Riemannian manifold M^n, let $\{x, y\}$ be an orthogonal basis for an oriented 2-plane π in the tangent space $T_p(M^n)$ at a point p of M^n.

The **Ivanov–Petrova metric** is a Riemannian metric on M^n, for which the eigenvalues of the antisymmetric curvature operator $\mathcal{R}(\pi) = R(x, y)$ ([IvSt95]) depend only on the point p of a Riemannian manifold M^n, but not upon the plane π.

- **Zoll metric**

A **Zoll metric** is a Riemannian metric on a smooth manifold M^n whose geodesics are all simple closed curves of an equal length. A two-dimensional sphere S^2 admits many such metrics, besides the obvious metrics of constant curvature. In terms of cylindrical coordinates (z, θ) $(z \in [-1, 1], \theta \in [0, 2\pi])$, the *line element*

$$ds^2 = \frac{(1 + f(z))^2}{1 - z^2} dz^2 + (1 - z^2) d\theta^2$$

defines a Zoll metric on S^2 for any smooth odd function $f : [-1, 1] \rightarrow (-1, 1)$ which vanishes at the end points of the interval.

- **Cycloidal metric**

The **cycloidal metric** is a Riemannian metric on the half-plane $\mathbb{R}^2_+ = \{x \in \mathbb{R}^2 : x_1 \geqslant 0\}$, defined by this *line element*

$$ds^2 = \frac{dx_1^2 + dx_2^2}{2x_1}.$$

It is called cycloidal because its geodesics are cycloid curves. The corresponding distance $d(x, y)$ between two points $x, y \in \mathbb{R}^2_+$ is equivalent to the distance

$$\rho(x, y) = \frac{|x_1 - y_1| + |x_2 - y_2|}{\sqrt{x_1} + \sqrt{x_2} + \sqrt{|x_2 - y_2|}}$$

in the sense that $d \leqslant C\rho$, and $\rho \leqslant Cd$ for some positive constant C.

- **Berger metric**

The **Berger metric** is a Riemannian metric on the *Berger sphere* (i.e., the three-sphere S^3 squashed in one direction), defined by the *line element*

$$ds^2 = d\theta^2 + \sin^2\theta\, d\phi^2 + \cos^2\alpha(d\psi + \cos\theta\, d\phi)^2,$$

where α is a constant, and θ, ϕ, ψ are *Euler angles*.

- **Carnot–Carathéodory metric**

A *distribution* (or *polarization*) on a manifold M^n is a subbundle of the tangent bundle $T(M^n)$ of M^n. Given a distribution $H(M^n)$, a vector field in $H(M^n)$ is called *horizontal*. A curve γ on M^n is called *horizontal* (or *distinguished*, *admissible*) with respect to $H(M^n)$ if $\gamma'(t) \in H_{\gamma(t)}(M^n)$ for any t. A distribution $H(M^n)$ is called *completely non-integrable* if the Lie brackets $[\cdots, [H(M^n), H(M^n)]]$ of $H(M^n)$ span the tangent bundle $T(M^n)$, i.e., for all $p \in M^n$ any tangent vector v from $T_p(M^n)$ can be presented as a linear combination of vectors of the following types: u, $[u, w]$, $[u, [w, t]]$, $[u, [w, [t, s]]], \ldots \in T_p(M^n)$, where all vector fields $u, w, t, s, \ldots$ are horizontal.

The **Carnot–Carathéodory metric** (or *C-C metric*) is a metric on a manifold M^n with a completely non-integrable horizontal distribution $H(M^n)$, defined as the section g_C of positive-definite *scalar products* on $H(M^n)$. The distance $d_C(p, q)$ between any points $p, q \in M^n$ is defined as the infimum of the g_C-lengths of the horizontal curves, joining the points p and q.

A *sub-Riemannian manifold* (or *polarized manifold*) is a manifold M^n equipped with a Carnot–Carathéodory metric. It is a generalization of a Riemannian manifold. Roughly, in order to measure distances in a sub-Riemannian manifold, one is allowed to go only along curves tangent to horizontal spaces.

- **Pseudo-Riemannian metric**

Consider a real n-dimensional differentiable manifold M^n in which every tangent space $T_p(M^n)$, $p \in M^n$, is equipped with a *scalar product* which varies smoothly from point to point and is non-degenerate, but indefinite.

A **pseudo-Riemannian metric** on M^n is a collection of scalar products $\langle , \rangle_p$ on the tangent spaces $T_p(M^n)$, $p \in M^n$, one for each $p \in M^n$.

Every scalar product $\langle , \rangle_p$ is completely defined by scalar products $\langle e_i, e_j \rangle_p = g_{ij}(p)$ of elements $e_1, \dots, e_n$ of standard basis in $\mathbb{E}^n$, i.e., by real symmetric indefinite $n \times n$ matrix $((g_{ij})) = ((g_{ij}(p)))$, called **metric tensor** (cf. **Riemannian metric** in which case the metric tensor is a real symmetric positive-definite $n \times n$ matrix). In fact, $\langle x, y \rangle_p = \sum_{i,j} g_{ij}(p) x_i y_j$, where $x = (x_1, \dots, x_n)$ and $y = (y_1, \dots, y_n) \in T_p(M^n)$. The smooth function g completely determines the pseudo-Riemannian metric.

The length ds of the vector $(dx_1, \dots, dx_n)$ is expressed by the quadratic differential form

$$ds^2 = \sum_{i,j} g_{ij} dx_i dx_j.$$

The length of a curve $\gamma : [0, 1] \to M^n$ is expressed by the formula

$$\int_\gamma \sqrt{\sum_{i,j} g_{ij}\, dx_i\, dx_j} = \int_0^1 \sqrt{\sum_{i,j} g_{ij} \frac{dx_i}{dt} \frac{dx_j}{dt}}\, dt.$$

In general case it may be real, purely imaginary or zero (an *isotropic curve*).

A pseudo-Riemannian metric on M^n is a metric with a fixed, but indefinite signature (p, q), $p + q = n$. A pseudo-Riemannian metric is non-degenerate, i.e., its metric discriminant $\det((g_{ij})) \neq 0$. Therefore, it is a **non-degenerate indefinite metric**.

A *pseudo-Riemannian manifold* (or *pseudo-Riemannian space*) is a real n-dimensional differentiable manifold M^n equipped with a pseudo-Riemannian metric. The theory of pseudo-Riemannian spaces is called *Pseudo-Riemannian Geometry*.

The model space of a pseudo-Riemannian space of signature (p, q) is the *pseudo-Euclidean space* $\mathbb{R}^{p,q}$, $p + q = n$, which is a real n-dimensional vector space $\mathbb{R}^n$ equipped with the metric tensor $((g_{ij}))$ of signature (p, q), defined by $g_{11} = \dots = g_{pp} = 1$, $g_{p+1,p+1} = \dots = g_{nn} = -1$, $g_{ij} = 0$ for $i \neq j$. The *line element* of the corresponding metric is given by

$$ds^2 = dx_1^2 + \dots + dx_p^2 - dx_{p+1}^2 - \dots - dx_n^2.$$

- **Lorentzian metric**

A **Lorentzian metric** (or **Lorentz metric**) is a pseudo-Riemannian metric of signature $(1, p)$.

A *Lorentzian manifold* is a manifold equipped with a Lorentzian metric. A principal assumption of the General Theory of Relativity is that *space-time* can be modeled as a Lorentzian manifold of signature $(1, 3)$. The *Minkowski space* $\mathbb{R}^{1,3}$ with the flat **Minkowski metric** is a model of Lorentzian manifold.

- **Osserman Lorentzian metric**

An **Osserman Lorentzian metric** is a **Lorentzian metric** for which the Riemannian curvature tensor R is *Osserman*. It means, that the eigenvalues of the *Jacobi operator* $\mathcal{J}(x) : y \to R(y, x)x$ are independent of the unit vectors x.

A *Lorentzian manifold* is *Osserman* if and only if it is a manifold of constant curvature.

- **Blaschke metric**

The **Blaschke metric** on a non-degenerated hypersurface is a pseudo-Riemannian metric, associated to the affine normal of the immersion $\phi : M^n \to \mathbb{R}^{n+1}$, where M^n is an n-dimensional manifold, and $\mathbb{R}^{n+1}$ is considered as an affine space.

- **Semi-Riemannian metric**

A **semi-Riemannian metric** on a real n-dimensional differentiable manifold M^n is a degenerate Riemannian metric, i.e., a collection of positive-semi-definite *scalar products* $\langle x, y\rangle_p = \sum_{i,j} g_{ij}(p)x_i y_j$ on the tangent spaces $T_p(M^n)$, $p \in M^n$; the metric discriminant $det((g_{ij})) = 0$.

A *semi-Riemannian manifold* (or *semi-Riemannian space*) is a real n-dimensional differentiable manifold M^n equipped with a semi-Riemannian metric.

The model space of a semi-Riemannian manifold is the *semi-Euclidean space* R^n_d, $d \geqslant 1$ (sometimes denoted also by $\mathbb{R}^n_{n-d}$), i.e., a real n-dimensional vector space $\mathbb{R}^n$ equipped with a semi-Riemannian metric. It means, that there exists a scalar product of vectors such that, relative to a suitably chosen basis, the scalar product $\langle x, x\rangle$ of any vector with itself has the form $\langle x, x\rangle = \sum_{i=1}^{n-d} x_i^2$. The number $d \geqslant 1$ is called *defect* (or *deficiency*) of the space.

- **Semi-pseudo-Riemannian metric**

A **semi-pseudo-Riemannian metric** on a real n-dimensional differentiable manifold M^n is a degenerate pseudo-Riemannian metric, i.e., a collection of degenerate indefinite *scalar products* $\langle x, y\rangle_p = \sum_{i,j} g_{ij}(p)x_i y_j$ on the tangent spaces $T_p(M^n)$, $p \in M^n$; the metric discriminant $det((g_{ij})) = 0$. In fact, a semi-pseudo-Riemannian metric is a **degenerate indefinite metric**.

A *semi-pseudo-Riemannian manifold* (or *semi-pseudo-Riemannian space*) is a real n-dimensional differentiable manifold M^n equipped with a semi-pseudo-Riemannian metric.

The model space of a semi-pseudo-Riemannian manifold is the *semi-pseudo-Euclidean space* $\mathbb{R}^n_{\substack{l_1,\dots,l_r\\ m_1,\dots,m_{r-1}}}$, i.e., a real n-dimensional vector space $\mathbb{R}^n$ equipped with a semi-pseudo-Riemannian metric. It means, that there exist r scalar products $\langle x, y\rangle_a = \sum \varepsilon_{i_a} x_{i_a} y_{i_a}$, where $a = 1, \dots, r$, $0 = m_0 < m_1 < \dots < m_r = n$, $i_a = m_{a-1} + 1$, $\dots, m_a$, $\varepsilon_{i_a} = \pm 1$, and -1 occurs l_a times among the numbers ε_{i_a}. The product $\langle x, y\rangle_a$ is defined for those vectors for which all coordinates x_i, $i \leqslant m_{a-1}$ or $i > m_a + 1$, are zero. The first scalar square of an arbitrary vector x is a degenerate quadratic form $\langle x, x\rangle_1 = -\sum_{i=1}^{l_1} x_i^2 + \sum_{j=l_1+1}^{n-d} x_j^2$. The number $l_1 \geqslant 0$ is called *index*, and the number $d = n - m_1$ is called *defect* of the space. If $l_1 = \cdots = l_r = 0$, we obtain a *semi-Euclidean space*. The spaces $\underset{m}{\mathbb{R}^n}$ and $\underset{m}{\mathbb{R}^n_{k,l}}$ are called *quasi-Euclidean spaces*.

The *semi-pseudo-non-Euclidean space* $\underset{m_1,\dots,m_{r-1}}{\mathbb{S}^n_{l_1,\dots,l_r}}$ can be defined as a hypersphere in $\underset{m_1,\dots,m_{r-1}}{\mathbb{R}^{n+1}_{l_1,\dots,l_r}}$ with identified antipodal points. If $l_1 = \cdots = l_r = 0$, the space $\underset{m_1,\dots,m_{r-1}}{\mathbb{S}^n}$ is called *semi-elliptic space* (or *semi-non-Euclidean space*). If there exist $l_i \neq 0$, the space $\underset{m_1,\dots,m_{r-1}}{\mathbb{S}^n_{l_1,\dots,l_r}}$ is called *semi-hyperbolic space*.

- **Finsler metric**

Consider a real n-dimensional differentiable manifold M^n in which every tangent space $T_p(M^n)$, $p \in M^n$, is equipped with a *Banach norm* $\|.\|$ such that the Banach norm as a function of position is smooth, and the matrix $((g_{ij}))$,

$$g_{ij} = g_{ij}(p, x) = \frac{1}{2}\frac{\partial^2 \|x\|^2}{\partial x_i \partial x_j},$$

is positive-definite for any $p \in M^n$ and any $x \in T_p(M^n)$.

A **Finsler metric** on M^n is a collection of Banach norms $\|.\|$ on the tangent spaces $T_p(M^n)$, one for each $p \in M^n$. The *line element* of this metric has the form

$$ds^2 = \sum_{i,j} g_{ij}\, dx_i\, dx_j.$$

The Finsler metric can be given by a real positive-definite convex function $F(p, x)$ of coordinates of $p \in M^n$ and components of vectors $x \in T_p(M^n)$ acting at the point p. $F(p, x)$ is positively homogeneous of degree one in x: $F(p, \lambda x) = \lambda F(p, x)$ for every $\lambda > 0$. The value of $F(p, x)$ is interpreted as the length of the vector x. The *Finsler metric tensor* has the form

$$\big((g_{ij})\big) = \left(\left(\frac{1}{2}\frac{\partial^2 F^2(p, x)}{\partial x_i \partial x_j}\right)\right).$$

The length of a curve $\gamma : [0, 1] \to M^n$ is given by $\int_0^1 F(p, \frac{dp}{dt})\, dt$. For each fixed p the Finsler metric tensor is Riemannian in the variables x.

The Finsler metric is a generalization of the Riemannian metric, where the general definition of the length $\|x\|$ of a vector $x \in T_p(M^n)$ is not necessarily given in the form of the square root of a symmetric bilinear form as in the Riemannian case.

A *Finsler manifold* (or *Finsler space*) is a real n-dimensional differentiable manifold M^n equipped with a Finsler metric. The theory of Finsler spaces is called *Finsler Geometry*. The difference between a Riemannian space and a Finsler space is that the former behaves locally like an Euclidean, and the latter locally like a *Minkowskian space*, or, analytically, that to an ellipsoid in the Riemannian case there corresponds an arbitrary convex surface which has the origin as the center.

A *generalized Finsler space* is a space with the **intrinsic metric**, subject to certain restrictions on the behavior of shortest curves, i.e., the curves with length equal to the distance between their ends. Such spaces include **spaces of geodesics**, Finsler spaces, etc. Generalized Finsler spaces differ from Finsler spaces not only in their greater generality, but also in the fact that they are defined and investigated starting from a metric, without coordinates.

- **Kropina metric**

The **Kropina metric** is a Finsler metric F_{Kr} on a real n-dimensional manifold M^n, defined by

$$\frac{\sum_{i,j} g_{ij} x_i x_j}{\sum_i b_i(p) y_i}$$

for any $p \in M^n$ and $x \in T_p(M^n)$, where $((g_{ij}))$ is a Riemannian metric tensor, and $b(p) = (b_i(p))$ is a vector field.

- **Randers metric**

The **Randers metric** is a Finsler metric F_{Ra} on a real n-dimensional manifold M^n, defined by

$$\sqrt{\sum_{i,j} g_{ij} x_i x_j} + \sum_i b_i(p) y_i$$

for any $p \in M^n$ and $x \in T_p(M^n)$, where $((g_{ij}))$ is a Riemannian metric tensor, and $b(p) = (b_i(p))$ is a vector field.

- **Funk metric**

The **Funk metric** is a Finsler metric F_{Fu} on the *open unit ball* $B^n = \{x \in \mathbb{R}^n : \|x\|_2 < 1\}$ in $\mathbb{R}^n$, defined by

$$\frac{\sqrt{\|y\|_2^2 - (\|x\|_2^2 \|y\|_2^2 - \langle x, y\rangle^2)} + \langle x, y\rangle}{1 - \|x\|_2^2}$$

for any $x \in B^n$ and $u \in T_x(B^n)$, where $\|.\|_2$ is the Euclidean norm on $\mathbb{R}^n$, and $\langle , \rangle$ is the ordinary *inner product* on $\mathbb{R}^n$. It is a **projective metric**.

- **Shen metric**

Given a vector $a \in \mathbb{R}^n$, $\|a\|_2 < 1$, the **Shen metric** is a Finsler metric F_{Sh} on the *open unit ball* $B^n = \{x \in \mathbb{R}^n : \|x\|_2 < 1\}$ in $\mathbb{R}^n$, defined by

$$\frac{\sqrt{\|y\|_2^2 - (\|x\|_2^2 \|y\|_2^2 - \langle x, y\rangle^2)} + \langle x, y\rangle}{1 - \|x\|_2^2} + \frac{\langle a, y\rangle}{1 + \langle a, x\rangle}$$

for any $x \in B^n$ and $y \in T_x(B^n)$, where $\|.\|_2$ is the Euclidean norm on $\mathbb{R}^n$, and $\langle , \rangle$ is the ordinary *inner product* on $\mathbb{R}^n$. It is a **projective metric**. For $a = 1$ it becomes the **Funk metric**.

- **Berwald metric**

The **Berwald metric** is a Finsler metric F_{Be} on the *open unit ball* $B^n = \{x \in \mathbb{R}^n : \|x\|_2 < 1\}$ in $\mathbb{R}^n$, defined by

$$\frac{(\sqrt{\|y\|_2^2 - (\|x\|_2^2\|y\|_2^2 - \langle x, y\rangle^2)} + \langle x, y\rangle)^2}{(1 - \|x\|_2^2)^2\sqrt{\|y\|_2^2 - (\|x\|_2^2\|y\|_2^2 - \langle x, y\rangle^2)}}$$

for any $x \in B^n$ and $u \in T_x(B^n)$, where $\|.\|_2$ is the Euclidean norm on $\mathbb{R}^n$, and $\langle , \rangle$ is the ordinary *inner product* on $\mathbb{R}^n$. It is a **projective metric**.

- **Bryant metric**

Let α is an angle with $|\alpha| < \frac{\pi}{2}$. Let, for any $x, y \in \mathbb{R}^n$, $A = \|y\|_2^4 \sin^2 2\alpha + (\|y\|_2^2 \cos 2\alpha + \|x\|_2^2\|y\|_2^2 - \langle x, y\rangle^2)^2$, $B = \|y\|_2^2 \cos 2\alpha + \|x\|_2^2\|y\|_2^2 - \langle x, y\rangle^2$, $C = \langle x, y\rangle \sin 2\alpha$, $D = \|x\|_2^4 + 2\|x\|_2^2 \cos 2\alpha + 1$. Then one obtains a (**projective**) Finsler metric F by

$$\sqrt{\frac{\sqrt{A} + B}{2D} + \left(\frac{C}{D}\right)^2} + \frac{C}{D}.$$

On the two-dimensional *unit sphere* S^2, it is the **Bryant metric**.

- **Kawaguchi metric**

The **Kawaguchi metric** is a metric on a smooth n-dimensional manifold M^n, given by the arc element ds of a regular curve $x = x(t)$, $t \in [t_0, t_1]$, expressed by the formula

$$ds = F\left(x, \frac{dx}{dt}, \dots, \frac{d^k x}{dt^k}\right) dt,$$

where the *metric function* F satisfies Zermelo's conditions: $\sum_{s=1}^{k} s x^{(s)} F_{(s)i} = F$, $\sum_{s=r}^{k} \binom{s}{k} x^{(s-r+1)i} F_{(s)i} = 0$, $x^{(s)i} = \frac{d^s x^i}{dt^s}$, $F_{(s)i} = \frac{\partial F}{\partial x^{(s)i}}$, and $r = 2, \dots, k$. These conditions ensure that the arc element ds is independent of the parametrization of the curve $x = x(t)$.

A *Kawaguchi manifold* (or *Kawaguchi space*) is a smooth manifold equipped with a Kawaguchi metric. It is a generalization of a *Finsler manifold*.

- **DeWitt supermetric**

The **DeWitt supermetric** (or *Wheeler–DeWitt supermetric*) $G = ((G_{ijkl}))$ is a generalization of a Riemannian (or pseudo-Riemannian) metric $g = ((g_{ij}))$ used to calculate distances between points of a given manifold, to the case of distances between metrics on this manifold.

More exactly, for a given connected smooth 3-dimensional manifold M^3, consider the space $\mathcal{M}(M^3)$ of all Riemannian (or pseudo-Riemannian) metrics on M^3. Identifying points of $\mathcal{M}(M^3)$ that are related by a diffeomorphism of M^3, one obtains the space $Geom(M^3)$ of 3-*geometries* (of fixed topology), points of which are the classes of diffeomorphically equivalent metrics. The space $Geom(M^3)$ is called *superspace*. It plays an important role in several formulations of Quantum Gravity.

A **supermetric**, i.e., a "metric of metrics", is a metric on $\mathcal{M}(M^3)$ (or on $Geom(M^3)$) which is used for measuring distances between metrics on M^3 (or between their equivalence classes). Given a metric $g = ((g_{ij})) \in \mathcal{M}(M^3)$, we obtain

$$\|\delta g\|^2 = \int_{M^3} d^3x G^{ijkl}(x)\delta g_{ij}(x)\delta g_{kl}(x),$$

where G^{ijkl} is the inverse of the **DeWitt supermetric**

$$G_{ijkl} = \frac{1}{2\sqrt{det((g_{ij}))}}(g_{ik}g_{jl} + g_{il}g_{jk} - \lambda g_{ij}g_{kl}).$$

The value λ parameterizes the distance between metrics in $\mathcal{M}(M^3)$, and may take any real value except $\lambda = \frac{2}{3}$, for which the supermetric is *singular*.

- **Lund–Regge supermetric**

The **Lund–Regge supermetric** (or **simplicial supermetric**) is an analog of the **DeWitt supermetric**, used to measure the distances between *simplicial* 3-*geometries* in a *simplicial configuration space*.

More exactly, given a closed *simplicial* 3-dimensional manifold M^3 consisting of several *tetrahedra* (i.e., 3-*simplices*), an *simplicial geometry* on M^3 is fixed by an assignment of values to the squared edge lengths of M^3, and a flat Riemannian Geometry to the interior of each tetrahedron consistent with those values. The squared edge lengths should be positive and constrained by the triangle inequalities and their analogs for the tetrahedra, i.e., all squared measures (lengths, areas, volumes) must be non-negative (cf. **tetrahedron inequality**). The set $\mathcal{T}(M^3)$ of all simplicial geometries on M^3 is called *simplicial configuration space*.

The **Lund–Regge supermetric** $((G_{mn}))$ on $\mathcal{T}(M^3)$ is induced from the DeWitt supermetric on $\mathcal{M}(M^3)$, using for representations of points in $\mathcal{T}(M^3)$ such metrics in $\mathcal{M}(M^3)$ which are piecewise flat in the tetrahedra.

7.2. RIEMANNIAN METRICS IN INFORMATION THEORY

Some special Riemannian metrics are commonly used in Information Theory. A list of such metrics is given below.

- **Fisher information metric**

In Statistic, Probability, and Information Geometry, the **Fisher information metric** (or **Fisher metric**, **Rao metric**) is a Riemannian metric for a statistical differential manifold (see, for example, [Amar85], [Frie98]). It addresses the differential geometry properties of families of classical probability densities.

Formally, let $p_\theta = p(x, \theta)$ be a family of densities, indexed by n parameters $\theta = (\theta_1, \dots, \theta_n)$ which form the *parameter manifold* P. The **Fisher information metric** $g = g_\theta$ on P is a Riemannian metric, defined by the *Fisher information matrix* $((I(\theta)_{ij}))$, where

$$I(\theta)_{ij} = \mathbb{E}_\theta\left[\frac{\partial \ln p_\theta}{\partial \theta_i} \cdot \frac{\partial \ln p_\theta}{\partial \theta_j}\right] = \int \frac{\partial \ln p(x,\theta)}{\partial \theta_i} \frac{\partial \ln p(x,\theta)}{\partial \theta_j} p(x,\theta)dx.$$

It is a symmetric bilinear form which gives a classical measure (*Rao measure*) for the statistical distinguishability of distribution parameters. Putting $i(x, \theta) = -\ln p(x, \theta)$, one obtains an equivalent formula

$$I(\theta)_{ij} = \mathbb{E}_\theta\left[\frac{\partial^2 i(x,\theta)}{\partial \theta_i \partial \theta_j}\right] = \int \frac{\partial^2 i(x,\theta)}{\partial \theta_i \partial \theta_j} p(x,\theta)dx.$$

In a coordinate free language, we get

$$I(\theta)(u, v) = \mathbb{E}_\theta\big[u(\ln p_\theta) \cdot v(\ln p_\theta)\big],$$

where u and v are vectors tangent to the parameter manifold P, and $u(\ln p_\theta) = \frac{d}{dt} \ln p_{\theta+tu|t=0}$ is the derivative of $\ln p_\theta$ along the direction u.

A *manifold of densities* M is the image of the parameter manifold P under the mapping $\theta \to p_\theta$ with certain regularity conditions. A vector u tangent to this manifold is of the form $u = \frac{d}{dt} p_{\theta+tu|t=0}$, and the Fisher metric $g = g_p$ on M, obtained from the metric g_θ on P, can be written as

$$g_p(u, v) = \mathbb{E}_p\left[\frac{u}{p} \cdot \frac{v}{p}\right].$$

- **Fisher–Rao metric**

Let $\mathcal{P}_n = \{p \in \mathbb{R}^n : \sum_{i=1}^n p_i = 1, p_i > 0\}$ be the simplex of strictly positive probability vectors. An element $p \in \mathcal{P}_n$ is a density of the n-point set $\{1, \dots, n\}$ with $p(i) = p_i$. An element u of the tangent space $T_p(\mathcal{P}_n) = \{u \in \mathbb{R}^n : \sum_{i=1}^n u_i = 0\}$ at a point $p \in \mathcal{P}_n$ is a function on $\{1, \dots, n\}$ with $u(i) = u_i$.

The **Fisher–Rao metric** g_p on $\mathcal{P}_n$ is a Riemannian metric, defined by

$$g_p(u, v) = \sum_{i=1}^n \frac{u_i v_i}{p_i}$$

for any $u, v \in T_p(\mathcal{P}_n)$, i.e., it is the **Fisher information metric** on $\mathcal{P}_n$. The Fisher–Rao metric is the unique (up to a constant factor) Riemannian metric on $\mathcal{P}_n$, contracting under stochastic maps ([Chen72]).

The Fisher–Rao metric is isometric, by $p \to 2(\sqrt{p_1}, \dots, \sqrt{p_n})$, with the standard metric on an open subset of the sphere of radius two in $\mathbb{R}^n$. This identification of $\mathcal{P}_n$ allows to obtain on $\mathcal{P}_n$ the **geodesic distance**, called **Fisher distance** (or **Bhattacharya distance 1**), by

$$2 \arccos\left(\sum_i p_i^{1/2} q_i^{1/2}\right).$$

The Fisher–Rao metric can be extended to the set $\mathcal{M}_n = \{p \in \mathbb{R}^n, p_i > 0\}$ of all finite strictly positive measures on the set $\{1, \dots, n\}$. In this case, the geodesic distance on $\mathcal{M}_n$ can be written as

$$2\left(\sum_i (\sqrt{p_i} - \sqrt{q_i})^2\right)^{1/2}$$

for any $p, q \in \mathcal{M}_n$ (cf. **Hellinger metric**).

- **Monotone metric**

Let M_n be the set of all complex $n \times n$ matrices. Let $\mathcal{M} \subset M_n$ be the manifold of all complex positive-definite $n \times n$ matrices. Let $\mathcal{D} \subset \mathcal{M}$, $\mathcal{D} = \{\rho \in \mathcal{M}: \ Tr\,\rho = 1\}$, be the manifold of all *density matrices*. The tangent space of $\mathcal{M}$ at $\rho \in \mathcal{M}$ is $T_\rho(\mathcal{M}) = \{x \in M_n: \ x = x^*\}$, i.e., the set of all $n \times n$ *Hermitian matrices*. The tangent space $T_\rho(\mathcal{D})$ at $\rho \in \mathcal{D}$ is the subspace of *traceless* (i.e., with trace 0) matrices in $T_\rho(\mathcal{M})$.

A Riemannian metric λ on $\mathcal{M}$ is called **monotone metric** if the inequality

$$\lambda_{h(\rho)}\big(h(u), h(u)\big) \leqslant \lambda_\rho(u, u)$$

holds for any $\rho \in \mathcal{M}$, any $u \in T_\rho(\mathcal{M})$, and any completely positive trace preserving mapping h, called *stochastic mapping*. In fact ([Petz96]), λ is monotone if and only if it can be written as

$$\lambda_\rho(u, v) = Tr\, u J_\rho(v),$$

where J_ρ is an operator of the form $J_\rho = \frac{1}{f(L_\rho/R_\rho)R_\rho}$. Here L_ρ and R_ρ are the left and the right multiplication operators, and $f : (0, \infty) \to \mathbb{R}$ is an operator monotone function which is *symmetric*, i.e., $f(t) = tf(t^{-1})$, and *normalized*, i.e., $f(1) = 1$. $J_\rho(v) = \rho^{-1}v$ if v and ρ are commute, i.e., any monotone metric is equal to the **Fisher information metric** on commutative submanifolds. Therefore, monotone metrics generalize the Fisher information metric on the class of probability densities (classical or commutative case) to the class of density matrices (quantum or non-commutative case) which are used in Quantum Statistics and Information Theory. In fact, $\mathcal{D}$ is the space of faithful states of an n-level quantum system.

A monotone metric $\lambda_\rho(u, v) = Tr\, u \frac{1}{f(L_\rho/R_\rho)R_\rho}(v)$ can be rewritten as $\lambda_\rho(u, v) = Tr\, uc(L_\rho, R_\rho)(v)$, where the function $c(x, y) = \frac{1}{f(x/y)y}$ is the *Morozova–Chentsov function*, related to λ.

The **Bures metric** is the smallest monotone metric, obtained for $f(t) = \frac{1+t}{2}$ (for $c(x, y) = \frac{2}{x+y}$). In this case $J_\rho(v) = g$, $\rho g + g\rho = 2v$, is the *symmetric logarithmic derivative*.

The **right logarithmic derivative metric** is the greatest monotone metric, corresponding to the function $f(t) = \frac{2t}{1+t}$ (to the function $c(x, y) = \frac{x+y}{2xy}$). In this case $J_\rho(v) = \frac{1}{2}(\rho^{-1}v + v\rho^{-1})$ is the *right logarithmic derivative*.

The **Bogolubov–Kubo–Mori metric** is obtained for $f(x) = \frac{x-1}{\ln x}$ (for $c(x, y) = \frac{\ln x - \ln x}{x-y}$). It can be written as $\lambda_\rho(u, v) = \frac{\partial^2}{\partial s \partial t} Tr(\rho + su)\ln(\rho + tv)|_{s,t=0}$.

The **Wigner–Yanase–Dyson metrics** λ_ρ^α are monotone for $\alpha \in [-3, 3]$. For $\alpha = \pm 1$, we obtain the Bogolubov–Kubo–Mori metric; for $\alpha = \pm 3$ we obtain the right logarithmic derivative metric. The smallest in the family is the **Wigner–Yanase metric**, obtained for $\alpha = 0$.

- **Bures metric**

The **Bures metric** (or **statistical metric**) is a **monotone metric** on the manifold $\mathcal{M}$ of all complex positive-definite $n \times n$ matrices, defined by

$$\lambda_\rho(u, v) = Tr\, u J_\rho(v),$$

where $J_\rho(v) = g$, $\rho g + g\rho = 2v$, is the *symmetric logarithmic derivative*. It is the smallest monotone metric.

For any $\rho_1, \rho_2 \in \mathcal{M}$ the **Bures distance**, i.e., the **geodesic distance**, defined by the Bures metric, can be written as

$$2\sqrt{Tr\,\rho_1 + Tr\,\rho_2 - 2\,Tr(\rho_1^{1/2}\rho_2\rho_1^{1/2})^{1/2}}.$$

On the submanifold $\mathcal{D} = \{\rho \in \mathcal{M} : \; Tr\,\rho = 1\}$ of density matrices it has the form

$$2 \arccos Tr\big(\rho_1^{1/2}\rho_2\rho_1^{1/2}\big)^{1/2}.$$

- **Right logarithmic derivative metric**

The **right logarithmic derivative metric** (or *RLD-metric*) is a **monotone metric** on the manifold $\mathcal{M}$ of all complex positive-definite $n \times n$ matrices, defined by

$$\lambda_\rho(u, v) = Tr\, u J_\rho(v),$$

where $J_\rho(v) = \frac{1}{2}(\rho^{-1}v + v\rho^{-1})$ is the *right logarithmic derivative*. It is the greatest monotone metric.

- **Bogolubov–Kubo–Mori metric**

  The **Bogolubov–Kubo–Mori metric** (or *BKM-metric*) is a **monotone metric** on the manifold $\mathcal{M}$ of all complex positive-definite $n \times n$ matrices, defined by

$$\lambda_\rho(u, v) = \frac{\partial^2}{\partial s \partial t} Tr(\rho + su) \ln(\rho + tv)|_{s,t=0}.$$

- **Wigner–Yanase–Dyson metrics**

  The **Wigner–Yanase–Dyson metrics** (or *WYD-metrics*) form a family of metrics on the manifold $\mathcal{M}$ of all complex positive-definite $n \times n$ matrices, defined by

$$\lambda_\rho^\alpha(u, v) = \frac{\partial^2}{\partial t \partial s} Tr\, f_\alpha(\rho + tu) f_{-\alpha}(\rho + sv)|_{s,t=0},$$

  where $f_\alpha(x) = \frac{2}{1-\alpha} x^{\frac{1-\alpha}{2}}$, if $\alpha \neq 1$, and is $\ln x$, if $\alpha = 1$. These metrics are monotone for $\alpha \in [-3, 3]$. For $\alpha = \pm 1$ one obtains the **Bogolubov–Kubo–Mori metric**; for $\alpha = \pm 3$ one obtains the **right logarithmic derivative metric**.

  The **Wigner–Yanase metric** (or *WY-metric*) λ_ρ is the Wigner–Yanase–Dyson metric λ_ρ^0, obtained for $\alpha = 0$. It can be written as

$$\lambda_\rho(u, v) = 4\, Tr\, u\left(\sqrt{L_\rho} + \sqrt{R_\rho}\right)^2(v),$$

  and is the smallest metric in the family. For any $\rho_1, \rho_2 \in \mathcal{M}$ the **geodesic distance**, defined by the WY-metric, has the form

$$2\sqrt{Tr\, \rho_1 + Tr\, \rho_2 - 2\, Tr(\rho_1^{1/2} \rho_2^{1/2})}.$$

  On the submanifold $\mathcal{D} = \{\rho \in \mathcal{M} : \; Tr\, \rho = 1\}$ of density matrices it is equal to

$$2 \arccos Tr\left(\rho_1^{1/2} \rho_2^{1/2}\right).$$

- **Connes metric**

  Roughly, the **Connes metric** is a generalization (from the space of all probability measures of a set X, to the *state space* of any *unital C^*-algebra*) of the **Kantorovich–Mallows–Monge–Wasserstein metric** defined as the **Lipschitz distance between measures**.

  Let M^n be a smooth n-dimensional manifold. Let $A = C^\infty(M^n)$ be the (commutative) algebra of smooth complex-valued functions on M^n, represented as multiplication operators on the Hilbert space $H = L^2(M^n, S)$ of square integrable sections of the spinor bundle on M^n by $(f\xi)(p) = f(p)\xi(p)$ for all $f \in A$ and for all $\xi \in H$. Let D be the *Dirac operator*. Let the commutator $[D, f]$ for $f \in A$ be the *Clifford multiplication* by the gradient ∇f so that its operator norm $\|.\|$ in H is given by $\|[D, f]\| = \sup_{p \in M^n} \|\nabla f\|$.

The **Connes metric** is the **intrinsic metric** on M^n, defined by

$$\sup_{f \in A,\ \|[D,f]\| \leqslant 1} \left| f(p) - f(q) \right|.$$

This definition can also be applied to discrete spaces, and even generalized to "noncommutative spaces" (*unital C^*-algebras*). In particular, for a labeled connected *locally finite* graph $G = (V, E)$ with the vertex-set $V = \{v_1, \dots, v_n, \dots\}$, the Connes metric on V is defined by

$$\sup_{\|[D,f]\| = \|df\| \leqslant 1} |f_{v_i} - f_{v_j}|$$

for any $v_i, v_j \in V$, where $\{f = \sum f_{v_i} v_i : \ \sum |f_{v_i}|^2 < \infty\}$ is the set of formal sums f forms a Hilbert space, and $\|[D, f]\|$ can be obtained by $\|[D, f]\| = \sup_i (\sum_{k=1}^{\deg(v_i)} (f_{v_k} - f_{v_i})^2)^{\frac{1}{2}}$.

7.3. HERMITIAN METRICS AND GENERALIZATIONS

A *vector bundle* is a geometrical construct where to every point of a *topological space* M we attach a vector space so that all those vector spaces "glued together" form another topological space E. A continuous mapping $\pi : E \to M$ is called *projection* E on M. For every $p \in M$, the vector space $\pi^{-1}(p)$ is called *fiber* of the vector bundle. A *real* (*complex*) *vector bundle* is a vector bundle $\pi : E \to M$ whose fibers $\pi^{-1}(p)$, $p \in M$, are real (complex) vector spaces.

In a real vector bundle, for every $p \in M$, the fiber $\pi^{-1}(p)$ locally looks like the vector space $\mathbb{R}^n$, i.e., there is an *open neighborhood* U of p, a natural number n, and a homeomorphism $\varphi : U \times \mathbb{R}^n \to \pi^{-1}(U)$ such that, for all $x \in U$, $v \in \mathbb{R}^n$, one has $\pi(\varphi(x, v)) = v$, and the mapping $v \to \varphi(x, v)$ yields an isomorphism between $\mathbb{R}^n$ and $\pi^{-1}(x)$. The set U, together with φ, is called *local trivialization* of the bundle. If there exists a "global trivialization", then a real vector bundle $\pi : M \times \mathbb{R}^n \to M$ is called *trivial*. Similarly, in a complex vector bundle, for every $p \in M$, the fiber $\pi^{-1}(p)$ locally looks like the vector space $\mathbb{C}^n$. The basic example of a complex vector bundle is the trivial bundle $\pi : U \times \mathbb{C}^n \to U$, where U is an open subset of $\mathbb{R}^k$.

Important special cases of a real vector bundle are the *tangent bundle* $T(M^n)$ and the *cotangent bundle* $T^*(M^n)$ of a *real n-dimensional manifold* $M^n_{\mathbb{R}} = M^n$. Important special cases of a complex vector bundle is the tangent bundle and the cotangent bundle of a *complex n-dimensional manifold*.

Namely, a *complex n-dimensional manifold* $M^n_{\mathbb{C}}$ is a *topological space* in which every point has an open neighborhood homeomorphic to an open set of the n-dimensional complex vector space $\mathbb{C}^n$, and there is an atlas of charts such that the change of coordinates between charts are analytic. The (complex) tangent bundle $T_{\mathbb{C}}(M^n_{\mathbb{C}})$ of a complex manifold $M^n_{\mathbb{C}}$ is a vector bundle of all (complex) *tangent spaces* of $M^n_{\mathbb{C}}$ at every point $p \in M^n_{\mathbb{C}}$. It can be obtained as a *complexification* $T_{\mathbb{R}}(M^n_{\mathbb{R}}) \otimes \mathbb{C} = T(M^n) \otimes \mathbb{C}$ of the corresponding real tangent bundle, and is called *complexified tangent bundle* of $M^n_{\mathbb{C}}$. The *complexified cotangent* *bundle* of $M^n_{\mathbb{C}}$ is obtained in similar manner as $T^*(M^n) \otimes \mathbb{C}$. Any complex n-dimensional manifold $M^n_{\mathbb{C}} = M^n$ can be regarded as a special case of a real $2n$-dimensional manifold equipped with a *complex structure* on each tangent space. A *complex structure* on a real vector space V is the structure of a complex vector space on V that is compatible with the original real structure. It is completely determined by the operator of multiplication by the number i, the role of which can be taken by an arbitrary linear transformation $J : V \to V$, $J^2 = -\,id$, where id is the *identity mapping*.

A *connection* (or *covariant derivative*) is a way of specifying a derivative of a *vector field* along another vector field in a vector bundle. A **metric connection** is a linear connection in a vector bundle $\pi : E \to M$, equipped with a bilinear form in the fibers, for which parallel displacement along an arbitrary piecewise-smooth curve in M preserves the form, that is, the *scalar product* of two vectors remains constant under parallel displacement. In the case of non-degenerative symmetric bilinear form, the metric connection is called *Euclidean connection*. In the case of non-degenerate antisymmetric bilinear form, the metric connection is called *symplectic connection*.

- **Bundle metric**

A **bundle metric** is a metric on a vector bundle.

- **Hermitian metric**

A **Hermitian metric** on a complex vector bundle $\pi : E \to M$ is a collection of *Hermitian inner products* (i.e., positive-definite symmetric sesquilinear forms) on every fiber $E_p = \pi^{-1}(p)$, $p \in M$, that varies smoothly with the point p in M. Any complex vector bundle has a Hermitian metric.

The basic example of a vector bundle is the trivial bundle $\pi : U \times \mathbb{C}^n \to U$, where U is an open set in $\mathbb{R}^k$. In this case a Hermitian inner product on $\mathbb{C}^n$, and hence, a Hermitian metric on the bundle $\pi : U \times \mathbb{C}^n \to U$, is defined by

$$\langle u, v \rangle = u^T H \overline{v},$$

where H is a *positive-definite Hermitian matrix*, i.e., a complex $n \times n$ matrix such that $H^* = \overline{H}^T = H$, and $\overline{v}^T H v > 0$ for all $v \in \mathbb{C}^n \backslash \{0\}$. In the simplest case, one has $\langle u, v \rangle = \sum_{i=1}^n u_i \overline{v}_i$.

An important special case is a Hermitian metric h on a complex manifold M^n, i.e., on the complexified tangent bundle $T(M^n) \otimes \mathbb{C}$ of M^n. This is the Hermitian analog of a Riemannian metric. In this case $h = g + iw$, its real part g is a Riemannian metric, and its imaginary part w is a non-degenerate antisymmetric bilinear form, called *fundamental form*. Here $g(J(x), J(y)) = g(x, y)$, $w(J(x), J(y)) = w(x, y)$, and $w(x, y) = g(x, J(y))$, where the operator J is an operator of complex structure on M^n, as the rule, $J(x) = ix$. Any of the forms g, w determines h uniquely. The term "Hermitian metric" can also refer to the corresponding Riemannian metric g, which gives M^n a Hermitian structure.

On a complex manifold a Hermitian metric h can be expressed in local coordinates by a *Hermitian symmetric tensor* $((h_{ij}))$:

$$h = \sum_{i,j} h_{ij} dz_i \otimes d\overline{z}_j,$$

where $((h_{ij}))$ is a positive-definite Hermitian matrix. The associated fundamental form w is then written as $w = \frac{i}{2} \sum_{i,j} h_{ij} dz_i \wedge d\overline{z}_j$.

A *Hermitian manifold* (or *Hermitian space*) is a complex manifold equipped with a Hermitian metric.

- **Kähler metric**

A **Kähler metric** (or *Kählerian metric*) is a Hermitian metric $h = g + iw$ on a complex manifold M^n, whose fundamental form w is *closed*, i.e., satisfies the condition $dw = 0$. A *Kähler manifold* is a complex manifold equipped with a Kähler metric.

If h is expressed in local coordinates, i.e., $h = \sum_{i,j} h_{ij} dz_i \otimes d\overline{z}_j$, then the associated fundamental form w can be written as $w = \frac{i}{2} \sum_{i,j} h_{ij} dz_i \wedge d\overline{z}_j$, where $\wedge$ is the *wedge product* which is antisymmetric, i.e., $dx \wedge dy = -dy \wedge dx$ (hence, $dx \wedge dx = 0$). In fact, w is a *differential 2-form* on M^n, i.e., a tensor of rank 2 that is antisymmetric under exchange of any pair of indices: $w = \sum_{i,j} f_{ij} dx^i \wedge dx^j$, where f_{ij} is a function on M^n. The *exterior derivative* dw of w is defined by $dw = \sum_{i,j} \sum_k \frac{\partial f_{ij}}{\partial x_k} dx_k \wedge dx_i \wedge dx_k$. If $dw = 0$, then w is a *symplectic* (i.e., closed non-degenerate) differential 2-form. Such differential 2-forms are called *Kähler forms*.

The term **Kähler metric** can also refer to the corresponding Riemannian metric g, which gives M^n a Kähler structure. Then a Kähler manifold is defined as a complex manifold which carries a Riemannian metric and a Kähler form on the underlying real manifold.

- **Calabi–Yau metric**

The **Calabi–Yau metric** is a **Kähler metric** which is **Ricci-flat**.

A *Calabi–Yau manifold* (or *Calabi–Yau space*) is a simply-connected complex manifold equipped with a Calabi–Yau metric. It can be considered as $2n$-dimensional (six-dimensional case being particularly interesting) smooth manifold with holonomy group (i.e., the set of linear transformations of tangent vectors arising from parallel transport along closed loops) in the special unitary group.

- **Kähler–Einstein metric**

A **Kähler–Einstein metric** (or **Einstein metric**) is a *Kähler metric* on a complex manifold M^n whose *Ricci curvature tensor* is proportional to the metric tensor. This proportionality is an analog of the *Einstein field equation* in the General Theory of Relativity.

A *Kähler–Einstein manifold* (or *Einstein manifold*) is a complex manifold equipped with a Kähler-Einstein metric. In this case the Ricci curvature tensor, considered as an operator on the tangent space, is just multiplication by a constant.

Such a metric exists on any *domain* $D \subset \mathbb{C}^n$ that is bounded and *pseudo-convex*. It can be given by the *line element*

$$ds^2 = \sum_{i,j} \frac{\partial^2 u(z)}{\partial z_i \partial \overline{z}_j} dz_i d\overline{z}_j,$$

where u is a solution to the *boundary value problem*: $det(\frac{\partial^2 u}{\partial z_i \partial \overline{z}_j}) = e^{2u}$ on D, and $u = \infty$ on ∂D.

The Kähler–Einstein metric is a **complete** metric. On the *unit disk* $\Delta = \{z \in \mathbb{C}: \ |z| < 1\}$ it is coincides with the **Poincaré metric**.

- **Hodge metric**

The **Hodge metric** is a **Kähler metric** whose *fundamental form* w defines an integral cohomology class or, equivalently, has integral periods.

A *Hodge manifold* (or *Hodge variety*) is a complex manifold equipped with a Hodge metric. A compact complex manifold is a Hodge manifold if and only if it is isomorphic to a smooth algebraic subvariety of some complex projective space.

- **Fubini–Study metric**

The **Fubini–Study metric** is a **Kähler metric** on a *complex projective space* $\mathbb{C}P^n$, defined by a *Hermitian inner product* $\langle , \rangle$ in $\mathbb{C}^{n+1}$. It is given by the *line element*

$$ds^2 = \frac{\langle x, x\rangle\langle dx, dx\rangle - \langle x, d\overline{x}\rangle\langle \overline{x}, dx\rangle}{\langle x, x\rangle^2}.$$

The distance between two points $(x_1 : \ldots : x_{n+1}), (y_1 : \ldots : y_{n+1}) \in \mathbb{C}P^n$, where $x = (x_1, \ldots, x_{n+1}), y = (y_1, \ldots, y_{n+1}) \in \mathbf{C}^{n+1} \backslash \{0\}$, is equal to

$$\arccos \frac{|\langle x, y\rangle|}{\sqrt{\langle x, x\rangle\langle y, y\rangle}}.$$

The Fubini–Study metric is a **Hodge metric**. The space $\mathbb{C}P^n$ endowed with the Fubini–Study metric is called *Hermitian elliptic space* (cf. **Hermitian elliptic metric**).

- **Bergman metric**

The **Bergman metric** is a **Kähler metric** on a bounded *domain* $D \subset \mathbb{C}^n$, defined by the *line element*

$$ds^2 = \sum_{i,j} \frac{\partial^2 \ln K(z, z)}{\partial z_i \partial \overline{z}_j} dz_i d\overline{z}_j,$$

where $K(z, u)$ is the *Bergman kernel function*. The Bergman metric is invariant under all automorphisms of D; it is **complete** if D is homogeneous. For the *unit disk* $\Delta = \{z \in \mathbb{C}: \ |z| < 1\}$ the Bergman metric coincides with the **Poincaré metric** (cf. also **Bergman p-metric**).

The Bergman kernel function is defined as follow. Consider a domain $D \subset \mathbb{C}^n$ in which there exists analytic functions $f \neq 0$ of class $L_2(D)$ with respect to the Lebesgue measure. The set of these functions forms the **Hilbert space** $L_{2,a}(D) \subset L_2(D)$ with an orthonormal basis $(\phi_i)_i$. The *Bergman kernel function* in the domain $D \times D \subset \mathbb{C}^{2n}$ is defined by $K_D(z, u) = K(z, u) = \sum_{i=1}^{\infty} \phi_i(z)\overline{\phi_i(u)}$.

- **Hyper-Kähler metric**

A **hyper-Kähler metric** is a Riemannian metric g on an $4n$-dimensional *Riemannian manifold* which is compatible with a quaternionic structure on the tangent bundle of the manifold. Thus, the metric g is Kählerian with respect to three Kähler structures (I, w_I, g), (J, w_J, g), and (K, w_K, g), corresponding to the complex structures, as endomorphisms of the tangent bundle which satisfy the quaternionic relationship

$$I^2 = J^2 = K^2 = IJK = -JIK = -1.$$

A *hyper-Kähler manifold* is a Riemannian manifold equipped with a hyper-Kähler metric. It is a special case of a *Kähler manifold*. All hyper-Kähler manifolds are Ricci-flat. Compact four-dimensional hyper-Kähler manifolds are called *K_3-surfaces*, they are studied in Algebraic Geometry.

- **Calabi metric**

The **Calabi metric** is a **hyper-Kähler metric** on the cotangent bundle $T^*(\mathbb{C}P^{n+1})$ of a *complex projective space* $\mathbb{C}P^{n+1}$. For $n = 4k + 4$, this metric can be given by the *line element*

$$ds^2 = \frac{dr^2}{1 - r^{-4}} + \frac{1}{4}r^2(1 - r^{-4})\lambda^2 + r^2(\nu_1^2 + \nu_2^2) + \frac{1}{2}(r^2 - 1)(\sigma_{1\alpha}^2 + \sigma_{2\alpha}^2) + \frac{1}{2}(r^2 + 1)(\Sigma_{1\alpha}^2 + \Sigma_{2\alpha}^2),$$

where $(\lambda, \nu_1, \nu_2, \sigma_{1\alpha}, \sigma_{2\alpha}, \Sigma_{1\alpha}, \Sigma_{2\alpha})$, with α running over k values, are left-invariant *one-forms* (i.e., linear real-valued functions) on the coset $SU(k+2)/U(k)$. Here $U(k)$ is the *unitary group* consisting of complex $k \times k$ *unitary matrices*, and $SU(k)$ is the *special unitary group* consisting of complex $k \times k$ unitary matrices with determinant 1.

For $k = 0$, the Calabi metric coincides with the **Eguchi–Hanson metric**.

- **Stenzel metric**

The **Stenzel metric** is a **hyper-Kähler metric** on the cotangent bundle $T^*(S^{n+1})$ of a sphere S^{n+1}.

- ***SO*(3)-invariant metric**

An ***SO*(3)-invariant metric** is an 4-dimensional hyper-Kähler metric with the *line element*, given, in the Bianchi-IX formalism, by

$$ds^2 = f^2(t)dt^2 + a^2(t)\sigma_1^2 + b^2(t)\sigma_2^2 + c^2(t)\sigma_3^2,$$

where the invariant *one-forms* $\sigma_1, \sigma_2, \sigma_3$ of $SO(3)$ are expressed in terms of *Eler angles* θ, ψ, ϕ as $\sigma_1 = \frac{1}{2}(\sin\psi d\theta - \sin\theta\cos\psi d\phi)$, $\sigma_2 = -\frac{1}{2}(\cos\psi d\theta + \sin\theta\sin\psi d\phi)$, $\sigma_3 = \frac{1}{2}(d\psi + \cos\theta d\phi)$, and the normalization has been chosen so that $\sigma_i \wedge \sigma_j = \frac{1}{2}\varepsilon_{ijk} d\sigma_k$. The coordinate t of the metric can always be chosen so that $f(t) = \frac{1}{2}abc$, using a suitable reparametrization.

- **Atiyah–Hitchin metric**

  The **Atiyah–Hitchin metric** is a **complete regular** $SO(3)$**-invariant metric** with the *line element*

$$ds^2 = \frac{1}{4}a^2b^2c^2\left(\frac{dk}{k(1-k^2)K^2}\right)^2 + a^2(k)\sigma_1^2 + b^2(k)\sigma_2^2 + c^2(k)\sigma_3^2,$$

  where a, b, c are functions of k, $ab = -K(k)(E(k) - K(k))$, $bc = -K(k)(E(k) - (1-k^2)K(k))$, $ac = -K(k)E(k)$, and $K(k)$, $E(k)$ are the complete elliptic integrals of the first and second kind, respectively, with $0 < k < 1$. The coordinate t is given by the change of variables $t = -\frac{2K(1-k^2)}{\pi K(k)}$ up to an additive constant.

- **Taub–NUT metric**

  The **Taub–NUT metric** is a **complete regular** $SO(3)$**-invariant metric** with the *line element*

$$ds^2 = \frac{1}{4}\frac{r+m}{r-m}dr^2 + (r^2 - m^2)(\sigma_1^2 + \sigma_2^2) + 4m^2\frac{r-m}{r+m}\sigma_3^2,$$

  where m is the relevant moduli parameter, and the coordinate r is related to t by $r = m + \frac{1}{2mt}$.

- **Eguchi–Hanson metric**

  The **Eguchi–Hanson metric** is a **complete regular** $SO(3)$**-invariant metric** with the *line element*

$$ds^2 = \frac{dr^2}{1-(\frac{a}{r})^4} + r^2\left(\sigma_1^2 + \sigma_2^2 + \left(1 - \left(\frac{a}{r}\right)^4\right)\sigma_3^2\right),$$

  where a is the moduli parameter, and the coordinate r is related to t by $r^2 = a^2\coth(a^2t)$.

  The Eguchi–Hanson metric coincides with the four-dimensional **Calabi metric**.

- **Complex Finsler metric**

  A **complex Finsler metric** is an upper semi-continuous function $F : T(M^n) \to \mathbb{R}_+$ on a complex manifold M^n with the analytic tangent bundle $T(M^n)$ satisfying the following conditions:

  1. F^2 is smooth on $\check{M}^n$, where $\check{M}^n$ is the complement in $T(M^n)$ of the zero section;
  2. $F(p, x) > 0$ for all $p \in M^n$ and $x \in \check{M}^n_p$;
  3. $F(p, \lambda x) = |\lambda|F(p, x)$ for all $p \in M^n$, $x \in T_p(M^n)$, and $\lambda \in \mathbb{C}$.

The function $G = F^2$ can be locally expressed in terms of the coordinates $(p_1, \dots, p_n, x_1, \dots, x_n)$; the *Finsler metric tensor* of the complex Finsler metric is given by the matrix$((G_{ij})) = ((\frac{1}{2}\frac{\partial^2 F^2}{\partial x_i \partial \overline{x}_j}))$, called *Levi matrix*. If the matrix $((G_{ij}))$ is positive-definite, the complex Finsler metric F is called *strongly pseudo-convex*.

- **Distance-decreasing semi-metric**

Let d be a semi-metric which can be defined on some class $\mathcal{M}$ of complex manifolds containing the *unit disk* $\Delta = \{z \in \mathbb{C}: |z| < 1\}$. It is called **distance-decreasing** for all analytic mappings if, for any analytic mapping $f: M_1 \to M_2$, $M_1, M_2 \in \mathcal{M}$, the inequality $d(f(p), f(q)) \leqslant d(p, q)$ holds for all $p, q \in M_1$ (cf. **Kobayashi metric**, **Carathéodory metric**, **Wu metric**).

- **Kobayashi metric**

Let D be a *domain* in $\mathbb{C}^n$. Let $\mathcal{O}(\Delta, D)$ be the set of all analytic mappings $f: \Delta \to D$, where $\Delta = \{z \in \mathbb{C}: |z| < 1\}$ is the *unit disk*.

The **Kobayashi metric** (or **Kobayashi–Royden metric**) F_K is a **complex Finsler metric**, defined by

$$F_K(z, u) = \inf\{\alpha > 0: \exists f \in \mathcal{O}(\Delta, D),\ f(0) = z,\ \alpha f'(0) = u\}$$

for all $z \in D$ and $u \in \mathbb{C}^n$. It is a generalization of the **Poincaré metric** to higher-dimensional domains. $F_K(z, u) \geqslant F_C(z, u)$, where F_C is the **Carathéodory metric**. If D is convex, and $d(z, u) = \inf\{\lambda:\ z + \frac{u}{\alpha} \in D \text{ if } |\alpha| > \lambda\}$, then $\frac{d(z,u)}{2} \leqslant F_K(z, u) = F_C(z, u) \leqslant d(z, u)$.

Given a complex manifold M^n, the **Kobayashi semi-metric** F_K is defined by

$$F_K(p, u) = \inf\{\alpha > 0: \exists f \in \mathcal{O}(\Delta, M^n),\ f(0) = p,\ \alpha f'(0) = u\}$$

for all $p \in M^n$ and $u \in T_p(M^n)$. $F_K(p, u)$ is a semi-norm of the tangent vector u, called *Kobayashi semi-norm*. F_K is a metric if M^n is *taut*, i.e., $\mathcal{O}(\Delta, M^n)$ is a *normal family*.

The Kobayashi semi-metric is an infinitesimal form of the **Kobayashi semi-distance** K_{M^n} on M^n, defined as follow. Given $p, q \in M^n$, a *chain of disks* α from p to q is a collection of points $p = p^0, p^1, \dots, p^k = q$ of M^n, pairs of points $a^1, b^1; \dots; a^k, b^k$ of the unit disk Δ, and analytic mappings $f_1, \dots f_k$ from Δ into M^n, such that $f_j(a^j) = p^{j-1}$ and $f_j(b^j) = p^j$ for all j. The length $l(\alpha)$ of a chain α is the sum $d_P(a^1, b^1) + \dots + d_P(a^k, b^k)$, where d_P is the Poincaré metric. The **Kobayashi semi-distance** (or *Kobayashi pseudo-distance*) K_{M^n} on M^n is a semi-metric on M^n, defined by

$$K_{M^n}(p, q) = \inf_{\alpha} l(\alpha),$$

where the infimum is taken over all lengths $l(\alpha)$ of chains of disks α from p to q.

The Kobayashi semi-distance is **distance-decreasing** for all analytic mappings. It is the greatest semi-metric among all semi-metrics on M^n, that are distance-decreasing for all analytic mappings from Δ into M^n, where distances on Δ are measured in the Poincaré metric. K_Δ coincides with the Poincaré metric, and $K_{\mathbb{C}^n} \equiv 0$.

A manifold is called *Kobayashi hyperbolic* if the Kobayashi semi-distance is a metric on it. In fact, a manifold is Kobayashi hyperbolic if and only if it is biholomorphic to a bounded homogeneous domain.

- **Kobayashi–Busemann metric**

Given a complex manifold M^n, the **Kobayashi–Busemann semi-metric** on M^n is the double dual of the **Kobayashi semi-metric** on M^n. It is a metric if M^n is *taut*.

- **Carathéodory metric**

Let D be a *domain* in $\mathbb{C}^n$. Let $\mathcal{O}(D, \Delta)$ be the set of all analytic mappings $f : D \to \Delta$, where $\Delta = \{z \in \mathbb{C}: |z| < 1\}$ is the *unit disk*.

The **Carathéodory metric** F_C is a **complex Finsler metric**, defined by

$$F_C(z, u) = \sup\{|f'(z)u|: f \in \mathcal{O}(D, \Delta)\}$$

for any $z \in D$ and $u \in \mathbb{C}^n$. It is a generalization of the **Poincaré metric** to higher-dimensional domains. $F_C(z, u) \leqslant F_K(z, u)$, where F_K is the Kobayashi metric. If D is convex and $d(z, u) = \inf\{\lambda: z + \frac{u}{\alpha} \in D \text{ if } |\alpha| > \lambda\}$, then $\frac{d(z,u)}{2} \leqslant F_C(z, u) = F_K(z, u) \leqslant d(z, u)$.

Given a complex manifold M^n, the **Carathéodory semi-metric** F_C is defined by

$$F_C(p, u) = \sup\{|f'(p)u|: f \in \mathcal{O}(M^n, \Delta)\}$$

for all $p \in M^n$ and $u \in T_p(M^n)$. F_C is a metric if M^n is *taut*.

The **Carathéodory semi-distance** (or *Carathéodory pseudo-distance*) C_{M^n} is a semi-metric on a complex manifold M^n, defined by

$$C_{M^n}(p, q) = \sup\{d_P(f(p), f(q)): f \in \mathcal{O}(M^n, \Delta)\},$$

where d_P is the Poincaré metric. In general, the integrated semi-metric of the infinitesimal Carathéodory semi-metric is **internal** for the Carathéodory semi-distance, but does not coincides with it.

The Carathéodory semi-distance is **distance-decreasing** for all analytic mappings. It is the smallest distance-decreasing semi-metric. C_Δ coincides with the Poincaré metric, and $C_{\mathbb{C}^n} \equiv 0$.

- **Azukawa metric**

Let D be a *domain* in $\mathbb{C}^n$. Let $g_D(z, u) = \sup\{f(u): f \in K_D(z)\}$, where $K_D(z)$ is the set of all *logarithmically plurisubharmonic* functions $f : D \to [0, 1)$ such that there exist $M, r > 0$ with $f(u) \leqslant M\|u - z\|_2$ for all $u \in B(z, r) \subset D$; here $\|.\|_2$ is the l_2-*norm* on $\mathbb{C}^n$, and $B(z, r) = \{x \in \mathbb{C}^n: \|z - x\|_2 < r\}$.

The **Azukawa metric** (in general, a semi-metric) F_A is a **complex Finsler metric**, defined by

$$F_A(z,u) = \lim_{\lambda \to 0} \sup \frac{1}{|\lambda|} g_D(z, z + \lambda u)$$

for all $z \in D$ and $u \in \mathbb{C}^n$. It "lies between" the **Carathéodory metric** F_C and the **Kobayashi metric** F_K: $F_C(z,u) \leqslant F_A(z,u) \leqslant F_K(z,u)$ for all $z \in D$ and $u \in \mathbb{C}^n$. If D is convex, then all these metrics coincide.

The Azukawa metric is an infinitesimal form of the **Azukawa semi-distance**.

- **Sibony metric**

Let D be a *domain* in $\mathbb{C}^n$. Let $K_D(z)$ be the set of all *logarithmically plurisubharmonic* functions $f : D \to [0, 1)$ such that there exist $M, r > 0$ with $f(u) \leqslant M\|u - z\|_2$ for all $u \in B(z,r) \subset D$; here $\|.\|_2$ is the l_2-*norm* on $\mathbb{C}^n$, and $B(z,r) = \{x \in \mathbb{C}^n : \|z - x\|_2 < r\}$. Let $C^2_{loc}(z)$ be the set of all functions of class C^2 on some open neighborhood of z.

The **Sibony metric** (in general, a semi-metric) F_S is a **complex Finsler metric**, defined by

$$F_S(z,u) = \sup_{f \in K_D(z) \cap C^2_{loc}(z)} \sqrt{\sum_{i,j} \frac{\partial^2 f}{\partial z_i \partial \overline{z}_j}(z) u_i \overline{u}_j}$$

for all $z \in D$ and $u \in \mathbb{C}^n$. It "lies between" the **Carathéodory metric** F_C and the **Kobayashi metric** F_K: $F_C(z,u) \leqslant F_S(z,u) \leqslant F_A(z,u) \leqslant F_K(z,u)$ for all $z \in D$ and $u \in \mathbb{C}^n$, where F_A is the **Azukawa metric**. If D is convex, then all these metrics coincide.

The Sibony metric is an infinitesimal form of the **Sibony semi-distance**.

- **Wu metric**

The **Wu metric** W_{M^n} is an upper-semi-continuous Hermitian metric on a complex manifold M^n, that is **distance-decreasing** for all analytic mappings. In fact, for two n-dimensional complex manifolds M_1^n and M_2^n, the inequality $W_{M_2^n}(f(p), f(q)) \leqslant \sqrt{n} W_{M_1^n}(p,q)$ holds for all $p, q \in M_1^n$.

The invariant metrics including the Carathéodory, Kobayashi, Bergman, and Kähler-Einstein metrics play an important role in the Complex Function Theory and Convex Geometry. The Carathéodory and Kobayashi metrics are used mostly because of the distance-decreasing property. But they are almost never Hermitian. On the other hand, the Bergman metric and the Kähler-Einstein metric are Hermitian (in fact, Kählerian), but the distance-decreasing property, in general, fails for them.

- **Teichmüller metric**

A *Riemann surface* R is an one-dimensional complex manifold. Two Riemann surfaces R_1 and R_2 are called *conformally equivalent* if there exists a bijective analytic function (i.e., a conformal homeomorphism) from R_1 into R_2. More precisely, consider a fixed closed Riemann surface R_0 of a given *genus* $g \geqslant 2$. For a closed Riemann surface R of genus g, construct a pair (R, f), where $f : R_0 \to R$ is a homeomorphism. Two pairs (R, f) and (R_1, f_1) are called conformally equivalent if there exists a conformal homeomorphism $h : R \to R_1$ such that the mapping $(f_1)^{-1} \cdot h \cdot f : R_0 \to R_0$ is homotopic to the identity. An *abstract Riemann surface* $R^* = (R, f)^*$ is the equivalence class of all Riemann surfaces, conformally equivalent to R. The set of all equivalence classes is called *Teichmüller space* $T(R_0)$ of the surface R_0. For closed surfaces R_0 of given genus g the spaces $T(R_0)$ are isometrically isomorphic, and one can speak of the *Teichmüller space* T_g *of surfaces of genus* g. T_g is a complex manifold. If R_0 is obtained from a compact surface of genus $g \geqslant 2$ by removing n points, then the dimension of T_g is $3g - 3 + n$.

The **Teichmüller metric** is a metric on T_g, defined by

$$\frac{1}{2} \ln \inf_h K(h)$$

for any $R_1^*, R_2^* \in T_g$, where $h : R_1 \to R_2$ is a quasi-conformal homeomorphism, homotopic to the identity, and $K(h)$ is the maximal dilatation of h.

In fact, there exists an unique extremal mapping, called *Teichmüller mapping*, which minimizes the maximal dilatation of all such h, and the distance between R_1^* and R_2^* is equal to $\frac{1}{2} \ln K$, where the constant K is the dilatation of the Teichmüller mapping.

In terms of the *extremal length* $ext_{R^*}(\gamma)$, the distance between R_1^* and R_2^* can be written as

$$\frac{1}{2} \ln \sup_\gamma \frac{ext_{R_1^*}(\gamma)}{ext_{R_2^*}(\gamma)},$$

where the supremum is taken over all simple closed curves on R_0.

The *moduli space* R_g of conformal classes of Riemann surfaces of genus g is obtained by factorization of T_g by some countable group of automorphisms of it, called *modular group*. Examples of metrics related to moduli and Teichmüller spaces are **Teichmüller metric**, **Carathéodory metric**, **Kobayashi metric**, *Cheng–Yau–Mok's–Kähler–Einstein metric*, *Mc-Mullen metric*, **Bergman metric**, *asymptotic Poincaré metric*, *Ricci metric*, *perturbed Ricci metric*, **Weyl–Petersson metric**, *VHS-metric*, **Quillen metric**, etc.

- **Weyl–Petersson metric**

The **Weyl–Peterson metric** is a **Kähler metric** on the Teichmüller space $T_{g,n}$ of abstract Riemann surfaces of genus g with n punctures and negative Euler characteristic.

- **Gibbons–Manton metric**

The **Gibbons–Manton metric** is an $4n$-dimensional **hyper-Kähler metric** on the moduli space of *n-monopoles*, admitted an isometric action of the n-dimensional torus T^n. It can be described also as a hyper-Kähler quotient of a flat quaternionic vector space.

- **Metrics on determinant lines**

Let M^n be an n-dimensional compact smooth manifold, and let F be a flat vector bundle over M^n. Let $H^\bullet(M^n, F) = \bigoplus_{i=0}^n H^i(M^n, F)$ be the *de Rham cohomology* of M^n with coefficients in F. Given an n-dimensional vector space V, the *determinant line det V* of V is defined as the top exterior power of V, i.e., $det\, V = \wedge^n V$. Given a finite-dimensional graded vector space $V = \bigoplus_{i=0}^n V_i$, the determinant line of V is defined as the tensor product $det\, V = \bigotimes_{i=0}^n (det\, V_i)^{(-1)^i}$. Thus, the determinant line $det\, H^\bullet(M^n, F)$ of the cohomology $H^\bullet(M^n, F)$ can be written as $det\, H^\bullet(M^n, F) = \bigotimes_{i=0}^n (det\, H^i(M^n, F))^{(-1)^i}$.

The **Reidemeister metric** is a metric on $det\, H^\bullet(M^n, F)$, defined by a given smooth triangulation of M^n, and the classical *Reidemeister–Franz torsion*.

Let g^F and $g^{T(M^n)}$ be smooth metrics on the vector bundle F and tangent bundle $T(M^n)$, respectively. These metrics induce a canonical L_2**-metric** $h^{H^\bullet(M^n,F)}$ on $H^\bullet(M^n, F)$. The **Ray–Singler metric** on $det\, H^\bullet(M^n, F)$ is defined as the product of the metric induced on $det\, H^\bullet(M^n, F)$ by $h^{H^\bullet(M^n,F)}$ with the *Ray–Singler analytic torsion*. The **Milnor metric** on $det\, H^\bullet(M^n, F)$ ca be defined in similar manner using the *Milnor analytic torsion*. If g^F is flat, the above two metrics coincide with the Reidemeister metric. Using a co-Euler structure, one can define a *modified Ray–Singler metric* on $det\, H^\bullet(M^n, F)$.

The **Poincaré–Reidemeister metric** is a metric on the cohomological determinant line $det\, H^\bullet(M^n, F)$ of a closed connected oriented odd-dimensional manifold M^n. It can be constructed using a combination of the Reidemeister torsion with the Poincaré duality. Equivalently, one can define the *Poincaré–Reidemeister scalar product* on $det H^\bullet(M^n, F)$ which completely determines the Poincaré–Reidemeister metric but contains an additional sign or phase information.

The **Quillen metric** is a metric on the inverse of the cohomological determinant line of a compact Hermitian one-dimensional complex manifold. It can be defined as the product of the L_2-metric with the Ray–Singler analytic torsion.

- **Kähler supermetric**

The **Kähler supermetric** is a generalization of the **Kähler metric** on the case of a *supermanifold*. A *supermanifold* is a generalization of an usual manifold with *fermionic* as well as *bosonic* coordinates. The bosonic coordinates are ordinary numbers, whereas the fermionic coordinates are *Grassmann numbers*.

- **Hofer metric**

A *symplectic manifold* (M^n, w), $n = 2k$, is a smooth even-dimensional manifold M^n equipped with a *symplectic form*, i.e, a closed non-degenerate 2-form, w.

A *Lagrangian manifold* is an k-dimensional smooth submanifold L^k of a symplectic manifold (M^n, w), $n = 2k$, such that the form w vanishes identically on L^k, i.e., for any $p \in L^k$ and any $x, y \in T_p(L^k)$, one has $w(x, y) = 0$.

Let $L(M^n, \Delta)$ be the set of all Lagrangian submanifolds of a closed symplectic manifold (M^n, w), diffeomorphic to a given Lagrangian submanifold Δ. A smooth family $\alpha = \{L_t\}_t$, $t \in [0, 1]$, of Lagrangian submanifolds $L_t \in L(M^n, \Delta)$ is called *exact path*, connecting L_0 and L_1, if there exists a smooth mapping $\Psi : \Delta \times [0, 1] \to M^n$ such that, for every $t \in [0, 1]$, one has $\Psi(\Delta \times \{t\}) = L_t$, and $\Psi * w = dH_t \wedge dt$ for some smooth function $H : \Delta \times [0, 1] \to \mathbb{R}$. The *Hofer length* $l(\alpha)$ of an exact path α is defined by $l(\alpha) = \int_0^1 \{\max_{p\in\Delta} H(p, t) - \min_{p\in\Delta} H(p, t)\}dt$.

The **Hofer metric** on the set $L(M^n, \Delta)$ is defined by

$$\inf_{\alpha} l(\alpha)$$

for any $L_0, L_1 \in L(M^n, \Delta)$, where the infimum is taken over all exact paths on $L(M^n, \Delta)$, that connect L_0 and L_1.

The Hofer metric can be defined in similar way on the group $Ham(M^n, w)$ of *Hamiltonian diffeomorphisms* of a closed symplectic manifold (M^n, w), whose elements are time one mappings of *Hamiltonian flows* ϕ_t^H: it is $\inf_\alpha l(\alpha)$, where the infimum is taken over all smooth paths $\alpha = \{\phi_t^H\}$, $t \in [0, 1]$, connecting ϕ and ψ.

- **Sasakian metric**

A **Sasakian metric** is a metric of positive scalar curvature on a *contact manifold*, naturally adapted to the *contact structure*. A contact manifold equipped with a Sasakian metric is called *Sasakian space*, and is an odd-dimensional analog of *Kähler manifolds*.

- **Cartan metric**

A *Killing form* (or *Cartan–Killing form*) on a finite-dimensional *Lie algebra* Ω over a field $\mathbb{F}$ is a symmetric bilinear form

$$B(x, y) = Tr(ad_x \cdot ad_y),$$

where Tr denotes the trace of a linear operator, and ad_x is the image of x under *the adjoint representation* of Ω, i.e., the linear operator on the vector space Ω defined by the rule $z \to [x, z]$, where [,] is the Lie bracket.

Let $e_1, \dots e_n$ be a basis for the Lie algebra Ω, and $[e_i, e_j] = \sum_{k=1}^n \gamma_{ij}^k e_k$, where γ_{ij}^k are corresponding *structure constants*. Then the Killing form is given by

$$B(x_i, x_j) = g_{ij} = \sum_{k,l=1}^{n} \gamma_{il}^k \gamma_{ik}^l.$$

The **metric tensor** $((g_{ij}))$ is called, especially in the Theoretical Physics, **Cartan metric**.

Chapter 8

Distances on Surfaces and Knots

8.1. GENERAL SURFACE METRICS

A *surface* is a real two-dimensional *manifold* M^2, i.e., a **Hausdorff space**, each point of which has a neighborhood which is homeomorphic to a plane $\mathbb{E}^2$, or a closed half-plane (cf. Chapter 7).

A compact orientable surface is called *closed* if it has no boundary, and it is called *surface with boundary*, otherwise. There are compact non-orientable surfaces (closed or with boundary); the simplest such surface is the *Möbius strip*. Non-compact surfaces without boundary are called *open*.

Any closed connected surface is homeomorphic to either a sphere with, say, g (cylindric) handles, or a sphere with, say, g *cross-caps* (i.e., caps with a twist like Möbius strip in them). In both cases the number g is called *genus* of the surface. In the case of handles, the surface is orientable; it is called *torus* (doughnut), *double torus*, and *triple torus* for $g = 1, 2$ and 3, respectively. In the case of cross-caps, the surface is non-orientable; it is called *real projective plane*, *Klein bottle*, and *Dyck's surface* for $g = 1, 2$ and 3, respectively. The genus is the maximal number of disjoint simple closed curves which can be cut from a surface without disconnecting it (the *Jordan curve theorem* for surfaces).

The *Euler–Poincaré characteristic* of a surface is (the same for all polyhedral decompositions of a given surface) the number $\chi = v - e + f$, where v, e and f are, respectively, the number of vertices, edges and faces of the decomposition. It holds $\chi = 2 - 2g$ if the surface is orientable, and $\chi = 2 - g$ if not. Every surface with boundary is homeomorphic to a sphere with appropriated number of (disjoint) *holes* (i.e., what remains if an open disk is removed) and handles or cross-caps. If h is the number of holes, then $\chi = 2 - 2g - h$ holds if the surface is orientable, and $\chi = 2 - g - h$ if not.

The *connectivity number* of a surface is the largest number of closed cuts that can be made on the surface without separating it into two or more parts. This number is equal to $3 - \chi$ for closed surfaces, and $2 - \chi$ for surfaces with boundaries. A surface with connectivity number 1, 2 and 3 is called, respectively, *simply, doubly* and *triply connected*. A sphere is simply connected, while a torus is triply connected.

A surface can be considered as a metric space with its own **intrinsic metric**, or as a figure in space. A surface in $\mathbb{E}^3$ is called *complete* if it is a **complete** metric space with respect to its intrinsic metric.

A surface is called *differentiable*, *regular*, or *analytic*, respectively, if in a neighborhood of each of its points it can be given by an expression

$$r = r(u, v) = r(x_1(u, v), x_2(u, v), x_3(u, v)),$$

where the *position vector* $r = r(u, v)$ is a differentiable, *regular* (i.e., a sufficient number of times differentiable), or *real analytic*, respectively, vector function satisfying the condition $r_u \times r_v \neq 0$.

Any regular surface has the intrinsic metric with the *line element* (or *first fundamental form*)

$$ds^2 = dr^2 = E(u, v)\,du^2 + 2F(u, v)\,du\,dv + G(u, v)\,dv^2,$$

where $E(u, v) = \langle r_u, r_u \rangle$, $F(u, v) = \langle r_u, r_v \rangle$, $G(u, v) = \langle r_v, r_v \rangle$. The length of a curve, defined on the surface by the equations $u = u(t)$, $v = v(t)$, $t \in [0, 1]$, is computed by

$$\int_0^1 \sqrt{Eu'^2 + 2Fu'v' + Gv'^2}\,dt,$$

and the distance between any points $p, q \in M^2$ is defined as the infimum of the lengths of all curves on M^2, connecting p and q. A **Riemannian metric** is a generalization of the first fundamental form of a surface.

For surfaces, two kinds of *curvature* are considered: *Gaussian curvature*, and *mean curvature*. To compute those curvatures at a given point of the surface, consider the intersection of the surface with a plane, containing a fixed *normal vector*, i.e., a vector which is perpendicular to the surface at this point. This intersection is a plane curve. The *curvature* k of this plane curve is called *normal curvature* of the surface at the given point. If we vary the plane, the normal curvature k will change, and there are two extremal values – the *maximal curvature* k_1, and the *minimal curvature* k_2, called *principal curvatures* of the surface. A curvature is taken to be *positive* if the curve turns in the same direction as the surface's chosen normal, otherwise it is taken to be *negative*. The *Gaussian curvature* is $K = k_1 k_2$ (it can be given entirely in terms of the first fundamental form). The *mean curvature* is $H = \frac{1}{2}(k_1 + k_2)$.

A *minimal surface* is a surface with mean curvature zero, or, equivalently, a surface of minimum area subject to constraints on the location of its boundary.

A *Riemann surface* is an one-dimensional *complex manifold*, or a two-dimensional real manifold with a complex structure, i.e., in which the local coordinates in neighborhoods of points are related by complex analytic functions. It can be thought as a deformed version of the complex plane. All Riemann surfaces are orientable. Closed Riemann surfaces are geometrical models of *complex algebraic curves*. Every connected Riemann surface can be turned into a *complete* two-dimensional *Riemannian manifold* with constant curvature -1, 0, or 1. The Riemann surfaces with the curvature -1 are called *hyperbolic*, the *unit disk* $\Delta = \{z \in \mathbb{C}: |z| < 1\}$ is the canonical example. The Riemann surfaces with the curvature 0 are called *parabolic*, $\mathbb{C}$ is a typical example. The Riemann surfaces with the curvature 1 are called *elliptic*, the *Riemann sphere* $\mathbb{C} \cup \{\infty\}$ is a typical example.

- **Regular metric**

The intrinsic metric of a surface is called **regular** if it can be specified using the *line element*

$$ds^2 = E\,du^2 + 2F\,du\,dv + G\,dv^2,$$

where the coefficients of the form ds^2 are regular functions.

Any regular surface, given by an expression $r = r(u, v)$, has a regular metric with the *line element* ds^2, where $E(u, v) = \langle r_u, r_u \rangle$, $F(u, v) = \langle r_u, r_v \rangle$, $G(u, v) = \langle r_v, r_v \rangle$.

- **Analytic metric**

The intrinsic metric on a surface is called **analytic** if it can be specified using the *line element*

$$ds^2 = E\,du^2 + 2F\,du\,dv + G\,dv^2,$$

where the coefficients of the form ds^2 are real analytic functions.

Any analytic surface, given by an expression $r = r(u, v)$, has an analytic metric with the *line element* ds^2, where $E(u, v) = \langle r_u, r_u \rangle$, $F(u, v) = \langle r_u, r_v \rangle$, $G(u, v) = \langle r_v, r_v \rangle$.

- **Metric of positive curvature**

A **metric of positive curvature** is the intrinsic metric on a *surface of positive curvature*.

A *surface of positive curvature* is a surface in $\mathbb{E}^3$ that has positive Gaussian curvature at every point.

- **Metric of negative curvature**

A **metric of negative curvature** is the intrinsic metric on a *surface of negative curvature*.

A *surface of negative curvature* is a surface in $\mathbb{E}^3$ that has negative Gaussian curvature at every point. A surface of negative curvature locally have a saddle-like structure. The intrinsic geometry of a surface of constant negative curvature (in particular, of a *pseudo-sphere*) locally coincides with the geometry of *Lobachevsky plane*. There exists no surface in $\mathbb{E}^3$, whose intrinsic geometry coincides completely with the geometry of Lobachevsky plane (i.e., a complete regular surface of constant negative curvature).

- **Metric of non-positive curvature**

A **metric of non-positive curvature** is the intrinsic metric on a *saddle-like surface*.

A *saddle-like surface* is a generalization of a surface of negative curvature: a twice continuously-differentiable surface is a saddle-like surface if and only if at each point of the surface its Gaussian curvature is non-positive. These surfaces can be seen as antipodes of *convex surfaces*, but they do not form such a natural class of surfaces as do convex surfaces.

- **Metric of non-negative curvature**

A **metric of non-negative curvature** is the intrinsic metric on a *convex surface*.

A *convex surface* is a *domain* (i.e., a connected open set) on the boundary of a *convex body* in $\mathbb{E}^3$ (in some sense, it is an antipode of saddle-like surface). The entire boundary of a convex body is called *complete convex surface*. If the body is finite (bounded), the complete convex surface is called *closed*. Otherwise, it is called *infinite* (an infinite convex surface is homeomorphic to a plane or to a circular cylinder).

Any convex surface M^2 in $\mathbb{E}^3$ is a *surface of bounded curvature*. The *total Gaussian curvature* $w(A) = \int\int_A K(x)\, d\sigma(x)$ of a set $A \subset M^2$ is always non-negative (here $\sigma(.)$ is the *area*, and $K(x)$ is the *Gaussian curvature* of M^2 at a point x), i.e., a convex surface can be seen as a *surface of non-negative curvature*.

The intrinsic metric of a convex surface is a **convex metric** in the sense of surface theory, i.e., it displays the *convexity condition*: the sum of the angles of any triangle whose sides are shortest curves is not less that π.

- **Metric with alternating curvature**

A **metric with alternating curvature** is the intrinsic metric on a surface with alternating (positive or negative) Gaussian curvature.

- **Flat metric**

A **flat metric** is the intrinsic metric on a *developable surface*, i.e., a surface, on which the Gaussian curvature is everywhere zero.

- **Metric of bounded curvature**

A **metric of bounded curvature** is the intrinsic metric ρ on a *surface of bounded curvature*.

A surface M^2 with an intrinsic metric ρ is called *surface of bounded curvature* if there exists a sequence of **Riemannian metrics** ρ_n, defined on M^2, such that for any compact set $A \subset M^2$ one has $\rho_n \to \rho$ uniformly, and the sequence $|w_n|(A)$ is bounded, where $|w|_n(A) = \int\int_A |K(x)|\, d\sigma(x)$ is *total absolute curvature* of the metric ρ_n (here $K(x)$ is the Gaussian curvature of M^2 at a point x, and $\sigma(.)$ is the *area*).

- **Λ-metric**

A **Λ-metric** (or *metric of type Λ*) is a **complete** metric on a surface with curvature bounded from above by a negative constant.

An Λ-metric does not have embeddings into $\mathbb{E}^3$. It is a generalization of the classical result of Hilbert (1901): no complete regular surface of constant negative curvature (i.e., a surface whose intrinsic geometry coincides completely with the geometry of Lobachevsky plane) exists in $\mathbb{E}^3$.

- **(h, Δ)-metric**

An **(h, Δ)-metric** is a metric on a surface with a slowly-changing negative curvature.

A **complete** (h, Δ)-metric does not permit a regular *isometric embedding* in three-dimensional Euclidean space (cf. **Λ-metric**).

- ***G*-distance**

A connected set G of points on a surface M^2 is called *geodesic region* if, for each point $x \in G$, there exists a *disk* $B(x, r)$ with center at x, such that $B_G = G \cap B(x, r)$ has one of the following forms: $B_G = B(x, r)$ (x is a *regular interior point* of G); B_G is a semi-disk of $B(x, r)$ (x is a *regular boundary point* of G); B_G is a sector of $B(x, r)$

other than a semi-disk (x is an *angular point* of G); B_G consists of a finite number of sectors of $B(x, r)$ with no common points except x (x is a *nodal point* of G).

The **G-distance** between any x and $y \in G$ is defined as the greatest lower bound of the lengths of all rectifiable curves connecting x and $y \in G$, and completely contained in G.

- **Conformally invariant metric**

Let R be a Riemann surface. A *local parameter* (or *local uniformizing parameter*, *local uniformizer*) is a complex variable z considered as a continuous function $z_{p_0} = \phi_{p_0}(p)$ of a point $p \in R$ which is defined everywhere in some neighborhood (*parametric neighborhood*) $V(p_0)$ of a point $p_0 \in R$ and which realizes a homeomorphic mapping (*parametric mapping*) of $V(p_0)$ onto the disk (*parametric disk*) $\Delta(p_0) = \{z \in \mathbb{C}: |z| < r(p_0)\}$, where $\phi_{p_0}(p_0) = 0$. Under a parametric mapping, any point function $g(p)$, defined in the parametric neighborhood $V(p_0)$, goes into a function of the local parameter z: $g(p) = g(\phi_{p_0}^{-1}(z)) = G(z)$.

A **conformally invariant metric** is a differential $\rho(z)|dz|$ on the Riemann surface R which is invariant with respect to the choice of the local parameter z. Thus, to each local parameter z ($z : U \to \overline{\mathbb{C}}$) a function $\rho_z : z(U) \to [0, \infty]$ is associated such that, for any local parameters z_1 and z_2, we have:

$$\frac{\rho_{z_2}(z_2(p))}{\rho_{z_1}(z_1(p))} = \left|\frac{dz_1(p)}{dz_2(p)}\right| \quad \text{for any } p \in U_1 \cap U_2.$$

Every linear differential $\lambda(z)dz$ and every *quadratic differential* $Q(z)dz^2$ induce conformally invariant metrics $|\lambda(z)||dz|$ and $|Q(z)|^{1/2}||dz|$, respectively (cf. **Q-metric**).

- **Q-metric**

An **Q-metric** is a **conformally invariant metric** $\rho(z)|dz| = |Q(z)|^{1/2}|dz|$ on a Riemann surface R, defined by a *quadratic differential* $Q(z)dz^2$.

A *quadratic differential* $Q(z)dz^2$ is a non-linear differential on a Riemann surface R which is invariant with respect to the choice of the local parameter z. Thus, to each local parameter z ($z : U \to \overline{\mathbb{C}}$) a function $Q_z : z(U) \to \overline{\mathbb{C}}$ is associated such that, for any local parameters z_1 and z_2, we have:

$$\frac{Q_{z_2}(z_2(p))}{Q_{z_1}(z_1(p))} = \left(\frac{dz_1(p)}{dz_2(p)}\right)^2 \quad \text{for any } p \in U_1 \cap U_2.$$

- **Extremal metric**

An **extremal metric** is a **conformally invariant metric** in the *modulus problem* for a family Γ of locally rectifiable curves on a Riemann surface R which realizes the infimum in the definition of the *modulus* $M(\Gamma)$.

Formally, let Γ be a family of locally rectifiable curves on a Riemann surface R, let P be a non-empty class of conformally invariant metrics $\rho(z)|dz|$ on R such that $\rho(z)$ is square-integrable in the z-plane for every local parameter z, and the integrals

$$A_\rho(R) = \iint_R \rho^2(z)\,dx\,dy \quad \text{and} \quad L_\rho(\Gamma) = \inf_{\gamma \in \Gamma} \int_\gamma \rho(z)|dz|$$

are not simultaneously equal to 0 or ∞ (each of the above integrals is understood as a Lebesgue integral). The *modulus of the family of curves* Γ is defined by

$$M(\Gamma) = \inf_{\rho \in P} \frac{A_\rho(R)}{(L_\rho(\Gamma))^2}.$$

The *extremal length of the family of curves* Γ is equal to $\sup_{\rho \in P} \frac{(L_\rho(\Gamma))^2}{A_\rho(R)}$, i.e., is the reciprocal of $M(\Gamma)$.

The modulus problem for Γ is defined as follows: let P_L be the subclass of P such that, for any $\rho(z)|dz| \in P_L$ and any $\gamma \in \Gamma$, one has $\int_\gamma \rho(z)|dz| \geqslant 1$. If $P_L \neq \emptyset$, then the modulus $M(\Gamma)$ of the family Γ can be written as $M(\Gamma) = \inf_{\rho \in P_L} A_\rho(R)$. Every metric from P_L is called **admissible metric** for the modulus problem on Γ. If there exists ρ^* for which

$$M(\Gamma) = \inf_{\rho \in P_L} A_\rho(R) = A_{\rho^*}(R),$$

the metric $\rho^*|dz|$ is called **extremal metric** for the modulus problem on Γ.

- **Fréchet surface metric**

  Let (X, d) be a metric space, M^2 be a compact two-dimensional manifold, f be a continuous mapping $f : M^2 \to X$, called *parameterized surface*, and $\sigma : M^2 \to M^2$ be a homeomorphism of M^2 onto itself. Two parameterized surfaces f_1 and f_2 are called *equivalent* if $\inf_\sigma \max_{p \in M^2} d(f_1(p), f_2(\sigma(p))) = 0$, where the infimum is taken over all possible homeomorphisms σ. A class f^* of parameterized surfaces, equivalent to f, is called *Fréchet surface*. It is a generalization of the notion of a surface in an Euclidean space to the case of an arbitrary metric space (X, d).

  The **Fréchet surface metric** is a metric on the set of all Fréchet surfaces, defined by

  $$\inf_\sigma \max_{p \in M^2} d\big(f_1(p), f_2(\sigma(p))\big)$$

  for any Fréchet surfaces f_1^* and f_2^*, where the infimum is taken over all possible homeomorphisms σ (cf. **Fréchet metric**).

8.2. INTRINSIC METRICS ON SURFACES

In this section we list intrinsic metrics, given by their *line elements* (which, in fact, are two-dimensional **Riemannian metrics**), for some selected surfaces.

- **Quadric metric**

A *quadric* (or *quadratic surface*, *surface of second order*) is a set of points in $\mathbb{E}^3$, whose coordinates in a Cartesian coordinate system satisfy an algebraic equation of degree two. There are 17 classes of such surfaces, among them are: *ellipsoids*, *one-sheet* and *two-sheet hyperboloids*, *elliptic paraboloids*, *hyperbolic paraboloids*, *elliptic*, *hyperbolic* and *parabolic cylinders*, and *conical surfaces*.

For example, a *cylinder* can be given by the following parametric equations:

$$x_1(u,v) = a\cos v, \quad x_2(u,v) = a\sin v, \quad x_3(u,v) = u.$$

The intrinsic metric on it is given by the *line element*

$$ds^2 = du^2 + a^2\,dv^2.$$

An *elliptic cone* (i.e., a cone with elliptical cross-section) has the following parametric equations:

$$x_1(u,v) = a\frac{h-u}{h}\cos v, \quad x_2(u,v) = b\frac{h-u}{h}\sin v, \quad x_3(u,v) = u,$$

where h is the *height*, a is the *semi-major axis*, and b is the *semi-minor axis* of the cone. The intrinsic metric on it is given by the *line element*

$$ds^2 = \frac{h^2 + a^2\cos^2 v + b^2\sin^2 v}{h^2}\,du^2 + 2\frac{(a^2-b^2)(h-u)\cos v\sin v}{h^2}\,du\,dv + \frac{(h-u)^2(a^2\sin^2 v + b^2\cos^2 v)}{h^2}\,dv^2.$$

- **Sphere metric**

A *sphere* is a *quadric*, given by the Cartesian equation $(x_1-a)^2+(x_2-b)^2+(x_3-c)^2 = r^2$, where the point (a,b,c) is the *center* of the sphere, and $r > 0$ is the *radius* of the sphere. The sphere of radius r, centered at the origin, can be given by the following parametric equations:

$$x_1(\theta,\phi) = r\sin\theta\cos\phi, \quad x_2(\theta,\phi) = r\sin\theta\sin\phi, \quad x_3(\theta,\phi) = r\cos\theta,$$

where the *azimuthal angle* $\phi \in [0, 2\pi)$, and the *polar angle* $\theta \in [0,\pi]$. The intrinsic metric on it (in fact, the two-dimensional **spherical metric**) is given by the *line element*

$$ds^2 = r^2\,d\theta^2 + r^2\sin^2\theta\,d\phi^2.$$

A sphere of radius r has constant positive Gaussian curvature equal to r.

- **Ellipsoid metric**

An *ellipsoid* is a *quadric* given by the Cartesian equation $\frac{x_1^2}{a^2} + \frac{x_2^2}{b^2} + \frac{x_3^2}{c^2} = 1$, or by the following parametric equations:

$$x_1(\theta, \phi) = a \cos\phi \sin\theta, \quad x_2(\theta, \phi) = b \sin\phi \sin\theta, \quad x_3(\theta, \phi) = c \cos\theta,$$

where the *azimuthal angle* $\phi \in [0, 2\pi)$, and the *polar angle* $\theta \in [0, \pi]$. The intrinsic metric on it is given by the *line element*

$$ds^2 = \left(b^2\cos^2\phi + a^2\sin^2\phi\right)\sin^2\theta\, d\phi^2 + \left(b^2 - a^2\right)\cos\phi \sin\phi \cos\theta \sin\theta\, d\theta\, d\phi + \left(\left(a^2\cos^2\phi + b^2\sin^2\phi\right)\cos^2\theta + c^2\sin^2\theta\right) d\theta^2.$$

- **Spheroid metric**

A *spheroid* is an *ellipsoid* having two axes of equal length. It is also a *rotation surface*, given by the following parametric equations:

$$x_1(u, v) = a \sin v \cos u, \quad x_2(u, v) = a \sin v \sin u, \quad x_3(u, v) = c \cos v,$$

where $0 \leqslant u < 2\pi$, and $0 \leqslant v \leqslant \pi$. The intrinsic metric on it is given by the *line element*

$$ds^2 = a^2 \sin^2 v\, du^2 + \frac{1}{2}\left(a^2 + c^2 + \left(a^2 - c^2\right)\cos(2v)\right) dv^2.$$

- **Hyperboloid metric**

A *hyperboloid* is a *quadric* which may be one- or two-sheeted. The one-sheeted hyperboloid is a *surface of revolution* obtained by rotating a hyperbola about the perpendicular bisector to the line between the foci, while the two-sheeted hyperboloid is a surface of revolution obtained by rotating a hyperbola about the line joining the foci. The one-sheeted circular hyperboloid, oriented along the x_3-axis, is given by the Cartesian equation $\frac{x_1^2}{a^2} + \frac{x_2^2}{a^2} - \frac{x_3^2}{c^2} = 1$, or by the following parametric equations:

$$x_1(u, v) = a\sqrt{1+u^2}\cos v, \quad x_2(u, v) = a\sqrt{1+u^2}\sin v, \quad x_3(u, v) = cu,$$

where $v \in [0, 2\pi)$. The intrinsic metric on it is given by the *line element*

$$ds^2 = \left(c^2 + \frac{a^2u^2}{u^2+1}\right) du^2 + a^2\left(u^2 + 1\right) dv^2.$$

- **Rotation surface metric**

A *rotation surface* (or *surface of revolution*) is a surface generated by rotating a two-dimensional curve about an axis. It is given by the following parametric equations:

$$x_1(u, v) = \phi(v) \cos u, \quad x_2(u, v) = \phi(v) \sin u, \quad x_3(u, v) = \psi(v).$$

The intrinsic metric on it is given by the *line element*

$$ds^2 = \phi^2\, du^2 + \left(\phi'^2 + \psi'^2\right) dv^2.$$

- **Pseudo-sphere metric**

A *pseudo-sphere* is a half of the *rotation surface*, generated by rotating a *tractrix* about its asymptote. It is given by the following parametric equations:

$$x_1(u, v) = \operatorname{sech} u \cos v, \quad x_2(u, v) = \operatorname{sech} u \sin v, \quad x_3(u, v) = u - \tanh u,$$

where $u \geqslant 0$, and $0 \leqslant v < 2\pi$. The intrinsic metric on it is given by the *line element*

$$ds^2 = \tanh^2 u\, du^2 + \operatorname{sech}^2 u\, dv^2.$$

The pseudo-sphere has constant negative Gaussian curvature equal to -1, and in this sense is an analog of the sphere which has constant positive Gaussian curvature.

- **Torus metric**

A *torus* is a surface having genus one. A torus azimuthally symmetric about the x_3-axis is given by the Cartesian equation $(c - \sqrt{x_1^2 + x_2^2})^2 + x_3^2 = a^2$, or by the following parametric equations:

$$x_1(u, v) = (c + a\cos v)\cos u, \quad x_2(u, v) = (c + a\cos v)\sin u, \quad x_3(u, v) = a\sin v,$$

where $c > a$, and $u, v \in [0, 2\pi)$. The intrinsic metric on it is given by the *line element*

$$ds^2 = (c + a\cos v)^2\, du^2 + a^2\, dv^2.$$

- **Helical surface metric**

A *helical surface* (or *surface of screw motion*) is a surface described by a plane curve γ which, while rotating around an axis at a uniform rate, also advances along that axis at a uniform rate. If γ is located in the plane of the axis of rotation x_3 and is defined by the equation $x_3 = f(u)$, the position vector of the helical surface is

$$r = (u\cos v,\ u\sin v,\ f(u) = hv), \quad h = \text{const},$$

and the intrinsic metric on it is given by the *line element*

$$ds^2 = \left(1 + f'^2\right) du^2 + 2hf'\, du\, dv + \left(u^2 + h^2\right) dv^2.$$

If $f = \text{const}$, one has a *helicoid*; if $h = 0$, one has a *rotation surface*.

- **Catalan surface metric**

The *Catalan surface* is a *minimal surface*, given by the following parametric equations:

$$x_1(u, v) = u - \sin u \cosh v, \quad x_2(u, v) = 1 - \cos u \cosh v,$$
$$x_3(u, v) = 4 \sin\left(\frac{u}{2}\right) \sinh\left(\frac{v}{2}\right).$$

The intrinsic metric on it is given by the *line element*

$$ds^2 = 2\cosh^2\left(\frac{v}{2}\right)(\cosh v - \cos u)\, du^2 + 2\cosh^2\left(\frac{v}{2}\right)\left(\cosh v - \cos u\right) dv^2.$$

- **Monkey saddle metric**

The *monkey saddle* is a surface, given by the Cartesian equation $x_3 = x_1(x_1^2 - 3x_2^2)$, or by the following parametric equations:

$$x_1(u, v) = u, \quad x_2(u, v) = v, \quad x_3(u, v) = u^3 - 3uv^2.$$

This is a surface which a monkey can straddle with both legs and his tail. The intrinsic metric on it is given by the *line element*

$$ds^2 = \left(1 + \left(su^2 - 3v^2\right)^2\right) du^2 - 2\left(18uv\left(u^2 - v^2\right)\right) du\, dv + \left(1 + 36u^2v^2\right) dv^2.$$

8.3. DISTANCES ON KNOTS

A *knot* is a closed, non-self-intersecting curve that is embedded in S^3. The *trivial knot* (or *unknot*) O is a closed loop that is not knotted. A knot can be generalized to a link which is a set of disjoint knots. Every link has its *Seifert surface*, i.e., a compact oriented surface with given link as boundary. Two knots (links) are called *equivalent* if one can be smoothly deformed into another. Formally, a link is defined as a smooth one-dimensional *submanifold* of the 3-sphere S^3; a knot is a link consisting of one component; links L_1 and L_2 are called *equivalent* if there exists an orientation-preserving homeomorphism $f : S^3 \to S^3$ such that $f(L_1) = L_2$.

All the information about a knot can be described using a *knot diagram*. It is a projection of a knot onto a plane such that no more than two points of the knot are projected to the same point on the plane, and at each such point it is indicated which strand is closest to the plane, usually by erasing part of the lower strand. Two different knot diagrams may both represent the same knot. Much of knot theory is devoted to telling when two knot diagrams represent the same knot.

An *unknotting operation* is an operation which changes the overcrossing and the undercrossing at a double point of a given knot diagram. The *unknotting number* of a knot K is the minimum number of unknotting operations needed to deform a diagram of K into that of the trivial knot, where the minimum is taken over all diagrams of K. Roughly, unknotting number is the smallest number of times a knot K must be passed through itself to untie it.

An *$\sharp$-unknotting operation* in a diagram of a knot K is an analog of the unknotting operation for a *$\sharp$-part* of the diagram consisting of two pairs of parallel strands with one of the pair overcrossing another. Thus, an $\sharp$-unknotting operation changes the overcrossing and the undercrossing at each vertex of obtained quadrangle.

- **Gordian distance**

  The **Gordian distance** is a metric on the set of all knots, defined, for given knots K and K', as the minimum number of unknotting operations needed to deform a diagram of K into that of K', where the minimum is taken over all diagrams of K from which one can obtain diagrams of K'. The unknotting number of K is equal to the Gordian distance between K and the trivial knot O.

  Let rK be the knot obtained from K by taking its mirror image, and let $-K$ be the knot with the reversed orientation. The **positive reflection distance** $Ref_+(K)$ is the Gordian distance between K and rK. The **negative reflection distance** $Ref_-(K)$ is the Gordian distance between K and $-rK$. The **inversion distance** $Inv(K)$ is the Gordian distance between K and $-K$.

  The Gordian distance is the case $k = 1$ of the **C_k-distance** which is the minimum number of *C_k-moves* needed to transform K into K'; Habiro and Goussarov proved that, for $k > 1$, it is finite if and only if both knots have the same *Vassiliev invariants of order less than k*. A C_1-move is a single crossing change, a C_2-move (or *delta-move*) is simultaneous crossing change for 3 arcs forming triangle. C_2- and C_3-distances are called **delta distance** and **clasp-pass distance**, respectively.

- **$\sharp$-Gordian distance**

  The **$\sharp$-Gordian distance** (see, for example, [Mura85]) is a metric on the set of all knots, defined, for given knots K and K', as the minimum number of $\sharp$-unknotting operations needed to deform a diagram of K into that of K', where the minimum is taken over all diagrams of K from which one can obtain diagrams of K'.

  Let rK be the knot obtained from K by taking its mirror image, and let $-K$ be the knot with the reversed orientation. The **positive $\sharp$-reflection distance** $Ref^\sharp_+(K)$ is the $\sharp$-Gordian distance between K and rK. The **negative $\sharp$-reflection distance** $Ref^\sharp_-(K)$ is the $\sharp$-Gordian distance between K and $-rK$. The **$\sharp$-inversion distance** $Inv^\sharp(K)$ is the $\sharp$-Gordian distance between K and $-K$.

Chapter 9

Distances on Convex Bodies, Cones, and Simplicial Complexes

9.1. DISTANCES ON CONVEX BODIES

A *convex body* in the n-dimensional Euclidean space $\mathbb{E}^n$ is a *compact convex* subset of $\mathbb{E}^n$. It is called *proper* if it has non-empty interior. Let K denote the space of all convex bodies in $\mathbb{E}^n$, and let K_p be the subspace of all proper convex bodies.

Any metric space (K, d) on K is called **metric space of convex bodies**. Metric spaces of convex bodies, in particular the metrization by the **Hausdorff metric**, or by the **symmetric difference metric**, play a basic role in the foundations of analysis in Convex Geometry (see, for example, [Grub93]).

For $C, D \in K\backslash\{\emptyset\}$ the *Minkowski addition* and the *Minkowski non-negative scalar multiplication* are defined by $C + D = \{x + y\colon\ x \in C, y \in D\}$, and $\alpha C = \{\alpha x\colon\ x \in C\}$, $\alpha \geqslant 0$, respectively. The Abelian semi-group $(K, +)$ equipped with non-negative scalar multiplication operators can be considered as a *convex cone*.

The *support function* $h_C : S^{n-1} \to \mathbb{R}$ of $C \in K$ is defined by $h_C(u) = \sup\{\langle u, x\rangle\colon\ x \in C\}$ for any $u \in S^{n-1}$, where S^{n-1} is the $(n-1)$-dimensional *unit sphere* in $\mathbb{E}^n$, and $\langle , \rangle$ is the *inner product* in $\mathbb{E}^n$.

Given a set $X \subset \mathbb{E}^n$, its *convex hull, conv*(X), is defined as the minimal *convex* set containing X.

- **Area deviation**

  The **area deviation** (or **template metric**) is a metric on the set K_p in $\mathbb{E}^2$ (i.e., on the set of plane convex disks), defined by

  $$A(C \triangle D),$$

  where $A(.)$ is the *area*, and $\triangle$ is the *symmetric difference*. If $C \subset D$, then it is equal to $A(D) - A(C)$.

- **Perimeter deviation**

  The **perimeter deviation** is a metric on K_p in $\mathbb{E}^2$, defined by

  $$2p\big(conv(C \cup D)\big) - p(C) - p(D),$$

  where $p(.)$ is the *perimeter*. In the case $C \subset D$, it is equal to $p(D) - p(C)$.

- **Mean width metric**

The **mean width metric** is a metric on K_p in $\mathbb{E}^2$, defined by

$$2W\big(conv(C \cup D)\big) - W(C) - W(D),$$

where $W(.)$ is the *mean width*: $W(C) = p(C)/\pi$, and $p(.)$ is the *perimeter*.

- **Pompeiu–Hausdorff–Blaschke metric**

The **Pompeiu–Hausdorff–Blaschke metric** (or **Hausdorff metric**) is a metric on K, defined by

$$\max\Big\{\sup_{x\in C}\inf_{y\in D}\|x-y\|_2, \sup_{y\in D}\inf_{x\in C}\|x-y\|_2\Big\},$$

where $\|.\|_2$ is the Euclidean norm on $\mathbb{E}^n$.

In terms of support functions, respectively, using Minkowski addition, it is

$$\begin{aligned}\sup_{u\in S^{n-1}}\big|h_C(u)-h_D(u)\big| &= \|h_C-h_D\|_\infty \\ &= \inf\big\{\lambda \geqslant 0\colon\ C \subset D+\lambda\overline{B}^n, D \subset C+\lambda\overline{B}^n\big\},\end{aligned}$$

where $\overline{B}^n$ is the *unit ball* of $\mathbb{E}^n$.

This metric can be defined using any norm on $\mathbb{R}^n$ instead of the Euclidean norm. More generally, it can be defined for the space of bounded closed subsets of an arbitrary metric space.

- **Pompeiu–Eggleston metric**

The **Pompeiu–Eggleston metric** is a metric on K, defined by

$$\sup_{x\in C}\inf_{y\in D}\|x-y\|_2 + \sup_{y\in D}\inf_{x\in C}\|x-y\|_2,$$

where $\|.\|_2$ is the Euclidean norm on $\mathbb{E}^n$.

In terms of support functions, respectively, using Minkowski addition, it is

$$\begin{aligned}&\max\Big\{0, \sup_{u\in S^{n-1}}\big(h_C(u)-h_D(u)\big)\Big\} + \max\Big\{0, \sup_{u\in S^{n-1}}\big(h_D(u)-h_C(u)\big)\Big\} \\ &\quad = \inf\big\{\lambda \geqslant 0\colon\ C \subset D+\lambda\overline{B}^n\big\} + \inf\big\{\lambda \geqslant 0\colon\ D \subset C+\lambda\overline{B}^n\big\},\end{aligned}$$

where $\overline{B}^n$ is the *unit ball* of $\mathbb{E}^n$.

This metric can be defined using any norm on $\mathbb{R}^n$ instead of the Euclidean norm. More generally, it can be defined for the space of bounded closed subsets of an arbitrary metric space.

• **McClure–Vitale metric**

Given $1 \leqslant p \leqslant \infty$, the **McClure–Vitale metric** is a metric on K, defined by

$$\left(\int_{S^{n-1}} |h_C(u) - h_D(u)|^p \, d\sigma(u) \right)^{\frac{1}{p}} = \|h_C - h_D\|_p.$$

• **Florian metric**

The **Florian metric** is a metric on K, defined by

$$\int_{S^{n-1}} |h_C(u) - h_D(u)| \, d\sigma(u) = \|h_C - h_D\|_1.$$

It can be expressed in the form $2S(conv(C \cup D)) - S(C) - S(D)$ for $n = 2$ (cf. **perimeter deviation**); it can be expressed also in the form $nk_n(2W(conv(C \cup D)) - W(C) - W(D))$ for $n \geqslant 2$ (cf. **mean width metric**). Here $S(.)$ is the *surface area*, k_n is the *volume* of the *unit ball* $\overline{B}^n$ of $\mathbb{E}^n$, and $W(.)$ is the *mean width*: $W(C) = \frac{1}{nk_n} \int_{S^{n-1}} (h_C(u) + h_C(-u)) \, d\sigma(u)$.

• **Sobolev distance**

The **Sobolev distance** is a metric on K, defined by

$$\|h_C - h_D\|_w,$$

where $\|.\|_w$ is the *Sobolev* 1*-norm* on the set $C_{S^{n-1}}$ of all continuous functions on the *unit sphere* S^{n-1} of $\mathbb{E}^n$.

The *Sobolev* 1*-norm* is defined by $\|f\|_w = \langle f, f \rangle_w^{1/2}$, where $\langle , \rangle_w$ is an *inner product* on $C_{S^{n-1}}$, given by

$$\langle f, g \rangle_w = \int_{S^{n-1}} \big(fg + \nabla_s(f, g)\big) \, dw_0, \quad w_0 = \frac{1}{n \cdot k_n} w,$$

$\nabla_s(f, g) = \langle grad_s\, f, grad_s\, g \rangle$, $\langle , \rangle$ is the *inner product* in $\mathbb{E}^n$, and $grad_s$ is the *gradient* on S^{n-1} (see [ArWe92]).

• **Shephard metric**

The **Shephard metric** is a metric on K_p, defined by

$$\ln\big(1 + 2 \inf\{\lambda \geqslant 0 \colon \; C \subset D + \lambda(D - D), \, D \subset C + \lambda(C - C)\}\big).$$

• **Nikodym metric**

The **Nikodym metric** (or **symmetric difference metric**) is a metric on K_p, defined by

$$V(C \triangle D),$$

where $V(.)$ is the *volume* (i.e., the Lebesgue n-dimensional measure), and $\triangle$ is the *symmetric difference*. For $n = 2$, one obtains the **area deviation**.

- **Steinhaus metric**

The **Steinhaus metric** (or **homogeneous symmetric difference metric**, **Steinhaus distance**) is a metric on K_p, defined by

$$\frac{V(C \triangle D)}{V(C \cup D)},$$

where $V(.)$ is the *volume*. So, it is $\frac{d_\triangle(C,D)}{V(C \cup D)}$, where $d_\triangle$ is the **Nikodym metric**. This metric is **bounded**; it is affine invariant, while the Nikodym metric is invariant only under volume-preserving affine transformations.

- **Eggleston distance**

The **Eggleston distance** (or **symmetric surface area deviation**) is a distance on K_p, defined by

$$S(C \cup D) - S(C \cap D),$$

where $S(.)$ is the *surface area*. The measure of surface deviation is not a metric.

- **Asplund metric**

The **Asplund metric** is a metric on the space $K_p/\approx$ of affine-equivalence classes in K_p, defined by

$$\ln \inf\{\lambda \geqslant 1 : \exists T : \mathbb{E}^n \to \mathbb{E}^n \text{ affine}, \; x \in \mathbb{E}^n, \; C \subset T(D) \subset \lambda C + x\}$$

for any equivalence classes C^* and D^* with the representatives C and D, respectively.

- **Macbeath metric**

The **Macbeath metric** is a metric on the space $K_p/\approx$ of affine-equivalence classes in K_p, defined by

$$\ln \inf\{|\det T \cdot P| : \exists T, P : \mathbb{E}^n \to \mathbb{E}^n \text{ regular affine}, \; C \subset T(D), \; D \subset P(C)\}$$

for any equivalence classes C^* and D^* with the representatives C and D, respectively.

Equivalently, it can be written as

$$\ln \delta_1(C, D) + \ln \delta_1(D, C),$$

where $\delta_1(C, D) = \inf_T\{\frac{V(T(D))}{V(C)} : C \subset T(D)\}$, and T is a regular affine mapping of $\mathbb{E}^n$ onto itself.

- **Banach–Mazur metric**

The **Banach–Mazur metric** is a metric on the space $K_{po}/\sim$ of the equivalence classes of proper 0-symmetric convex bodies with respect to linear transformations, defined by

$$\ln \inf\{\lambda \geqslant 1:\ \exists T : \mathbb{E}^n \to \mathbb{E}^n \text{ linear},\ C \subset T(D) \subset \lambda C\}$$

for any equivalence classes C^* and D^* with the representatives C and D, respectively.

It is a special case of the **Banach–Mazur distance** between n-dimensional *normed spaces*.

- **Separation distance**

The **separation distance** is the minimum Euclidean distance between two disjoint convex bodies C and D in $\mathbb{E}^n$ (in general, the **set–set distance** between any two disjoint subsets of $\mathbb{E}^n$): $\inf\{\|x - y\|_2:\ x \in C, y \in D\}$, while $\sup\{\|x - y\|_2:\ x \in C, y \in D\}$ is called the **spanning distance**.

- **Penetration depth distance**

The **penetration depth distance** between two inter-penetrating convex bodies C and D in $\mathbb{E}^n$ (in general, between any two inter-penetrating subsets of $\mathbb{E}^n$) is defined as the minimum *translation distance* that one body undergoes to make the interiors of C and D disjoint:

$$\min\{\|t\|_2:\ \mathit{interior}(C + t) \cap D = \emptyset\}.$$

This distance is a natural extension of the Euclidean **separation distance** for disjoint objects to overlapping objects. This distance can be defined by $\inf\{d(C, D + x):\ x \in \mathbb{E}^n\}$, or by $\inf_s d(C, s(D))$, where the infimum is taken over all similarities $s : \mathbb{E}^n \to \mathbb{E}^n$, or ..., where d is one of the metrics above.

- **Growth distance**

For convex polyhedra, the **growth distance** (see [GiOn96] for details) is defined as the amount objects must be grown from their internal seed points until their surfaces touch.

- **Minkowski difference**

The **Minkowski difference** on the set of all compact subsets, in particular, on the set of all *sculptured objects* (or *free form objects*), of $\mathbb{R}^3$ is defined by

$$A - B = \{x - y:\ x \in A, y \in B\}.$$

If we consider object B to be free to move with fixed orientation, the Minkowski difference is a set containing all the translations that bring B to intersect with A. The closest point from the Minkowski difference boundary, $\partial(A - B)$, to the origin gives the **separation distance** between A and B. If both objects intersect, the origin is inside the Minkowski difference, and the obtained distance can be interpreted as a **penetration distance**.

- **Maximum polygon distance**

  The **maximum polygon distance** is a distance between two convex polygons $P = (p_1, \dots, p_n)$ and $Q = (q_1, \dots, q_m)$, defined by

$$\max_{i,j} \|p_i - q_j\|_2, \; i \in \{1, \dots, n\}, \; j \in \{1, \dots, m\},$$

  where $\|.\|_2$ is the Euclidean norm.

- **Grenander distance**

  Let $P = (p_1, \dots, p_n)$ and $Q = (q_1, \dots, q_m)$ be two disjoint convex polygons, and $l(p_i, q_j), l(p_m, q_l)$ be two intersecting *critical support lines* for P and Q. Then the **Grenander distance** between P and Q is defined by

$$\|p_i - q_j\|_2 + \|p_m - q_l\|_2 - \Sigma(p_i, p_m) - \Sigma(g_j, q_l),$$

  where $\|.\|_2$ is the Euclidean norm, and $\Sigma(p_i, p_m)$ is the sum of the edges lengths of the polynomial chain $p_i, \dots, p_m$.

  Here $P = (p_1, \dots, p_n)$ is a convex polygon with the vertices in standard form, i.e., the vertices are specified according to Cartesian coordinates in a clockwise order, and no three consecutive vertices are collinear. A line l is a *line of support* of P if the interior of P lies completely to one side of l. Given two disjoint polygons P and Q, the line $l(p_i, q_j)$ is a *critical support line* if it is a line of support for P at p_i, a line of support for Q at q_j, and P and Q lie on opposite sides of $l(p_i, q_j)$.

9.2. DISTANCES ON CONES

A *convex cone* C in a real vector space V is a subset C of V such that $C + C \subset C$, $\lambda C \subset C$ for any $\lambda \geqslant 0$, and $C \cap (-C) = \{0\}$. A cone C induces a *partial order* on V by

$$x \preceq y \quad \text{if and only if} \quad y - x \in C.$$

The order $\preceq$ respects the vector structure of V, i.e., if $x \preceq y$ and $z \preceq u$, then $x + z \preceq y + u$, and if $x \preceq y$, then $\lambda x \preceq \lambda y$, $\lambda \in \mathbb{R}$, $\lambda \geqslant 0$. Elements $x, y \in V$ are called *comparable* and denoted by $x \sim y$ if there exist positive real numbers α and β such that $\alpha y \preceq x \preceq \beta y$. Comparability is an equivalence relation; its equivalence classes (which belong to C or to $-C$) are called *parts* (or *components, constituents*).

Given a convex cone C, a subset $S = \{x \in C\colon\; T(x) = 1\}$, where $T : V \to \mathbb{R}$ is some positive linear functional, is called *cross-section* of C.

A convex cone C is called *almost Archimedean* if the closure of its restriction to any two-dimensional subspace is also a cone.

- **Thompson part metric**

Given a convex cone C in a real vector space V, the **Thompson part metric** on a *part* $K \subset C \backslash \{0\}$ is defined by

$$\ln \max\{m(x, y), m(y, x)\}$$

for any $x, y \in K$, where $m(x, y) = \inf\{\lambda \in \mathbb{R}: \ y \preceq \lambda x\}$.

If C is *almost Archimedean*, then K equipped with the Thompson part metric is a **complete** metric space. If C is finite-dimensional, then one obtains a *chord space*, i.e., a metric space in which there is a distinguished set of geodesics, satisfying certain axioms. The *positive cone* $\mathbb{R}^n_+ = \{(x_1, \dots, x_n): \ x_i \geqslant 0 \text{ for } 1 \leqslant i \leqslant n\}$ equipped with the Thompson part metric is isometric to a *normed space* which one may think as being flat.

If C is a closed cone in $\mathbb{R}^n$ with non-empty interior, then *int* C can be considered as an n-dimensional manifold M^n. If for any tangent vector $v \in T_p(M^n)$, $p \in M^n$, we define a norm $\|v\|^T_p = \inf\{\alpha > 0: \ -\alpha p \preceq v \preceq \alpha p\}$, then the length of any piecewise differentiable curve $\gamma : [0, 1] \to M^n$ can be written as $l(\gamma) = \int_0^1 \|\gamma'(t)\|^T_{\gamma(t)}\, dt$, and the distance between x and y is equal to $\inf_\gamma l(\gamma)$, where the infimum is taken over all such curves γ with $\gamma(0) = x$ and $\gamma(1) = y$.

- **Hilbert projective semi-metric**

Given a convex cone C in a real vector space V, the **Hilbert projective semi-metric** is a semi-metric on $C \backslash \{0\}$, defined by

$$\ln\big(m(x, y) \cdot m(y, x)\big)$$

for any $x, y \in C\backslash\{0\}$, where $m(x, y) = \inf\{\lambda \in \mathbb{R}: \ y \preceq \lambda x\}$. It is equal to 0 if and only if $x = \lambda y$ for some $\lambda > 0$, and becomes a metric on the space of rays of the cone.

If C is finite-dimensional, and S is a *cross-section* of C (in particular, $S = \{x \in C: \|x\| = 1\}$, where $\|.\|$ is a norm on V), then, for any distinct points $x, y \in S$, the distance between them is equal to $|\ln(x, y, z, t)|$, where z, t is the points of the intersection of the line $l_{x,y}$ with the boundary of S, and (x, y, z, t) is the *cross-ratio* of x, y, z, t.

If C is *almost Archimedean* and finite-dimensional, then each part of C is a *chord space* under the Hilbert projective metric. The *Lorentz cone* $\{(t, x_1, \dots, x_n) \in \mathbb{R}^{n+1} : t^2 > x_1^2 + \dots + x_n^2\}$ equipped with the Hilbert projective metric is isometric to the n-dimensional *hyperbolic space*. The *positive cone* $\mathbb{R}^n_+ = \{(x_1, \dots, x_n): \ x_i \geqslant 0 \text{ for } 1 \leqslant i \leqslant n\}$ with the Hilbert projective metric is isometric to a *normed space* which one may think as being flat.

If C is a closed cone in $\mathbb{R}^n$ with non-empty interior, then *int* C can be considered as an n-dimensional manifold M^n. If for any tangent vector $v \in T_p(M^n)$, $p \in M^n$, we define a semi-norm $\|v\|^H_p = m(p, v) - m(v, p)$, then the length of any piecewise differentiable curve $\gamma : [0, 1] \to M^n$ can be written as $l(\gamma) = \int_0^1 \|\gamma'(t)\|^H_{\gamma(t)}\, dt$, and the distance between x and y is equal to $\inf_\gamma l(\gamma)$, where the infimum is taken over all such curves γ with $\gamma(0) = x$ and $\gamma(1) = y$.

- **Bushell metric**

Given a convex cone C in a real vector space V, the **Bushell metric** on the set $S = \{x \in C: \sum_{i=1}^{n} |x_i| = 1\}$ (in general, on any *cross-section* of C) is defined by

$$\frac{1 - m(x, y) \cdot m(y, x)}{1 + m(x, y) \cdot m(y, x)}$$

for any $x, y \in S$, where $m(x, y) = \inf\{\lambda \in \mathbb{R}: y \preceq \lambda x\}$. In fact, it is equal to $\tanh(\frac{1}{2}h(x, y))$, where h is the **Hilbert projective semi-metric**.

- **k-oriented distance**

A *simplicial cone* C in $\mathbb{R}^n$ is defined as the intersection of n (open or closed) half-spaces, each of whose supporting planes contain the origin 0. For any set X of n points on the *unit sphere*, there is a unique simplicial cone C that contains these points. The *axes* of the cone C can be constructed as the set of the n rays, where each ray originates at the origin, and contains one of the points from X.

Given a *partition* $\{C_1, \dots, C_k\}$ of $\mathbb{R}^n$ into a set of simplicial cones $C_1, \dots, C_k$, the **k-oriented distance** is a metric on $\mathbb{R}^n$, defined by

$$d_k(x - y)$$

for all $x, y \in \mathbb{R}^n$, where, for any $x \in C_i$, the value $d_k(x)$ is the length of the shortest path from the origin 0 to x traveling only in directions parallel to the axes of C_i.

- **Cone metric**

A *cone* $Con(X)$ over a metric space (X, d) is the quotient of the product $X \times [0, \infty)$ obtained by identifying all points in the fiber $X \times \{0\}$. This point is called *apex* of the cone.

The **cone metric** is a metric on $Con(X)$, defined, for any $(x, t), (y, s) \in Con(X)$, by

$$\sqrt{t^2 + s^2 - 2ts\cos(\min\{d(x, y), \pi\})}.$$

- **Suspension metric**

A *spherical cone* (or *suspension*) $\Sigma(X)$ over a metric space (X, d) is the quotient of the product $X \times [0, a]$ obtained by identifying all points in the fibers $X \times \{0\}$ and $X \times \{a\}$.

If (X, d) is a **length space** with diameter $diam(X) \leqslant \pi$, and $a = \pi$, the **suspension metric** is a metric on $\Sigma(X)$, defined, for any $(x, t), (y, s) \in \Sigma(X)$, by

$$\arccos\big(\cos t \cos s + \sin t \sin s \cos d(x, y)\big).$$

9.3. DISTANCES ON SIMPLICIAL COMPLEXES

An r-dimensional *simplex* (or *geometrical simplex*, *hypertetrahedron*) is the *convex hull* of $r + 1$ points of $\mathbb{E}^n$ which do not lie in any $(r - 1)$-plane. The simplex is so-named because it represents the simplest possible polytope in any given space. The boundary of an r-simplex has $r+1$ 0-*faces* (polytope vertices), $\frac{r(r+1)}{2}$ 1-*faces* (polytope edges), and $\binom{r+1}{i+1}$ i-*faces*, where $\binom{r}{i}$ is a binomial coefficient. The *content* (i.e., the *hypervolume*) of a simplex can be computed using the *Cayley–Menger determinant*. The regular simplex in r dimensions with is denoted by α_r.

Roughly, a *geometrical simplicial complex* is a space with a *triangulation*, i.e., a decomposition of it into closed simplices such that any two simplices either do not intersect or intersect along a common face.

An *abstract simplicial complex* S is a set, whose elements are called vertices, in which a family of finite non-empty subsets, called *simplices*, is distinguished, such that every non-empty subset of a simplex s is a simplex, called *face* of s, and every one-element subset is a simplex. A simplex is called i-dimensional if it consists of $i+1$ vertices. The *dimension* of S is the maximal dimension of its simplices. For every simplicial complex S there exists a triangulation of a polyhedron, whose simplicial complex is S. This geometric simplicial complex, denoted by GS, is called *geometric realization* of S.

- **Simplicial metric**

Let S be an abstract simplicial complex, and GS be a geometric simplicial complex which is a geometric realization of S. The points of GS can be identified with the functions $\alpha : S \to [0, 1]$ for which the set $\{x \in S: \alpha(x) \neq 0\}$ is a simplex in S, and $\sum_{x\in S}\alpha(x) = 1$. The number $\alpha(x)$ is called x-*th barycentric coordinate* of α.

The **simplicial metric** is a metric on GS, defined by

$$\sqrt{\sum_{x\in S}(\alpha(x) - \beta(x))^2}.$$

- **Polyhedral metric**

A **polyhedral metric** is the **intrinsic metric** of a connected geometric simplicial complex in $\mathbb{E}^n$ in which identified boundaries are isometric. In fact, it is defined as the infimum of the lengths of the polygonal lines joining the points x and y such that each link is within one of the simplices.

An example of a polyhedral metric is the intrinsic metric on the surface of a convex polyhedron in $\mathbb{E}^3$. A polyhedral metric can be considered on a complex of simplices in a *space of constant curvature*. In general, polyhedral metrics are considered for complexes which are *manifolds* or *manifolds with boundary*.

- **Polyhedral chain metric**

An r-dimensional *polyhedral chain* A in $\mathbb{E}^n$ is a linear expression

$$\sum_{i=1}^{m} d_i t_i^r,$$

where, for any i, the value t_i^r is an r-dimensional simplex of $\mathbb{E}^n$. The *boundary* of a chain is the linear combination of boundaries of the simplices in the chain. The boundary of an r-dimensional chain is an $(r-1)$-dimensional chain.

A **polyhedral chain metric** is a **norm metric**

$$\|A - B\|$$

on the set $C_r(\mathbb{E}^n)$ of all r-dimensional polyhedral chains. As a norm $\|.\|$ on $C_r(\mathbb{E}^n)$ one can take:

1. The *mass* of a polyhedral chain, i.e., $|A| = \sum_{i=1}^{m} |d_i||t_i^r|$, where $|t^r|$ is the volume of the cell t_i^r;
2. The *flat norm* of a polyhedral chain, i.e., $|A|^\flat = \inf_D\{|A - \partial D| + |D|\}$, where $|D|$ is the mass of D, ∂D is the boundary of D, and the infimum is taken over all $(r+1)$-dimensional polyhedral chains; the completion of the metric space $(C_r(\mathbb{E}^n), |.|^\flat)$ by the flat norm is a **separable Banach space**, denoted by $C_r^\flat(\mathbb{E}^n)$, its elements are known as r-dimensional *flat chains*;
3. The *sharp norm* of a polyhedral chain, i.e.,

$$|A|^\sharp = \inf\left(\frac{\sum_{i=1}^{m} |d_i||t_i^r||v_i|}{r+1} + \left|\sum_{i=1}^{m} d_i T_{v_i} t_i^r\right|^\flat\right),$$

   where $|A|^\flat$ is the flat norm of A, and the infimum is taken over all *shifts* v (here $T_v t^r$ is the sell obtained by shifting t^r by a vector v of length $|v|$); the completion of the metric space $(C_r(\mathbb{E}^n), |.|^\sharp)$ by the sharp norm is a separable Banach space, denoted by $C_r^\sharp(\mathbb{E}^n)$, its elements are called *r-dimensional sharp chains*. A flat chain of finite mass is a sharp chain. If $r = 0$, than $|A|^\flat = |A|^\sharp$.

The metric space of *polyhedral co-chains* (i.e., linear functions of polyhedral chains) can be defined in similar way. As a norm of a polyhedral co-chain X one can take:

1. The *co-mass* of a polyhedral co-chain, i.e., $|X| = \sup_{|A|=1} |X(A)|$, where $X(A)$ is the value of the co-chain X on a chain A;
2. The *flat co-norm* of a polyhedral co-chain, i.e., $|X|^\flat = \sup_{|A|^\flat=1} |X(A)|$;
3. The *sharp co-norm* of a polyhedral co-chain, i.e., $|X|^\sharp = \sup_{|A|^\sharp=1} |X(A)|$.

Part III

Distances in Classical Mathematics

Chapter 10. Distances in Algebra
Chapter 11. Distances on Strings and Permutations
Chapter 12. Distances on Numbers, Polynomials, and Matrices
Chapter 13. Distances in Functional Analysis
Chapter 14. Distances in Probability Theory

Chapter 10

Distances in Algebra

10.1. GROUP METRICS

A *group* $(G, \cdot, e)$ is a set G of elements with a binary operation $\cdot$, called *group operation*, that together satisfy the four fundamental properties of *closure* ($x \cdot y \in G$ for any $x, y \in G$), *associativity* ($x \cdot (y \cdot z) = (x \cdot y) \cdot z$ for any $x, y, z \in G$), the *identity property* ($x \cdot e = e \cdot x = x$ for any $x \in G$), and the *inverse property* (for any $x \in G$, there exists $x^{-1} \in G$ such that $x \cdot x^{-1} = x^{-1} \cdot x = e$). In additive notation, a group $(G, +, 0)$ is a set G with a binary operation $+$ such that the following properties hold: $x + y \in G$ for any $x, y \in G$, $x + (y + z) = (x + y) + z$ for any $x, y, z \in G$, $x + 0 = 0 + x = x$ for any $x \in G$, and, for any $x \in G$, there exists $-x \in G$ such that $x + (-x) = (-x) + x = 0$. A group $(G, \cdot, e)$ is called *finite* if the set G is finite. A group $(G, \cdot, e)$ is called *Abelian* if it is *commutative*, i.e., $x \cdot y = y \cdot x$ holds for any $x, y \in G$.

Most of metrics, considered in this section, are **group norm metrics** on a group $(G, \cdot, e)$, defined by

$$\|x \cdot y^{-1}\|$$

(or, sometimes, by $\|y^{-1} \cdot x\|$), where $\|.\|$ is a *group norm*, i.e., a function $\|.\| : G \to \mathbb{R}$ such that, for any $x, y \in G$, we have the following properties:

1. $\|x\| \geqslant 0$, with $\|x\| = 0$ if and only if $x = e$;
2. $\|x\| = \|x^{-1}\|$;
3. $\|x \cdot y\| \leqslant \|x\| + \|y\|$ (*triangle inequality*).

In additive notation, a group norm metric on a group $(G, +, 0)$ is defined by $\|x + (-y)\| = \|x - y\|$, or, sometimes, by $\|(-y) + x\|$.

The simplest example of a group norm metric is the **bi-invariant ultrametric** (sometimes called *Hamming metric*) $\|x \cdot y^{-1}\|_H$, where $\|x\|_H = 1$ for $x \neq e$, and $\|e\|_H = 0$.

- **Bi-invariant metric**

  A metric (in general, a semi-metric) d on a group $(G, \cdot, e)$ is called **bi-invariant** if

  $$d(x, y) = d(x \cdot z, y \cdot z) = d(z \cdot x, z \cdot y)$$

  holds for any $x, y, z \in G$ (cf. **translation invariant metric**). Any **group norm metric** on an Abelian group is bi-invariant.

  A metric (in general, a semi-metric) d on a group $(G, \cdot, e)$ is called **right-invariant metric** if $d(x, y) = d(x \cdot z, y \cdot z)$ holds for any $x, y, z \in G$, i.e., the operation of right multiplication by an element z is a **motion** of the metric space (G, d). Any group norm metric, defined by $\|x \cdot y^{-1}\|$, is right-invariant.

A metric (in general, a semi-metric) d on a group $(G, \cdot, e)$ is called **left-invariant metric** if $d(x, y) = d(z \cdot x, z \cdot y)$ holds for any $x, y, z \in G$, i.e., the operation of left multiplication by an element z is a motion of the metric space (G, d). Any group norm metric, defined by $\|y^{-1} \cdot x\|$, is left-invariant.

Any right-invariant, as well as any left-invariant, in particular, bi-invariant, metric d on G is a group norm metric, since one can define a group norm on G by $\|x\| = d(x, 0)$.

- **Positively homogeneous metric**

A metric (in general, a distance) d on an Abelian group $(G, +, 0)$ is called **positively homogeneous** if

$$d(mx, my) = md(x, y)$$

holds for all $x, y \in G$ and for all $m \in \mathbb{N}$, where mx is the sum of m terms all equal to x.

- **Translation discrete metric**

A **group norm metric** (in general, a group norm semi-metric) on a group $(G, \cdot, e)$ is called **translation discrete** if the *translation distances* (or *translation numbers*)

$$\tau_G(x) = \lim_{n \to \infty} \frac{\|x^n\|}{n}$$

of the *non-torsion elements* x (i.e., such that $x^n \neq e$ for any $n \in \mathbb{N}$) of the group with respect to that metric are bounded away from zero.

If the numbers $\tau_G(x)$ are just non-zero, such group norm metric is called **translation proper metric**.

- **Word metric**

Let $(G, \cdot, e)$ be a finitely-generated group with a set A of generators. The *word length* $w_W^A(x)$ of an element $x \in G \backslash \{e\}$ is defined by

$$w_W^A(x) = \inf\{r : \ x = a_1^{\varepsilon_1} \cdots a_r^{\varepsilon_r}, \ a_i \in A, \ \varepsilon_i \in \{\pm 1\}\},$$

and $w_W^A(e) = 0$.

The **word metric** d_W^A *associated with* A is a **group norm metric** on G, defined by

$$w_W^A(x \cdot y^{-1}).$$

As the word length w_W^A is a *group norm* on G, then d_W^A is **right-invariant**. Sometimes it is defined as $w_W^A(y^{-1} \cdot x)$, and then it is **left-invariant**. In fact, d_W^A is the maximal metric on G that is right-invariant, and such that the distance of any element of A or A^{-1} to the identity element e is equal to one.

If A and B are two finite sets of generators of the group $(G, \cdot, e)$, then the identity mapping between the metric spaces (G, d_W^A) and (G, d_W^B) is a **quasi-isometry**, i.e., the word metric is unique up to quasi-isometry.

The word metric is the **path metric** of the *Cayley graph* Γ of $(G, \cdot, e)$, constructed with respect to A. Namely, Γ is a graph with the vertex-set G in which two vertices x and $y \in G$ are connected by an edge if and only if $y = a^{\varepsilon}x$, $\varepsilon = \pm 1$, $a \in A$.

- **Weighted word metric**

Let $(G, \cdot, e)$ be a finitely-generated group with a set A of generators. Given a bounded *weight function* $w : A \to (0, \infty)$, the *weighted word length* $w_{WW}^A(x)$ of an element $x \in G \backslash \{e\}$ is defined by

$$w_{WW}^A(x) = \inf\left\{\sum_{i=1}^{t} w(a_i),\ t \in \mathbb{N}:\ x = a_1^{\varepsilon_1} \dots a_t^{\varepsilon_t},\ a_i \in A,\ \varepsilon_i \in \{\pm 1\}\right\},$$

and $w_{WW}^A(e) = 0$.

The **weighted word metric** d_{WW}^A *associated with* A is a **group norm metric** on G, defined by

$$w_{WW}^A(x \cdot y^{-1}).$$

As the weighted word length w_{WW}^A is a *group norm* on G, then d_{WW}^A is **right-invariant**. Sometimes it is defined as $w_{WW}^A(y^{-1} \cdot x)$, and then it is **left-invariant**.

The metric d_{WW}^A is the supremum of semi-metrics d on G with the property that $d(e, a) \leqslant w(a)$ for any $a \in A$.

The metric d_{WW}^A is a **coarse-path metric**, and every right-invariant coarse path metric is a weighted word metric up to **coarse isometry**.

The metric d_{WW}^A is the **path metric** of the *weighted Cayley graph* Γ_W of $(G, \cdot, e)$ constructed with respect to A. Namely, Γ_W is a weighted graph with the vertex-set G in which two vertices x and $y \in G$ are connected by an edge with the weight $w(a)$ if and only if $y = a^{\varepsilon}x$, $\varepsilon = \pm 1$, $a \in A$.

- **Interval norm metric**

An **interval norm metric** is a **group norm metric** on a finite group $(G, \cdot, e)$, defined by

$$\|x \cdot y^{-1}\|_{int},$$

where $\|.\|_{int}$ is an *interval norm* on G, i.e., a *group norm* such that the values of $\|.\|_{int}$ form a set of consecutive integers starting with 0.

To each interval norm $\|.\|_{int}$ corresponds an ordered *partition* $\{B_0, \dots, B_m\}$ of G with $B_i = \{x \in G:\ \|x\|_{int} = i\}$ (cf. **Sharma–Kaushik distance**). The *Hamming norm* and the *Lee norm* are special cases of interval norms. A *generalized Lee norm* is an interval norm for which each class has a form $B_i = \{a, a^{-1}\}$.

- **C-metric**

A **C-metric** d is a metric on a group $(G, \cdot, e)$, satisfying the following conditions:

1. The values of d form a set of consecutive integers starting with 0;
2. The cardinality of the sphere $S(x, r) = \{y \in G: d(x, y) = r\}$ is independent of the particular choice of $x \in G$.

The **word metric**, the **Hamming metric**, and the **Lee metric** are C-metrics. Any **interval norm metric** is a C-metric.

- **Order norm metric**

Let $(G, \cdot, e)$ be a finite Abelian group. Let $ord(x)$ is the *order* of an element $x \in G$, i.e., the smallest positive integer n such that $x^n = e$. Then the function $\|.\|_{ord} : G \to \mathbb{R}$, defined by $\|x\|_{ord} = \ln ord(x)$, is a *group norm* on G, called *order norm.*

The **order norm metric** is a **group norm metric** on G, defined by

$$\|x \cdot y^{-1}\|_{ord}.$$

- **Monomorphism norm metric**

Let $(G, +, 0)$ be a group. Let $(H, \cdot, e)$ be a group with a *group norm* $\|.\|_H$. Let $f : G \to H$ be a *monomorphism* of groups G and H, i.e., an injective function such that $f(x + y) = f(x) \cdot f(y)$ for any $x, y \in G$. Then the function $\|.\|_G^f : G \to \mathbb{R}$, defined by $\|x\|_G^f = \|f(x)\|_H$, is a *group norm* on G, called *monomorphism norm.*

The **monomorphism norm metric** is a **group norm metric** on G, defined by

$$\|x - y\|_G^f.$$

- **Product norm metric**

Let $(G, +, 0)$ be a group with a *group norm* $\|.\|_G$. Let $(H, \cdot, e)$ be a group with a group norm $\|.\|_H$. Let $G \times H = \{\alpha = (x, y): x \in G, y \in H\}$ be the Cartesian product of G and H, and $(x, y) \cdot (z, t) = (x + z, y \cdot t)$. Then the function $\|.\|_{G\times H} : G \times H \to \mathbb{R}$, defined by $\|\alpha\|_{G\times H} = \|(x, y)\|_{G\times H} = \|x\|_G + \|y\|_H$, is a group norm on $G \times H$, called *product norm.*

The **product norm metric** is a **group norm metric** on $G \times H$, defined by

$$\|\alpha \cdot \beta^{-1}\|_{G\times F}.$$

On the Cartesian product $G \times H$ of two finite groups with the *interval norms* $\|.\|_G^{int}$ and $\|.\|_H^{int}$ an interval norm $\|.\|_{G\times H}^{int}$ can be defined. In fact, $\|\alpha\|_{G\times H}^{int} = \|(x, y)\|_{G\times H}^{int} = \|x\|_G + (m + 1)\|y\|_H$, where $m = \max_{a\in G} \|a\|_G^{int}$.

- **Quotient norm metric**

Let $(G, \cdot, e)$ be a group with a *group norm* $\|.\|_G$. Let $(N, \cdot, e)$ be a *normal subgroup* of $(G, \cdot, e)$, i.e., $xN = Nx$ for any $x \in G$. Let $(G/N, \cdot, eN)$ be the *quotient group* of G, i.e., $G/N = \{xN\colon\ x \in G\}$ with $xN = \{x \cdot a\colon\ a \in N\}$, and $xN \cdot yN = xyN$. Then the function $\|.\|_{G/N} : G/N \to \mathbb{R}$, defined by $\|xN\|_{G/N} = \min_{a\in N} \|xa\|_X$, is a *group norm* on G/N, called *quotient norm*.

A **quotient norm metric** is a **group norm metric** on G/N, defined by

$$\left\|xN \cdot (yN)^{-1}\right\|_{G/N} = \left\|xy^{-1}N\right\|_{G/N}.$$

If $G = \mathbb{Z}$ with the norm being the absolute value, and $N = m\mathbb{Z}$, $m \in \mathbb{N}$, then the quotient norm on $\mathbb{Z}/m\mathbb{Z} = \mathbb{Z}_m$ coincides with the *Lee norm*.

If a metric d on a group $(G, \cdot, e)$ is **right-invariant**, then for any normal subgroup $(N, \cdot, e)$ of $(G, \cdot, e)$ the metric d induced a right-invariant metric (in fact, the **Hausdorff metric**) d^* on G/N by

$$d^*(xN, yN) = \max\left\{\max_{b\in yN}\min_{a\in xN} d(a, b),\ \max_{a\in xN}\min_{b\in yN} d(a, b)\right\}.$$

- **Commutation distance**

Let $(G, \cdot, e)$ be a finite non-Abelian group. Let $Z(G) = \{c \in G\colon\ x\cdot c = c\cdot x$ for any $x \in G\}$ be the *center* of G. The *commutation graph* of G is defined as a graph with the vertex-set G in which distinct elements $x, y \in G$ are connected by an edge whenever they *commute*, i.e., $x \cdot y = y \cdot x$. Obviously, any two distinct elements $x, y \in G$ that are not commute, are connected in this graph by the path x, c, y, where c is any element of $Z(G)$ (for example, e). A path $x = x^1, x^2, \dots, x^k = y$ in the commutation graph is called $(x - y)$ *N-path* if $x^i \notin Z(G)$ for any $i \in \{1, \dots, k\}$. In this case elements $x, y \in G\backslash Z(G)$ are called *N-connected*.

The **commutation distance** (see [DeHu98]) d is an extended distance on G, defined by the following conditions:

1. $d(x, x) = 0$;
2. $d(x, y) = 1$ if $x \neq y$, and $x \cdot y = y \cdot x$;
3. $d(x, y)$ is the minimum length of an $(x - y)$ N-path for any N-connected elements x and $y \in G\backslash Z(G)$;
4. $d(x, y) = \infty$ if $x, y \in G\backslash Z(G)$ are not connected by any N-path.

- **Modular distance**

Let $(\mathbb{Z}_m, +, 0)$, $m \geqslant 2$, be a finite *cyclic group*. Let $r \in \mathbb{N}$, $r \geqslant 2$. The *modular r-weight* $w_r(x)$ of an element $x \in \mathbb{Z}_m = \{0, 1, \dots, m\}$ is defined as $w_r(x) = \min\{w_r(x), w_r(m - x)\}$, where $w_r(x)$ is the *arithmetic r-weight* of the integer x. The value $w_r(x)$ can be obtained as the number of non-zero coefficients in the *generalized non-adjacent form* $x = e_n r^n + \cdots e_1 r + e_0$ with $e_i \in \mathbb{Z}$, $|e_i| < r$, $|e_i + e_{i+1}| < r$, and $|e_i| < |e_{i+1}|$ if $e_i e_{i+1} < 0$ (cf. **arithmetic r-norm metric**).

The **modular distance** is a distance on $\mathbb{Z}_m$, defined by

$$w_r(x - y).$$

The modular distance is a metric for $w_r(m) = 1$, $w_r(m) = 2$, and for several special cases with $w_r(m) = 3$ or 4. In particular, it is a metric for $m = r^n$ or $m = r^n - 1$; if $r = 2$, it is a metric also for $m = 2^n + 1$ (see, for example, [Ernv85]).

The most popular metric on $\mathbb{Z}_m$ is the **Lee metric**, defined by $\|x-y\|_{Lee}$, where $\|x\|_{Lee} = \min\{x, m-x\}$ is the *Lee norm* of an element $x \in \mathbb{Z}_m$.

- **G-norm metric**

  Consider a finite field $\mathbb{F}_{p^n}$ for a prime p and a natural number n. Given a compact convex centrally-symmetric body G in $\mathbb{R}^n$, define the *G-norm* of an element $x \in \mathbb{F}_{p^n}$ by $\|x\|_G = \inf\{\mu \geqslant 0\colon\ x \in p\mathbb{Z}^n + \mu G\}$.

  The **G-norm metric** is a **group norm metric** on $\mathbb{F}_{p^n}$, defined by

$$\|x \cdot y^{-1}\|_G.$$

- **Permutation norm metric**

  Given a finite metric space (X, d), the **permutation norm metric** is a **group norm metric** on the group $(Sym_X, \cdot, id)$ of all permutations of X (id is the *identity mapping*), defined by

$$\|f \cdot g^{-1}\|_{Sym},$$

  where the *group norm* $\|.\|_{Sym}$ on Sym_X is given by $\|f\|_{Sym} = \max_{x \in X} d(x, f(x))$.

- **Metric of motions**

  Let (X, d) be a metric space, and let $p \in X$ be a fixed element of X.

  The **metric of motions** (see [Buse55]) is a metric on the group $(\Omega, \cdot, id)$ of all **motions** of X (id is the *identity mapping*), defined by

$$\sup_{x \in X} d\big(f(x), g(x)\big) \cdot e^{-d(p,x)}$$

  for any $f, g \in \Omega$ (cf. **Busemann metric of sets**). If the space (X, d) is bounded, the similar metric on Ω can be defined as

$$\sup_{x \in X} d\big(f(x), g(x)\big).$$

  Given a semi-metric space (X, d), the **semi-metric of motions** on $(\Omega, \cdot, id)$ is defined by

$$d\big(f(p), g(p)\big).$$

- **General linear group semi-metric**

  Let $\mathbb{F}$ be a locally compact non-discrete *topological field*. Let $(\mathbb{F}^n, \|.\|_{\mathbb{F}^n})$, $n \geqslant 2$, be a *normed vector space* over $\mathbb{F}$. Let $\|.\|$ be the *operator norm* associated with the normed vector space $(\mathbb{F}^n, \|.\|_{\mathbb{F}^n})$. Let $GL(n, \mathbb{F})$ be the *general linear group* over $\mathbb{F}$. Then the function $|.|_{op} : GL(n, \mathbb{F}) \to \mathbb{R}$, defined by $|g|_{op} = \sup\{|\ln \|g\||, |\ln \|g^{-1}\||\}$, is a semi-norm on $GL(n, \mathbb{F})$.

The **general linear group semi-metric** is a semi-metric on the group $GL(n, \mathbb{F})$, defined by

$$|g \cdot h^{-1}|_{op}.$$

It is a **right-invariant** semi-metric which is unique, up to **coarse isometry**, since any two norms on $\mathbb{F}^n$ are **bi-Lipschitz equivalent**.

- **Generalized torus semi-metric**

Let $(T, \cdot, e)$ be a *generalized torus*, i.e., a *topological group* which is isomorphic to a direct product of n multiplicative groups $\mathbb{F}_i^*$ of locally compact non-discrete *topological fields* $\mathbb{F}_i$; then there is a proper continuous homomorphism $v : T \to \mathbb{R}^n$, namely, $v(x_1, \dots, x_n) = (v_1(x_1), \dots, v_n(x_n))$, where $v_i : \mathbb{F}_i^* \to \mathbb{R}$ are proper continuous homomorphisms from the $\mathbb{F}_i^*$ to the additive group $\mathbb{R}$, given by the logarithm of the *valuation*. Every other proper continuous homomorphism $v' : T \to \mathbb{R}^n$ is of the form $v' = \alpha \cdot v$ with $\alpha \in GL(n, \mathbb{R})$. If $\|.\|$ is a norm on $\mathbb{R}^n$, one obtains the corresponding semi-norm $\|x\|_T = \|v(x)\|$ on T.

The **generalized torus semi-metric** is a semi-metric on the group $(T, \cdot, e)$, defined by

$$\|xy^{-1}\|_T = \|v(xy^{-1})\| = \|v(x) - v(y)\|.$$

- **Heisenberg metric**

Let $(H, \cdot, e)$ be the first *Heisenberg group*, i.e., a group on the set $H = \mathbb{C} \otimes \mathbb{R}$ with the group law $x \cdot y = (z, t) \cdot (u, s) = (z+u, t+s+2\Im(z\bar{u}))$, and the identity $e = (0, 0)$. Let $|.|_{Heis}$ be the *Heisenberg norm* on H, defined by $|x|_{Heis} = |(z, t)|_{Heis} = (|z|^4 + t^2)^{1/4}$.

The **Heisenberg metric** (or **gauge metric**, **Korányi metric**) d_{Heis} is a **group norm metric** on H, defined by

$$|x^{-1} \cdot y|_H.$$

The second natural metric on $(H, \cdot, e)$ is the **Carnot–Carathéodory metric** (or *C-C metric*, *control metric*) d_C, defined as the **intrinsic metric** using horizontal vector fields on H. The metrics d_{Heis} and d_C are **bi-Lipschitz equivalent**; in fact, $\frac{1}{\sqrt{\pi}} d_{Heis}(x, y) \leqslant d_C(x, y) \leqslant d_{Heis}(x, y)$.

The Heisenberg metric can be defined, in a similar manner, on any Heisenberg group $(H^n, \cdot, e)$ with $H^n = \mathbb{C}^n \otimes \mathbb{R}$.

- **Metric between intervals**

Let G be the set of all intervals $[a, b]$ of $\mathbb{R}$. The set G forms semi-groups $(G, +)$ and $(G, \cdot)$ under addition $I + J = \{x + y : x \in I, y \in J\}$ and under multiplication $I \cdot J = \{x \cdot y : x \in I, y \in J\}$, respectively.

The **metric between intervals** is a metric on G, defined by

$$\max\{|I|, |J|\}$$

for all $I, J \in G$, where, for $I = [a, b]$, one has $|I| = |a - b|$.

- **Ring semi-metric**

  Let $(A, +, \cdot)$ be a *factorial ring*, i.e., a ring with unique factorization.

  The **ring semi-metric** is a semi-metric on the set $A \backslash \{0\}$, defined by

$$\ln \frac{l.c.m.(x, y)}{g.c.d.(x, y)},$$

  where $l.c.m.(x, y)$ is the *least common multiple*, and $g.c.d.(x, y)$ is the *greatest common divisor* of elements $x, y \in A \backslash \{0\}$.

10.2. METRICS ON BINARY RELATIONS

A *binary relation* R on a set X is a subset of $X \times X$; it is the arc-set of the directed graph (X, R) with the vertex-set X.

A binary relation R which is *symmetric* ($(x, y) \in R$ implies $(y, x) \in R$), *reflexive* (all $(x, x) \in R$), and *transitive* ($(x, y), (y, z) \in R$ imply $(x, z) \in R$) is called *equivalence relation* or a *partition* (of X into equivalence classes). Any *q-ary sequence* $x = (x_1, \dots, x_n)$, $q \geqslant 2$ (i.e., with $0 \leqslant x_i \leqslant q - 1$ for $1 \leqslant i \leqslant n$), corresponds to the partition $\{B_0, \dots, B_{q-1}\}$ of $V_n = \{1, \dots, n\}$, where $B_j = \{1 \leqslant i \leqslant n\colon x_i = j\}$ are the equivalence classes.

A binary relation R which is *antisymmetric* ($(x, y), (y, x) \in R$ imply $x = y$), reflexive, and transitive is called *partial order*, and the pair (X, R) is called *poset* (partially ordered set). A partial order R on X is denoted also by $\preceq$ with $x \preceq y$ if and only if $(x, y) \in R$. The order $\preceq$ is called *linear* if any two elements $x, y \in X$ are *compatible*, i.e., $x \preceq y$ or $y \preceq x$.

A poset $(L, \preceq)$ is called *lattice* if every two elements $x, y \in L$ have the *join* $x \vee y$ and the *meet* $x \wedge y$. All partitions of X form a lattice by refinement; it is a sublattice of the lattice (by set-inclusion) of all binary relations.

- **Kemeny distance**

  The **Kemeny distance** between binary relations R_1 and R_2 on a set X is the **Hamming metric** $|R_1 \triangle R_2|$. The Kemeny distance is twice the minimal number of inversions of pairs of adjacent elements of X which is necessary to obtain R_2 from R_1.

  If R_1, R_2 are *partitions*, then the Kemeny distance coincides with the **Mirkin–Tcherny distance**, and

$$1 - \frac{|R_1 \triangle R_2|}{n(n-1)}$$

  is the *Rand index*.

The **Drapal–Kepka distance** between distinct quasigroups $(X, +)$ and $(X, \cdot)$ is defined by $|\{(x, y) : x + y \neq x \cdot y\}|$.

If binary relations R_1, R_2 are *linear orders* (or *rankings, permutations*) on the set X, then the Kemeny distance coincides with the **inversion metric** on permutations.

- **Metrics between partitions**

Let X be a finite set of cardinality $n = |X|$, and let A, B be non-empty subsets of X. Let P_X be the set of partitions of X, and $P, Q \in P_X$. Let $B_1, \dots, B_q$ are *blocks* in the partition P, i.e., the pairwise disjoint sets such that $X = B_1 \cup \dots \cup B_q$, $q \geqslant 2$. Let $P \vee Q$ be the *join* of P and Q, and $P \wedge Q$ be the *meet* of P and Q in the *lattice* $\mathbb{P}_X$ *of partitions* of X.

Consider the following *editing operations* on partitions:

- An *augmentation* transforms a partition P of $A \backslash \{B\}$ into a partition of A by either including the objects of B in a block, or including B itself as a new block;
- An *removal* transforms a partition P of A into a partition of $A \backslash \{B\}$ by deleting the objects in B from each block that contains them;
- A *division* transforms one partition P into another by the simultaneous removal of B from B_i (where $B \subset B_i$, $B \neq B_i$), and augmentation of B as a new block;
- A *merging* transforms one partition P into another by the simultaneous removal of B from B_i (where $B = B_i$), and augmentation of B to B_j (where $j \neq i$);
- A *transfer* transforms one partition P into another by the simultaneous removal of B from B_i (where $B \subset B_i$), and augmentation of B to B_j (where $j \neq i$).

Define (see, for example, [Day81]), in terms of above operations, the following **editing metrics** on P_X:

1. The minimum number of augmentations and removals of single objects needed to transform P into Q;
2. The minimum number of divisions, mergings, and transfers of single objects needed to transform P into Q;
3. The minimum number of divisions, mergings, and transfers needed to transform P into Q;
4. The minimum number of divisions and mergings needed to transform P into Q; in fact, it is equal to $|P| + |Q| - 2|P \vee Q|$;
5. $\sigma(P) + \sigma(Q) - 2\sigma(P \wedge Q)$, where $\sigma(P) = \sum_{P_i \in P} |P_i|(|P_i| - 1)$;
6. $e(P) + e(Q) - 2e(P \wedge Q)$, where $e(P) = \log_2 n + \sum_{P_i \in P} \frac{|P_i|}{n} \log_2 \frac{|P_i|}{n}$.

The **Reignier distance** is the minimum number of elements that must be moved between the blocks of partition P in order to transform it into Q. (Cf. **Earth Mover distance** and above metric 2.)

10.3. LATTICE METRICS

Consider a poset $(L, \preceq)$. The *meet* (or *infimum*) $x \wedge y$ (if it exists) of two elements x and y is the unique element satisfying $x \wedge y \preceq x, y$, and $z \preceq x \wedge y$ if $z \preceq x, y$; similarly, the *join* (or *supremum*) $x \vee y$ (if it exists) is the unique element such that $x, y \preceq x \vee y$, and $x \vee y \preceq z$ if $x, y \preceq z$.

A poset $(L, \preceq)$ is called *lattice* if every two elements $x, y \in L$ have the join $x \vee y$ and the meet $x \wedge y$. A poset $(L, \preceq)$ is called *meet semi-lattice* (or *lower semi-lattice*) if only meet-operation is defined. A poset $(L, \preceq)$ is called *join semi-lattice* (or *upper semi-lattice*) if only join-operation is defined.

A lattice $\mathbb{L} = (L, \preceq, \vee, \wedge)$ is called *semi-modular lattice* (or *semi-Dedekind lattice*) if the *modularity relation* xMy is symmetric: xMy implies yMx for any $x, y \in L$. The *modularity relation* here is defined as follows: two elements x and y are said to constitute a *modular pair*, in symbols xMy, if $x \wedge (y \vee z) = (x \wedge y) \vee z$ for any $z \preceq x$. A lattice $\mathbb{L}$ in which every pair of elements is modular, is called *modular lattice* (or *Dedekind lattice*). A lattice is modular if and only if the *modular law* is valid: if $z \preceq x$, then $x \wedge (y \vee z) = (x \wedge y) \vee z$ for any y. A lattice is called *distributive* if $x \wedge (y \vee z) = (x \wedge y) \vee (x \wedge z)$ holds for any $x, y, z \in L$.

Given a lattice $\mathbb{L}$, a function $v : L \to \mathbb{R}_{\geqslant 0}$, satisfying $v(x \vee y) + v(x \wedge y) \leqslant v(x) + v(y)$ for all $x, y \in L$, is called *subvaluation* on $\mathbb{L}$. A subvaluation v is called *isotone* if $v(x) \leqslant v(y)$ whenever $x \preceq y$, and it is called *positive* if $v(x) < v(y)$ whenever $x \preceq y, x \neq y$.

A subvaluation v is called *valuation* if it is isotone and $v(x \vee y) + v(x \wedge y) = v(x) + v(y)$ holds for all $x, y \in L$. Integer-valued valuation is called *height* (or *length*) of $\mathbb{L}$.

- **Lattice valuation metric**

  Let $\mathbb{L} = (L, \preceq, \vee, \wedge)$ be a lattice, and let v be an isotone subvaluation on $\mathbb{L}$. The *lattice subvaluation semi-metric* d_v on L is defined by

$$2v(x \vee y) - v(x) - v(y).$$

  (It can be defined also on some semi-lattices.) If v is a positive subvaluation on L, one obtains a metric, called **lattice subvaluation metric**. If v is a valuation, d_v can be written as

$$v(x \vee y) - v(x \wedge y) = v(x) + v(y) - 2v(x \wedge y),$$

  and is called *valuation semi-metric*. If v is a positive valuation on L, one obtains a metric, called **lattice valuation metric**.

  If $L = \mathbb{N}$ (the set of positive integers), $x \vee y = l.c.m.(x, y)$ (least common multiple), $x \wedge y = g.c.d.(x, y)$ (greatest common divisor), and the positive valuation $v(x) = \ln x$, then $d_v(x, y) = \ln \frac{l.c.m.(x,y)}{g.c.d(x,y)}$. This metric can be generalized on any *factorial* ring (i.e., a ring with unique factorization) equipped with a positive valuation v such that $v(x) \geqslant 0$ with equality only for the multiplicative unit of the ring, and $v(xy) = v(x) + v(y)$.

- **Finite subgroup metric**

  Let $(G, \cdot, e)$ be a group. Let $\mathbb{L} = (L, \subset, \cap)$ be the meet semi-lattice of all finite subgroups of the group $(G, \cdot, e)$ with the meet $X \cap Y$ and the valuation $v(X) = \ln |X|$.

  The **finite subgroup metric** is a **valuation metric** on L, defined by

$$v(X) + v(Y) - 2v(X \wedge Y) = \ln \frac{|X||Y|}{(|X \cap Y|)^2}.$$

- **Scalar and vectorial metrics**

  Let $\mathbb{L} = (L, \leqslant, \max, \min)$ be a lattice with the join $\max\{x, y\}$, and the meet $\min\{x, y\}$ on a set $L \subset [0, \infty)$ which has a fixed number a as the greatest element and is closed under *negation*, i.e., for any $x \in L$, one has $\overline{x} = a - x \in L$.

  The **scalar metric** d on L is defined, for $x \neq y$, by

$$d(x, y) = \max\bigl\{\min\{x, \overline{y}\}, \min\{\overline{x}, y\}\bigr\}.$$

  The **scalar metric** d^* on $L^* = L \cup \{*\}$, $* \notin L$, is defined, for $x \neq y$, by

$$d^*(x, y) = \begin{cases} d(x, y), & \text{if } x, y \in L, \\ \max\{x, \overline{x}\}, & \text{if } y = *, x \neq *, \\ \max\{y, \overline{y}\}, & \text{if } x = *, y \neq *. \end{cases}$$

  Given a norm $\|.\|$ on $\mathbb{R}^n$, $n \geqslant 2$, the **vectorial metric** on L^n is defined by

$$\big\|\bigl(d(x_1, y_1), \dots, d(x_n, y_n)\bigr)\big\|,$$

  and the **vectorial metric** on $(L^*)^n$ is defined by

$$\big\|\bigl(d^*(x_1, y_1), \dots, d^*(x_n, y_n)\bigr)\big\|.$$

  The vectorial metric on $L_2^n = \{0, 1\}^n$ with l_1-*norm* on $\mathbb{R}^n$ is the **Fréchet–Nikodym–Aronszyan distance**. The vectorial metric on $L_m^n = \{0, \frac{1}{m-1}, \dots, \frac{m-2}{m-1}, 1\}^n$ with l_1-*norm* on $\mathbb{R}^n$ is the **Sgarro m-valued metric**. The vectorial metric on $[0, 1]^n$ with l_1-*norm* on $\mathbb{R}^n$ is the **Sgarro fuzzy metric**. If L is L_m or $[0, 1]$, and $x = (x_1, \dots, x_n, x_{n+1}, \dots, x_{n+r})$, $y = (y_1, \dots, y_n, *, \dots, *)$, where $*$ stands in r places, then the vectorial metric between x and y is the **Sgarro metric** (see, for example, [CSY01]).

- **Metrics on Riesz space**

  A *Riesz space* (or *vector lattice*) is a partially ordered vector space $(V_{Ri}, \preceq)$ in which the following conditions hold:

  1. The vector space structure and the partial order structure are compatible: from $x \preceq y$ follows that $x + z \preceq y + z$, and from $x \succ 0$, $\lambda \in \mathbb{R}$, $\lambda > 0$ follows that $\lambda x \succ 0$;

2. For any two elements $x, y \in V_{Ri}$ there exists join $x \vee y \in V_{Ri}$ (in particular, the join and the meet of any finite set exist).

The **Riesz norm metric** is a **norm metric** on V_{Ri}, defined by

$$\|x - y\|_{Ri},$$

where $\|.\|_{Ri}$ is a *Riesz norm*, i.e., a *norm* on V_{Ri} such that, for any $x, y \in V_{Ri}$, the inequality $|x| \leqslant |y|$, where $|x| = (-x) \vee (x)$, implies $\|x\|_{Ri} \leqslant \|y\|_{Ri}$. The space $(V_{Ri}, \|.\|_{Ri})$ is called *normed Riesz space*. In the case of completeness it is called *Banach lattice*. All Riesz norms on a Banach lattice are equivalent.

An element $e \in V_{Ri}^+ = \{x \in V_{Ri}\colon\ x \succ 0\}$ is called *strong unit* of V_{Ri} if for each $x \in V_{Ri}$ there exists $\lambda \in \mathbb{R}$ such that $|x| \preceq \lambda e$. If a Riesz space V_{Ri} has a *strong unit* e, then $\|x\| = \inf\{\lambda \in \mathbb{R}\colon\ |x| \preceq \lambda e\}$ is a Riesz norm, and one obtains on V_{Ri} a Riesz norm metric

$$\inf\bigl\{\lambda \in \mathbb{R}\colon\ |x - y| \preceq \lambda e\bigr\}.$$

A *weak unit* of V_{Ri} is an element e of V_{Ri}^+ such that $e \wedge |x| = 0$ implies $x = 0$. A Riesz space V_{Ri} is called *Archimedean* if, for any two $x, y \in V_{Ri}^+$, there exists a natural number n, such that $nx \preceq y$. The **uniform metric** on an Archimedean Riesz space with a weak unit e is defined by

$$\inf\bigl\{\lambda \in \mathbb{R}\colon\ |x - y| \wedge e \preceq \lambda e\bigr\}.$$

- **Gallery distance of flags**

Let $\mathbb{L}$ be a lattice. A *chain* C in $\mathbb{L}$ is a subset of L which is *linearly ordered*, i.e., any two elements of C are compatible. A *flag* is a chain in $\mathbb{L}$ which is maximal with respect to inclusion. If $\mathbb{L}$ is a semi-modular lattice, containing a finite flag, then $\mathbb{L}$ has an unique minimal and an unique maximal element, and any two flags C, D in $\mathbb{L}$ have the same cardinality, $n + 1$. Then n is the height of the lattice $\mathbb{L}$. Two flags C, D in $\mathbb{L}$ are called *adjacent* if they are equal or D contains exactly one element not in C. A *gallery* from C to D of length m is a sequence of flags $C = C_0, C_1, \dots, C_m = D$ such that C_{i-1} and C_i are adjacent for $i = 1, \dots, m$.

A **gallery distance of flags** (see [Abel91]) is a distance on the set of all flags of a semi-modular lattice $\mathbb{L}$ with finite height, defined as the minimum of lengths of galleries from C to D. It can be written as

$$|C \vee D| - |C| = |C \vee D| - |D|,$$

where $C \vee D = \{c \vee d\colon\ c \in C,\ d \in D\}$ is the upper sub-semi-lattice generated by C and D.

The gallery distance of flags is a special case of the **gallery metric** (of the *chamber system* consisting of flags).

Chapter 11

Distances on Strings and Permutations

An *alphabet* is a finite set $\mathcal{A}$, $|\mathcal{A}| \geqslant 2$, elements of which are called *characters* (or *symbols*). A *string* (or *word*) is a sequence of characters over a given finite alphabet $\mathcal{A}$. The set of all finite strings over the alphabet $\mathcal{A}$ is denoted by $W(\mathcal{A})$. The strings below are finite except for **Baire**, **Duncan**, and **Fréchet permutation metrics**.

A *substring* (or *factor*, *chain*, *block*) of the string $x = x_1 \dots x_n$ is any its contiguous subsequence $x_i x_{i+1} \dots x_k$ with $1 \leqslant i \leqslant k \leqslant n$. A *prefix* of a string $x_1 \dots x_n$ is any its substring starting with x_1; a *suffix* is any its substring finishing with x_n. If a string is a part of a text, then the *delimiters* (a space, a dot, a comma, etc.) are added to the alphabet $\mathcal{A}$.

A *vector* is any finite sequence consisting of real numbers, i.e., a finite string over *infinite alphabet* $\mathbb{R}$. A *frequency vector* (or *discrete probability distribution*) is any string $x_1 \dots x_n$ with all $x_i \geqslant 0$ and $\sum_{i=1}^{n} x_i = 1$. A *permutation* (or *ranking*) is any string $x_1 \dots x_n$ with all x_i being different numbers from $\{1, \dots, n\}$.

An *editing operation* is an operation on strings, i.e., a *symmetric binary relation* on the set of all considered strings. Given a set of editing operations $\mathcal{O} = \{O_1, \dots, O_m\}$, the corresponding **unit cost edit distance** between strings x and y is the minimum number of editing operations from $\mathcal{O}$ needed to obtain y from x. It is a metric; moreover, it is the **path metric** of a graph with the vertex-set $W(\mathcal{A})$ and xy being an edge if y can be obtained from x by one of the operations from $\mathcal{O}$. In some applications, a *cost function* is assigned to each type of editing operation; then the distance is the minimal total cost of transforming x into y.

Main editing operations on strings are:

- *Character indel*, i.e., insertion or deletion of a character;
- *Character replacement*;
- *Substring move*, i.e., transforming, say, the string $x = x_1 \dots x_n$ into the string $x_1 \dots x_{i-1} \mathbf{x_j} \dots \mathbf{x_{k-1}} x_i \dots x_{j-1} x_k \dots x_n$;
- *Substring copy*, i.e., transforming, say, $x = x_1 \dots x_n$ into $x_1 \dots x_{i-1} \mathbf{x_j} \dots \mathbf{x_{k-1}} x_i \dots x_n$;
- *Substring uncopy*, i.e., the removal of a substring provided that a copy of it remains in the string.

We list below main distances on strings. However, some string distances will appear in Chapters 15, 21 and 23, where they fit better, with respect to the needed level of generalization or specification.

11.1. DISTANCES ON GENERAL STRINGS

- **Levenstein metric**

The **Levenstein metric** (or *shuffle-Hamming distance, Hamming+Gap metric, the editing metric*) is an editing metric on $W(\mathcal{A})$, obtained for $\mathcal{O}$ consisting of only character replacements and indels.

The Levenstein metric between strings $x = x_1 \dots x_m$ and $y = y_1 \dots y_n$ is equal to

$$\min\{d_H(x^*, y^*)\},$$

where x^*, y^* are strings of length k, $k \geqslant \max\{m, n\}$, over alphabet $\mathcal{A}^* = \mathcal{A} \cup \{*\}$, so that after deleting all new characters $*$, strings x^* and y^* shrink to x and y, respectively. Here, the *gap* is the new symbol $*$, and x^*, y^* are *shuffles* of strings x and y with strings consisting of only $*$.

- **Editing metric with moves**

The **editing metric with moves** is an editing metric on $W(\mathcal{A})$ ([Corm03]), obtained for $\mathcal{O}$ consisting of only substring moves and indels.

- **Editing compression metric**

The **editing compression metric** is an editing metric on $W(\mathcal{A})$ ([Corm03]), obtained for $\mathcal{O}$ consisting of only indels, copy and uncopy operations.

- **Indel metric**

The **indel metric** is an editing metric on $W(\mathcal{A})$, obtained for $\mathcal{O}$ consisting of only indels.

It is an analog of the **Hamming metric** $|X \Delta Y|$ between sets X and Y. For strings $x = x_1 \dots x_m$ and $y = y_1 \dots y_n$ it is equal to $m + n - 2LCS(x, y)$, where the similarity $LCS(x, y)$ is the length of the longest common subsequence of x and y.

The **factor distance** on $W(\mathcal{A})$ is defined by $m + n - 2LCF(x, y)$, where the similarity $LCF(x, y)$ is the length of the longest common substring (factor) of x and y.

- **Multiset metric**

The **multiset metric** is a metric on $W(\mathcal{A})$, defined by

$$\max\{|X - Y|, |Y - X|\}$$

for any strings x and y, where X, Y are *bags of symbols* (multisets of characters) in strings x, y, respectively.

- **Normalized information distance**

The **normalized information distance** d is a symmetric function on $W(\{0, 1\})$ ([LCLM04]), defined by

$$\frac{\max\{K(x|y^*), K(y|x^*)\}}{\max\{K(x), K(y)\}}$$

for every two binary strings x and y. Here, for binary strings u and v, u^* is a shortest binary program to compute u on an appropriated universal computer, the *Kolmogorov complexity* (or *algorithmic entropy*) $K(u)$ is the length of u^* (the ultimate compressed version of u), and $K(u|v)$ is the length of the shortest program to compute u if v is provided as an auxiliary input.

The function $d(x, y)$ is a metric up to small error term: $d(x, x) = O((K(x))^{-1})$, and $d(x, z) - d(x, y) - d(y, z) = O((\max\{K(x), K(y), K(z)\})^{-1})$. (Cf. $d(x, y)$ with the following **information metric** (or *entropy metric*) $H(X|Y) + H(Y|X)$ between stochastic sources X and Y.)

The **normalized compression distance** is a distance on $W(\{0, 1\})$ ([LCLM04], [BGLVZ98]), defined by

$$\frac{C(xy) - \min\{C(x), C(y)\}}{\max\{C(x), C(y)\}}$$

for any binary strings x and y, where $C(x)$, $C(y)$, and $C(xy)$ denote the size of compressed (by fixed compressor C, such as gzip, bzip2, or PPMZ) of strings x, y, and their *concatenation* xy. This distance is not a metric. It is an approximation of the normalized information distance. A similar distance is defined by $\frac{C(xy)}{C(x)+C(y)} - \frac{1}{2}$.

- **Marking metric**

The **marking metric** is a metric on $W(\mathcal{A})$ ([EhHa88]), defined by

$$\ln_2\big((\mathit{diff}(x, y) + 1)(\mathit{diff}(y, x) + 1)\big)$$

for any strings $x = x_1 \dots x_m$ and $y = y_1 \dots y_n$, where $\mathit{diff}(x, y)$ is the minimal size $|M|$ of a subset $M \subset \{1, \dots, m\}$ such that any substring of x, not containing any x_i with $i \in M$, is a substring of y.

Another metric, defined in [EhHa88], is $\ln_2(\mathit{diff}(x, y) + \mathit{diff}(y, x) + 1)$.

- **Jaro similarity**

Given strings $x = x_1 \dots x_m$ and $y = y_1 \dots y_n$, call a character x_i *common with* y if $x_i = y_j$, where $|i - j| \leqslant \frac{\min\{m,n\}}{2}$. Let $x' = x'_1 \dots x'_{m'}$ be the all characters of x, which are common with y (in the same order as they appear in x), and let $y' = y'_1 \dots y'_{n'}$ be the analogous string for y.

The **Jaro similarity** $\mathit{Jaro}(x, y)$ between strings x and y is defined by

$$\frac{1}{3}\left(\frac{m'}{m} + \frac{n'}{n} + \frac{|\{1 \leqslant i \leqslant \min\{m', n'\}:\ x'_i = y'_i\}|}{\min\{m', n'\}}\right).$$

This and following two similarities are used in Record Linkage.

- **Jaro–Winkler similarity**

  The **Jaro–Winkler similarity** between strings x and y is defined by

$$Jaro(x, y) + \frac{\max\{4, LCP(x, y)\}}{10}\bigl(1 - Jaro(x, y)\bigr),$$

  where $Jaro(x, y)$ is the **Jaro similarity**, and $LCP(x, y)$ is the length of the longest common prefix of x and y.

- **q-gram similarity**

  The **q-gram similarity** between strings x and y is defined by

$$\frac{q(x, y) + q(y, x)}{2},$$

  where $q(x, y)$ is the number of substrings of length q in the string y, which occur also as substrings in x, divided by the number of all substrings of length q in y.

  This similarity is an example of **token-based similarities**, i.e., ones defined in terms of *tokens* (selected substrings or words). Here tokens are *q-grams*, i.e., substrings of length q. Examples of other token-based similarities on strings, used in Record Linkage, are **Jaccard similarity of community** and **TF-IDF** (a version of **cosine similarity**).

- **Prefix-Hamming metric**

  The **prefix-Hamming metric** between strings $x = x_1 \dots x_m$ and $y = y_1 \dots y_n$ is defined by

$$\bigl(\max\{m, n\} - \min\{m, n\}\bigr) + \bigl|\{1 \leqslant i \leqslant \min\{m, n\}: \ x_i \neq y_i\}\bigr|.$$

- **Weighted Hamming metric**

  If $(\mathcal{A}, d)$ is a metric space, the **weighted Hamming metric** $d_{wH}(x, y)$ between strings $x = x_1 \dots x_m$, $y = y_1 \dots y_m \in W(\mathcal{A})$ is defined by

$$\sum_{i=1}^{m} d(x_i, y_i).$$

- **Needleman–Wunsch–Sellers metric**

  If $(\mathcal{A}, d)$ is a metric space, the **Needleman–Wunsch–Sellers metric** (or *Levenstein metric with costs, global alignment metric*) is an **editing metric with costs** on $W(\mathcal{A})$ ([NeWu70]), obtained for $\mathcal{O}$ consisting of only indels, each of fixed cost $q > 0$, and character replacements, where the cost of replacement of i by j is $d(i, j)$. This metric is the minimal total cost of transforming x into y by those operations.

  Equivalently, it is equal to

$$\min\bigl\{d_{wH}(x^*, y^*)\bigr\},$$

where x^*, y^* are strings of length k, $k \geqslant \max\{m, n\}$, over alphabet $\mathcal{A}^* = \mathcal{A} \cup \{*\}$, so that after deleting all new characters $*$ strings x^* and y^* shrink to x and y, respectively. Here $d_{wH}(x^*, y^*)$ is the **weighted Hamming metric** between x^* and y^* with weight $d(x_i^*, y_i^*) = q$ (i.e., the editing operation is an indel) if one of x_i^*, y_i^* is $*$, and $d(x_i^*, y_i^*) = d(i, j)$, otherwise.

The **Gotoh–Smith–Waterman distance** (or *string distance with affine gaps*) is a more specialized editing metric with costs (see [Coto82]). It discounts mismatching parts in the beginning and in the end of the strings x, y, and introduces two indel costs: one for starting an *affine gap* (contiguous block of indels), and another one (lower) for extending a gap.

- **Martin metric**

The **Martin metric** d^a between strings $x = x_1 \dots x_m$ and $y = y_1 \dots y_n$ is defined by

$$\left|2^{-m} - 2^{-n}\right| + \sum_{t=1}^{\max\{m,n\}} \frac{a_t}{|\mathcal{A}|^t} \sup_z \left|k(z, x) - k(z, y)\right|,$$

where z is any string of the length t, $k(z, x)$ is the *Martin kernel* ([MaSt99]) of a *Markov chain* $M = \{M_t\}_{t=0}^{\infty}$, and the sequence $a \in \{a = \{a_t\}_{t=0}^{\infty}\colon\ a_t > 0,\ \sum_{t=1}^{\infty} a_t < \infty\}$ is a parameter.

- **Baire metric**

The **Baire metric** is an ultrametric between finite or infinite strings $x = x_1 \dots x_m \dots$ and $y = y_1 \dots y_n \dots$, defined, for $x \neq y$, by

$$\frac{1}{1 + LCP(x, y)},$$

where $LCP(x, y)$ is the length of the longest common prefix of x and y.

Moreover, the function $a^{LCP(x,y)}$ is an ultrametric, for any a with $0 < a < 1$, on the set of all infinite strings.

- **Duncan metric**

Consider the set X of all strictly increasing infinite sequences $x = \{x_n\}_n$ of positive integers. Define $N(n, x)$ as the number of elements in $x = \{x_n\}_n$ which are less than n, and $\delta(x)$ as the *density* of x, i.e., $\delta(x) = \lim_{n\to\infty} \frac{N(n,x)}{n}$. Let Y be the subset of X consisting of all sequences $x = \{x_n\}_n$ for which $\delta(x) < \infty$.

The **Duncan metric** is a metric on Y, defined, for $x \neq y$, by

$$\frac{1}{1 + LCP(x, y)} + \left|\delta(x) - \delta(y)\right|,$$

where $LCP(x, y)$ is the length of the longest common prefix of x and y. The metric space (Y, d) is called *Duncan space*.

11.2. DISTANCES ON PERMUTATIONS

A *permutation* (or *ranking*) is any string $x_1 \dots x_n$ with all x_i being different numbers from $\{1, \dots, n\}$; a *signed permutation* is any string $x_1 \dots x_n$ with all $|x_i|$ being different numbers from $\{1, \dots, n\}$. Denote by $(Sym_n, \cdot, id)$ the group of all permutations of the set $\{1, \dots, n\}$, where *id* is the *identity mapping*.

The restriction, on the set Sym_n of all n-permutation vectors, of any metric on $\mathbb{R}^n$ is a metric on Sym_n; main example is the **l_p-metric** $(\sum_{i=1}^{n} |x_i - y_i|^p)^{\frac{1}{p}}$, $p \geqslant 1$.

Main editing operations on permutations are:

- *Block transposition*, i.e., a substring move;
- *Character move*, i.e., a transposition of a block consisting of only one character;
- *Character swap*, i.e., a move of character only on one position to the right or the left (it interchanges adjacent characters);
- *Character exchange*, i.e., interchanging of any two characters (in Group Theory, it is called *transposition*);
- *One-level character exchange*, i.e., exchange of characters x_i and x_j, $i < j$, such that, for any k with $i < k < j$, it holds either $\min\{x_i, x_j\} > x_k$, or $x_k > \max\{x_i, x_j\}$;
- *Block reversal*, i.e., transforming, say, the permutation $x = x_1 \dots x_n$ into the permutation $x_1 \dots x_{i-1}\mathbf{x_j}x_{j-1} \dots \mathbf{x_{i+1}x_i}x_{j+1} \dots x_n$ (so, a swap is a reversal of a block consisting only of two characters);
- *Signed reversal*, i.e., a reversal in signed permutation, followed by multiplication on -1 all characters of reversed block.

Below we list most used editing and other metrics on Sym_n.

- **Hamming metric on permutations**

  The **Hamming metric on permutations** d_H is an editing metric on Sym_n, obtained for $\mathcal{O}$ consisting of only character replacements. It is a **bi-invariant** metric. Also, $n - d_H(x, y)$ is the number of fixed points of xy^{-1}.

- **Spearman ρ distance**

  The **Spearman ρ distance** is the Euclidean metric on Sym_n:

$$\sqrt{\sum_{i=1}^{n}(x_i - y_i)^2}.$$

  (Cf. **Spearman ρ rank correlation similarity** in Statistics.)

- **Spearman footrule distance**

  The **Spearman footrule distance** is the l_1**-metric** on Sym_n:

$$\sum_{i=1}^{n} |x_i - y_i|.$$

  (Cf. **Spearman footrule similarity** in Statistics.)

  Both Spearman distances are **bi-invariant**.

- **Kendall τ distance**

  The **Kendall τ distance** (or *inversion metric, swap metric*) I is an editing metric on Sym_n, obtained for $\mathcal{O}$ consisting of only character swaps.

  In terms of Group Theory, $I(x, y)$ is the number of adjacent transpositions needed to obtain x from y. Also, $I(x, y)$ is the number of *relative inversions* of x and y, i.e., pairs (i, j), $1 \leqslant i < j \leqslant n$, with $(x_i - x_j)(y_i - y_j) < 0$. (Cf. **Kendall τ rank correlation similarity** in Statistics.)

  In [BCFS97] were also given the following metrics, associated with metric $I(x, y)$:

  1. $\min_{z \in Sym_n}(I(x, z) + I(z^{-1}, y^{-1}))$;
  2. $\max_{z \in Sym_n} I(zx, zy)$;
  3. $\min_{z \in Sym_n} I(zx, zy) = T(x, y)$, where T is the **Cayley metric**;
  4. Editing metric, obtained for $\mathcal{O}$ consisting of only one-level character exchanges.

- **Daniels–Guilbaud semi-metric**

  The **Daniels–Guilbaud semi-metric** is a semi-metric on Sym_n, defined, for any $x, y \in Sym_n$, as the number of triples (i, j, k), $1 \leqslant i < j < k \leqslant n$, such that (x_i, x_j, x_k) is not a cyclic shift of (y_i, y_j, y_k); so, it is 0 if and only if x is a cyclical shift of y (see [Monj98]).

- **Cayley metric**

  The **Cayley metric** T is an editing metric on Sym_n, obtained for $\mathcal{O}$ consisting of only character exchanges.

  In terms of Group Theory, $T(x, y)$ is the minimum number of transpositions needed to obtain x from y. Also, $n - T(x, y)$ is the number of cycles in xy^{-1}. The metric T is **bi-invariant**.

- **Ulam metric**

  The **Ulam metric** (or **permutation editing metric**) U is an editing metric on Sym_n, obtained for $\mathcal{O}$ consisting of only character moves.

  Equivalently, it is an editing metric, obtained for $\mathcal{O}$ consisting of only indels. Also, $n - U(x, y) = LCS(x, y) = LIS(xy^{-1})$, where $LCS(x, y)$ is the length of longest common subsequence (not necessarily a substring) of x and y, while $LIS(z)$ is the length of longest increasing subsequence of $z \in Sym_n$.

  This and above six metrics are **right-invariant**.

- **Reversal metric**

The **reversal metric** is an editing metric on Sym_n, obtained for $\mathcal{O}$ consisting of only block reversals.

- **Signed reversal metric**

The **signed reversal metric** is an editing metric on the set of all $2^n n!$ signed permutations of the set $\{1, \dots, n\}$, obtained for $\mathcal{O}$ consisting of only signed reversals.

This metric is used in Biology, where a signed permutation represents a single-chromosome genome, seen as a permutation of genes (along the chromosome) having each a direction (so, a sign $+$ or $-$).

- **Chain metric**

The **chain metric** (or *rearrangement metric*) is a metric on Sym_n ([Page65]), defined, for any $x, y \in Sym_n$, as the minimum number, minus 1, of chains (substrings) $y'_1, \dots, y'_t$ of y, so that x can be *parsed* (concatenated) into, i.e., $x = y'_1 \dots y'_t$.

- **Lexicographic metric**

The **lexicographic metric** is a metric on Sym_n, defined by

$$\left|N(x) - N(y)\right|,$$

where $N(x)$ is the ordinal number of the position (among $1, \dots, n!$) occupied by the permutation x in the *lexicographic ordering* of the set Sym_n.

In the *lexicographic ordering* of Sym_n, $x = x_1 \dots x_n \prec y = y_1 \dots y_n$ if there exists $1 \leqslant i \leqslant n$ such that $x_1 = x_1, \dots, x_{i-1} = y_{i-1}$, but $x_i < y_i$.

- **Fréchet permutation metric**

The **Fréchet permutation metric** is the **Fréchet product metric** on the set Sym_∞ of permutations of positive integers, defined by

$$\sum_{i=1}^{\infty} \frac{1}{2^i} \frac{|x_i - y_i|}{1 + |x_i - y_i|}.$$

Chapter 12

Distances on Numbers, Polynomials, and Matrices

12.1. METRICS ON NUMBERS

Here we consider some most important metrics on the classical number systems: the semi-ring $\mathbb{N}$ of natural numbers, the ring $\mathbb{Z}$ of integers, and the fields $\mathbb{Q}$, $\mathbb{R}$, and $\mathbb{C}$ of rational, real, and complex numbers, respectively. We consider also the algebra $\mathcal{Q}$ of quaternions.

- **Metrics on natural numbers**

  There are several well-known metrics on the set $\mathbb{N}$ of natural numbers:

  1. $|n - m|$; the restriction of the **natural metric** (from $\mathbb{R}$) on $\mathbb{N}$;
  2. $p^{-\alpha}$, where α is the highest power of a given prime number p dividing $m - n$, for $m \neq n$ (and equal to 0 for $m = n$); the restriction of the **p-adic metric** (from $\mathbb{Q}$) on $\mathbb{N}$;
  3. $\ln \frac{l.c.m.(m,n)}{g.c.d.(m,n)}$; an example of the **lattice valuation metric**;
  4. $w_r(n - m)$, where $w_r(n)$ is the *arithmetic r-weight* of n; the restriction of the **arithmetic r-norm metric** (from $\mathbb{Z}$) on $\mathbb{N}$;
  5. $\frac{|n-m|}{mn}$ (cf. **M-relative metric**);
  6. $1 + \frac{1}{m+n}$ for $m \neq n$ (and equal to 0 for $m = n$); the **Sierpinski metric**.

  Most of these metrics on $\mathbb{N}$ can be extended on $\mathbb{Z}$. Moreover, any above metric can be used in the case of an arbitrary countable set X. For example, the **Sierpinski metric** is defined, in general, on a countable set $X = \{x_n : n \in \mathbb{N}\}$ by $1 + \frac{1}{m+n}$ for all $x_m, x_n \in X$ with $m \neq n$ (and is equal to 0, otherwise).

- **Arithmetic r-norm metric**

  Let $r \in \mathbb{N}$, $r \geqslant 2$. The *modified r-ary form* of an integer x is a representation

  $$x = e_n r^n + \cdots + e_1 r + e_0,$$

  where $e_i \in \mathbb{Z}$, and $|e_i| < r$ for all $i = 0, \ldots, n$. An r-ary form is called *minimal* if the number of non-zero coefficients is minimal. The minimal form is not unique, in general. But if the coefficients e_i, $0 \leqslant i \leqslant n - 1$, satisfy the conditions $|e_i + e_{i+1}| < r$, and $|e_i| < |e_{i+1}|$ if $e_i e_{i+1} < 0$, then the above form is unique and minimal; it is called *generalized non-adjacent form*. The *arithmetic r-weight* $w_r(x)$ of an integer x is the number of non-zero coefficients in a *minimal r-ary form* of x, in particular, in the generalized non-adjacent form.

The **arithmetic r-norm metric** (see, for example, [Ernv85]) is a metric on $\mathbb{Z}$, defined by

$$w_r(x-y).$$

- ***p*-adic metric**

  Let p be a prime number. Any non-zero rational number x can be represented as $x = p^\alpha \frac{c}{d}$, where c and d are integers not divisible by p, and α is an unique integer. The *p-adic norm* of x is defined by $|x|_p = p^{-\alpha}$. Moreover, $|0|_p = 0$ holds.

  The **p-adic metric** is a **norm metric** on the set $\mathbb{Q}$ of rational numbers, defined by

$$|x-y|_p.$$

  This metric forms the basis for the algebra of p-adic numbers. In fact, the **Cauchy completion** of the metric space $(\mathbb{Q}, |x-y|_p)$ gives the field $\mathbb{Q}_p$ of *p-adic numbers*, as well as the Cauchy completion of the metric space $(\mathbb{Q}, |x-y|)$ with the **natural metric** $|x-y|$ gives the field $\mathbb{R}$ of real numbers.

- **Natural metric**

  The **natural metric** (or **absolute value metric**) is a metric on $\mathbb{R}$, defined by

$$|x-y| = \begin{cases} y-x, & \text{if } x-y<0, \\ x-y, & \text{if } x-y \geqslant 0. \end{cases}$$

  On $\mathbb{R}$ all **l_p-metrics** coincide with it. The metric space $(\mathbb{R}, |x-y|)$ is called *real line* (or *Euclidean line*).

  There exist many other useful metrics on $\mathbb{R}$. In particular, for a given $0 < \alpha < 1$, the **generalized absolute value metric** on $\mathbb{R}$ is defined by $|x-y|^\alpha$.

- **Zero bias metric**

  The **zero bias metric** is a metric on $\mathbb{R}$, defined by

$$1 + |x-y|$$

  if one and only one of x and y is strictly positive, and by

$$|x-y|,$$

  otherwise, where $|x-y|$ is the **natural metric** (see, for example, [Gile87]).

- **Extended real line metric**

  An **extended real line metric** is a metric on $\mathbb{R} \cup \{+\infty\} \cup \{-\infty\}$. Main example (see, for example, [Cops68]) of such metric is given by

$$\left|f(x) - f(y)\right|,$$

where $f(x) = \frac{x}{1+|x|}$ for $x \in \mathbb{R}$, $f(+\infty) = 1$, and $f(-\infty) = -1$. Another metric, commonly used on $\mathbb{R} \cup \{+\infty\} \cup \{-\infty\}$, is defined by

$$|\arctan x - \arctan y|,$$

where $-\frac{1}{2}\pi < \arctan x < \frac{1}{2}\pi$ for $-\infty < x < \infty$, and $\arctan(\pm\infty) = \pm\frac{1}{2}\pi$.

- **Complex modulus metric**

  The **complex modulus metric** is a metric on the set $\mathbb{C}$ of complex numbers, defined by

$$|z - u|,$$

  where, for any $z \in \mathbb{C}$, the real number $|z| = |z_1 + z_2 i| = \sqrt{z_1^2 + z_2^2}$ is the *complex modulus*. The metric space $(\mathbb{C}, |z - u|)$ is called *complex plane* (or *Argand plane*).

  Examples of other useful metrics on $\mathbb{C}$ are: the **British Rail metric**, defined by

$$|z| + |u|$$

  for $z \neq u$ (and is equal to 0, otherwise); the ***p*-relative metric**, $1 \leqslant p \leqslant \infty$ (cf. **(p, q)-relative metric**), defined by

$$\frac{|z - u|}{(|z|^p + |u|^p)^{\frac{1}{p}}}$$

  for $|z| + |u| \neq 0$ (and is equal to 0, otherwise); for $p = \infty$ one obtains the **relative metric**, written for $|z| + |u| \neq 0$ as

$$\frac{|z - u|}{\max\{|z|, |u|\}}.$$

- **Chordal metric**

  The **chordal metric** (or *spherical metric*) d_χ is a metric on the set $\overline{\mathbb{C}} = \mathbb{C} \cup \{\infty\}$, defined by

$$d_\chi(z, u) = \frac{2|z - u|}{\sqrt{1 + |z|^2}\sqrt{1 + |u|^2}}$$

  for all $z, u \in \mathbb{C}$, and by

$$d_\chi(z, \infty) = \frac{2}{\sqrt{1 + |z|^2}}$$

  for all $z \in \mathbb{C}$ (cf. ***M*-relative metric**). The metric space $(\overline{\mathbb{C}}, d_\chi)$ is called *extended complex plane*. It is homeomorphic and conformally equivalent to the *Riemann sphere*.

  In fact, a *Riemann sphere* is a sphere in the Euclidean space $\mathbb{E}^3$, considered as a metric subspace of $\mathbb{E}^3$, onto which the extended complex plane is one-to-one mapped under stereographic projection. The *unit sphere* $S^2 = \{(x_1, x_2, x_3) \in \mathbb{E}^3 : x_1^2 + x_2^2 + x_3^2 = 1\}$ can be taken as the Riemann sphere, and the plane $\overline{\mathbb{C}}$ can be identified with the plane $x_3 = 0$ such the real axis coincides with the x_1-axis, and the imaginary axis with the x_2-axis. Under stereographic projection, each point $z \in \mathbb{C}$ corresponds to the point $(x_1, x_2, x_3) \in S^2$ obtained as the point of intersection of the ray drawn from the "north pole" (0, 0, 1) of the sphere to the point z with the sphere S^2; the "north pole" corresponds to the point at infinity ∞. The chordal (spherical) distance between two points $p, q \in S^2$ is taken to be the distance between their preimages $z, u \in \overline{\mathbb{C}}$.

The chordal metric can be defined equivalently on $\overline{\mathbb{R}}^n = \mathbb{R}^n \cup \{\infty\}$. Thus, for any $x, y \in \mathbb{R}^n$, one has

$$d_\chi(x, y) = \frac{2\|x - y\|_2}{\sqrt{1 + \|x\|_2^2}\sqrt{1 + \|y\|_2^2}},$$

and for any $x \in \mathbb{R}^n$, one has

$$d_\chi(x, \infty) = \frac{2}{\sqrt{1 + \|x\|_2^2}},$$

where $\|.\|_2$ is the ordinary Euclidean norm on $\mathbb{R}^n$. The metric space $(\overline{\mathbb{R}}^n, d_\chi)$ is called *Möbius space*. It is a *Ptolemaic* metric space (cf. **Ptolemaic metric**).

Given $\alpha > 0$, $\beta \geqslant 0$, $p \geqslant 1$, the **generalized chordal metric** is a metric on $\mathbb{C}$ (in general, on $(\mathbb{R}^n, \|.\|_2)$ and even on any *Ptolemaic* space $(V, \|.\|)$), defined by

$$\frac{|z - u|}{(\alpha + \beta|z|^p)^{\frac{1}{p}} \cdot (\alpha + \beta|u|^p)^{\frac{1}{p}}}.$$

It can be easy generalized on $\overline{\mathbb{C}}$ (on $\overline{\mathbb{R}}^n$).

- **Quaternion metric**

Quaternions are members of a non-commutative division algebra $\mathcal{Q}$ over the field $\mathbb{R}$, geometrically realizable in a four-dimensional space ([Hami66]). The quaternions can be written in the form $q = q_1 + q_2 i + q_3 j + q_4 k$, $q_i \in \mathbb{R}$, where the quaternions i, j, and k, called *basic units*, satisfy the following identities, known as *Hamilton's rules*: $i^2 = j^2 = k^2 = -1$, and $ij = -ji = k$.

The *quaternion norm* $\|q\|$ of $q = q_1 + q_2 i + q_3 j + q_4 k \in \mathcal{Q}$ is defined by

$$\|q\| = \sqrt{q\overline{q}} = \sqrt{q_1^2 + q_2^2 + q_3^2 + q_4^2}, \quad \overline{q} = q_1 - q_2 i - q_3 j - q_4 k.$$

The **quaternion metric** is a **norm metric** on the set $\mathcal{Q}$ of all quaternions, defined by $\|x - y\|$.

12.2. METRICS ON POLYNOMIALS

A *polynomial* is an expression involving a sum of powers in one or more variables multiplied by coefficients. A *polynomial in one variable* (or *univariate polynomial*) with constant real (complex) coefficients is given by $P = P(z) = \sum_{k=0}^{n} a_k z^k$, $a_k \in \mathbb{R}$ ($a_k \in \mathbb{C}$). The set $\mathcal{P}$ of all real (complex) polynomials forms a ring $(\mathcal{P}, +, \cdot, 0)$. It is also a vector space over $\mathbb{R}$ (over $\mathbb{C}$).

- **Polynomial norm metric**

  A **polynomial norm metric** (or **polynomial bar metric**) is a **norm metric** on the set $\mathcal{P}$ of all real (complex) polynomials, defined by

$$\|P - Q\|,$$

  where $\|.\|$ is a *polynomial norm*, i.e., a function $\|.\| : \mathcal{P} \to \mathbb{R}$ such that, for all $P, Q \in \mathcal{P}$ and for any scalar k, we have the following properties:

  1. $\|P\| \geqslant 0$, with $\|P\| = 0$ if and only if $P \equiv 0$;
  2. $\|kP\| = |k|\|P\|$;
  3. $\|P + Q\| \leqslant \|P\| + \|Q\|$ (*triangle inequality*).

  For the set $\mathcal{P}$ several classes of norms are commonly used. The l_p*-norm*, $1 \leqslant p \leqslant \infty$, of a polynomial $P(z) = \sum_{k=0}^{n} a_k z^k$ is defined by

$$\|P\|_p = \left(\sum_{k=0}^{n} |a_k|^p\right)^{1/p},$$

  giving the special cases $\|P\|_1 = \sum_{k=0}^{n} |a_k|$, $\|P\|_2 = \sqrt{\sum_{k=0}^{n} |a_k|^2}$, and $\|P\|_\infty = \max_{0 \leqslant k \leqslant n} |a_k|$. The value $\|P\|_\infty$ is called *polynomial height*. The L_p*-norm*, $1 \leqslant p \leqslant \infty$, of a polynomial $P(z) = \sum_{k=0}^{n} a_k z^k$ is defined by

$$\|P\|_{L_p} = \left(\int_0^{2\pi} |P(e^{i\theta})|^p \frac{d\theta}{2\pi}\right)^{\frac{1}{p}},$$

  giving the special cases $\|P\|_{L_1} = \int_0^{2\pi} |P(e^{i\theta})| \frac{d\theta}{2\pi}$, $\|P\|_{L_2} = \sqrt{\int_0^{2\pi} |P(e^{i\theta})|^2 \frac{d\theta}{2\pi}}$, and $\|P\|_{L_\infty} = \sup_{|z|=1} |P(z)|$.

- **Bombieri metric**

  The **Bombieri metric** (or **polynomial bracket metric**) is a **polynomial norm metric** on the set $\mathcal{P}$ of all real (complex) polynomials, defined by

$$[P - Q]_p,$$

where $[.]_p$, $0 \leqslant p \leqslant \infty$, is the *Bombieri p-norm*. For a polynomial $P(z) = \sum_{k=0}^{n} a_k z^k$ it is defined by

$$[P]_p = \left(\sum_{k=0}^{n} \binom{n}{k}^{1-p} |a_k|^p \right)^{\frac{1}{p}},$$

where $\binom{n}{k}$ is a binomial coefficient.

12.3. METRICS ON MATRICES

An $m \times n$ matrix $A = ((a_{ij}))$ over a field $\mathbb{F}$ is a table consisting of m rows and n columns with the entries a_{ij} from $\mathbb{F}$. The set of all $m \times n$ matrices with real (complex) entries is denoted by $M_{m,n}$. It forms a *group* $(M_{m,n}, +, 0_{m,n})$, where $((a_{ij}))+((b_{ij})) = ((a_{ij}+b_{ij}))$, and the matrix $0_{m,n} \equiv 0$, i.e., all its entries are equal to 0. It is also an mn-dimensional vector space over $\mathbb{R}$ (over $\mathbb{C}$). The *transpose* of a matrix $A = ((a_{ij})) \in M_{m,n}$ is the matrix $A^T = ((a_{ji})) \in M_{n,m}$. The *conjugate transpose* (or *adjoint*) of a matrix $A = ((a_{ij})) \in M_{m,n}$ is the matrix $A^* = ((\overline{a}_{ji})) \in M_{n,m}$.

A matrix is called *square matrix* if $m = n$. The set of all square $n \times n$ matrices with real (complex) entries is denoted by M_n. It forms a *ring* $(M_n, +, \cdot, 0_n)$, where $+$ and 0_n are defined as above, and $((a_{ij})) \cdot ((b_{ij})) = ((\sum_{k=1}^{n} a_{ik} b_{kj}))$. It is also an n^2-dimensional vector space over $\mathbb{R}$ (over $\mathbb{C}$). A matrix $A = ((a_{ij})) \in M_n$ is called *symmetric* if $a_{ij} = a_{ji}$ for all $i, j \in \{1, \dots, n\}$, i.e., if $A = A^T$. Special types of square $n \times n$ matrices include the *identity matrix* $1_n = ((c_{ij}))$ with $c_{ii} = 1$, and $c_{ij} = 0$, $i \neq j$. An *unitary matrix* $U = ((u_{ij}))$ is a square matrix, defined by $U^{-1} = U^*$, where U^{-1} is the *inverse matrix* for U, i.e., $U \cdot U^{-1} = 1_n$. An *orthonormal matrix* is a matrix $A \in M_{m,n}$ such that $A^* A = 1_n$.

If for a matrix $A \in M_n$ there is a vector x such that $Ax = \lambda x$ for some scalar λ, then λ is called *eigenvalue* of A with corresponding *eigenvector* x. Given a complex matrix $A \in M_{m,n}$, its *singular values* $s_i(A)$ are defined as the square roots of the *eigenvalues* of the matrix A^*A, where A^* is the conjugate transpose of A. They are non-negative real numbers $s_1(A) \geqslant s_2(A) \geqslant \cdots$.

- **Matrix norm metric**

A **matrix norm metric** is a **norm metric** on the set $M_{m,n}$ of all real (complex) $m \times n$ matrices, defined by

$$\|A - B\|,$$

where $\|.\|$ is a *matrix norm*, i.e., a function $\|.\| : M_{m,n} \to \mathbb{R}$ such that, for all $A, B \in M_{m,n}$, and for any scalar k, we have the following properties:

1. $\|A\| \geqslant 0$, with $\|A\| = 0$ if and only if $A = 0_{m,n}$;
2. $\|kA\| = |k|\|A\|$;
3. $\|A + B\| \leqslant \|A\| + \|B\|$ (*triangle inequality*).

All matrix norm metrics on $M_{m,n}$ are equivalent. A matrix norm $\|.\|$ on the set M_n of all real (complex) square $n \times n$ matrices is called *sub-multiplicative* if it is *compatible* with matrix multiplication, i.e., $\|AB\| \leqslant \|A\| \cdot \|B\|$ for all $A, B \in M_n$. The set M_n with a sub-multiplicative norm is a *Banach algebra*.

The simplest example of a matrix norm metric is the **Hamming metric** on $M_{m,n}$ (in general, on the set $M_{m,n}(\mathbb{F})$ of all $m \times n$ matrices with entries from a field $\mathbb{F}$), defined by $\|A - B\|_H$, where $\|A\|_H$ is the *Hamming norm* of $A \in M_{m,n}$, i.e., the number of non-zero entries in A.

- **Natural norm metric**

  A **natural norm metric** (or **induced norm metric**, **subordinate norm metric**) is a **matrix norm metric** on the set M_n of all real (complex) square $n \times n$ matrices, defined by

$$\|A - B\|_{nat},$$

  where $\|.\|_{nat}$ is a *natural norm* on M_n. The *natural norm* $\|.\|_{nat}$ on M_n, induced by the vector norm $\|x\|$, $x \in \mathbb{R}^n$ $(x \in \mathbb{C}^n)$, is a *sub-multiplicative matrix norm*, defined by

$$\|A\|_{nat} = \sup_{\|x\| \neq 0} \frac{\|Ax\|}{\|x\|} = \sup_{\|x\|=1} \|Ax\| = \sup_{\|x\| \leqslant 1} \|Ax\|.$$

  The natural norm metric can be defined in similar way on the set $M_{m,n}$ of all $m \times n$ real (complex) matrices: given vector norms $\|.\|_{\mathbb{R}^m}$ on $\mathbb{R}^m$ and $\|.\|_{\mathbb{R}^n}$ on $\mathbb{R}^n$, the *natural norm* $\|A\|_{nat}$ of a matrix $A \in M_{m,n}$, induced by $\|.\|_{\mathbb{R}^n}$ and $\|.\|_{\mathbb{R}^m}$, is a matrix norm, defined by $\|A\|_{nat} = \sup_{\|x\|_{\mathbb{R}^n}=1} \|Ax\|_{\mathbb{R}^m}$.

- **Matrix p-norm metric**

  A **matrix p-norm metric** is a **natural norm metric** on M_n, defined by

$$\|A - B\|^p_{nat},$$

  where $\|.\|^p_{nat}$ is the *matrix p-norm*, i.e., a *natural norm*, induced by the vector *l_p-norm*, $1 \leqslant p \leqslant \infty$:

$$\|A\|^p_{nat} = \max_{\|x\|_p=1} \|Ax\|_p, \quad \text{where } \|x\|_p = \left(\sum_{i=1}^n |x_i|^p\right)^{1/p}.$$

  The **maximum absolute column metric** (more exactly, **maximum absolute column sum norm metric**) is the **matrix 1-norm metric** $\|A - B\|^1_{nat}$ on M_n. The *matrix 1-norm* $\|.\|^1_{nat}$, induced by the vector *l_1-norm*, is called also *maximum absolute column sum norm*. For a matrix $A = ((a_{ij})) \in M_n$ it can be written as

$$\|A\|^1_{nat} = \max_{1 \leqslant j \leqslant n} \sum_{i=1}^n |a_{ij}|.$$

The **maximum absolute row metric** (more exactly, **maximum absolute row sum norm metric**) is the **matrix ∞-norm metric** $\|A - B\|_{nat}^{\infty}$ on M_n. The *matrix ∞-norm* $\|.\|_{nat}^{\infty}$, induced by the vector *l_∞-norm*, is called also *maximum absolute row sum norm*. For a matrix $A = ((a_{ij})) \in M_n$ it can be written as

$$\|A\|_{nat}^{\infty} = \max_{1 \leqslant j \leqslant n} \sum_{j=1}^{n} |a_{ij}|.$$

The **spectral norm metric** is the **matrix 2-norm metric** $\|A - B\|_{nat}^{2}$ on M_n. The matrix 2-norm $\|.\|_{nat}^{2}$, induced by the vector *l_2-norm*, is called also *spectral norm* and denoted by $\|.\|_{sp}$. For a matrix $A = ((a_{ij})) \in M_n$, it can be written as

$$\|A\|_{sp} = (\text{maximum eigenvalue of } A^*A)^{\frac{1}{2}},$$

where $A^* = ((\overline{a}_{ji})) \in M_n$ is the conjugate transpose of A (cf. **Ky-Fan norm metric**).

- **Frobenius norm metric**

The **Frobenius norm metric** is a **matrix norm metric** on $M_{m,n}$, defined by

$$\|A - B\|_{Fr},$$

where $\|.\|_{Fr}$ is the *Frobenius norm*. For a matrix $A = ((a_{ij})) \in M_{m,n}$, it is defined by

$$\|A\|_{Fr} = \sqrt{\sum_{i=1}^{m}\sum_{j=1}^{n} |a_{ij}|^2}.$$

It is also equal to the square root of the matrix trace of A^*A, where $A^* = ((\overline{a}_{ji}))$ is the conjugate transpose of A, or, equivalently, to the square root of the sum of *eigenvalues* λ_i of A^*A: $\|A\|_{Fr} = \sqrt{Tr(A^*A)} = \sqrt{\sum_{i=1}^{\min\{m,n\}} \lambda_i}$ (cf. **Schatten norm metric**). This norm comes from an *inner product* on the space $M_{m,n}$, but it is not *sub-multiplicative* for $m = n$.

- **(c, p)-norm metric**

Let $k \in \mathbb{N}$, $k \leqslant \min\{m, n\}$, $c \in \mathbb{R}^k$, $c_1 \geqslant c_2 \geqslant \cdots \geqslant c_k > 0$, and $1 \leqslant p < \infty$.

The (c, p)**-norm metric** is a **matrix norm metric** on $M_{m,n}$, defined by

$$\|A - B\|_{(c,p)}^{k},$$

where $\|.\|_{(c,p)}^{k}$ is the *(c, p)-norm* on $M_{m,n}$. For a matrix $A \in M_{m,n}$, it is defined by

$$\|A\|_{(c,p)}^{k} = \left(\sum_{i=1}^{k} c_i s_i^p(A)\right)^{\frac{1}{p}},$$

where $s_1(A) \geqslant s_2(A) \geqslant \cdots \geqslant s_k(A)$ are the first k *singular values* of A. If $p = 1$, one obtains the *c-norm*. If, moreover, $c_1 = \cdots = c_k = 1$, one obtains the *Ky-Fan k-norm*.

- **Ky-Fan norm metric**

Given $k \in \mathbb{N}$, $k \leqslant \min\{m, n\}$, the **Ky-Fan norm metric** is a **matrix norm metric** on $M_{m,n}$, defined by

$$\|A - B\|_{KF}^k,$$

where $\|.\|_{KF}^k$ is the *Ky-Fan k-norm* on $M_{m,n}$. For a matrix $A \in M_{m,n}$, it is defined as the sum of its first k *singular values*:

$$\|A\|_{KF}^k = \sum_{i=1}^{k} s_i(A).$$

For $k = 1$, one obtains the *spectral norm*. For $k = \min\{m, n\}$, one obtains the *trace norm*.

- **Schatten norm metric**

Given $1 \leqslant p < \infty$, the **Schatten norm metric** is a **matrix norm metric** on $M_{m,n}$, defined by

$$\|A - B\|_{Sch}^p,$$

where $\|.\|_{Sch}^p$ is the *Schatten p-norm* on $M_{m,n}$. For a matrix $A \in M_{m,n}$, it is defined as the p-th root of the sum of the p-th powers of all its singular values:

$$\|A\|_{Sch}^p = \left(\sum_{i=1}^{\min\{m,n\}} s_i^p(A) \right)^{\frac{1}{p}}.$$

For $p = 2$, one obtains the *Frobenius norm*, and, for $p = 1$, one obtains the *trace norm*.

- **Trace norm metric**

The **trace norm metric** is a **matrix norm metric** on $M_{m,n}$, defined by

$$\|A - B\|_{tr},$$

where $\|.\|_{tr}$ is the *trace norm* on $M_{m,n}$. For a matrix $A \in M_{m,n}$, it is defined as the sum of all its *singular values*:

$$\|A\|_{tr} = \sum_{i=1}^{\min\{m,n\}} s_i(A).$$

- **Rosenbloom–Tsfasman metric**

Let $M_{m,n}(\mathbb{F}_q)$ be the set of all $m \times n$ matrices with entries from a finite field $\mathbb{F}_q$. The *Rosenbloom–Tsfasman norm* $\|.\|_{RT}$ on $M_{m,n}(\mathbb{F}_q)$ is defined as follow: if $m = 1$ and

$a = (\xi_1, \xi_2, \dots, \xi_n) \in M_{1,n}(\mathbb{F}_q)$, then $\|0_{1,n}\|_{RT} = 0$, and $\|a\|_{RT} = \max\{i|\xi_i \neq 0\}$ for $a \neq 0_{1,n}$; if $A = (a_1, \dots, a_m)^T \in M_{m,n}(\mathbb{F}_q)$, $a_j \in M_{1,n}(\mathbb{F}_q)$, $1 \leqslant j \leqslant m$, then

$$\|A\|_{RT} = \sum_{j=1}^{m} \|a_j\|_{RT}.$$

The **Rosenbloom–Tsfasman metric** ([RoTs96]) is a **matrix norm metric** (in fact, an **ultrametric**) on $M_{m,n}(\mathbb{F}_q)$, defined by

$$\|A - B\|_{RT}.$$

- **Angle distances between subspaces**

Consider the *Grassmannian space* $G(m, n)$ of all n-dimensional subspaces of Euclidean space $\mathbb{E}^m$; it is a compact *Riemannian manifold* of dimension $n(m - n)$.

Given two subspaces $A, B \in G(m, n)$, the *principal angles* $\frac{\pi}{2} \geqslant \theta_1 \geqslant \cdots \geqslant \theta_n \geqslant 0$ between them are defined, for $k = 1, \dots, n$, inductively by

$$\cos\theta_k = \max_{x \in A} \max_{y \in B} x^T y = \left(x^k\right)^T y^k$$

subject to the conditions $\|x\|_2 = \|y\|_2 = 1$, $x^T x^i = 0$, $y^T y^i = 0$, for $1 \leqslant i \leqslant k - 1$, where $\|.\|_2$ is the Euclidean norm. The principal angles can also be defined in terms of orthonormal matrices Q_A and Q_B spanning subspaces A and B, respectively: in fact, n ordered *singular values* of the matrix $Q_A Q_B \in M_n$ can be expressed as cosines $\cos\theta_1, \dots, \cos\theta_n$.

The **geodesic distance** between subspaces A and B is (Wong, 1967) defined by

$$\sqrt{2\sum_{i=1}^{n} \theta_i^2}.$$

The **Martin distance** between subspaces A and B is defined by

$$\sqrt{\ln \prod_{i=1}^{n} \frac{1}{\cos^2\theta_i}}.$$

In the case, when subspaces represent *autoregressive models*, the Martin distance can be expressed in terms of *cepstrum* of the autocorrelation functions of the models (cf. **Martin cepstrum distance**).

The **Asimov distance** between subspaces A and B is defined by

$$\theta_1.$$

It can be expressed also in terms of the **Finsler metric** on the manifold $G(m, n)$.

The **gap distance** between subspaces A and B is defined by

$$\sin\theta_1.$$

It can be expressed also in terms of *orthogonal projectors* as the l_2*-norm* of the difference of the projectors onto A and B, respectively. Many versions of this distance are used in Control Theory (cf. **gap metric**).

The **Frobenius distance** between subspaces A and B is defined by

$$\sqrt{2\sum_{i=1}^{n}\sin^2\theta_i}.$$

It can be expressed also in terms of *orthogonal projectors* as the *Frobenius norm* of the difference of the projectors onto A and B, respectively. A similar distance $\sqrt{\sum_{i=1}^{n}\sin^2\theta_i}$ is called **chordal distance**.

- **Semi-metrics on resemblances**

The following two semi-metrics are defined for any two **resemblances** d_1 and d_2 on a given finite set X (moreover, for any two real symmetric matrices).

The **Lerman semi-metric** (cf. **Kendall τ distance** on permutations) is defined by

$$\frac{|\{(\{x,y\},\{u,v\}):\ (d_1(x,y)-d_1(u,v))(d_2(x,y)-d_2(u,v))<0\}|}{\binom{|X|+1}{2}^2},$$

where $(\{x,y\},\{u,v\})$ is any pair of unordered pairs $\{x,y\},\{u,v\}$ of elements x,y,u,v from X.

The **Kaufman semi-metric** is defined by

$$\frac{|\{(\{x,y\},\{u,v\}):\ (d_1(x,y)-d_1(u,v))(d_2(x,y)-d_2(u,v))<0\}|}{|\{(\{x,y\},\{u,v\}):\ (d_1(x,y)-d_1(u,v))(d_2(x,y)-d_2(u,v))\neq 0\}|}.$$

Chapter 13

Distances in Functional Analysis

Functional Analysis is the branch of Mathematics, concerned with the study of spaces of functions. This usage of the word *functional* goes back to the calculus of variations, implying a function whose argument is a function. In the modern view, Functional Analysis is seen as the study of complete *normed vector spaces*, i.e., **Banach spaces**. For any real number $p \geqslant 1$, an example of a Banach space is given by **L_p-space** of all Lebesgue-measurable functions whose absolute value's p-th power has finite integral. A **Hilbert space** is a Banach space in which the norm arises from an *inner product*. Also, in Functional Analysis are considered the *continuous linear operators* defined on Banach and Hilbert spaces.

13.1. METRICS ON FUNCTION SPACES

Let $I \subset \mathbb{R}$ be an *open interval* (i.e., a non-empty connected open set) in $\mathbb{R}$. A real function $f : I \to \mathbb{R}$ is called *real analytic* on I if it agrees with its *Taylor series* in an *open neighborhood* U_{x_0} of every point $x_0 \in I$:

$$f(x) = \sum_{n=0}^{\infty} \frac{f^{(n)}(x_0)}{n!}(x - x_0)^n \quad \text{for any } x \in U_{x_0}.$$

Let $D \subset \mathbb{C}$ be a *domain* (i.e., a *convex* open set) in $\mathbb{C}$. A complex function $f : D \to \mathbb{C}$ is called *complex analytic* (or, simply, *analytic*) on D if it agrees with its Taylor series in an open neighborhood of every point $z_0 \in D$. A complex function f is analytic on D if and only if it is *holomorphic* on D, i.e., if it has a complex derivative

$$f'(z_0) = \lim_{z \to z_0} \frac{f(z) - f(z_0)}{z - z_0}$$

at every point $z_0 \in D$.

- **Integral metric**

  The **integral metric** is the **L_1-metric** on the set $C_{[a,b]}$ of all continuous real (complex) functions on a given segment $[a, b]$, defined by

$$\int_a^b |f(x) - g(x)| dx.$$

The corresponding metric space is abbreviated by $C^1_{[a,b]}$. It is a Banach space.

In general, for any **compact** (or *countably compact*) topological space X the integral metric can be defined on the set of all continuous functions $f : X \to \mathbb{R}$ $(\mathbb{C})$ by $\int_X |f(x) - g(x)|dx$.

- **Uniform metric**

The **uniform metric** (or **sup metric**) is the L_∞**-metric** on the set $C_{[a,b]}$ of all real (complex) continuous functions on a given segment $[a, b]$, defined by

$$\sup_{x\in[a,b]} \left|f(x) - g(x)\right|.$$

The corresponding metric space is abbreviated by $C^\infty_{[a,b]}$. It is a Banach space.

A generalization of $C^\infty_{[a,b]}$ is the *space of continuous functions* $C(X)$, i.e., a metric space on the set of all continuous (more generally, bounded) functions $f : X \to \mathbb{C}$ of a topological space X with the L_∞-metric $\sup_{x\in X} |f(x) - g(x)|$.

In the case of the metric space $C(X, Y)$ of continuous (more generally, bounded) functions $f : X \to Y$ from one **metric compactum** (X, d_X) to another (Y, d_Y), the sup metric between two functions $f, g \in C(X, Y)$ is defined by $\sup_{x\in X} d_Y(f(x), g(x))$.

The metric space $C^\infty_{[a,b]}$, as well as the metric space $C^1_{[a,b]}$, are two of the most important cases of the metric space $C^p_{[a,b]}$, $1 \leqslant p \leqslant \infty$, on the set $C_{[a,b]}$ with the L_p-metric $(\int_a^b |f(x) - g(x)|^p dx)^{\frac{1}{p}}$. The space $C^p_{[a,b]}$ is an example of L_p-space.

- **Dogkeeper distance**

Given a metric space (X, d), the **dogkeeper distance** is a metric on the set of all functions $f : [0, 1] \to X$, defined by

$$\inf_\sigma \sup_{t\in[0,1]} d\big(f(t), g\big(\sigma(t)\big)\big),$$

where $\sigma : [0, 1] \to [0, 1]$ is a continuous, monotone increasing function such that $\sigma(0) = 0$, $\sigma(1) = 1$. This metric is a special case of the **Fréchet metric**. It is used for measuring the distances between curves.

- **Bohr metric**

Let $\mathbb{R}$ be a metric space with a metric ρ. A continuous function $f : \mathbb{R} \to \mathbb{R}$ is called *almost-periodic* if, for every $\varepsilon > 0$, there exists $l = l(\varepsilon) > 0$ such that every interval $[t_0, t_0 + l(\varepsilon)]$ contains at least one number τ for which $\rho(f(t), f(t + \tau)) < \varepsilon$ for $-\infty < t < +\infty$.

The **Bohr metric** is the **norm metric** $\|f - g\|$ on the set AP of all almost-periodic functions, defined by the norm

$$\|f\| = \sup_{-\infty<t<+\infty} \left|f(t)\right|.$$

It makes AP a Banach space. Some generalizations of almost-periodic functions were obtained using other norms by Besicovitch, Stepanov, Weyl, von Neumann, Turing, Bochner, and others (cf. **Stepanov distance**, **Weyl distance**, and **Besicovitch distance**).

- **Stepanov distance**

The **Stepanov distance** is a distance on the set of all measurable functions $f : \mathbb{R} \to \mathbb{C}$ with summable p-th power on each bounded integral, defined by

$$\sup_{x \in \mathbb{R}} \left(\frac{1}{l} \int_{x}^{x+l} \left| f(x) - g(x) \right|^{p} dx \right)^{1/p}.$$

The **Weyl distance** is a distance on the same set, defined by

$$\lim_{l \to \infty} \sup_{x \in \mathbb{R}} \left(\frac{1}{l} \int_{x}^{x+l} \left| f(x) - g(x) \right|^{p} dx \right)^{1/p}.$$

Corresponding to these distances one has the *generalized Stepanov* and *Weyl almost-periodic functions*.

- **Besicovitch distance**

The **Besicovitch distance** is a distance on the set of all measurable functions $f : \mathbb{R} \to \mathbb{C}$ with summable p-th power on each bounded integral, defined by

$$\left(\overline{\lim}_{T \to \infty} \frac{1}{2T} \int_{-T}^{T} \left| f(x) - g(x) \right|^{p} dx \right)^{1/p}.$$

Corresponding to this distance one has the *generalized Besicovitch almost-periodic functions*.

- **Bergman p-metric**

Given $1 \leqslant p < \infty$, let $L_p(\Delta)$ be the L_p-space of Lebesgue measurable functions f on the *unit disk* $\Delta = \{z \in \mathbb{C}\colon \ |z| < 1\}$ with

$$\|f\|_p = \left(\int_{\Delta} |f(z)|^p \mu(dz) \right)^{\frac{1}{p}} < \infty.$$

The *Bergman space* $L_p^a(\Delta)$ is the subspace of $L_p(\Delta)$ consisting of analytic functions, and the **Bergman p-metric** is the **L_p-metric** on $L_p^a(\Delta)$ (cf. **Bergman metric**). Any Bergman space is a Banach space.

- **Bloch metric**

The *Bloch space* B on the *unit disk* $\Delta = \{z \in \mathbb{C}\colon \ |z| < 1\}$ is the set of all analytic functions f on Δ such that $\|f\|_B = \sup_{z \in \Delta}(1 - |z|^2)|f'(z)| < \infty$. Using complete *semi-norm* $\|.\|_B$, a norm on B is defined by

$$\|f\| = \left|f(0)\right| + \|f\|_B.$$

The **Bloch metric** is the **norm metric** $\|f - g\|$ on B. It makes B a Banach space.

- **Besov metric**

Given $1 < p < \infty$, the *Besov space* B_p on the *unit disk* $\Delta = \{z \in \mathbb{C}\colon\ |z| < 1\}$ is the set of all analytic functions f in Δ such that

$$\|f\|_{B_p} = \left(\int_\Delta \left(1 - |z|^2\right)^p \left|f'(z)\right|^p d\lambda(z)\right)^{\frac{1}{p}} < \infty, \quad \text{where } d\lambda(z) = \frac{\mu(dz)}{(1 - |z|^2)^2}$$

is the Möbius invariant measure on Δ. Using complete *semi-norm* $\|.\|_{B_p}$, a norm on B_p is defined by

$$\|f\| = \left|f(0)\right| + \|f\|_{B_p}.$$

The **Besov metric** is the **norm metric** $\|f - g\|$ on B_p. It makes B_p a Banach space.

The set B_2 is the classical *Dirichlet space* of analytic on Δ functions with square integrable derivative, equipped with the **Dirichlet metric**. The *Bloch space* B can be considered as B_∞.

- **Hardy metric**

Given $1 \leqslant p < \infty$, the *Hardy space* $H^p(\Delta)$ is the class of functions, analytic on the *unit disk* $\Delta = \{z \in \mathbb{C}\colon\ |z| < 1\}$, and satisfying the following growth condition for the *Hardy norm* $\|.\|_{H^p}$:

$$\|f\|_{H^p(\Delta)} = \sup_{0<r<1}\left(\frac{1}{2\pi}\int_0^{2\pi} \left|f\left(re^{i\theta}\right)\right|^p d\theta\right)^{\frac{1}{p}} < \infty.$$

The **Hardy metric** is the **norm metric** $\|f - g\|_{H^p(\Delta)}$ on $H^p(\Delta)$. It makes $H^p(\Delta)$ a Banach space.

In Complex Analysis, the Hardy spaces are analogs of the L_p-spaces of Functional Analysis. Such spaces are applied in Mathematical Analysis itself, and also to Scattering Theory and Control Theory (cf. Chapter 17).

- **Part metric**

The **part metric** is a metric on a *domain* D of $\mathbb{R}^2$, defined by

$$\sup_{f \in H^+}\left|\ln\left(\frac{f(x)}{f(y)}\right)\right|$$

for any $x, y \in \mathbb{R}^2$, where H^+ is the set of all positive *harmonic functions* on the domain D.

A twice-differentiable real function $f : D \to \mathbb{R}$ is called *harmonic* on D if its *Laplacian* $\Delta f = \frac{\partial^2 f}{\partial x_1^2} + \frac{\partial^2 f}{\partial x_2^2}$ vanishes on D.

- **Orlicz metric**

Let $M(u)$ be an even convex function of a real variable which is increasing for u positive, and $\lim_{u\to 0} u^{-1}M(u) = \lim_{u\to\infty} u(M(u))^{-1} = 0$. In this case the function $p(v) = M'(v)$ does not decrease on $[0, \infty)$, $p(0) = \lim_{v\to 0} p(v) = 0$, and $p(v) > 0$ when $v > 0$. Writing $M(u) = \int_0^{|u|} p(v)dv$, and defining $N(u) = \int_0^{|u|} p^{-1}(v)dv$, one obtains a pair $(M(u), N(u))$ of *complementary functions*.

Let $(M(u), N(u))$ be a pair of complementary functions, and let G be a bounded closed set in $\mathbb{R}^n$. The *Orlicz space* $L_M^*(G)$ is the set of Lebesgue-measurable functions f on G satisfying the following growth condition for the *Orlicz norm* $\|f\|_M$:

$$\|f\|_M = \sup\left\{\int_G f(t)g(t)\,dt\colon \int_G N\big(g(t)\big)\,dt \leqslant 1\right\} < \infty.$$

The **Orlicz metric** is the norm metric $\|f - g\|_M$ on $L_M^*(G)$. It makes $L_M^*(G)$ a Banach space ([Orli32]).

When $M(u) = u^p$, $1 < p < \infty$, $L_M^*(G)$ coincides with the space $L_p(G)$, and, up to scalar factor, the L_p*-norm* $\|f\|_p$ coincides with $\|f\|_M$. Orlicz norm is equivalent to the *Luxemburg norm* $\|f\|_{(M)} = \inf\{\lambda > 0\colon \int_G M(\lambda^{-1}f(t))\,dt \leqslant 1\}$; in fact, $\|f\|_{(M)} \leqslant \|f\|_M \leqslant 2\|f\|_{(M)}$.

- **Orlicz–Lorentz metric**

Let $w : (0, \infty) \to (0, \infty)$ be a non-increasing function. Let $M : [0, \infty) \to [0, \infty)$ be a non-decreasing and convex function with $M(0) = 0$. Let G be a bounded closed set in $\mathbb{R}^n$.

The *Orlicz–Lorentz space* $L_{w,M}(G)$ is the set of all Lebesgue-measurable functions f on G satisfying the following growth condition for the *Orlicz–Lorentz norm* $\|f\|_{w,M}$:

$$\|f\|_{w,M} = \inf\left\{\lambda > 0\colon \int_0^\infty w(x)M\left(\frac{f^*(x)}{\lambda}\right)dx \leqslant 1\right\} < \infty,$$

where $f^*(x) = \sup\{t\colon \mu(|f| \geqslant t) \geqslant x\}$ is the *non-increasing rearrangement* of f.

The **Orlicz–Lorentz metric** is the **norm metric** $\|f - g\|_{w,M}$ on $L_{w,M}(G)$. It makes $L_{w,M}(G)$ a Banach space.

The Orlicz–Lorentz space is a generalization of the *Orlicz space* $L_M^*(G)$ (cf. **Orlicz metric**), and the *Lorentz space* $L_{w,q}(G)$, $1 \leqslant q < \infty$, of all Lebesgue-measurable functions f on G satisfying the following growth condition for the *Lorentz norm* $\|f\|_{w,q}$:

$$\|f\|_{w,q} = \left(\int_0^\infty w(x)\big(f^*(x)\big)^q\right)^{\frac{1}{q}} < \infty.$$

- **Hölder metric**

Let $L^{\alpha}(G)$ be the set of all bounded continuous functions f, defined on a subset G of $\mathbb{R}^n$, and satisfying the *Hölder condition* on G. Here, a function f satisfies the *Hölder condition* at a point $y \in G$ with *index* (or *order*) α, $0 < \alpha \leqslant 1$, and with coefficient $A(y)$, if $|f(x) - f(y)| \leqslant A(y)|x - y|^{\alpha}$ for all $x \in G$ sufficiently close to y. If $A = \sup_{y \in G}(A(y)) < \infty$, the Hölder condition is called *uniform* on G, and A is called *Hölder coefficient* of G. The quantity $|f|_{\alpha} = \sup_{x,y \in G} \frac{|f(x)-f(y)|}{|x-y|^{\alpha}}$, $0 \leqslant \alpha \leqslant 1$, is called *Hölder α-semi-norm* of f, and the *Hölder norm* of f is defined by

$$\|f\|_{L^{\alpha}(G)} = \sup_{x \in G}\bigl|f(x)\bigr| + |f|_{\alpha}.$$

The **Hölder metric** is the **norm metric** $\|f - g\|_{L^{\alpha}(G)}$ on $L^{\alpha}(G)$. It makes $L^{\alpha}(G)$ a Banach space.

- **Sobolev metric**

The *Sobolev space* $W^{k,p}$ is a subset of an L_p-space such that f and its derivatives up to order k have a finite L_p-*norm*. Formally, given a subset G of $\mathbb{R}^n$, define

$$W^{k,p} = W^{k,p}(G) = \bigl\{f \in L_p(G):\ f^{(i)} \in L_p(G), 1 \leqslant i \leqslant k\bigr\},$$

where $f^{(i)} = \partial_{x_1}^{\alpha_1} \dots \partial_{x_n}^{\alpha_n} f$, $\alpha_1 + \dots + \alpha_n = i$, and the derivatives are taken in a weak sense. The *Sobolev norm* on $W^{k,p}$ is defined by

$$\|f\|_{k,p} = \sum_{i=0}^{k} \bigl\|f^{(i)}\bigr\|_p.$$

In fact, it is enough to take only the first and last in the sequence, i.e., the norm defined by $\|f\|_{k,p} = \|f\|_p + \|f^{(k)}\|_p$ is equivalent to the norm above. For $p = \infty$, the Sobolev norm is equal to the *essential supremum* of $|f|$: $\|f\|_{k,\infty} = \operatorname{ess\,sup}_{x \in G} |f(x)|$, i.e., it is the infimum of all numbers $a \in \mathbb{R}$ for which $|f(x)| > a$ holds on a set of measure zero.

The **Sobolev metric** is the **norm metric** $\|f - g\|_{k,p}$ on $W^{k,p}$. It makes $W^{k,p}$ a Banach space.

The Sobolev space $W^{k,2}$ is denoted by H^k. It is a Hilbert space for the *inner product* $\langle f, g\rangle_k = \sum_{i=1}^{k}\langle f^{(i)}, g^{(i)}\rangle_{L_2} = \sum_{i=1}^{k}\int_G f^{(i)}\overline{g}^{(i)}\mu(d\omega)$.

Sobolev spaces are the modern replacement for the space C^1 (of functions having continuous derivatives) for solutions of *partial differential equations*.

- **Variable exponent space metrics**

Let G be a non-empty open subset of $\mathbb{R}^n$, and let $p : G \to [1, \infty)$ be a measurable bounded function, called *variable exponent*. The *variable exponent Lebesgue space* $L_{p(.)}(G)$ is the set of all measurable functions $f : G \to \mathbb{R}$ for which the *modular*

$\varrho_{p(.)}(f) = \int_G |f(x)|^{p(x)}\,dx$ is finite. The *Luxemburg norm* on this space is defined by

$$\|f\|_{p(.)} = \inf\{\lambda > 0:\ \varrho_{p(.)}(f/\lambda) \leqslant 1\}.$$

The **variable exponent Lebesgue space metric** is the **norm metric** $\|f - g\|_{p(.)}$ on $L_{p(.)}(G)$.

A *variable exponent Sobolev space* $W^{1,p(.)}(G)$ is a subspace of $L_{p(.)}(G)$ consisting of functions f whose distributional gradient exists almost everywhere and satisfies the condition $|\nabla f| \in L_{p(.)}(G)$. The norm

$$\|f\|_{1,p(.)} = \|f\|_{p(.)} + \|\nabla f\|_{p(.)}$$

makes $W^{1,p(.)}(G)$ a Banach space. The **variable exponent Sobolev space metric** is the norm metric $\|f - g\|_{1,p(.)}$ on $W^{1,p(.)}$.

- **Schwartz metric**

The *Schwartz space* (or *space of rapidly decreasing functions*) $S(\mathbb{R}^n)$ is the class of all *Schwartz functions*, i.e., infinitely-differentiable functions $f : \mathbb{R}^n \to \mathbb{C}$ that decrease at infinity, as do all their derivatives, faster than any inverse power of x. More precisely, f is a Schwartz function if we have the following growth condition:

$$\|f\|_{\alpha\beta} = \sup_{x\in\mathbb{R}^n}\left|x_1^{\beta_1}\dots x_n^{\beta_n}\frac{\partial^{\alpha_1+\dots+\alpha_n} f(x_1,\dots,x_n)}{\partial x_1^{\alpha_1}\dots\partial x_n^{\alpha_n}}\right| < \infty$$

for any non-negative integer vectors α and β. The family of *semi-norms* $\|.\|_{\alpha\beta}$ defines a locally convex topology of $S(\mathbb{R}^n)$ which is metrizable and complete. The **Schwartz metric** is a metric on $S(\mathbb{R}^n)$ which can be obtained using this topology (cf. **countably normed space**).

The corresponding metric space on $S(\mathbb{R}^n)$ is a *Fréchet space* in the sense of Functional Analysis, i.e., a locally convex F-*space*.

- **Bregman quasi-distance**

Let $G \subset \mathbb{R}^n$ be a closed set with the non-empty interior G^0. Let f be a *Bregman function with zone* G.

The **Bregman quasi-distance** $D_f : G \times G^0 \to \mathbb{R}_{\geqslant 0}$ is defined by

$$D_f(x, y) = f(x) - f(y) - \langle\nabla f(y), x - y\rangle,$$

where $\nabla f = (\frac{\partial f}{\partial x_1}, \dots \frac{\partial f}{\partial x_n})$. $D_f(x, y) = 0$ if and only if $x = y$, $D_f(x, y) + D_f(y, z) - D_f(x, z) = \langle\nabla f(z) - \nabla f(y), x - y\rangle$, but, in general, D_f does not satisfy the triangle inequality, and is not symmetric.

A real-valued function f whose effective domain contains G is called *Bregman function with zone* G if the following conditions hold:

1. f is continuously differentiable on G^0;
2. f is strictly convex and continuous on G;
3. For all $\delta \in \mathbb{R}$ the *partial level sets* $\Gamma(x, \delta) = \{y \in G^0 : D_f(x, y) \leqslant \delta\}$ are bounded for all $x \in G$;
4. If $\{y_n\}_n \subset G^0$ converges to y^*, then $D_f(y^*, y_n)$ converges to 0;
5. If $\{x_n\}_n \subset G$ and $\{y_n\}_n \subset G^0$ are sequences such that $\{x_n\}_n$ is bounded, $\lim_{n\to\infty} = y^*$, and $\lim_{n\to\infty} D_f(x_n, y_n) = 0$, then $\lim_{n\to\infty} x_n = y^*$.

When $G = \mathbb{R}^n$, a sufficient condition for a strictly convex function to be a Bregman function has the form: $\lim_{\|x\|\to\infty} \frac{f(x)}{\|x\|} = \infty$.

13.2. METRICS ON LINEAR OPERATORS

A *linear operator* is a function $T : V \to W$ between two vector spaces V, W over a field $\mathbb{F}$, that is compatible with their linear structures, i.e., for any $x, y \in V$ and any scalar $k \in \mathbb{F}$, we have the following properties: $T(x + y) = T(x) + T(y)$, and $T(kx) = kT(x)$.

- **Operator norm metric**

  Consider the set of all linear operators from a *normed space* $(V, \|.\|_V)$ into a normed space $(W, \|.\|_W)$. The *operator norm* $\|T\|$ of a *linear operator* $T : V \to W$ is defined as the largest value by which T stretches an element of V, i.e.,

$$\|T\| = \sup_{\|v\|_V \neq 0} \frac{\|T(v)\|_W}{\|v\|_V} = \sup_{\|v\|_V = 1} \|T(v)\|_W = \sup_{\|v\|_V \leqslant 1} \|T(v)\|_W.$$

  A linear operator $T : V \to W$ from a normed space V into a normed space W is called *bounded* if its operator norm is finite. For normed spaces, a linear operator is bounded if and only if it is *continuous*.

  The **operator norm metric** is a **norm metric** on the set $B(V, W)$ of all bounded linear operators from V into W, defined by

$$\|T - P\|.$$

  The space $(B(V, W), \|.\|)$ is called *space of bounded linear operators*. This metric space is **complete** if W is. If $V = W$ is complete, the space $B(V, V)$ is a *Banach algebra*, as the operator norm is a *sub-multiplicative norm*.

  A linear operator $T : V \to W$ from a Banach space V into another Banach space W is called *compact* if the image of any bounded subset of V is a relatively compact subset of W. Any compact operator is bounded (and, hence, continuous). The space $(K(V, W), \|.\|)$ on the set $K(V, W)$ of all compact operators from V into W with the operator norm $\|.\|$ is called *space of compact operators*.

- **Nuclear norm metric**

Let $B(V, W)$ be the space of all bounded linear operators mapping a Banach space $(V, \|.\|_V)$ into another Banach space $(W, \|.\|_W)$. Let the *Banach dual* of V be denoted by V', and the value of a functional $x' \in V'$ at a vector $x \in V$ by $\langle x, x' \rangle$. A linear operator $T \in B(V, W)$ is called *nuclear operator* if it can be represented in the form $x \mapsto T(x) = \sum_{i=1}^{\infty} \langle x, x'_i \rangle y_i$, where $\{x'_i\}_i$ and $\{y_i\}_i$ are sequences in V' and W, respectively, such that $\sum_{i=1}^{\infty} \|x'_i\|_{V'} \|y_i\|_W < \infty$. This representation is called *nuclear*, and can be regarded as an expansion of T as a sum of operators of rank 1 (i.e., with one-dimensional range). The *nuclear norm* of T is defined as

$$\|T\|_{nuc} = \inf \sum_{i=1}^{\infty} \|x'_i\|_{V'} \|y_i\|_W,$$

where the infimum is taken over all possible nuclear representations of T.

The **nuclear norm metric** is the **norm metric** $\|T - P\|_{nuc}$ on the set $N(V, W)$ of all nuclear operators mapping V into W. The space $(N(V, W), \|.\|_{nuc})$, called *space of nuclear operators*, is a Banach space.

A *nuclear space* is defined as a **locally convex** space for which all continuous linear functions into an arbitrary Banach space are nuclear operators. A nuclear space is constructed as a projective limit of Hilbert spaces H_α with the property that, for each $\alpha \in I$, one can find $\beta \in I$ such that $H_\beta \subset H_\alpha$, and the embedding operator $H_\beta \ni x \to x \in H_\alpha$ is a *Hilbert–Schmidt operator*. A *normed space* is nuclear if and only if it is finite-dimensional.

- **Finite nuclear norm metric**

Let $F(V, W)$ be the space of all *linear operators of finite rank* (i.e., with finite-dimensional range) mapping a Banach space $(V, \|.\|_V)$ into another Banach space $(W, \|.\|_W)$. A linear operator $T \in F(V, W)$ can be represented in the form $x \mapsto T(x) = \sum_{i=1}^{n} \langle x, x'_i \rangle y_i$, where $\{x'_i\}_i$ and $\{y_i\}_i$ are sequences in V' (*Banach dual* of V) and W, respectively, and $\langle x, x' \rangle$ is the value of a functional $x' \in V'$ at a vector $x \in V$. The *finite nuclear norm* of T is defined as

$$\|T\|_{fnuc} = \inf \sum_{i=1}^{n} \|x'_i\|_{V'} \|y_i\|_W,$$

where the infimum is taken over all possible finite representations of T.

The **finite nuclear norm metric** is the **norm metric** $\|T - P\|_{fnuc}$ on $F(V, W)$. The space $(F(V, W), \|.\|_{fnuc})$ is called *space of operators of finite rank*. It is a dense linear subspace of the *space of nuclear operators* $N(V, W)$.

- **Hilbert–Schmidt norm metric**

Consider the set of all linear operators from a Hilbert space $(H_1, \|.\|_{H_1})$ into a Hilbert space $(H_2, \|.\|_{H_2})$. The *Hilbert–Schmidt norm* $\|T\|_{HS}$ of a linear operator $T : H_1 \to H_2$

is defined by

$$\|T\|_{HS} = \left(\sum_{\alpha \in I} \|T(e_\alpha)\|^2_{H_2}\right)^{1/2},$$

where $(e_\alpha)_{\alpha \in I}$ is an orthonormal basis in H_1. A linear operator $T : H_1 \to H_2$ is called *Hilbert–Schmidt operator* if $\|T\|^2_{HS} < \infty$.

The **Hilbert–Schmidt norm metric** is the **norm metric** $\|T - P\|_{HS}$ on the set $S(H_1, H_2)$ of all Hilbert–Schmidt operators from H_1 into H_2.

For $H_1 = H_2 = H$, the algebra $S(H, H) = S(H)$ with the Hilbert–Schmidt norm is a *Banach algebra*. It contains operators of finite rank as a dense subset, and is contained in the space $K(H)$ of compact operators. An *inner product* $\langle , \rangle_{HS}$ on $S(H)$ is defined by $\langle T, P\rangle_{HS} = \sum_{\alpha \in I}\langle T(e_\alpha), P(e_\alpha)\rangle$, and $\|T\|_{HS} = \langle T, T\rangle^{1/2}_{HS}$. Therefore, $S(H)$ is a Hilbert space (independent on the choice basis $(e_\alpha)_{\alpha \in I}$).

- **Trace-class norm metric**

  Given a Hilbert space H, the *trace-class norm* of a linear operator $T : H \to H$ is defined by

  $$\|T\|_{tc} = \sum_{\alpha \in I}\big\langle |T|(e_\alpha), e_\alpha\big\rangle,$$

  where $|T|$ is the *absolute value* of T in the *Banach algebra* $B(H)$ of all bounded operators from H into itself, and $(e_\alpha)_{\alpha \in I}$ is an orthonormal basis of H. An operator $T : H \to H$ is called *trace-class operator* if $\|T\|_{tc} < \infty$. Any such operator is a product of two *Hilbert–Schmidt operators*.

  The **trace-class norm metric** is the **norm metric** $\|T - P\|_{tc}$ on the set $L(H)$ of all trace-class operators from H into itself. The set $L(H)$ with the norm $\|.\|_{tc}$ forms a Banach algebra which is contained in the algebra $K(H)$ (of all compact operators from H into itself), and contains the algebra $S(H)$ (of all Hilbert–Schmidt operators from H into itself).

- **Schatten p-class norm metric**

  Let $1 \leqslant p < \infty$. Given a separable Hilbert space H, the *Schatten p-class norm* of a compact linear operator $T : H \to H$ is defined by

  $$\|T\|^p_{Sch} = \left(\sum_n |s_n|^p\right)^{\frac{1}{p}},$$

  where $\{s_n\}_n$ is the sequence of *singular values* of T. A compact operator $T : H \to H$ is called *Schatten p-class operator*, if $\|T\|^p_{Sch} < \infty$.

  The **Schatten p-class norm metric** is the norm metric $\|T - P\|^p_{Sch}$ on the set $S_p(H)$ of all Schatten p-class operators from H onto itself. The set $S_p(H)$ with the norm $\|.\|^p_{Sch}$ forms a Banach space. $S_1(H)$ is the *trace-class* of H, and $S_2(H)$ is the *Hilbert–Schmidt class* of H (cf. also **Schatten norm metric**).

- **Continuous dual space**

Let $(V, \|.\|)$ be a *normed vector space*. Let V' be the set of all *continuous* linear functionals T from V into the base field ($\mathbb{R}$ or $\mathbb{C}$). Let $\|.\|'$ be the *operator norm* on V', defined by

$$\|T\|' = \sup_{\|x\| \leqslant 1} |T(x)|.$$

The space $(V', \|.\|')$ is a Banach space which is called **continuous dual** (or *Banach dual*) of $(V, \|.\|)$.

In fact, the continuous dual of the **metric space** l_p^n (l_p^∞) is l_q^n (l_q^∞, respectively). The continuous dual of l_1^n (l_1^∞) is l_∞^n (l_∞^∞, respectively). The continuous duals of the Banach spaces C (consisting of all convergent sequences, with the l_∞**-metric**) and C_0 (consisting of all sequences converging to zero, with the l_∞**-metric**) are both naturally identified with l_1^∞.

- **Distance constant of operator algebra**

Let $\mathcal{A}$ be an operator algebra contained in $B(H)$, the set of all bounded operators on a Hilbert space H. For any operator $T \in B(H)$ let $\beta(T, \mathcal{A}) = \sup\{\|P^\perp TP\|: P$ is a projection, and $P^\perp \mathcal{A} P = (0)\}$. Let $dist(T, \mathcal{A})$ be the *distance* of T from the algebra $\mathcal{A}$, i.e., the smallest norm of an operator $T - A$, where A runs over $\mathcal{A}$. The smallest positive constant C (if it exists) such that, for any operator $T \in B(H)$,

$$dist(T, \mathcal{A}) \leqslant C\beta(T, \mathcal{A})$$

holds, is called **distance constant** for the algebra $\mathcal{A}$.

Chapter 14

Distances in Probability Theory

A *probability space* is a *measurable space* $(\Omega, \mathcal{A}, P)$, where $\mathcal{A}$ is the set of all measurable subsets of Ω, and P is a measure on $\mathcal{A}$ with $P(\Omega) = 1$. The set Ω is called *sample space*. An element $a \in \mathcal{A}$ is called an *event*, in particular, an *elementary event* is a subset of Ω that contains only one element; $P(a)$ is called *probability* of the event a. The measure P on $\mathcal{A}$ is called *probability measure*, or *(probability) distribution law*, or simply *(probability) distribution*.

A *random variable* X is a measurable function from a probability space $(\Omega, \mathcal{A}, P)$ into a measurable space, called a *state space* of possible values of the variable; it is usually taken to be the real numbers with the *Borel σ-algebra*, so $X : \Omega \to \mathbb{R}$. The range $\mathcal{X}$ of random variable X is called *support* of distribution P; an element $x \in \mathcal{X}$ is called a *state*.

A distribution law can be uniquely described via a *cumulative distribution function* (CDF, *distribution function, cumulative density function*) $F(x)$ which describes the probability that a random value X takes on a value at most x: $F(x) = P(X \leqslant x) = P(\omega \in \Omega : X(\omega) \leqslant x)$.

So, any random variable X gives rise to a *probability distribution* which assigns to the interval $[a, b]$ the probability $P(a \leqslant X \leqslant b) = P(\omega \in \Omega : a \leqslant X(\omega) \leqslant b)$, i.e., the probability that the variable X will take a value in the interval $[a, b]$.

A distribution is called *discrete* if $F(x)$ consists of a sequence of finite jumps at x_i; a distribution is called *continuous* if $F(x)$ is continuous. We consider (as in majority of applications) only discrete or *absolutely continuous* distributions, i.e., CDF function $F : \mathbb{R} \to \mathbb{R}$ is *absolutely continuous*. It means that, for every number $\varepsilon > 0$, there is a number $\delta > 0$ such that, for any sequence of pairwise disjoint intervals $[x_k, y_k]$, $1 \leqslant k \leqslant n$, the inequality $\sum_{1 \leqslant k \leqslant n}(y_k - x_k) < \delta$ implies the inequality $\sum_{1 \leqslant k \leqslant n}|F(y_k) - F(x_k)| < \varepsilon$.

A distribution law also can be uniquely defined via a *probability density function* (PDF, *density function, probability function*) $p(x)$ of the underlying random variable. For an absolutely continuous distribution, CDF is almost everywhere differentiable, and PDF is defined as the derivative $p(x) = F'(x)$ of the CDF; so, $F(x) = P(X \leqslant x) = \int_{-\infty}^{x} p(t)\,dt$, and $\int_a^b p(t)\,dt = P(a \leqslant X \leqslant b)$. In the discrete case, PDF (the density of the random variable X) is defined by its values $p(x_i) = P(X = x)$; so $F(x) = \sum_{x_i \leqslant x} p(x_i)$. In contrast, each elementary events has probability zero in any continuous case.

The random variable X is used to "push-forward" the measure P on Ω to a measure dF on $\mathbb{R}$. The underlying probability space is a technical device used to guarantee the existence of random variables and sometimes to construct them.

Probability metrics between distributions are called *simple metrics*, while metrics between random variables are called *compound metrics*; see [Rach91]. For simplicity, we usually present the discrete version of probability metrics, but many of them are defined on any measurable space. For probability metrics d, the condition $P(X = Y) = 1$ implies (and characterizes) $d(X, Y) = 0$. In many cases, some *ground* distance d is given on the state space $\mathcal{X}$ and presented distance is a lifting of it to a distance on distributions.

In Statistics, many of distances below, between distributions P_1 and P_2, are used as measures of *goodness of fit* between estimated, P_2, and theoretical, P_1, distributions.

Below we use notation $\mathbb{E}[X]$ for the *expected value* (or *mean*) of the random variable X: in discrete case $\mathbb{E}[X] = \sum_x xp(x)$, in continuous case $\mathbb{E}[X] = \int xp(x)\,dx$. The *variance* of X is $\mathbb{E}[(X - \mathbb{E}[X])^2]$. Also we denote $p_X = p(x) = P(X = x)$, $F_X = F(x) = P(X \leqslant x)$, $p(x, y) = P(X = x, Y = y)$.

14.1. DISTANCES ON RANDOM VARIABLES

All distances in this section are defined on the set $\mathbf{Z}$ of all random variables with the same support $\mathcal{X}$; here $X, Y \in \mathbf{Z}$.

- **L_p-metric between variables**

  The **L_p-metric between variables** is a metric on $\mathbf{Z}$ with $\mathcal{X} \subset \mathbb{R}$ and $\mathbb{E}[|Z|^p] < \infty$ for all $Z \in \mathbf{Z}$, defined by

$$\left(\mathbb{E}\left[|X - Y|^p\right]\right)^{1/p} = \left(\sum_{(x,y)\in\mathcal{X}\times\mathcal{X}} |x - y|^p p(x, y)\right)^{1/p}.$$

  For $p = 1, 2$ and ∞, it is called, respectively, *engineer metric*, *mean-square distance* and *essential supremum distance between variables*.

- **Indicator metric**

  The **indicator metric** is a metric on $\mathbf{Z}$, defined by

$$\mathbb{E}[1_{X\neq Y}] = \sum_{(x,y)\in\mathcal{X}\times\mathcal{X}} 1_{x\neq y}\, p(x, y) = \sum_{(x,y)\in\mathcal{X}\times\mathcal{X},\, x\neq y} p(x, y).$$

  (Cf. **Hamming metric**.)

- **Ky-Fan metric K**

  The **Ky-Fan metric** K is a metric K on $\mathbf{Z}$, defined by

$$\inf\left\{\varepsilon > 0:\ P\left(|X - Y| > \varepsilon\right) < \varepsilon\right\}.$$

  It is the case $d(x, y) = |X - Y|$ of the **probability distance** below.

- **Ky-Fan metric K^***

  The **Ky-Fan metric** K^* is a metric K^* on **Z**, defined by

$$\mathbb{E}\left[\frac{|X-Y|}{1+|X-Y|}\right] = \sum_{(x,y)\in\mathcal{X}\times\mathcal{X}} \frac{|x-y|}{1+|x-y|}p(x,y).$$

- **Probability distance**

  Given a metric space $(\mathcal{X}, d)$, the **probability distance** on **Z** is defined by

$$\inf\{\varepsilon:\ P\big(d(X,Y)>\varepsilon\big)<\varepsilon\}.$$

14.2. DISTANCES ON DISTRIBUTION LAWS

All distances in this section are defined on the set $\mathcal{P}$ of all distribution laws such that corresponding random variables have the same range $\mathcal{X}$; here $P_1, P_2 \in \mathcal{P}$.

- **L_p-metric between densities**

  The **L_p-metric between densities** is a metric on $\mathcal{P}$ (for a countable $\mathcal{X}$), defined, for any $p > 0$, by

$$\left(\sum_x |p_1(x)-p_2(x)|^p\right)^{\min(1,\frac{1}{p})}.$$

  For $p = 1$, its half is called **total variation metric** (or *distance in variation*, *trace-distance*). The *point metric* $\sup_x |p_1(x) - p_2(x)|$ corresponds to $p = \infty$.

- **Mahalanobis semi-metric**

  The **Mahalanobis semi-metric** (or *quadratic distance*, *quadratic metric*) is a semi-metric on $\mathcal{P}$ (for $\mathcal{X} \subset \mathbb{R}^n$), defined by

$$\sqrt{(\mathbb{E}_{P_1}[X]-\mathbb{E}_{P_2}[X])^T A^{-1}(\mathbb{E}_{P_1}[X]-\mathbb{E}_{P_2}[X])}$$

  for a given positive-definite matrix A.

- **Engineer semi-metric**

  The **engineer semi-metric** is a semi-metric on $\mathcal{P}$ (for $\mathcal{X} \subset \mathbb{R}$), defined by

$$\big|\mathbb{E}_{P_1}[X]-\mathbb{E}_{P_2}[X]\big| = \left|\sum_x x\big(p_1(x)-p_2(x)\big)\right|.$$

- **Stop-loss metric of order m**

  The **stop-loss metric of order** m is a metric on $\mathcal{P}$ (for $\mathcal{X} \subset \mathbb{R}$), defined by

$$\sup_{t\in\mathbb{R}} \sum_{x\geqslant t} \frac{(x-t)^m}{m!}\big(p_1(x)-p_2(x)\big).$$

- **Kolmogorov–Smirnov metric**

The **Kolmogorov–Smirnov metric** (or *Kolmogorov metric*, *uniform metric*) is a metric on $\mathcal{P}$ (for $\mathcal{X} \subset \mathbb{R}$), defined by

$$\sup_{x \in \mathbb{R}} \bigl| P_1(X \leqslant x) - P_2(X \leqslant x) \bigr|.$$

The **Kuiper distance** on $\mathcal{P}$ is defined by

$$\sup_{x \in \mathbb{R}} \bigl(P_1(X \leqslant x) - P_2(X \leqslant x) \bigr) + \sup_{x \in \mathbb{R}} \bigl(P_2(X \leqslant x) - P_1(X \leqslant x) \bigr).$$

(Cf. **Pompeiu–Eggleston metric** on convex bodies.)

The **Anderson–Darling distance** on $\mathcal{P}$ is defined by

$$\sup_{x \in \mathbb{R}} \frac{|(P_1(X \leqslant x) - P_2(X \leqslant x)|}{\ln \sqrt{(P_1(X \leqslant x)(1 - P_1(X \leqslant x))}}.$$

The **Crnkovic–Drachma distance** is defined by

$$\sup_{x \in \mathbb{R}} \bigl(P_1(X \leqslant x) - P_2(X \leqslant x) \bigr) \ln \frac{1}{\sqrt{(P_1(X \leqslant x)(1 - P_1(X \leqslant x))}}$$
$$+ \sup_{x \in \mathbb{R}} \bigl(P_2(X \leqslant x) - P_1(X \leqslant x) \bigr) \ln \frac{1}{\sqrt{(P_1(X \leqslant x)(1 - P_1(X \leqslant x))}}.$$

Above three distances are used in Statistics as measures of *goodness of fit*, especially, for VaR (Value at Risk) measurements in Finance.

- **Cramer–von Mises distance**

The **Cramer–von Mises distance** is a distance on $\mathcal{P}$ (for $\mathcal{X} \subset \mathbb{R}$), defined by

$$\int_{-\infty}^{+\infty} \bigl(P_1(X \leqslant x) - P_2(X \leqslant x) \bigr)^2 \, dx.$$

This is the squared L_2**-metric** between cumulative density functions.

- **Levy metric**

The **Levy metric** is a metric on $\mathcal{P}$ (for $\mathcal{X} \subset \mathbb{R}$ only), defined by

$$\inf \bigl\{ \varepsilon > 0: \ P_1(X \leqslant x - \varepsilon) - \varepsilon \leqslant P_2(X \leqslant x) \leqslant P_1(X \leqslant x + \varepsilon) + \varepsilon \text{ for any } x \in \mathbb{R} \bigr\}.$$

It is a special case of the **Prokhorov metric** for $(\mathcal{X}, d) = (\mathbb{R}, |x - y|)$.

- **Prokhorov metric**

Given a metric space $(\mathcal{X}, d)$, the **Prokhorov metric** on $\mathcal{P}$ is defined by

$$\inf\{\varepsilon > 0: \ P_1(X \in B) \leqslant P_2(X \in B^{\varepsilon}) + \varepsilon \text{ and } P_2(X \in B) \leqslant P_1(X \in B^{\varepsilon}) + \varepsilon\},$$

where B is any Borel subset of $\mathcal{X}$, and $B^{\varepsilon} = \{x: \ d(x, y) < \varepsilon, y \in B\}$.

It is the smallest (over all joint distributions of pairs (X, Y) of random variables X, Y such that marginal distributions of X and Y are P_1 and P_2, respectively) **probability distance** between random variables X and Y.

- **Dudley metric**

Given a metric space $(\mathcal{X}, d)$, the **Dudley metric** on $\mathcal{P}$ is defined by

$$\sup_{f \in F} \left|\mathbb{E}_{P_1}[f(X)] - \mathbb{E}_{P_2}[f(X)]\right| = \sup_{f \in F} \left|\sum_{x \in \mathcal{X}} f(x)\big(p_1(x) - p_2(x)\big)\right|,$$

where

$$F = \{f: \ \mathcal{X} \to \mathbb{R}, \|f\|_{\infty} + Lip_d(f) \leqslant 1\}, \text{ and } Lip_d(f) = \sup_{x, y \in \mathcal{X}, x \neq y} \frac{|f(x) - f(y)|}{d(x, y)}.$$

- **Szulga metric**

Given a metric space $(\mathcal{X}, d)$, the **Szulga metric** on $\mathcal{P}$ is defined by

$$\sup_{f \in F} \left|\left(\sum_{x \in \mathcal{X}} |f(x)|^p p_1(x)\right)^{1/p} - \left(\sum_{x \in \mathcal{X}} |f(x)|^p p_2(x)\right)^{1/p}\right|,$$

where $F = \{f: \ X \to \mathbb{R}, Lip_d(f) \leqslant 1\}$, and $Lip_d(f) = \sup_{x, y \in \mathcal{X}, x \neq y} \frac{|f(x) - f(y)|}{d(x, y)}$.

- **Zolotarev semi-metric**

The **Zolotarev semi-metric** is a semi-metric on $\mathcal{P}$, defined by

$$\sup_{f \in F} \left|\sum_{x \in \mathcal{X}} f(x)\big(p_1(x) - p_2(x)\big)\right|,$$

where F is any set of functions $f: \mathcal{X} \to \mathbb{R}$ (in the continuous case, F is any set of such bounded continuous functions); cf. **Szulga metric**, **Dudley metric**.

- **Convolution metric**

Let G be a separable locally compact Abelian group, and let $C(G)$ be the set of all real bounded continuous function on G vanishing at infinity. Fix a function $g \in C(G)$ such that $|g|$ is integrable with respect to Haar measure on G, and $\{\beta \in G^*: \ \widehat{g}(\beta) = 0\}$ has empty interior; here G^* is the dual group of G, and $\widehat{g}$ is the Fourier transform of g.

Yukich's **convolution metric** (or *smoothing metric*) is defined, for any two finite signed Baire measures, P_1 and P_2, on G, by

$$\sup_{x\in G}\left|\int_{y\in G} g(xy^{-1})(dP_1 - dP_2)(y)\right|.$$

This metric can also be seen as the difference $T_{P_1}(g) - T_{P_2}(g)$ of *convolution operators* on $C(G)$, where, for any $f \in C(G)$, the operator $T_P f(x)$ is $\int_{y\in G} f(xy^{-1})\,dP(y)$.

- **Discrepancy metric**

  Given a metric space $(\mathcal{X}, d)$, the **discrepancy metric** on $\mathcal{P}$ is defined by

$$\sup\bigl\{\left|P_1(X \in B) - P_2(X \in B)\right|:\ B \text{ is any closed ball}\bigr\}.$$

- **Bi-discrepancy semi-metric**

  The **bi-discrepancy semi-metric** is a semi-metric, evaluating the proximity of distributions P_1, P_2 defined over different collections $\mathcal{A}_1$, $\mathcal{A}_2$ of measurable sets in the following way:

$$D(P_1, P_2) + D(P_2, P_1),$$

  where $D(P_1, P_2) = \sup\{\inf\{P_2(C):\ B \subset C \in \mathcal{A}_2\} - P_1(B):\ B \in \mathcal{A}_1\}$ (*discrepancy*).

- **Le Cam distance**

  The **Le Cam distance** is a semi-metric, evaluating the proximity of probability distributions P_1, P_2 defined on different spaces $\mathcal{X}_1$, $\mathcal{X}_2$ in the following way:

$$\max\bigl\{\delta(P_1, P_2), \delta(P_2, P_1)\bigr\},$$

  where $\delta(P_1, P_2) = \inf_B \sum_{x_2\in\mathcal{X}_2} |BP_1(X_2 = x_2) - BP_2(X_2 = x_2)|$ is the *Le Cam deficiency*. Here $BP_1(X_2 = x_2) = \sum_{x_1\in\mathcal{X}_1} p_1(x_1)b(x_2|x_1)$, where B is a probability distribution over $\mathcal{X}_1 \times \mathcal{X}_2$, and

$$b(x_2|x_1) = \frac{B(X_1 = x_1, X_2 = x_2)}{B(X_1 = x_1)} = \frac{B(X_1 = x_1, X_2 = x_2)}{\sum_{x\in\mathcal{X}_2} B(X_1 = x_1, X_2 = x)}.$$

  So, $BP_2(X_2 = x_2)$ is a probability distribution over $\mathcal{X}_2$, since $\sum_{x_2\in\mathcal{X}_2} b(x_2|x_1) = 1$.

  Le Cam distance is not a probability distance, since P_1 and P_2 are defined over different spaces; it is a distance between statistical experiments (models).

- **Skorokhod–Billingsley metric**

  The **Skorokhod–Billingsley metric** is a metric on $\mathcal{P}$, defined by

$$\inf_f \max\left\{\sup_x\left|P_1(X \leqslant x) - P_2\bigl(X \leqslant f(x)\bigr)\right|, \sup_x\left|f(x) - x\right|, \sup_{x\neq y}\left|\ln\frac{f(y) - f(x)}{y - x}\right|\right\},$$

  where $f : \mathbb{R} \to \mathbb{R}$ is any strictly increasing continuous function.

- **Skorokhod metric**

The **Skorokhod metric** is a metric on $\mathcal{P}$, defined by

$$\inf\Bigl\{\varepsilon > 0: \ \max\Bigl\{\sup_x \bigl|P_1(X < x) - P_2\bigl(X \leqslant f(x)\bigr)\bigr|, \sup_x \bigl|f(x) - x\bigr|\Bigr\} < \varepsilon\Bigr\},$$

where $f : \mathbb{R} \to \mathbb{R}$ is a strictly increasing continuous function.

- **Birnbaum–Orlicz distance**

The **Birnbaum–Orlicz distance** is a distance on $\mathcal{P}$, defined by

$$\sup_{x \in \mathbb{R}} f\bigl(\bigl|P_1(X \leqslant x) - P_2(X \leqslant x)\bigr|\bigr),$$

where $f : \mathbb{R}_+ \to \mathbb{R}_+$ is any non-decreasing continuous function with $f(0) = 0$, and $f(2t) \leqslant Kf(t)$ for any $t > 0$ and some fixed K. It is a **near-metric**, since it holds $d(P_1, P_2) \leqslant K(d(P_1, P_3) + d(P_3, P_2))$.

Birnbaum–Orlicz distance is also used, in Functional Analysis, on the set of all integrable functions on the segment [0, 1], where it is defined by $\int_0^1 H(|f(x) - g(x)|)\,dx$, where H is a non-decreasing continuous function from $[0, \infty)$ onto $[0, \infty)$ which vanishes at the origin and satisfies the *Orlicz condition*: $\sup_{t>0} \frac{H(2t)}{H(t)} < \infty$.

- **Kruglov distance**

The **Kruglov distance** is a distance on $\mathcal{P}$, defined by

$$\int f\bigl(P_1(X \leqslant x) - P_2(X \leqslant x)\bigr)\,dx,$$

where $f : \mathbb{R}_+ \to \mathbb{R}_+$ is any even strictly increasing function with $f(0) = 0$, and $f(s+t) \leqslant K(f(s) + f(t))$ for any $s, t \geqslant 0$ and some fixed $K \geqslant 1$. It is a **near-metric**, since it holds $d(P_1, P_2) \leqslant K(d(P_1, P_3) + d(P_3, P_2))$.

- **Burbea–Rao distance**

Consider a continuous convex function $\phi(t) : (0, \infty) \to \mathbb{R}$ and put $\phi(0) = \lim_{t \to 0} \phi(t) \in (-\infty, \infty]$. The convexity of ϕ implies non-negativity of the function $\delta_\phi : [0, 1]^2 \to (-\infty, \infty]$, defined by $\delta_\phi(x, y) = \frac{\phi(x)+\phi(y)}{2} - \phi(\frac{x+y}{2})$ if $(x, y) \neq (0, 0)$, and $\delta_\phi(0, 0) = 0$.

The corresponding **Burbea–Rao distance** on $\mathcal{P}$ is defined by

$$\sum_x \delta_\phi\bigl(p_1(x), p_2(x)\bigr).$$

- **Bregman distance**

Consider a differentiable convex function $\phi(t) : (0, \infty) \to \mathbb{R}$, and put $\phi(0) = \lim_{t \to 0} \phi(t) \in (-\infty, \infty]$. The convexity of ϕ implies that the function $\delta_\phi : [0, 1]^2 \to$

$(-\infty, \infty]$ defined by continuous extension of $\delta_\phi(u, v) = \phi(u) - \phi(v) - \phi'(v)(u - v)$, $0 < u, v \leqslant 1$, on $[0, 1]^2$ is non-negative.

The corresponding **Bregman distance** on $\mathcal{P}$ is defined by

$$\sum_1^m \delta_\phi(p_i, q_i).$$

(Cf. **Bregman quasi-distance**.)

- **f-divergence of Csizar**

The **f-divergence of Csizar** is a function on $\mathcal{P} \times \mathcal{P}$, defined by

$$\sum_x p_2(x) f\left(\frac{p_1(x)}{p_2(x)}\right),$$

where f is a convex function $f : \mathbb{R}_+ \to \mathbb{R}$.

The cases $f(t) = t \ln t$ and $f(t) = (t - 1)^2/2$ correspond to the **Kullback–Leibler distance** and to the **χ^2-distance** below, respectively. The case $f(t) = |t - 1|$ corresponds to the **L_1-metric between densities**, and the case $f(t) = 4(1 - \sqrt{t})$ (as well as $f(t) = 2(t + 1) - 4\sqrt{t}$) corresponds to the squared **Hellinger metric**.

Semi-metrics can also be obtained, as the square root of the f-divergence of Csizar, in the cases $f(t) = (t - 1)^2/(t + 1)$ (the **Vajda–Kus semi-metric**), $f(t) = |t^a - 1|^{1/a}$ with $0 < a \leqslant 1$ (the **Matusita semi-metric**), and

$$f(t) = \frac{(t^a + 1)^{1/a} - 2^{(1-a)/a}(t + 1)}{1 - 1/\alpha}$$

(the **Osterreicher semi-metric**).

- **Fidelity similarity**

The **fidelity similarity** (or *Bhattacharya coefficient*, *Hellinger affinity*) on $\mathcal{P}$ is defined by

$$\rho(P_1, P_2) = \sum_x \sqrt{p_1(x) p_2(x)}.$$

- **Hellinger metric**

In terms of **fidelity similarity**, the **Hellinger metric** (or *Hellinger–Kakutani metric*) on $\mathcal{P}$ is defined by

$$\left(2 \sum_x (\sqrt{p_1(x)} - \sqrt{p_2(x)})^2\right)^{\frac{1}{2}} = 2(1 - \rho(P_1, P_2))^{\frac{1}{2}}.$$

This is the **L_2-metric** between the square roots of density functions.

- **Harmonic mean similarity**

  The **harmonic mean similarity** is a similarity on $\mathcal{P}$, defined by

$$2\sum_x \frac{p_1(x)p_2(x)}{p_1(x)+p_2(x)}.$$

- **Bhattacharya distance 1**

  In terms of **fidelity similarity**, the **Bhattacharya distance** 1 on $\mathcal{P}$ is defined by

$$\left(\arccos \rho(P_1, P_2)\right)^2.$$

  Twice this distance is used also in Statistics and Machine Learning, where it is called **Fisher distance**.

- **Bhattacharya distance 2**

  In terms of **fidelity similarity**, the **Bhattacharya distance** 2 on $\mathcal{P}$ is defined by

$$-\ln \rho(P_1, P_2).$$

- **χ^2-distance**

  The **χ^2-distance** (or **Neyman χ^2-distance**) is a quasi-distance on $\mathcal{P}$, defined by

$$\sum_x \frac{(p_1(x)-p_2(x))^2}{p_2(x)}.$$

  The **Pearson χ^2-distance** is

$$\sum_x \frac{(p_1(x)-p_2(x))^2}{p_1(x)}.$$

  Probabilistic **symmetric χ^2-measure** is a distance on $\mathcal{P}$, defined by

$$2\sum_x \frac{(p_1(x)-p_2(x))^2}{p_1(x)+p_2(x)}.$$

- **Separation distance**

  The **separation distance** is a quasi-distance on $\mathcal{P}$ (for any countable $\mathcal{X}$), defined by

$$\max_x\left(1-\frac{p_1(x)}{p_2(x)}\right).$$

  (Not to be confused with **separation distance** between convex bodies.)

- **Kullback–Leibler distance**

The **Kullback–Leibler distance** (or *relative entropy*, *information deviation*, *KL-distance*) is a quasi-distance on $\mathcal{P}$, defined by

$$KL(P_1, P_2) = \mathbb{E}_{P_1}[\ln L] = \sum_x p_1(x) \ln \frac{p_1(x)}{p_2(x)},$$

where $L = \frac{p_1(x)}{p_2(x)}$ is the *likelihood ratio*. Therefore,

$$KL(P_1, P_2) = -\sum_x \big(p_1(x) \ln p_2(x)\big) + \sum_x \big(p_1(x) \ln p_1(x)\big) = H(P_1, P_2) - H(P_1),$$

where $H(P_1)$ is the *entropy* of P_1, and $H(P_1, P_2)$ is the *cross-entropy* of P_1 and P_2.

If P_2 is the product of marginals of P_1, the KL-distance $KL(P_1, P_2)$ is called **Shannon information quantity** and is equal to $\sum_{(x,y)\in\mathcal{X}\times\mathcal{X}} p_1(x, y) \ln \frac{p_1(x,y)}{p_1(x)p_1(y)}$ (cf. **Shannon distance**).

- **Skew divergence**

The **skew divergence** is a quasi-distance on $\mathcal{P}$, defined by

$$KL(P_1, aP_2 + (1 - a)P_1),$$

where $a \in [0, 1]$ is a constant, and KL is the **Kullback–Leibler distance**. So, the case $a = 1$ corresponds to $KL(P_1, P_2)$. The skew divergence with $a = \frac{1}{2}$ is called *K-divergence*.

- **Jeffrey divergence**

The **Jeffrey divergence** (or *J-divergence*) is a symmetric version of the **Kullback–Leibler distance**, defined by

$$KL(P_1, P_2) + KL(P_2, P_1) = \sum_x \left(p_1(x) \ln \frac{p_1(x)}{p_2(x)} + p_2(x) \ln \frac{p_2(x)}{p_1(x)} \right).$$

For $P_1 \to P_2$, the Jeffrey divergence behaves like the **χ^2-distance**.

- **Jensen–Shannon divergence**

The **Jensen–Shannon divergence** is defined by

$$aKL(P_1, P_3) + (1 - a)KL(P_2, P_3),$$

where $P_3 = aP_1 + (1 - a)P_2$, and $a \in [0, 1]$ is a constant (cf. **clarity similarity**).

In terms of *entropy* $H(P) = \sum_x p(x) \ln p(x)$, the Jensen–Shannon divergence is equal to $H(aP_1 + (1 - a)P_2) - aH(P_1) - (1 - a)H(P_2)$.

The **Topsøe distance** is a symmetric version of the **Kullback–Leibler distance** on $\mathcal{P}$, defined by

$$KL(P_1, P_3) + KL(P_2, P_3) = \sum_x \left(p_1(x) \ln \frac{p_1(x)}{p_3(x)} + p_2(x) \ln \frac{p_2(x)}{p_3(x)} \right),$$

where $P_3 = \frac{1}{2}(P_1 + P_2)$. The Topsøe distance is twice the Jensen–Shannon divergence with $a = \frac{1}{2}$. Some authors use term "Jensen–Shannon divergence" only for this value of a. It is not a metric, but its square root is a metric.

- **Resistor-average distance**

Johnson–Simanović's **resistor-average distance** is a symmetric version of the **Kullback–Leibler distance** on $\mathcal{P}$ which is defined by the harmonic sum

$$\left(\frac{1}{KL(P_1, P_2)} + \frac{1}{KL(P_2, P_1)} \right)^{-1}.$$

(Cf. **resistance metric** for graphs.)

- **Ali–Silvey distance**

The **Ali–Silvey distance** is a quasi-distance on $\mathcal{P}$, defined by the functional

$$f\left(\mathbb{E}_{P_1}[g(L)]\right),$$

where $L = \frac{p_1(x)}{p_2(x)}$ is the *likelihood ratio*, f is a non-decreasing function, and g is a continuous convex function (cf. **f-divergence of Csizar**).

The case $f(x) = x$, $g(x) = x \ln x$ corresponds to the **Kullback–Leibler distance**; the case $f(x) = -\ln x$, $g(x) = x^t$ corresponds to the **Chernoff distance**.

- **Chernoff distance**

The **Chernoff distance** (or *Rényi cross-entropy*) is a distance on $\mathcal{P}$, defined by

$$\max_{t \in [0,1]} D_t(P_1, P_2),$$

where $D_t(P_1, P_2) = -\ln \sum_x (p_1(x))^t (p_2(x))^{1-t}$, which is proportional to the **Rényi distance**.

The case $t = \frac{1}{2}$ corresponds to the **Bhattacharya distance** 2.

- **Rényi distance**

The **Rényi distance** (or *order t Rényi entropy*) is a quasi-distance on $\mathcal{P}$, defined by

$$\frac{1}{t-1} \ln \sum_x p_2(x) \left(\frac{p_1(x)}{p_2(x)} \right)^t,$$

where $t \geqslant 0$, $t \neq 1$.

The limit of the Rényi distance, for $t \to 1$, is the **Kullback–Leibler distance**. For $t = \frac{1}{2}$, the half of the Rényi distance is the **Bhattacharya distance** 2 (cf. **f-divergence of Csizar** and **Chernoff distance**).

- **Clarity similarity**

The **clarity similarity** is a similarity on $\mathcal{P}$, defined by

$$\begin{aligned}&\big(KL(P_1, P_3) + KL(P_2, P_3)\big) - \big(KL(P_1, P_2) + KL(P_2, P_1)\big)\\&\quad = \sum_x \left(p_1(x) \ln \frac{p_2(x)}{p_3(x)} + p_2(x) \ln \frac{p_1(x)}{p_3(x)} \right),\end{aligned}$$

where KL is the **Kullback–Leibler distance**, and P_3 is a fixed referential probability law. It was introduced in [CCL01] with P_3 being the probability distribution of General English.

- **Shannon distance**

Given a *measure space* $(\Omega, \mathcal{A}, P)$, where the set Ω is finite, and P is a probability measure, the *entropy* of a function $f : \Omega \to X$, where X is a finite set, is defined by

$$H(f) = \sum_{x \in X} P(f = x) \ln(P(f = x));$$

so, f can be seen as a *partition* of the measure space. For any two such partitions $f : \Omega \to X$ and $g : \Omega \to Y$, denote by $H(f, g)$ the entropy of the partition $(f, g) : \Omega \to X \times Y$ (*joint entropy*), and by $H(f|g)$ the *conditional entropy*; then the **Shannon distance** between f and g is defined by

$$2H(f, g) - H(f) - H(g) = H(f|g) + H(g|f).$$

It is a metric. The **Shannon information quantity** is defined by

$$H(f, g) - H(f) - H(g) = \sum_{(x,y)} p(f = x, g = y) \ln \frac{p(f = x, g = y)}{p(f = x) p(g = y)}.$$

If P is uniform probability law, then V. Goppa showed that the Shannon distance can be obtained as a limit case of the **finite subgroup metric**.

In general, the **information metric** (or **entropy metric**) between two random variables (information sources) X and Y is defined by

$$H(X|Y) + H(Y|X),$$

where the *conditional entropy* $H(X|Y)$ is defined by $\sum_{x \in X} \sum_{y \in Y} p(x, y) \ln p(x|y)$, and $p(x|y) = P(X = x|Y = y)$ is the conditional probability.

The **normalized information metric** is defined by

$$\frac{H(X|Y)+H(Y|X)}{H(X,Y)}.$$

It is equal to 1 if X and Y are independent. (Cf. a different one, **normalized information distance**).

- **Kantorovich–Mallows–Monge–Wasserstein metric**

  Given a metric space $(\mathcal{X},d)$, the **Kantorovich–Mallows–Monge–Wasserstein metric** is defined by

$$\inf \mathbb{E}_S\big[d(X,Y)\big],$$

  where the infimum is taken over all joint distributions S of pairs (X,Y) of random variables X,Y such that marginal distributions of X and Y are P_1 and P_2.

  For any **separable** metric space $(\mathcal{X},d)$, this is equivalent to the **Lipschitz distance between measures** $\sup_f \int f d(P_1-P_2)$, where the supremum is taken over all functions f with $|f(x)-f(y)|\leqslant d(x,y)$ for any $x,y\in\mathcal{X}$.

  More generally, the L_p**-Wasserstein distance** for $\mathcal{X}=\mathbb{R}^n$ is defined by

$$\big(\inf \mathbb{E}_S\big[d^p(X,Y)\big]\big)^{1/p},$$

  and, for $p=1$, it is called also $\overline{\rho}$*-distance*. For $(\mathcal{X},d)=(\mathbb{R},|x-y|)$, it is called also L_p*-metric between distribution functions* (CDF), and can be written as

$$\begin{aligned}\big(\inf \mathbb{E}\big[|X-Y|^p\big]\big)^{1/p} &= \left(\int_{\mathbb{R}}\big|F_1(x)-F_2(x)\big|^p\,dx\right)^{1/p}\\ &= \left(\int_0^1\big|F_1^{-1}(x)-F_2^{-1}(x)\big|^p\,dx\right)^{1/p}\end{aligned}$$

  with $F_i^{-1}(x)=\sup_u(P_i(X\leqslant x)<u)$.

  The case $p=1$ of this metric is called **Monge–Kantorovich metric** (or *Hutchinson metric* in Fractal theory), **Wasserstein metric** (or *Fortet–Mourier metric*).

- **Ornstein $\overline{d}$-metric**

  The **Ornstein $\overline{d}$-metric** is a metric on $\mathcal{P}$ (for $\mathcal{X}=\mathbb{R}^n$), defined by

$$\frac{1}{n}\inf\int_{x,y}\left(\sum_{i=1}^n 1_{x_i\neq y_i}\right)dS,$$

  where the infimum is taken over all joint distributions S of pairs (X,Y) of random variables X,Y such that marginal distributions of X and Y are P_1 and P_2.

  This metric is used in Stationary Stochastic Processes, Dynamic Systems, and Coding Theory.

Part IV

Distances in Applied Mathematics

Chapter 15. Distances in Graph Theory
Chapter 16. Distances in Coding Theory
Chapter 17. Distances and Similarities in Data Analysis
Chapter 18. Distances in Mathematical Engineering

Chapter 15

Distances in Graph Theory

A *graph* is a pair $G = (V, E)$, where V is a set, called set of *vertices* of the graph G, and E is a set of unordered pairs of vertices, called *edges* of the graph G. A *directed graph* (or *digraph*) is a pair $D = (V, E)$, where V is a set, called set of *vertices* of the digraph D, and E is a set of ordered pairs of vertices, called *arcs* of the digraph D.

A graph in which at most one edge may connect any two vertices, is called *simple graph*. If multiple edges are allowed between vertices, the graph is called *multi-graph*. The graph is called *finite* (*infinite*) if the set V of its vertices is finite (infinite, respectively). The *order* of a finite graph is the number of its vertices; the *size* of a finite graph is the number of its edges.

A graph or directed graph, together with a function which assigns a positive weight to each edge, is called *weighted graph* or *network*. A network also called **framework** if the weights are interpreted as edge-lengths of a putative embedding into an Euclidean space. In terms of Rigidity Theory, the edges of a framework are *graph bars* (usually, of equal length); a **tensegrity** is a framework in which graph bars are either *cables* (i.e., cannot get further apart), or *struts* (i.e., cannot get closer together).

A *subgraph* of a graph G is a graph G' whose vertices and edges form subsets of the vertices and edges of G. If G' is a subgraph of G, then G is called *supergraph* of G'. An *induced subgraph* is a subset of the vertices of a graph G together with all edges whose endpoints are both in this subset.

A graph $G = (V, E)$ is called *connected* if, for any vertices $u, v \in V$, there exists an *$(u - v)$ path*, i.e., a sequence of edges $uw_1 = w_0w_1, w_1w_2, \ldots, w_{n-1}w_n = w_{n-1}v$ from E such that $w_i \neq w_j$ for $i \neq j, i, j \in \{0, 1, \ldots, n\}$. A digraph $D = (V, E)$ is called *strongly connected* if, for any vertices $u, v \in V$, both, the *directed $(u - v)$ path* and the *directed $(v - u)$ path*, exist. A maximal connected subgraph of a graph G is called its *connected component*.

Vertices connected by an edge are called *adjacent*. The *degree* $deg(v)$ of a vertex $v \in V$ of a graph $G = (V, E)$ is the number of its vertices adjacent to v.

A *complete graph* is a graph in which each pair of vertices is connected by an edge. A *bipartite graph* is a graph in which the set V of vertices is decomposed into two disjoint subsets so that no two vertices within the same subset are adjacent. A *path* is a simple connected graph in which two vertices have degree one, and other vertices (if they exist) have the degree two; the *length* of a path is the number of its edges. A *cycle* is a *closed path*, i.e., a simple connected graph in which every vertex has degree two. A *tree* is a simple connected graph without cycles.

Two graphs which contain the same number of vertices connected in the same way called *isomorphic*. Formally, two graphs $G = (V(G), E(G))$ and $H = (V(H), E(H))$ are called *isomorphic* if there is a bijection $f : V(G) \to V(H)$ such that, for any $u, v \in V(G)$, $uv \in E(G)$ if and only if $f(u)f(v) \in E(H)$.

We will consider only simple finite graphs and digraphs, more exactly, the equivalence classes of such isomorphic graphs.

15.1. DISTANCES ON VERTICES OF A GRAPH

- **Path metric**

The **path metric** (or **graphic metric**, *shortest path metric*) d_{path} is a metric on the vertex-set V of a connected graph $G = (V, E)$, defined, for any $u, v \in V$, as the length of a shortest $(u - v)$ path in G. A shortest $(u - v)$ path is called *geodesic*. The corresponding metric space is called *graphic metric space*, associated with the graph G.

The path metric of the *Cayley graph* Γ of a finitely-generated group $(G, \cdot, e)$ is called **word metric**. The path metric of a graph $G = (V, E)$, such that V can be cyclically ordered in a *Hamiltonian cycle*, is called **Hamiltonian metric**. The **hypercube metric** is the path metric of a *hypercube graph* $H(m, 2)$ with the vertex-set $V = \{0, 1\}^m$, and whose edges are the pairs of vectors $x, y \in \{0, 1\}^m$ such that $|\{i \in \{1, \dots, n\}: x_i \neq y_i\}| = 1$; it is equal to $|\{i \in \{1, \dots, n\}: x_i = 1\}\Delta\{i \in \{1, \dots, n\}: y_i = 1\}|$. The graphic metric space associated with a hypercube graph is called *hypercube metric space*. It coincides with the metric space $(\{0, 1\}^m, d_{l_1})$.

- **Weighted path metric**

The **weighted path metric** d_{wpath} is a metric on the vertex-set V of a connected weighted graph $G = (V, E)$ with positive edge-weights $(w(e))_{e \in E}$, defined by

$$\min_P \sum_{e \in P} w(e),$$

where the minimum is taken over all $(u - v)$ paths P in G.

- **Detour distance**

The **detour distance** is a distance on the vertex-set V of a connected graph $G = (V, E)$, defined as the length of a longest *induced path* (i.e., a path, that is an induced subgraph of G) from u to $v \in V$.

In general, it is not a metric. A graph is called *detour graph* if its detour distance coincides with its **path metric** (see, for example, [CJT93]).

- **Path quasi-metric in digraphs**

The **path quasi-metric in digraphs** d_{dpath} is a quasi-metric on the vertex-set V of a strongly connected directed graph $D = (V, E)$, defined, for any $u, v \in V$, as the length of a shortest directed $(u - v)$ path in D.

- **Circular metric in digraphs**

  The **circular metric in digraphs** is a metric on the vertex-set V of a strongly connected directed graph $D = (V, E)$, defined by

  $$d_{dpath}(u, v) + d_{dpath}(v, u),$$

  where d_{dpath} is the **path quasi-metric in digraphs**.

- **Υ-metric**

  Given a class Υ of connected graphs, the metric d of a metric space (X, d) is called **Υ-metric** if (X, d) is isometric to a subspace of a metric space (V, d_{wpath}), where $G = (V, E) \in \Upsilon$, and d_{wpath} is the **weighted path metric** on the vertex-set V of G with positive edge-weight function w (see **tree-like metric**).

- **Tree-like metric**

  A **tree-like metric** (or **weighted tree metric**) d on a set X is an **Υ-metric** for the class Υ of all trees, i.e., the metric space (X, d) is isometric to a subspace of a metric space (V, d_{wpath}), where $T = (V, E)$ is a tree, and d_{wpath} is the **weighted path metric** on the vertex-set V of T with a positive weight function w. A metric is a tree-like metric if and only if it satisfied the **four-point inequality**.

  A metric d on a set X is called **relaxed tree-like metric** if the set X can be embedding in some (not necessary positively) edge-weighted tree such that, for any $x, y \in X$, $d(x, y)$ is equal to the sum of all edge's weights along the (unique) path between corresponding vertices x and y in the tree. A metric is a relaxed tree-like metric if and only if it satisfied the **relaxed four-point inequality**.

- **Resistance metric**

  Given a connected graph $G = (V, E)$ with positive edge-weight function $w = (w(e))_{e \in E}$, let us interpret the edge-weights as resistances. For any two different vertices u and v, suppose that a battery is connected across them, so that one unit of a current flows in at u and out in v. The voltage (potential) difference, required for this, is, by Ohm's law, the effective resistance between u and v in an electrical network; it is called **resistance metric** $\Omega(u, v)$ between them ([KlRa93], cf. **resistor-average distance**). The number $\frac{1}{\Omega(u,v)}$ can be seen, like electrical *conductance*, as a measure of *connectivity* between u and v. In fact, it holds $\Omega(u, v) \leqslant \min_P \sum_{e \in P} \frac{1}{w(e)}$, where P is any $(u - v)$ path, with equality if and only if such path P is unique; so, if $w(e) = 1$ for all edges, the equality means that G is a tree. The resistance metric is applied (in Physics, Chemistry, and Networks) when the number of paths between any two vertices should be taken into account.

  If $w(e) = 1$ for all edges, then

  $$\Omega(u, v) = (g_{uu} + g_{vv}) - (g_{uv} + g_{vu}),$$

  where $((g_{ij}))$ is the *generalized inverse* of the *Laplacian matrix* $((l_{ij}))$ of the graph G: here l_{ii} is the degree of vertex i, while, for $i \neq j$, $l_{ij} = 1$ if the vertices i and

j are adjacent, and $l_{ij} = 0$, otherwise. A probabilistic interpretation is: $\Omega(u, v) = (deg(u)Pr(u \to v))^{-1}$, where $deg(u)$ is the degree of the vertex u, and $Pr(u \to v)$ is the probability for a random walk leaving u to arrive to v before returning to u.

- **Truncated metric**

  The **truncated metric** is a metric on the vertex-set of a graph, which is equal to 1 for any two adjacent vertices, and is equal to 2 for any non-adjacent different vertices. It is the 2-**truncated metric** for the **path metric** of the graph. It is the $(1, 2)$-B-**metric** if the degree of any vertex is at most B.

- **Multiply-sure distance**

  The **multiply-sure distance** is a distance on the vertex-set V of an *m-connected* weighted graph $G = (V, E)$, defined, for any $u, v \in V$, as the minimum weighted sum of lengths of m disjoint $(u - v)$ paths. It is a generalization of the concept of distance to situations in which one wishes to find several disjoint paths between two points, for example, in a communication networks, where $m - 1$ of $(u - v)$ paths are used to code the message sent by the remaining $(u - v)$ path (see [McCa97]).

  A graph G is called *m-connected* if there is no set of $m - 1$ edges whose removal disconnects the graph. A connected graph is 1-connected.

A *cut* is a *partition* of a set into two parts. Given a subset S of $V_n = \{1, \dots, n\}$, we obtain the partition $\{S, V_n \backslash S\}$ of V_n. The cut-semi-metric on V_n, defined by this partition, can be seen as a special semi-metric on the vertex-set of the *complete bipartite graph* $K_{S, V_n \backslash S}$, where the distance between vertices is equal to 1 if they belong to different parts of this graph, and is equal to 0, otherwise.

- **Cut semi-metric**

  Given a subset S of $V_n = \{1, \dots, n\}$, the **cut semi-metric** (or **split semi-metric**) δ_S is a semi-metric on V_n, defined by

$$\delta_S(i, j) = \begin{cases} 1, & \text{if } i \neq j,\ |S \cap \{i, j\}| = 1, \\ 0, & \text{otherwise.} \end{cases}$$

  Usually, it is considered as a vector in $\mathbb{R}^{|E_n|}$, $E(n) = \{\{i, j\}\colon\ 1 \leqslant i < j \leqslant n\}$.

  A *circular cut* of V_n is defined by a subset $S_{[k+1,l]} = \{k + 1, \dots, l\}(mod\, n) \subset V_n$: if we consider the points $\{1, \dots, n\}$ as being ordered along a circle in that circular order, then $S_{[k+1,l]}$ is the set of its consecutive vertices from $k + 1$ to l. For a circular cut, the corresponding cut-semi-metric is called **circular cut semi-metric**.

  An **even cut semi-metric** is δ_S on V_n with even $|S|$. An **odd cut semi-metric** is δ_S on V_n with odd $|S|$. An k-**uniform cut semi-metric** is δ_S on V_n with $|S| \in \{k, n - k\}$. An **equicut semi-metric** is δ_S on V_n with $|S| \in \{\lfloor \frac{n}{2} \rfloor, \lceil \frac{n}{2} \rceil\}$. An **inequicut semi-metric** is δ_S on V_n with $|S| \notin \{\lfloor \frac{n}{2} \rfloor, \lceil \frac{n}{2} \rceil\}$ (see, for example, [DeLa97]).

- **Decomposable semi-metric**

A **decomposable semi-metric** is a semi-metric on $V_n = \{1, \dots, n\}$ which can be represented as a non-negative linear combination of **cut semi-metrics**. The set of all decomposable semi-metrics on V_n is a *convex cone*, called *cut cone* CUT_n.

A semi-metric on V_n is decomposable if and only if it is a **finite** l_1**-semi-metric**.

A **circular decomposable semi-metric** is a semi-metric on $V_n = \{1, \dots, n\}$ which can be represented as a non-negative linear combination of **circular cut semi-metrics**.

A semi-metric on V_n is circular decomposable if and only if it is a **Kalmanson semi-metric** with respect to the same ordering (see [ChFi98]).

- **Finite l_p-semi-metric**

Given a finite set X, the **finite** l_p**-semi-metric** is a semi-metric d on X such that the metric space (X, d) is a semi-metric subspace of the l_p^m*-space* $(\mathbb{R}^m, d_{l_p})$ for some $m \in \mathbb{N}$. If $X = \{0, 1\}^n$, the metric space (X, d) is called l_p^n*-cube*. The l_1^n-cube is called *Hamming cube*.

- **Kalmanson semi-metric**

A **Kalmanson semi-metric** d is a semi-metric on $V_n = \{1, \dots, n\}$ which satisfies the condition

$$\max\{d(i, j) + d(r, s), d(i, s) + d(j, r)\} \leqslant d(i, r) + d(j, s)$$

for all $1 \leqslant i \leqslant j \leqslant r \leqslant s \leqslant n$. In this definition the ordering of the elements is important; so, d is a Kalmanson semi-metric *with respect to the ordering* $1, \dots, n$.

Equivalently, if we consider the points $\{1, \dots, n\}$ as being ordered along a circle C_n in that circular order, then the distance d on V_n is a Kalmanson semi-metric if the inequality

$$d(i, r) + d(j, s) \leqslant d(i, j) + d(r, s)$$

holds for all $i, j, r, s \in V_n$ such that the segments $[i, j]$ and $[r, s]$ are crossing chords of C_n.

A **tree-like metric** is a Kalmanson metric for some ordering of the vertices of the tree. The Euclidean metric, restricted to the points that form a convex polygon in the plane, is a Kalmanson metric.

- **Multi-cut semi-metric**

Let $\{S_1, \dots, S_q\}$, $q \geqslant 2$, be a *partition* of the set $V_n = \{1, \dots, n\}$, i.e., a collection $S_1, \dots, S_q$ of pairwise disjoint subsets of V_n such that $S_1 \cup \dots \cup S_q = V_n$.

The **multi-cut semi-metric** $\delta_{S_1,\dots,S_q}$ is a semi-metric on V_n, defined by

$$\delta_{S_1,\dots,S_q}(i, j) = \begin{cases} 0, & \text{if } i, j \in S_h \text{ for some } h,\ 1 \leqslant h \leqslant q, \\ 1, & \text{otherwise.} \end{cases}$$

- **Oriented cut quasi-semi-metric**

  Given a subset S of $V_n = \{1, \dots, n\}$, the **oriented cut quasi-semi-metric** δ'_S is a quasi-semi-metric on V_n, defined by

$$\delta'_S(i, j) = \begin{cases} 1, & \text{if } i \in S,\ j \notin S, \\ 0, & \text{otherwise.} \end{cases}$$

  Usually, it is considered as the vector of $\mathbb{R}^{|I_n|}$, $I(n) = \{(i, j):\ 1 \leqslant i \neq j \leqslant n\}$. The **cut semi-metric** δ_S is $\delta'_S + \delta'_{V_n \backslash S}$.

- **Oriented multi-cut quasi-semi-metric**

  Given a *partition* $\{S_1, \dots, S_q\}$, $q \geqslant 2$, of V_n, the **oriented multi-cut quasi-semi-metric** $\delta'_{S_1,\dots,S_q}$ is a quasi-semi-metric on V_n, defined by

$$\delta'_{S_1,\dots,S_n}(i, j) = \begin{cases} 1, & \text{if } i \in S_h,\ j \in S_m,\ h < m, \\ 0, & \text{otherwise .} \end{cases}$$

15.2. DISTANCE-DEFINED GRAPHS

- **Geodetic graph**

  A connected graph is called **geodetic** if there exists exactly one shortest path between any two its vertices. Every tree is a geodetic graph.

- **Isometric subgraph**

  A subgraph H of a graph $G = (V, E)$ is called **isometric subgraph** if the **path metric** between any two points of H is the same as their path metric in G.

- **Retract subgraph**

  A subgraph H of a graph $G = (V, E)$ is called **retract subgraph** if it is induced by an idempotent **short mapping** of G into itself, i.e., $f^2 = f : V \to V$ with $d_{path}(f(u), f(v)) \leqslant d_{path}(u, v)$ for all $u, v \in V$. Any retract subgraph is **isometric**.

- **Distance-regular graph**

  A connected graph $G = (V, E)$ of diameter T is called **distance-regular** if, for any its vertices u, v and any integers $0 \leqslant i, j \leqslant T$, the number of vertices w, such that $d_{path}(u, w) = i$ and $d_{path}(v, w) = j$, depends only on i, j and $k = d_{path}(u, v)$, but not on the choice of vertices u and v.

  A special case of it is a **distance-transitive graph**, i.e., such that its group of automorphisms is transitive, for any $0 \leqslant i \leqslant T$, on the pairs of vertices (u, v) with $d_{path}(u, v) = i$.

  For any $2 \leqslant i \leqslant T$, denote by G_i the graph with the same vertex-set as G, and with edges uv such that $d_{path}(u, v) = i$. The graph G is called **distance-polynomial graph** if the adjacency matrix of any G_i, $2 \leqslant i \leqslant T$, can be expressed as a polynomial in terms of the adjacency matrix of G. Any distance-regular graph is distance-polynomial.

Any distance regular-graph is also **distance-balanced graph**, i.e., $|\{x \in V: d(x,u) < d(x,v)\}| = |\{x \in V: d(x,v) < d(x,u)\}|$ for any its edge uv, and **distance degree regular graph**, i.e., $|\{x \in V: d(x,u) = i\}|$ depends only on i but not on $u \in V$.

Another name for a distance-regular graph is a *P-polynomial association scheme*. A **finite polynomial metric space** is a *P- and Q-polynomial association scheme*. The term **infinite polynomial metric spaces** is used for *compact connected two-point homogeneous spaces*; Wang classified them as the Euclidean unit spheres, the real, complex, and quaternionic projective spaces or the Cayley projective plane.

- **Distance-hereditary graph**

  A connected graph is called **distance-hereditary** if each of its connected induced subgraphs is isometric. A graph is distance-hereditary if each of its induced paths is isometric. Any *co-graph*, i.e., a graph containing no induced path of four vertices, is distance-hereditary. A graph is distance-hereditary if and only if its **path metric** satisfy the **relaxed four-point inequality**. A graph is: distance-hereditary, bipartite distance-hereditary, **block graph**, or tree if and only if its path metric is a **relaxed tree-like metric** for edge-weights being, respectively, non-zero half-integers, non-zero integers, positive half-integers, or positive integers.

- **Block graph**

  A graph is called **block graph** if each its *block*, i.e., a maximal 2-connected induced subgraph, is a complete graph. Any tree is a block graph. A graph is a block graph if and only if its **path metric** is a **tree-like metric** or, equivalently, satisfies the **four-point inequality**.

- **Ptolemaic graph**

  A graph is called **Ptolemaic** if its **path metric** satisfies the **Ptolemaic inequality**

  $$d(x,y)d(u,z) \leqslant d(x,u)d(y,z) + d(x,z)d(y,u).$$

  A graph is Ptolemaic if and only if it is distance-hereditary and *chordal*, i.e., every cycle of length greater than 3 has a chord. In particular, any **block graph** is Ptolemaic.

- ***D*-distance graph**

  Given a set D of positive numbers containing 1 and a metric space (X,d), the **D-distance graph** $D(X,d)$ is a graph with the vertex-set X and the edge-set $\{uv: d(u,v) \in D\}$ (cf. **D-chromatic number**).

  An D-distance graph is called *unit-distance graph* if $D = \{1\}$, *ε-unit graph* if $D = [1-\varepsilon, 1+\varepsilon]$, *unit-neighborhood graph* if $D = (0,1]$, *integral-distance graph* if $D = \mathbb{Z}_+$, *rational-distance graph* if $D = \mathbb{Q}_+$, *prime-distance graph* if D is the set of prime numbers (with 1).

  Usually, the metric space (X,d) is a subspace of an Euclidean space $\mathbb{E}^n$. Moreover, every finite graph $G = (V,E)$ can be represented by an D-distance graph in some $\mathbb{E}^n$. The minimum dimension of such Euclidean space is called *D-dimension* of G.

- **t-spanner**

A subgraph $H = (V, E(H))$ of a connected graph $G = (V, E)$ is called **t-spanner** of G if, for every $u, v \in V$, the inequality $d^H_{path}(u, v)/d^G_{path}(u, v) \leqslant t$ holds. The value t is called *stretch factor* of H.

A graph is **distance-hereditary** if and only if every its induced subgraph is 1-spanner.

A *spanning tree* of a connected graph $G = (V, E)$ is a subset of $|V| - 1$ edges that form a tree on the vertex-set V.

- **Steiner distance of a set**

The **Steiner distance of a set** $S \subset V$ of vertices in a connected graph $G = (V, E)$ is the minimum number of edges of a connected subgraph of G, containing S. Such a subgraph is, obviously, a tree, and is called *Steiner tree* for S. The vertices of Steiner tree, that are not in S, are called *Steiner points*.

- **Distance labelling scheme**

A graph family A is said (D. Peleg, 2000) to have an $l(n)$ **distance labelling scheme** if there is a function L labelling the vertices of each n-vertex graph in A with distinct labels of up to $l(n)$ bits, and there exists an algorithm, called **distance decoder**, that decides the distance between any two vertices u, v in a graph from A in time polynomial in the length of their labels $L(u), L(v)$.

15.3. DISTANCES ON GRAPHS

- **Subgraph-supergraph distances**

A *common subgraph* of graphs G and H is a graph which is isomorphic to induced subgraphs of both G and H. A *common supergraph* of graphs G and H is a graph which contains induced subgraphs isomorphic to G and H.

The **Zelinka distance** d_Z on the set **G** of all graphs (more exactly, on the set of all equivalence classes of isomorphic graphs) is defined by

$$\max\{n(G_1), n(G_2)\} - n(G_1, G_2)$$

for any $G_1, G_2 \in \mathbf{G}$, where $n(G_i)$ is the number of vertices in G_i, $i = 1, 2$, and $n(G_1, G_2)$ is the maximum number of vertices of a common subgraph of G_1 and G_2.

Given an arbitrary set **M** of graphs, the **common subgraph distance** d_M on **M** is defined by

$$\max\{n(G_1), n(G_2)\} - n(G_1, G_2),$$

and the **common supergraph distance** d^*_M on **M** is defined by

$$N(G_1, G_2) - \min\{n(G_1), n(G_2)\}$$

for any $G_1, G_2 \in \mathbf{M}$, where $n(G_i)$ is the number of vertices in G_i, $i = 1, 2$, $n(G_1, G_2)$ is the maximum number of vertices of a common subgraph $G \in \mathbf{M}$ of G_1 and G_2, and

$N(G_1, G_2)$ is the minimum number of vertices of a common supergraph $H \in \mathbf{M}$ of G_1 and G_2.

d_M is a metric on $\mathbf{M}$ if the following condition (i) holds: if $H \in \mathbf{M}$ is a common supergraph of $G_1, G_2 \in \mathbf{M}$, then there exists a common subgraph $G \in \mathbf{M}$ of G_1 and G_2 with $n(G) \geqslant n(G_1) + n(G_2) - n(H)$. d^*_M is a metric on $\mathbf{M}$ if the following condition (ii) holds: if $G \in \mathbf{M}$ is a common subgraph of $G_1, G_2 \in \mathbf{M}$, then there exists a common supergraph $H \in \mathbf{M}$ of G_1 and G_2 with $n(H) \leqslant n(G_1) + n(G_2) - n(G)$. One has $d_M \leqslant d^*_M$ if the condition (i) holds, and $d_M \geqslant d^*_M$ if the condition (ii) holds.

The distance d_M is a metric on the set $\mathbf{G}$ of all graphs, the set of all cycle-free graphs, the set of all bipartite graphs, and the set of all trees. The distance d^*_M is a metric on the set $\mathbf{G}$ of all graphs, the set of all connected graphs, the set of all connected bipartite graphs, and the set of all trees. The Zelinka distance d_Z coincides with d_M and d^*_M on the set $\mathbf{G}$ of all graphs. On the set $\mathbf{T}$ of all trees the distances d_M and d^*_M are identical, but different from the Zelinka distance restricted to $\mathbf{T}$.

The Zelinka distance d_Z is a metric on the set $\mathbf{G}(n)$ of all graphs with n vertices, and is equal to $n - k$ or to $K - n$ for all $G_1, G_2 \in \mathbf{G}(n)$, where k is the maximum number of vertices of a common subgraph of G_1 and G_2, and K is the minimum number of vertices of a common supergraph of G_1 and G_2. On the set $\mathbf{T}(n)$ of all trees with n vertices the distance d_Z is called **Zelinka tree distance** (see, for example, [Zeli75]).

- **Edge distance**

The **edge distance** is a distance on the set $\mathbf{G}$ of all graphs, defined by

$$|E_1| + |E_2| - 2|E_{12}| + \big||V_1| - |V_2|\big|$$

for any graphs $G_1 = (V_1, E_1)$ and $G_2 = (V_2, E_2)$, where $G_{12} = (V_{12}, E_{12})$ is a common subgraph of G_1 and G_2 with maximal number of edges. This distance has many applications in Organic and Medical Chemistry.

- **Contraction distance**

The **contraction distance** is a distance on the set $\mathbf{G}(n)$ of all graphs with n vertices, defined by

$$n - k$$

for any $G_1, G_2 \in \mathbf{G}(n)$, where k is the maximum number of vertices of a graph which is isomorphic simultaneously to a graph, obtained from each of graphs G_1, G_2 by a finite number of *edge contractions*.

To perform the *contraction* of the edge $uv \in E$ of a graph $G = (V, E)$ means to replace u and v by one vertex that is adjacent to all vertices of $V \backslash \{u, v\}$ which were adjacent to u or to v.

- **Edge move distance**

The **edge move distance** is a metric on the set $\mathbf{G}(n, m)$ of all graphs with n vertices and m edges, defined, for any $G_1, G_2 \in \mathbf{G}(m, n)$, as the minimum number of *edge moves* necessary for transforming the graph G_1 into the graph G_2. It is equal to $m - k$, where k is the maximum number of edges in a common subgraph of G_1 and G_2.

An *edge move* is one of the *edges transformations*, defined as follow: H can be obtained from G by an edge move if there exist (not necessarily distinct) vertices u, v, w, and x in G such that $uv \in E(G)$, $wx \notin E(G)$, and $H = G - uv + wx$.

- **Edge jump distance**

The **edge jump distance** is an extended metric (which in general can take value ∞) on the set $\mathbf{G}(n, m)$ of all graphs with n vertices and m edges, defined, for any $G_1, G_2 \in \mathbf{G}(m, n)$, as the minimum number of *edge jumps* necessary for transforming the graph G_1 into the graph G_2.

An *edge jump* is one of the *edges transformations*, defined as follow: H can be obtained from G by an edge jump if there exist four distinct vertices u, v, w, and x in G, such that $uv \in E(G)$, $wx \notin E(G)$, and $H = G - uv + wx$.

- **Edge rotation distance**

The **edge rotation distance** is a metric on the set $\mathbf{G}(n, m)$ of all graphs with n vertices and m edges, defined, for any $G_1, G_2 \in \mathbf{G}(m, n)$, as the minimum number of *edge rotations* necessary for transforming the graph G_1 into the graph G_2.

An *edge rotation* is one of the *edges transformations*, defined as follow: H can be obtained from G by an edge rotation if there exist distinct vertices u, v, and w in G, such that $uv \in E(G)$, $uw \notin E(G)$, and $H = G - uv + uw$.

- **Tree rotation distance**

The **tree rotation distance** is a metric on the set $\mathbf{T}(n)$ of all trees with n vertices, defined, for all $T_1, T_2 \in \mathbf{T}(n)$, as the minimum number of *tree edge rotations* necessary for transforming T_1 into T_2. For $\mathbf{T}(n)$ the tree rotation distance and the edge rotation distance may differ.

A *tree edge rotation* is an *edge rotation* performed on a tree, and resulting in a tree.

- **Edge shift distance**

The **edge shift distance** (or **edge slide distance**) is a metric on the set $\mathbf{G}_c(n, m)$ of all connected graphs with n vertices and m edges, defined, for any $G_1, G_2 \in \mathbf{G}_c(m, n)$, as the minimum number of *edge shifts* necessary for transforming the graph G_1 into the graph G_2.

An *edge shift* is one of the *edges transformations*, defined as follow: H can be obtained from G by an edge shift if there exist distinct vertices u, v, and w in G such that $uv, vw \in E(G)$, $uw \notin E(G)$, and $H = G - uv + uw$. Edge shift is a special kind of the *edge rotation* in the case when the vertices v, w are adjacent in G.

The edge shift distance can be defined between any graphs G and H with components G_i $(1 \leqslant i \leqslant k)$ and H_i $(1 \leqslant i \leqslant k)$, respectively, such that G_i and H_i have the same order and same size.

- **F-rotation distance**

  The F**-rotation distance** is a distance on the set $\mathbf{G}_F(n, m)$ of all graphs with n vertices and m edges, containing a subgraph isomorphic to a given graph F of order at least 2, defined, for all $G_1, G_2 \in \mathbf{G}_F(m, n)$, as the minimum number of *F-rotations* necessary for transforming the graph G_1 into the graph G_2.

  An *F-rotation* is one of the *edges transformations*, defined as follow: let F' be a subgraph of a graph G, isomorphic to F, let u, v, w be three distinct vertices of the graph G such that $u \notin V(F')$, $v, w \in V(F')$, $uv \in E(G)$, and $uw \notin E(G)$; H can be obtained from G by the F-rotation of the edge uv into the position uw if $H = G - uv + uw$.

- **Binary relation distance**

  Let R be a non-reflexive *binary relation* between graphs, i.e., $R \subset \mathbf{G} \times \mathbf{G}$, and there exists $G \in \mathbf{G}$ such that $(G, G) \notin R$.

  The **binary relation distance** is an extended metric (which in general can take value ∞) on the set $\mathbf{G}$ of all graphs, defined, for any graphs G_1 and G_2, as the minimum number of *R-transformations* necessary for transforming the graph G_1 into the graph G_2.

  We say, that a graph H can be obtained from a graph G by an *R-transformation* if $(H, G) \in R$.

  An example is the distance between two *triangular embeddings of a complete graph* (i.e., its cellular embeddings in a surface with only 3-gonal faces) defined as the minimal number t such that, up to replacing t faces, the embeddings are isomorphic.

- **Crossing-free transformation metrics**

  Given a set S of n points in $\mathbb{R}^2$, a *non-crossing spanning tree* of S is a tree whose vertices are points of S, and whose edges are pairwise non-crossing straight line segments.

  The **crossing-free edge move metric** (see [AAH00]) is a metric on the set $\mathbf{T}_S$ of all *non-crossing spanning trees* of a set S, defined, for any $T_1, T_2 \in \mathbf{T}_S$, as the minimum number of *crossing-free edge moves* necessary for transforming T_1 into T_2. A *crossing-free edge move* is a *edges transformation* which consists of adding some edge e in $T \in \mathbf{T}_S$ and removing some edge f from the induced cycle so that e and f do not cross.

  The **crossing-free edge slide metric** is a metric on the set $\mathbf{T}_S$ of all *non-crossing spanning trees* of a set S, defined, for any $T_1, T_2 \in \mathbf{T}_S$, as the minimum number of *crossing-free edge slides* necessary for transforming T_1 into T_2. A *crossing-free edge slide* is one of the *edges transformations* which consists of taking some edge e in $T \in \mathbf{T}_S$ and moving one of its endpoints along some edge adjacent to e in T, without introducing edge crossings and without sweeping across points in S (that gives a new edge f instead of e). The edge slide is a special kind of crossing-free edge move: the new tree is obtained by closing with f a cycle C of length three in T, and removing e from C, in a way such that f avoids the interior of the triangle C.

- **Traveling salesman tours distances**

  The *traveling salesman problem* is the problem of finding the shortest tour that visits a set of cities. We shall consider only traveling salesman problem with undirected links.

For an N-city traveling salesman problem, the space $\mathcal{T}_N$ of tours is the set of $\frac{(N-1)!}{2}$ cyclic permutations of the cities $1, 2, \dots, N$.

The metric D on $\mathcal{T}_N$ is defined in terms of the difference in form: if tours $T, T' \in \mathcal{T}_N$ differ in m links, then $D(T, T') = m$.

A *k-OPT transformation* of a tour T is obtained by deleting k links from T, and reconnecting. A tour T', obtained from T by an k-OPT transformation, is called *k-OPT* of T. The distance d on the set $\mathcal{T}_N$ is defined in terms of the 2-OPT transformations: $d(T, T')$ is the minimal i, for which there exists a sequence of i *2-OPT transformations* which transforms T to T'.

In fact, $d(T, T') \leqslant D(T, T')$ for any $T, T' \in \mathcal{T}_N$ (see, for example, [MaMo95]).

- **Subgraphs distances**

The standard distance on the set of all subgraphs of a connected graph $G = (V, E)$ is defined by

$$\min\{d_{path}(u, v) : u \in V(F), v \in V(H)\}$$

for any subgraphs F, H of G. For any subgraphs F, H of a strongly connected digraph $D = (V, E)$, the standard quasi-distance is defined by

$$\min\{d_{dpath}(u, v) : u \in V(F), v \in V(H)\}.$$

The **edge rotation distance** on the set $\mathbf{S}^k(G)$ of all edge-induced subgraphs with k edges in a connected graph G is defined as the minimum number of *edge rotations* required to transform $F \in \mathbf{S}^k(G)$ into $H \in \mathbf{S}^k(G)$. We say, that H can be obtained from F by an *edge rotation* if there exist distinct vertices u, v, and w in G such that $uv \in E(F)$, $uw \in E(G)\backslash E(F)$, and $H = F - uv + uw$.

The **edge shift distance** on the set $\mathbf{S}^k(G)$ of all edge-induced subgraphs with k edges in a connected graph G is defined as the minimum number of *edge shifts* required to transform $F \in \mathbf{S}^k(G)$ into $H \in \mathbf{S}^k(G)$. We say, that H can be obtained from F by an *edge shift* if there exist distinct vertices u, v and w in G such that $uv, vw \in E(F)$, $uw \in E(G)\backslash E(F)$, and $H = F - uv + uw$.

The **edge move distance** on the set $\mathbf{S}^k(G)$ of all edge-induced subgraphs with k edges of a graph G (not necessary connected) is defined as the minimum number of *edge moves* required to transform $F \in \mathbf{S}^k(G)$ into $H \in \mathbf{S}^k(G)$. We say, that H can be obtained from F by an *edge move* if there exist (not necessarily distinct) vertices u, v, w, and x in G such that $uv \in E(F)$, $wx \in E(G)\backslash E(F)$, and $H = F - uv + wx$. The edge move distance is a metric on $\mathbf{S}^k(G)$. If F and H have s edges in common, then it is equal to $k - s$.

The **edge jump distance** (which in general can take value ∞) on the set $\mathbf{S}^k(G)$ of all edge-induced subgraphs with k edges of a graph G (not necessary connected) is defined as the minimum number of *edge jumps* required to transform $F \in \mathbf{S}^k(G)$ into $H \in \mathbf{S}^k(G)$. We say, that H can be obtained from F by an *edge jump* if there exist four distinct vertices u, v, w, and x in G such that $uv \in E(F)$, $wx \in E(G)\backslash E(F)$, and $H = F - uv + wx$.

15.4. DISTANCES ON TREES

Let T be a *rooted tree*, i.e., a tree with one of its vertices being chosen as the *root*. The *depth* of a vertex v, $depth(v)$, is the number of edges on the path from v to the root. A vertex v is called *parent* of a vertex u, $v = par(u)$, if they are adjacent, and $depth(u) = depth(v)+1$; in this case u is called *child* of v. Two vertices are *siblings* if they have the same parent. The *in-degree* of a vertex is the number of its children. $T(v)$ is the subtree of T, rooted at a node $v \in V(T)$. If $w \in V(T(v))$, then v is an *ancestor* of w, and w is a *descendant* of v; $nca(u, v)$ is the *nearest common ancestor* of the vertices u and v. T is called *labeled tree* if a symbol from a fixed finite alphabet $\mathcal{A}$ is assigned to each node. T is called *ordered tree* if a left-to-right order among siblings in T is given.

On the set $\mathbb{T}_{rlo}$ of all rooted labeled ordered trees there are three *editing operations*:

- *Relabel* – change the label of a vertex v;
- *Deletion* – delete a non-rooted vertex v with parent v', making the children of v become the children of v'; the children are inserted in the place of v as a subsequence in the left-to-right order of the children of v';
- *Insertion* – the complement of deletion; insert a vertex v as a child of a v' making v the parent of a consecutive subsequence of the children of v'.

For unordered trees the editing operations can be defined similarly, but insert and delete operations work on a subset instead of a subsequence.

We assume that there is a *cost function* defined on each editing operation, and the *cost* of a sequence of editing operations is the sum of costs of these operations.

The *ordered edit distance mapping* is a representation of the editing operations. Formally, define the triple (M, T_1, T_2) to be an *ordered edit distance mapping* from T_1 to T_2, $T_1, T_2 \in \mathbb{T}_{rlo}$, if $M \subset V(T_1) \times V(T_2)$ and, for any $(v_1, w_1), (v_2, w_2) \in M$, the following conditions hold: $v_1 = v_2$ if and only if $w_1 = w_2$ (*one-to-one condition*), v_1 is an ancestor of v_2 if and only if w_1 is an ancestor of w_2 (*ancestor condition*), v_1 is to the left of v_2 if and only if w_1 is to the left of w_2 (*sibling condition*).

We say that a vertex v in T_1 and T_2 is *touched by a line* in M if v occurs in some pair in M. Let N_1 and N_2 be the set of vertices in T_1 and T_2, respectively, not touched by any line in M. The *cost* of M is given by $\gamma(M) = \sum_{(v,w)\in M} \gamma(v \to w) + \sum_{v\in N_1} \gamma(v \to \lambda) + \sum_{w\in N_2} \gamma(\lambda \to w)$, where $\gamma(a \to b) = \gamma(a, b)$ is the *cost* of an editing operation $a \to b$ which is a relabel if $a, b \in \mathcal{A}$, a deletion if $b = \lambda$, and an insertion if $a = \lambda$. Here $\lambda \notin \mathcal{A}$ is a special *blank symbol*, and γ is a metric on the set $\mathcal{A} \cup \lambda$ (excepting the value $\gamma(\lambda, \lambda)$).

- **Tree edit distance**

  The **tree edit distance** (see [Tai79]) is a distance on the set $\mathbb{T}_{rlo}$ of all rooted labeled ordered trees, defined, for any $T_1, T_2 \in \mathbb{T}_{rlo}$, as the minimum cost of a sequence of editing operations (relabels, insertions, and deletions) turning T_1 into T_2.

  In terms of ordered edit distance mappings, it is equal to $\min_{(M,T_1,T_2)} \gamma(M)$, where the minimum is taken over all ordered edit distance mappings (M, T_1, T_2).

The edit tree distance can be defined in similar way on the set of all rooted labeled unordered trees.

- **Selkow distance**

The **Selkow distance** (or **top-down edit distance**, **1-degree edit distance**) is a distance on the set $\mathbb{T}_{rlo}$ of all rooted labeled ordered trees, defined, for any $T_1, T_2 \in \mathbb{T}_{rlo}$, as the minimum cost of a sequence of editing operations (relabels, insertions, and deletions) turning T_1 into T_2 if insertions and deletions are restricted to leaves of the trees (see [Selk77]). The root of T_1 must be mapped to the root of T_2, and if a node v is to be deleted (inserted) in M, then subtree rooted at v, if any, is to be deleted (inserted).

In terms of ordered edit distance mappings, it is equal to $\min_{(M,T_1,T_2)} \gamma(M)$, where the minimum is taken over all ordered edit distance mappings (M, T_1, T_2) satisfying the following condition: if $(v, w) \in M$, where neither v, nor w is the root, then $(par(v), par(w)) \in M$.

- **Constrained edit distance**

The **constrained edit distance** (or **restricted edit distance**) is a distance on the set $\mathbb{T}_{rlo}$ of all rooted labeled ordered trees, defined, for any $T_1, T_2 \in \mathbb{T}_{rlo}$, as the minimum cost of a sequence of editing operations (relabels, insertions, and deletions) turning T_1 into T_2 with the restriction that disjoint subtrees should be mapped to disjoint subtrees.

In terms of ordered edit distance mappings, it is equal to $\min_{(M,T_1,T_2)} \gamma(M)$, where the minimum is taken over all ordered edit distance mappings (M, T_1, T_2) satisfying the following condition: for all $(v_1, w_1), (v_2, w_2), (v_3, w_3) \in M$, $nca(v_1, v_2)$ is a proper ancestor of v_3 if and only if $nca(w_1, w_2)$ is a proper ancestor of w_3.

This distance is equivalent to the **structure respecting edit distance**, defined by $\min_{(M,T_1,T_2)} \gamma(M)$, where the minimum is taken over all ordered edit distance mappings (M, T_1, T_2), satisfying the following condition: for all $(v_1, w_1), (v_2, w_2), (v_3, w_3) \in M$, such that none of v_1, v_2, and v_3 is an ancestor of the others, $nca(v_1, v_2) = nca(v_1, v_3)$ if and only if $nca(w_1, w_2) = nca(w_1, w_3)$.

- **Unit cost edit distance**

The **unit cost edit distance** is a distance on the set $\mathbb{T}_{rlo}$ of all rooted labeled ordered trees, defined, for any $T_1, T_2 \in \mathbb{T}_{rlo}$, as the minimum number of editing operations (relabels, insertions, and deletions) turning T_1 into T_2.

- **Alignment distance**

The **alignment distance** (see [JWZ94]) is a distance on the set $\mathbb{T}_{rlo}$ of all rooted labeled ordered trees, defined, for any $T_1, T_2 \in \mathbb{T}_{rlo}$, as the minimum *cost* of an *alignment* of T_1 and T_2. It corresponds to a restricted edit distance, where all insertions must be performed before any deletions.

Thus, one inserts *spaces*, i.e., vertices labeled with a *blank symbol* λ, into T_1 and T_2 so they become isomorphic when labels are ignored; the resulting trees are overlayed on top of each other giving the *alignment* $T_{\mathcal{A}}$ which is a tree, where each vertex is labeled by a pair of labels. The *cost* of $T_{\mathcal{A}}$ is the sum of costs of all pairs of opposite labels in $T_{\mathcal{A}}$.

- **Splitting-merging distance**

  The **splitting-merging distance** (see [ChLu85]) is a distance on the set $\mathbb{T}_{rlo}$ of all rooted labeled ordered trees, defined, for any $T_1, T_2 \in \mathbb{T}_{rlo}$, as the minimum number of vertex splittings and mergings needed to transform T_1 into T_2.

- **Degree-2 distance**

  The **degree-2 distance** is a metric on the set $\mathbb{T}_l$ of all labeled trees (*labeled free trees*), defined, for any $T_1, T_2 \in \mathbb{T}_l$, as the minimum weighted number of editing operations (relabels, insertions, and deletions) turning T_1 into T_2 if any vertex to be inserted (deleted) has no more than two neighbors. This metric is a natural extension of the **tree edit distance**, and the **Selkow distance**.

A *phylogenetic X-tree* is an unordered, unrooted tree with the labeled leaf set X and no vertices of degree two. If every interior vertex has degree three, the tree is called *binary* (or *fully resolved*).

A *cut* $A|B$ of X is a *partition* of X into two subsets A and B (see **cut semi-metric**). Removing an edge e from a phylogenetic X-tree induces a cut of the leaf set X which is called *cut associated with e*.

- **Robinson–Foulds metric**

  The **Robinson–Foulds metric** (or **closest partition metric**, *cut distance*) is a metric on the set $\mathbb{T}(X)$ of all phylogenetic X-trees, defined by

$$\frac{1}{2}\left|\Sigma(T_1)\triangle\Sigma(T_2)\right| = \frac{1}{2}\left|\Sigma(T_1) - \Sigma(T_2)\right| + \frac{1}{2}\left|\Sigma(T_2) - \Sigma(T_1)\right|$$

  for all $T_1, T_2 \in \mathbb{T}(X)$, where $\Sigma(T)$ is the collection of all cuts of X associated with edges of T.

- **Robinson–Foulds weighted metric.**

  The **Robinson–Foulds weighted metric** is a metric on the set $\mathbb{T}(X)$ of all phylogenetic X-trees, defined by

$$\sum_{A|B\in\Sigma(T_1)\cup\Sigma(T_2)} \left|w_1(A|B) - w_2(A|B)\right|$$

  for all $T_1, T_2 \in \mathbb{T}(X)$, where $w_i = (w(e))_{e\in E(T_i)}$ is the collection of positive weights, associated with the edges of the X-tree T_i, $\Sigma(T_i)$ is the collection of all cuts of X, associated with edges of T_i, and $w_i(A|B)$ is the weight of the edge, corresponding to the cut $A|B$ of X, $i = 1, 2$.

- **Nearest neighbor interchange metric**

  The **nearest neighbor interchange metric** (or **crossover metric**) is a metric on the set $\mathbb{T}(X)$ of all phylogenetic X-trees, defined, for all $T_1, T_2 \in \mathbb{T}(X)$, as the minimum number of *nearest neighbor interchanges* required to transform T_1 into T_2.

A *nearest neighbor interchange* consists of swapping two subtrees in a tree that are adjacent to the same internal edge; the remainder of the tree is unchanged.

- **Subtree prune and regraft distance**

The **subtree prune and regraft distance** is a metric on the set $\mathbb{T}(X)$ of all phylogenetic X-trees, defined, for all $T_1, T_2 \in \mathbb{T}(X)$, as the minimum number of *subtree prune and regraft transformations* required to transform T_1 into T_2.

A *subtree prune and regraft transformation* proceeds in three steps: one selects and removes an edge uv of the tree, thereby dividing the tree into two subtrees T_u (containing u) and T_v (containing v); then one selects and subdivides an edge of T_v, giving a new vertex w; finally, one connects u and w by an edge, and removes all vertices of degree two.

- **Tree bisection-reconnection metric**

The **tree bisection-reconnection metric** is a metric on the set $\mathbb{T}(X)$ of all phylogenetic X-trees, defined, for all $T_1, T_2 \in \mathbb{T}(X)$, as the minimum number of *tree bisection and reconnection* transformations required to transform T_1 into T_2.

A *tree bisection and reconnection transformation* proceeds in three steps: one selects and removes an edge uv of the tree, thereby dividing the tree into two subtrees T_u (containing u) and T_v (containing v); then one selects and subdivides an edge of T_v, giving a new vertex w, and an edge of T_u, giving a new vertex z; finally one connects w and z by an edge, and removes all vertices of degree two.

- **Quartet distance**

The **quartet distance** (see [EMM85]) is a distance of the set $\mathbb{T}_b(X)$ of all binary phylogenetic X-trees, defined, for all $T_1, T_2 \in \mathbb{T}_b(X)$, as the number mismatched *quartets* (from the total number $\binom{n}{4}$ possible quartets) for T_1 and T_2.

This distance is based on the fact that given four leaves $\{1, 2, 3, 4\}$ of a tree, they can only be combined in a binary subtree in 3 different ways: (12|34), (13|24), or (14|23): a notation (12|34) refers to the binary tree with the leaf set $\{1, 2, 3, 4\}$ in which removing the inner edge yields the trees with the leave sets $\{1, 2\}$ and $\{3, 4\}$.

- **Triples distance**

The **triples distance** (see [CPQ96]) is a distance of the set $\mathbb{T}_b(X)$ of all binary phylogenetic X-trees, defined, for all $T_1, T_2 \in \mathbb{T}_b(X)$, as the number of triples (from the total number $\binom{n}{3}$ possible triples) that differ (for example, by which leaf is the outlier) for T_1 and T_2.

- **Perfect matching distance**

The **perfect matching distance** is a distance on the set $\mathbb{T}_{br}(X)$ of all rooted binary phylogenetic X-trees with the set X of n labeled leaves, defined, for any $T_1, T_2 \in \mathbb{T}_{br}(X)$, as the minimum number of interchanges necessary to bring the perfect matching of T_1 to the perfect matching of T_2.

Given set $A = \{1, \dots, 2k\}$ of $2k$ points, a *perfect matching* of A is a *partition* A into k pairs. A rooted binary phylogenetic tree with n labeled leaves has a root and $n - 2$ internal vertices distinct from the root. It can be identified with a perfect matching on $2n - 2$, different from the root, vertices by following construction: label the internal vertices with numbers $n + 1, \dots, 2n - 2$ by putting the smallest available label as the parent of the pair of labeled children of which one has the smallest label among pairs of labeled children; now a matching is formed by peeling off the children, or sibling pairs, two by two.

- **Attributed tree metrics**

An *attributed tree* is a triple (V, E, α), where $T = (V, E)$ is the underlying tree, and α is a function which assigns an *attribute vector* $\alpha(v)$ to every vertex $v \in V$. Given two attributed trees (V_1, E_1, α) and (V_2, E_2, β), consider the set of all *subtree isomorphisms* between them, i.e., the set of all isomorphisms $f : H_1 \to H_2$, $H_1 \subset V_1$, $H_2 \subset V_2$, between their *induced subtrees*. Given a similarity s on the set of attributes, the similarity between isomorphic induced subtrees is defined as $W_s(f) = \sum_{v \in H_1} s(\alpha(v), \beta(f(v)))$. The isomorphism ϕ with maximal similarity $W_s(\phi) = W(\phi)$ is called *maximum similarity subtree isomorphism*.

The following semi-metrics on the set $\mathbf{T}_{att}$ of all attributed trees are used:

1. $\max\{|V_1|, |V_2|\} - W(\phi)$;
2. $|V_1| + |V_2| - 2W(\phi)$;
3. $1 - \frac{W(\phi)}{\max\{|V_1|, |V_2|\}}$;
4. $1 - \frac{W(\phi)}{|V_1| + |V_2| - W(\phi)}$.

They become metrics on the set of equivalences classes of attributed trees: two attributed trees (V_1, E_1, α) and (V_2, E_2, β) are called *equivalent* if they are *attribute-isomorphic*, i.e., if there exists an isomorphism $g : V_1 \to V_2$ between the trees T_1 and T_2, such that, for any $v \in V_1$, we have $\alpha(v) = \beta(g(v))$. In this case $|V_1| = |V_2| = W(g)$.

- **Greatest agreement subtree distance**

The **greatest agreement subtree distance** is a distance of the set $\mathbf{T}$ of all trees, defined, for all $T_1, T_2 \in \mathbf{T}$, as the minimum number of leaves removed to obtain a *(greatest) agreement subtree*.

An *agreement subtree* (or *common pruned tree*) of two trees is an identical subtree that can be obtained from both trees by pruning leaves with the same label.

Chapter 16

Distances in Coding Theory

Coding Theory deals with the design and the properties of *error-correcting codes* for the reliable transmission of information across noisy channels in transmission lines and storage devices. The aim of Coding Theory is to find codes which transmit and decode fast, contain many valid code words, and can correct, or at least detect, many errors. These aims are mutually exclusive, however; so, each application has its own good code.

In communications, a *code* is a rule for converting a piece of information (for example, a letter, word, or phrase) into another form or representation, not necessarily of the same sort. *Encoding* is the process by which a source (object) performs this conversion of information into data, which is then sent to a receiver (observer), such as a data processing system. *Decoding* is the reverse process of converting data, which has been sent by a source, into information understandable by a receiver.

An *error-correcting code* is a code in which every data signal conforms to specific rules of construction so that departures from this construction in the received signal can generally be automatically detected and corrected. It is used in computer data storage, for example in dynamic RAM, and in data transmission. Error detection is much simpler than error correction, and one or more "check" digits are commonly embedded in credit card numbers in order to detect mistakes. The two main classes of error-correcting codes are *block codes*, and *convolutional codes*.

A *block code* (or *uniform code*) of length n over an alphabet $\mathcal{A}$, usually, over a finite field $\mathbb{F}_q = \{0, \dots, q-1\}$, is a subset $C \subset \mathcal{A}^n$; every vector $x \in C$ is called *codeword*, $M = |C|$ is called *size* of the code; given metric d on $\mathbb{F}_q^n$ (usually, the **Hamming metric** d_H), the value $d^* = d^*(C) = \min_{x,y \in C, x \neq y} d(x, y)$ is called **minimum distance** of the code C. The *weight* $w(x)$ of a codeword $x \in C$ is defined as $w(x) = d(x, 0)$. An (n, M, d^*)*-code* is an q-ary block code of length n, size M, and minimum distance d^*. A *binary code* is a code over $\mathbb{F}_2$.

When codewords are chosen such that the distance between them is maximized, the code is called *error-correcting*, since slightly garbled vectors can be recovered by choosing the nearest codeword. A code C is an t*-error-correcting code* (and an $2t$*-error-detecting code*) if $d^*(C) \geqslant 2t+1$. In this case each *neighborhood* $U_t(x) = \{y \in C : d(x, y) \leqslant t\}$ of $x \in C$ is disjoint with $U_t(y)$ for any $y \in C$, $y \neq x$. A *perfect code* is an q-ary $(n, M, 2t+1)$-code for which the M spheres $U_t(x)$ or radius t centered on the codewords fill the whole space $\mathbb{F}_q^n$ completely, without overlapping.

A block code $C \subset \mathbb{F}_q^n$ is called *linear* if C is a vector subspace of $\mathbb{F}_q^n$. An $[n, k]$-code is an k-dimensional linear code $C \subset \mathbb{F}_q^n$ (with the minimum distance d^*); it has size q^k, i.e., is an (n, q^k, d^*)-code. The *Hamming code* is the linear perfect one-error correcting $(\frac{q^r-1}{q-1}, \frac{q^r-1}{q-1} - r, 3)$-code.

An $k \times n$ matrix G with rows that are basis vectors for a linear $[n, k]$-code C is called *generator matrix* of C. In *standard form* it can be written as $(1_k|A)$, where 1_k is the $k \times k$ identity matrix. Each *message* (or *information symbol, source symbol*) $u = (u_1, \dots, u_k) \in \mathbb{F}_q^k$ can be encoded by multiplying it (on the right) with the generator matrix: $uG \in C$. The matrix $H = (-A^T|1_{n-k})$ is called *parity-check matrix* of C. The number $r = n - k$ corresponds to the number of parity check digits in the code, and is called *redundancy* of the code C. The *information rate* (or *code rate*) of a code C is the number $R = \frac{\log_2 M}{n}$. For an q-ary $[n, k]$-code $R = \frac{k}{n} \log_2 q$; for a binary $[n, k]$-code $R = \frac{k}{n}$.

A *convolutional code* is a type of error-correction code in which each k-bit information symbol to be encoded is transformed into an n-bit codeword, where $R = \frac{k}{n}$ is the code rate ($n \geqslant k$), and the transformation is a function of the last m information symbols, where m is the *constraint length* of the code. Convolutional codes are often used to improve the performance of radio and satellite links. A *variable length code* is a code with codewords of different lengths.

In contrast to error-correcting codes which are designed only to increase the reliability of data communications, *cryptographic codes* are designed to increase their security. In Cryptography, the sender uses a *key* to encrypt a message before it is sent through a insecure channel, and an authorized receiver at the other end then uses a key to decrypt the received data to a message. Often, data compression algorithms and error-correcting codes are used in tandem with cryptographic codes to yield communications that are both efficient, robust to data transmission errors, and secure to eavesdropping and tampering. Encrypted messages which are, moreover, hidden in text, image, etc., are called *stenographic messages*.

16.1. MINIMUM DISTANCE AND RELATIVES

- **Minimum distance**

  Given a code $C \subset V$, where V is an n-dimensional vector space equipped with a metric d, the **minimum distance** $d^* = d^*(C)$ of the code C is defined by

$$\min_{x,y \in C,\, x \neq y} d(x, y).$$

  The metric d depends on the nature of the errors for the correction of which the code is intended. For a prescribed correcting capacity it is necessary to use codes with maximum number of codewords. The most widely investigated such codes are the *q-ary block codes* in the **Hamming metric** $d_H(x, y) = |\{i : x_i \neq y_i, i = 1, \dots, n\}|$.

  For a *linear code* C the minimum distance $d^*(C) = w(C)$, where $w(C) = \min\{w(x) : x \in C\}$ is a *minimum weight* of the code C. As there are $rank(H) \leqslant n - k$ independent columns in the parity check matrix H of an $[n, k]$-code C, then $d^*(C) \leqslant n - k + 1$ (*Singleton upper bound*).

- **Dual distance**

The **dual distance** $d^{\perp}$ of a *linear* $[n, k]$-code $C \subset \mathbb{F}_q^n$ is the **minimum distance** of the *dual code* $C^{\perp}$ of C.

The *dual code* $C^{\perp}$ of C is defined as the set of all vectors of $\mathbb{F}_q^n$, that are orthogonal to every codeword of C: $C^{\perp} = \{v \in \mathbb{F}_q^n\colon \langle v, u\rangle = 0 \text{ for any } u \in C\}$. The code $C^{\perp}$ is a linear $[n, n-k]$-code. The $(n-k) \times n$ generator matrix of $C^{\perp}$ is the parity-check matrix of C.

- **Bar product distance**

Given linear codes C_1 and C_2 of length n with $C_2 \subset C_1$, their *bar product* $C_1|C_2$ is a linear code of length $2n$, defined by $C_1|C_2 = \{x|x + y\colon\ x \in C_1, y \in C_2\}$.

The **bar product distance** is the minimum distance $d^*(C_1|C_2)$ of the bar product $C_1|C_2$.

- **Design distance**

A *linear code* is called *cyclic code* if all cyclic shifts of a codeword also belong to C, i.e., if for any $(a_0, \dots, a_{n-1}) \in C$ the vector $(a_{n-1}, a_0, \dots, a_{n-2}) \in C$. It is convenient to identify a codeword $(a_0, \dots, a_{n-1})$ with the polynomial $c(x) = a_0 + a_1x + \cdots + a_{n-1}x^{n-1}$; then every cyclic $[n, k]$-code can be represented as the principal ideal $\langle g(x)\rangle = \{r(x)g(x)\colon\ r(x) \in \mathbb{R}_n\}$ of the ring $\mathbb{R}_n = \mathbb{F}_q(x)/(x^n - 1)$, generated by the polynomial $g(x) = g_0 + g_1x + \cdots + x^{n-k}$, called *generator polynomial* of the code C.

Given an element α of order n in a finite field $\mathbb{F}_{q^s}$, a *Bose–Chaudhuri–Hocquenghem* $[n, k]$*-code* of **design distance** d is a cyclic code of length n, generated by a polynomial $g(x)$ in $\mathbb{F}_q(x)$ of degree $n - k$, that has roots at $\alpha, \alpha^2, \dots, \alpha^{d-1}$. The minimum distance d^* of a Bose–Chaudhuri–Hocquenghem code of odd design distance d is at least d.

A *Reed–Solomon code* is a Bose–Chaudhuri–Hocquenghem code with $s = 1$. The generator polynomial of a Reed–Solomon code of design distance d is $g(x) = (x-\alpha)\cdots(x-\alpha^{d-1})$ with degree $n - k = d - 1$; that is, for a Reed–Solomon code the design distance $d = n - k + 1$, and the minimum distance $d^* \geqslant d$. Since for a linear $[n, k]$-code the minimum distance $d^* \leqslant n - k + 1$ (*Singleton upper bound*), a Reed–Solomon code has the minimum distance $d^* = n-k+1$ and achieves the Singleton upper bound. Compact disk players use a double-error correcting $(255, 251, 5)$ Reed–Solomon code over $\mathbb{F}_{256}$.

- **Goppa designed minimum distance**

Goppa designed minimum distance ([Gopp71]) is a lower bound $d^\star(m)$ for the minimum distance of *one-point geometric Goppa codes* (or *algebraic geometry codes*) $G(m)$. For $G(m)$, associated to the divisors D and mP, $m \in \mathbb{N}$, of a smooth projective absolutely irreducible algebraic curve of genus $g > 0$ over a finite field $\mathbb{F}_q$, one has $d^\star(m) = m + 2 - 2g$ if $2g - 2 < m < n$.

In fact, for a Goppa code $C(m)$ the structure of the gap sequence at P may allow one to give a better lower bound of the minimum distance (cf. **Feng–Rao distance**).

- **Feng–Rao distance**

The **Feng–Rao distance** $\delta_{FR}(m)$ is a lower bound for the minimum distance of *one-point geometric Goppa codes* $G(m)$ which is better than the **Goppa designed minimum**

distance. The method of Feng and Rao for encoding the code $C(m)$ decodes errors up to half the Feng–Rao distance $\delta_{FR}(m)$, and gives an improvement of the number of errors that one can correct for one-point geometric Goppa codes.

Formally, the Feng–Rao distance is defined as follow. Let S be a *numerical semi-group*, i.e., a sub-semi-group S of $\mathbb{N} \cup \{0\}$ such that the *genus* $g = |\mathbb{N} \cup \{0\} \backslash S|$ of S is finite, and $0 \in S$. The **Feng–Rao distance** on S is a function $\delta_{FR} : S \to \mathbb{N} \cup \{0\}$ such that $\delta_{FR}(m) = \min\{\nu(r)\colon\ r \geqslant m, r \in S\}$, where $\nu(r) = |\{(a, b) \in S^2\colon\ a + b = r\}|$. The generalized r**-th Feng–Rao distance** on S is defined by $\delta^r_{FR}(m) = \min\{\nu[m_1, \dots, m_r]\colon\ m \leqslant m_1 < \cdots < m_r, m_i \in S\}$, where $\nu[m_1, \dots, m_r] = |\{a \in S\colon\ m_i - a \in S \text{ for some } i = 1, \dots, r\}|$. Then $\delta_{FR}(m) = \delta^1_{FR}(m)$. (See, for example, [FaMu03].)

- **Free distance**

The **free distance** is the minimum non-zero *Hamming weight* of any codeword in a *convolutional code* or a *variable length code*.

Formally, the k**-th minimum distance** d^*_k of a convolutional code or a variable length code is the smallest Hamming distance between any two initial codeword segments k frame long that disagree in the initial frame. The sequence $d^*_1, d^*_2, d^*_3, \dots$ $(d^*_1 \leqslant d^*_2 \leqslant d^*_3 \leqslant \cdots)$ is called *distance profile* of the code. The free distance of a convolutional code or a variable length code is $\max_l d^*_l = \lim_{l \to \infty} d^*_l = d^*_\infty$.

- **Effective free distance**

A *turbo code* is a long *block code* in which there are L input bits, and each of these bits is encoded q times. In the j-th encoding, the L bits are sent through a permutation box P_j, and then encoded via an $[N_j, L]$ block encoder (*code fragment encoder*) which can be thought of as an $L \times N_j$ matrix. The overall turbo code is then a *linear* $[N_1 + \cdots + N_q, L]$-code (see, for example, [BGT93]).

The *weight-i input minimum distance* $d^i(C)$ of a turbo-code C is the minimum weight among codewords corresponding to input words of weight i. The **effective free distance** of C is its *weight-2 input minimum distance* $d^2(C)$, i.e., the minimum *weight* among codewords corresponding to input words of weight 2.

- **Distance distribution**

Given a code C over a finite metric space (X, d) with the diameter $diam(X, d) = D$, the **distance distribution** of C is an $(D + 1)$-vector $(A_0, \dots, A_D)$, where $A_i = \frac{1}{|C|} |\{(c, c') \in C^2\colon\ d(c, c') = i\}|$. That is, one considers $A_i(c)$ as the number of code words at distance i from the codeword c, and takes A_i as the average of $A_i(c)$ over all $c \in C$. $A_0 = 1$, and if $d^* = d^*(C)$ is the minimum distance of C, then $A_1 = \cdots = A_{d^*-1} = 0$.

The distance distribution of a code with given parameters is important, in particular, for bounding the probability of decoding error under different decoding procedures from maximum likelihood decoding to error detection. Apart from this, it can be helpful in revealing structural properties of codes and establish nonexistence of some codes.

- **Unicity distance**

The **unicity distance** of a cryptosystem is the minimal length of cyphertext that is required in order to expect that there exists only one meaningful decryption for it. For classic cryptosystems with fixed key space, the unicity distance is approximated by the formula $H(K)/D$, where $H(K)$ is the *key space entropy* (roughly $\log_2 N$, where N is the number of keys), and D measures the *redundancy* of the plain text source language in bits per letter.

A cryptosystem offers perfect secrecy if its unicity distance is infinite. For example, the *one-time pads* offer perfect secrecy; they were used for the "red telephone" between Kremlin and White House.

16.2. MAIN CODING DISTANCES

- **Arithmetic codes distance**

An *arithmetic code* (or *code with correction of arithmetic errors*) C is a finite subset of the set $\mathbb{Z}$ of integers (usually, non-negative integers). It is intended for the control of the functioning of an *adder* (a module performing addition). When adding numbers represented in the binary number system, a single slip in the functioning of the adder leads to a change in the result by some power of 2, thus, to a single *arithmetic error*. Formally, a single *arithmetic error* on $\mathbb{Z}$ is defined as a transformation of a number $n \in \mathbb{Z}$ to a number $n' = n \pm 2^i$, $i = 1, 2, \dots$.

The **arithmetic codes distance** is a metric on $\mathbb{Z}$, defined, for any $n_1, n_2 \in \mathbb{Z}$, as the minimum number of *arithmetic errors* taking n_1 to n_2. It can be written as $w_2(n_1 - n_2)$, where $w_2(n)$ is the *arithmetic* 2*-weight* of n, i.e., the smallest possible number of non-zero coefficients in representations $n = \sum_{i=0}^{k} e_i 2^i$, where $e_i = 0, \pm 1$, and k is some non-negative integer. In fact, for each n there is an unique such representation with $e_k \neq 0$, $e_i e_{i+1} = 0$ for all $i = 0, \dots, k-1$, which has the smallest number of non-zero coefficients (cf. **arithmetic** r**-norm metric**).

- **Sharma–Kaushik distance**

Let $q \geqslant 2$, $m \geqslant 2$. A *partition* $\{B_0, B_1, \dots B_{q-1}\}$ of $\mathbb{Z}_m$ is called *Sharma–Kaushik partition* if the following conditions hold:

1. $B_0 = \{0\}$;
2. For any $i \in Z_m$, $i \in B_s$ if and only if $m - i \in B_s$, $s = 1, 2, \dots, q-1$;
3. If $i \in B_s$, $j \in B_t$, and $s > t$, then $\min\{i, m-i\} > \min\{j, m-j\}$;
4. If $s \geqslant t$, $s, t = 0, 1, \dots, q-1$, $|B_s| \geqslant |B_t|$ except for $s = q-1$ in which case $|B_{q-1}| \geqslant \frac{1}{2}|B_{q-2}|$.

Given a Sharma–Kaushik partition of $\mathbb{Z}_m$, the *Sharma–Kaushik weight* $w_{SK}(x)$ of any element $x \in \mathbb{Z}_m$ is defined by $w_{SK}(x) = i$ if $x \in B_i$, $i \in \{0, 1, \dots, q-1\}$.

The **Sharma–Kaushik distance** (see, for example, [ShKa97]) is a metric on $\mathbb{Z}_m$, defined by

$$w_{SK}(x - y).$$

The Sharma–Kaushik distance on Z_m^n is defined by $w_{SK}^n(x - y)$, where, for $x = (x_1, \ldots x_n) \in \mathbb{Z}_m^n$, one has $w_{SK}^n(x) = \sum_{i=1}^n w_{SK}(x_i)$.

The **Hamming metric** and the **Lee metric** arise from two specific partitions of the above type: $P_H = \{B_0, B_1\}$, where $B_1 = \{1, 2, \ldots, q-1\}$, and $P_L = \{B_0, B_1, \ldots, B_{\lfloor q/2 \rfloor}\}$, where $B_i = \{i, m-i\}$, $i = 1, \ldots, \lfloor \frac{q}{2} \rfloor$.

- **Absolute summation distance**

The **absolute summation distance** (or *Lee distance*) is the **Lee metric** on the set $\mathbb{Z}_m^n$, defined by

$$w_{Lee}(x - y),$$

where $w_{Lee}(x) = \sum_{i=1}^n \min\{x_i, m - x_i\}$ is the *Lee weight* of $x = (x_1, \ldots, x_n) \in \mathbb{Z}_m^n$. If $\mathbb{Z}_m^n$ is equipped with the absolute summation distance, then a subset C of $\mathbb{Z}_m^n$ is called *Lee distance code*. Lee distance codes are used for phase-modulated and multilevel quantized-pulse-modulated channels, and have several applications to the toroidal interconnection networks. Most important Lee distance codes are *negacyclic codes*.

- **Mannheim distance**

Let $\mathbb{Z}[i] = \{a + bi : a, b \in \mathbb{Z}\}$ be the set of *Gaussian integers*. Let $\pi = a + bi$ $(a > b > 0)$ be a *Gaussian prime*. It means, that $(a+bi)(a-bi) = a^2 + b^2 = p$, where $p \equiv 1 \pmod 4$ is a prime number, or that $\pi = p + 0 \cdot i = p$, where $p \equiv 3 \pmod 4$ is a prime number.

The **Mannheim distance** is a distance on $\mathbb{Z}[i]$, defined, for any two Gaussian integers x and y, as the sum of the absolute values of real and imaginary part of the difference $x - y \pmod \pi$. The modulo reduction, before summing the absolute values of real and imaginary part, is the difference between the **Manhattan metric** and the Mannheim distance.

The elements of the finite field $\mathbb{F}_p = \{0, 1, \ldots, p-1\}$ for $p \equiv 1 \pmod 4$, $p = a^2 + b^2$, and the elements of the finite field $\mathbb{F}_{p^2}$ for $p \equiv 3 \pmod 4$, $p = a$, can be mapped on a subset of the Gaussian integers using the complex modulo function $\mu(k) = k - [\frac{k(a-bi)}{p}](a + bi)$, $k = 0, \ldots, p-1$, where $[.]$ denotes rounding to the closest Gaussian integer. The set of the selected Gaussian integers with the minimal Galois norms is called *constellation*. This representation gives a new way to construct codes for two-dimensional signals. Mannheim distance was introduced to make *QAM*-like signals more susceptible for algebraic decoding methods. For codes over hexagonal signal constellations a similar metric can be introduced over the set of the *Eisenstein–Jacobi integers*. It is useful for block codes over tori. (See, for example, [Hube93], [Hube94].)

- **Poset distance**

Let $(V_n, \preceq)$ be a *poset* on $V_n = \{1, \ldots, n\}$. A subset I of V_n is called *ideal* if $x \in I$ and $y \preceq x$ imply that $y \in I$. If $J \subset V_n$, then $\langle J \rangle$ denotes the smallest ideal of V_n which contains J. Consider the vector space $\mathbb{F}_q^n$ over a finite field $\mathbb{F}_q$. The *P-weight* of an element $x = (x_1, \ldots, x_n) \in \mathbb{F}_q^n$ is defined as the cardinality of the smallest ideal of V_n containing the *support* of x: $w_P(x) = |\langle supp(x) \rangle|$, where $supp(x) = \{i : x_i \neq 0\}$.

The **poset distance** (see [BGL95]) is a metric on $\mathbb{F}_q^n$, defined by

$$w_P(x - y).$$

If $\mathbb{F}_q^n$ is equipped with a poset distance, then a subset C of $\mathbb{F}_q^n$ is called *poset code*. If V_n forms the chain $1 \leqslant 2 \leqslant \cdots \leqslant n$, then the linear code C of dimension k consisting of all vectors $(0, \dots, 0, a_{n-k+1}, \dots, a_n) \in \mathbb{F}_q^n$ is a perfect poset code with the minimum (poset) distance $d_P^*(C) = n - k + 1$. If V_n forms an antichain, then the poset distance coincides with the **Hamming metric**.

- **Rank distance**

Let $\mathbb{F}_q$ be a finite field, $\mathbb{K} = \mathbb{F}_{q^m}$ be an extension of degree m of $\mathbb{F}_q$, and $\mathbb{E} = \mathbb{K}^n$ be a vector space of dimension n over $\mathbb{K}$. For any $a = (a_1, \dots a_n) \in \mathbb{E}$ define its *rank*, $rank(a)$, as the dimension of the vector space over $\mathbb{F}_q$, generated by $\{a_1, \dots, a_n\}$.

The **rank distance** is a metric on $\mathbb{E}$, defined by

$$rank(a - b).$$

Since the rank distance between two codewords is at most the Hamming distance between them, for any code $C \subset \mathbb{E}$ its minimum (rank) distance $d_{RK}^*(C) \leqslant \min\{m, n - \log_{q^m} |C| + 1\}$. A code C with $d_{RK}^*(C) = n - \log_{q^m} |C| + 1$, $n < m$, is called *Gabidulin code* (see [Gabi85]). A code C with $d_{RK}^*(C) = m$, $m \leqslant n$, is called *full rank distance code*. Such code has at most q^n elements. A *maximal full rank distance code* is a full rank distance code with q^n elements; it exists if and only if m divides n.

- **Gabidulin–Simonis metrics**

Consider the vector space $\mathbb{F}_q^n$ (over a finite field $\mathbb{F}_q$) and a finite family $F = \{F_i : i \in I\}$ of its subsets such that $\cup_{i \in I} F_i = \mathbb{F}_q^n$. Without loss of generality, F can be an antichain of linear subspaces of $\mathbb{F}_q^n$. The *F-weight* w_F of a vector $x = (x_1, \dots, x_n) \in \mathbb{F}_q^n$ is defined as the cardinality of the smallest subset J of I such that $x \in \cup_{i \in J} F_i$.

A **Gabidulin–Simonis metric** (or **F-distance**, see [GaSi98]) is a metric on $\mathbb{F}_q^n$, defined by

$$w_F(x - y).$$

The **Hamming metric** corresponds to the case of F_i, $i \in I$, forming the standard basis. The **Vandermonde metric** is F-distance with F_i, $i \in I$, being the columns of a generalized Vandermonde matrix. Amongst other coding Gabidulin–Simonis metrics are: **rank distance**, *b-burst distance*, Gabidulin's *combinatorial metrics* (cf. **poset distance**), etc.

- **Rosenbloom–Tsfasman distance**

Let $M_{m,n}(\mathbb{F}_q)$ be the set of all $m \times n$ matrices with entries from a finite field $\mathbb{F}_q$ (in general, from any finite alphabet $\mathcal{A} = \{a_1, \dots, a_q\}$). The *Rosenbloom–Tsfasman norm* $\|.\|_{RT}$ on $M_{m,n}(\mathbb{F}_q)$ is defined as follow: if $m = 1$ and $a = (\xi_1, \xi_2, \dots, \xi_n) \in M_{1,n}(\mathbb{F}_q)$, then $\|0_{1,n}\|_{RT} = 0$, and $\|a\|_{RT} = \max\{i | \xi_i \neq 0\}$ for $a \neq 0_{1,n}$; if $A = (a_1, \dots, a_m)^T \in M_{m,n}(\mathbb{F}_q)$, $a_j \in M_{1,n}(\mathbb{F}_q)$, $1 \leqslant j \leqslant m$, then $\|A\|_{RT} = \sum_{j=1}^m \|a_j\|_{RT}$.

The **Rosenbloom–Tsfasman distance** ([RoTs96]) is a metric (in fact, an **ultrametric**) on $M_{m,n}(\mathbb{F}_q)$, defined by

$$\|A - B\|_{RT}.$$

For every matrix code $C \subset M_{m,n}(F_q)$ with q^k elements the minimum (Rosenbloom–Tsfasman) distance $d^*_{RT}(C) \leqslant mn-k+1$. Codes meeting this bound are called *maximum distance separable codes*.

The most used distance between codewords of a matrix code $C \subset M_{m,n}(F_q)$ is the **Hamming metric** on $M_{m,n}(F_q)$, defined by $\|A - B\|_H$, where $\|A\|_H$ is the *Hamming weight* of a matrix $A \in M_{m,n}(\mathbb{F}_q)$, i.e., the number of non-zero entries of A.

- **Interchange distance**

  The **interchange distance** is a metric on the code $C \subset \mathcal{A}^n$ over an alphabet $\mathcal{A}$, defined, for any $x, y \in C$, as the minimum number of *transpositions*, i.e., interchanges of adjacent pairs of symbols, converting x into y.

- **ACME distance**

  The **ACME distance** is a metric on a code $C \subset \mathcal{A}^n$ over an alphabet $\mathcal{A}$, defined by

  $$\min\{d_H(x, y), d_I(x, y)\},$$

  where d_H is the **Hamming metric**, and d_I is the **interchange distance**.

- **Indel distance**

  Let W be the set of all words over an alphabet $\mathcal{A}$. A *deletion* of a letter in a word $\beta = b_1 \dots b_n$ of the length n is a transformation of β into a word $\beta' = b_1 \dots b_{i-1}b_{i+1} \dots b_n$ of the length $n-1$. An *insertion* of a letter in a word $\beta = b_1 \dots b_n$ of the length n is a transformation of β into a word $\beta'' = b_1 \dots b_i b b_{i+1} \dots b_n$, of the length $n+1$.

  The **indel distance** (or **distance of codes with correction of deletions and insertions**) is a metric on W, defined, for any $\alpha, \beta \in W$, as the minimum number of deletions and insertions of letters converting α into β.

  A *code C with correction of deletions and insertions* is an arbitrary finite subset of W. An example of such code is the set of words $\beta = b_1 \dots b_n$ of length n over the alphabet $\mathcal{A} = \{0, 1\}$ for which $\sum_{i=1}^n ib_i \equiv 0 \ (mod\ n+1)$. The number of words in this code is equal to $\frac{1}{2(n+1)} \sum_k \phi(k)2^{(n+1)/k}$, where the sum is taken over all odd divisors k of $n+1$, and ϕ is the *Euler function*.

- **Interval distance**

  The **interval distance** (see, for example, [Bata95]) is a metric on a finite group $(G, +, 0)$, defined by

  $$w_{int}(x - y),$$

  where $w_{int}(x)$ is an *interval weight* on G, i.e., a *group norm* which values are consecutive non-negative integers $0, \dots, m$. This distance is used for *group codes* $C \subset G$.

- **Fano metric**

The **Fano metric** is a *decoding metric* with the goal to find the best sequence estimate used for the *Fano algorithm* of *sequential decoding* of *convolutional codes*.

A *convolutional code* is a type of error-correction code in which each k-bit information symbol to be encoded is transformed into an n-bit codeword, where $R = \frac{k}{n}$ is the code rate ($n \geqslant k$), and the transformation is a function of the last m information symbols. The linear time-invariant decoder (*fixed convolutional decoder*) maps an information symbol $u_i \in \{u_1, \dots, u_N\}$, $u_i = (u_{i1}, \dots u_{ik})$, $u_{ij} \in \mathbb{F}_2$, into a codeword $x_i \in \{x_1, \dots, x_N\}$, $x_i = (x_{i1}, \dots, x_{in})$, $x_{ij} \in \mathbb{F}_2$, so one has a code $\{x_1, \dots, x_N\}$ with N codewords which occur with probabilities $\{p(x_1), \dots, p(x_N)\}$. A sequence of l codewords form a *stream* (or *path*) $x = x_{[1,l]} = \{x_1, \dots, x_l\}$ which is transmitted through a *discrete memoryless channel*, resulting in the received sequence $y = y_{[1,l]}$. The task of a decoder which minimizes the sequence error probability is to find a sequence which maximizes the joint probability of input and output channel sequences $p(y, x) = p(y|x) \cdot p(x)$. Usually it is sufficient to find a procedure that maximizes $p(y|x)$, and a decoder that always chooses as its estimate one of the sequences that maximizes it or, equivalently, the **Fano metric**, is called *max-likelihood decoder*.

Roughly, we consider each code as a tree, where each branch represents one codeword. The decoder begins at the first vertex in the tree, and computes the branch metric for each possible branch, determined the best branch to be the one corresponding to the codeword x_j resulting in the largest branch metric, $\mu_F(x_j)$. This branch is added to the path, and the algorithm continues from the new node which represents the sum of the previous node and the number of bits in the current best codeword. Through iterating until a terminal node of the tree is reached, the algorithm traces the most likely path. In this construction, the **bit Fano metric** is defined by

$$\log_2 \frac{p(y_i|x_i)}{p(y_i)} - R,$$

the **branch Fano metric** is defined by

$$\mu_F(x_j) = \sum_{i=1}^{n} \left(\log_2 \frac{p(y_i|x_{ji})}{p(y_i)} - R \right),$$

and the **path Fano metric** is defined by

$$\mu_F(x_{[1,l]}) = \sum_{j=1}^{l} \mu_F(x_j),$$

where $p(y_i|x_{ji})$ are the channel transition probabilities,

$$p(y_i) = \sum_{x_m} p(x_m) p(y_i|x_m)$$

is the probability distribution of the output given the input symbols averaged over all input symbols, and $R = \frac{k}{n}$ is the code rate.

For a hard-decision decoder $p(y_j = 0|x_j = 1) = p(y_j = 1|x_j = 0) = p, 0 < p < \frac{1}{2}$, the Fano metric for a path $x_{[1,l]}$ can be written as

$$\mu_F(x_{[1,l]}) = -\alpha d_H(y_{[1,l]}, x_{[1,l]}) + \beta \cdot l \cdot n,$$

where $\alpha = -\log_2 \frac{p}{1-p} > 0$, $\beta = 1 - R + \log_2(1-p)$, and d_H is the **Hamming metric**. The **generalized Fano metric** for sequential decoding is defined by

$$\mu_F^w(x_{[1,l]}) = \sum_{j=1}^{ln} \left(\log_2 \frac{p(y_j|x_j)^w}{p(y_j)^{1-w}} - wR \right),$$

$0 \leqslant w \leqslant 1$. When $w = 1/2$, the generalized Fano metric reduces to the Fano metric with a multiplicative constant $1/2$.

- **Metric recursion of a MAP decoding**

  Maximum a posteriori sequence estimation, or *MAP decoding* for *variable length codes*, used the *Viterbi algorithm*, is based on the **metric recursion**

$$\Lambda_k^{(m)} = \Lambda_{k-1}^{(m)} + \sum_{n=1}^{l_k^{(m)}} x_{k,n}^{(m)} \log_2 \frac{p(y_{k,n}|x_{k,n}^{(m)} = +1)}{p(y_{k,n}|x_{k,n}^{(m)} = -1)} + 2\log_2 p\big(u_k^{(m)}\big),$$

  where $\Lambda_k^{(m)}$ is the **branch metric** of branch m at time (level) k, $x_{k,n}$ is the n-th bit of the codeword having $l_k^{(m)}$ bits labeled at each branch, $y_{k,n}$ is the respective received soft-bit, u_k^m is the source symbol of branch m at time k, and assuming statistical independence of the source symbols, the probability $p(u_k^{(m)})$ is equivalent to the probability of the source symbol labeled at branch m, that may be known or estimated. The metric increment is computed for each branch, and the largest value, when using log-likelihood-values, of each state is used for further recursion. The decoder first computes the metric of all branches, and then the branch sequence with largest metric starting from the final state backward is selected.

Chapter 17

Distances and Similarities in Data Analysis

A *data set* is a finite set comprised of m sequences $(x_1^j, \ldots, x_n^j)$, $j \in \{1, \ldots, m\}$, of length n. The values $x_i^1, \ldots, x_i^m$ represent *attribute* S_i. It can be *numerical*, including *continuous* (real numbers) and *binary* (presence/absence expressed by $1/0$), *ordinal* (numbers expressing rank only), or *nominal* (which are not ordered).

Cluster Analysis (or *Classification, Taxonomy, Pattern Recognition*) consists mainly of partition of data A into relatively small number of *clusters*, i.e., such sets of objects, that (with respect of selected measure of distance) the objects, at best possible degree, are "close" if they belong to the same cluster, "far" if they belong to different clusters, and further subdivision into clusters will impair above two conditions.

We give three typical examples. In *Information Retrieval* applications, nodes of peer-to-peer database network export a data (collection of text documents); each document is characterized by a vector from $\mathbb{R}^n$. An user *query* consists of a vector $x \in \mathbb{R}^n$, and user needs all documents in database which are *relevant* to it, i.e., belong to the *ball* in $\mathbb{R}^n$, centered in x, of fixed radius and with convenient distance function. In *Record Linkage*, each document (database record) is represented by a term-frequency vector $x \in \mathbb{R}^n$ or a string, and one wants to measure semantic relevancy of syntactically different records. In *Ecology*, let x, y be *species abundance distributions*, obtained by two sample methods (i.e., x_j, y_j are the numbers of individuals of species j, observed in corresponding sample); one needs a measure of distance between x and y, in order to compare two methods.

Once a distance d between objects is selected, the **linkage metric**, i.e., a distance between clusters $A = \{a_1, \ldots, a_m\}$ and $B = \{b_1, \ldots, b_n\}$ is usually one of the following:

average linkage: the average of the distances between the all members of those clusters, i.e., $\frac{\sum_i \sum_j d(a_i, b_j)}{mn}$;

single linkage: the distance between the nearest members of those clusters, i.e., $\min_{i,j} d(a_i, b_j)$;

complete linkage: the distance between the furthest members of those clusters, i.e., $\max_{i,j} d(a_i, b_j)$;

centroid linkage: the distance between the *centroids* of those clusters, i.e, $\|\tilde{a} - \tilde{b}\|_2$, where $\tilde{a} = \frac{\sum_i a_i}{m}$, and $\tilde{b} = \frac{\sum_j b_j}{n}$;

Ward linkage: the distance $\sqrt{\frac{mn}{m+n}} \|\tilde{a} - \tilde{b}\|_2$.

Multi-dimensional Scaling is a technique developed in the behavioral and social sciences for studying the structure of objects or people. Together with Cluster Analysis, it is based on distance methods. But in Multi-dimensional Scaling, as opposite to Cluster Analysis, one starts only with some $m \times m$ matrix D of distances of the objects and (iteratively) looks for a representation of objects in $\mathbb{R}^n$ with low n, so that their Euclidean distance matrix has minimal square deviation from the original matrix D.

There are many **similarities** used in Data Analysis; the choice depends on the nature of data and is not an exact science. We list below main such similarities and distances.

Given two objects, represented by non-zero vectors $x = (x_1, \dots, x_n)$ and $y = (y_1, \dots, y_n)$ from $\mathbb{R}^n$, the following notation are used in this chapter.

$\sum x_i$ means $\sum_{i=1}^n x_i$.

1_F is the *characteristic function* of event F: $1_F = 1$ if F happens, and $1_F = 0$, otherwise.

$\|x\|_2 = \sqrt{\sum x_i^2}$ is the ordinary Euclidean norm on $\mathbb{R}^n$.

By $\overline{x}$ is denoted $\frac{\sum x_i}{n}$, i.e., the *mean value* of components of x. So, $\overline{x} = \frac{1}{n}$ if x is a *frequency vector* (*discrete probability distribution*), i.e., all $x_i \geqslant 0$, and $\sum x_i = 1$, and $\overline{x} = \frac{n+1}{2}$ if x is a *ranking* (*permutation*), i.e., all x_i are different numbers from $\{1, \dots, n\}$.

In the binary case $x \in \{0, 1\}^n$ (i.e., when x is a binary n-sequence), let $X = \{1 \leqslant i \leqslant n\colon\ x_i = 1\}$ and $\overline{X} = \{1 \leqslant i \leqslant n\colon\ x_i = 0\}$. Let $|X \cap Y|$, $|X \cup Y|$, $|X \backslash Y|$ and $|X \Delta Y|$ denote the cardinality of intersection, union, difference and symmetric difference $(X \backslash Y) \cup (Y \backslash X)$ of the sets X and Y, respectively.

17.1. SIMILARITIES AND DISTANCES FOR NUMERICAL DATA

- **Ruzicka similarity**

  The **Ruzicka similarity** is a similarity on $\mathbb{R}^n$, defined by

$$\frac{\sum \min\{x_i, y_i\}}{\sum \max\{x_i, y_i\}}.$$

- **Roberts similarity**

  The **Roberts similarity** is a similarity on $\mathbb{R}^n$, defined by

$$\frac{\sum (x_i + y_i)\frac{\min\{x_i, y_i\}}{\max\{x_i, y_i\}}}{\sum (x_i + y_i)}.$$

- **Ellenberg similarity**

  The **Ellenberg similarity** is a similarity on $\mathbb{R}^n$, defined by

$$\frac{\sum (x_i + y_i) 1_{x_i \cdot y_i \neq 0}}{\sum (x_i + y_i)(1 + 1_{x_i y_i = 0})}.$$

  Binary cases of Ellenberg and **Ruzicka similarities** coincide; it is called **Tanimoto similarity** (or **Jaccard similarity of community**):

$$\frac{|X \cap Y|}{|X \cup Y|}.$$

The **Tanimoto distance** (or **biotope distance**) is a distance on $\{0, 1\}^n$, defined by

$$1 - \frac{|X \cap Y|}{|X \cup Y|} = \frac{|X \Delta Y|}{|X \cup Y|}.$$

- **Gleason similarity**

The **Gleason similarity** is a similarity on $\mathbb{R}^n$, defined by

$$\frac{\sum (x_i + y_i) 1_{x_i \cdot y_i \neq 0}}{\sum (x_i + y_i)}.$$

Binary cases of Cleason, **Motyka** and **Bray–Curtis similarities** coincide; it is called **Dice similarity** (or *Sorensen similarity*, *Czekanowsky similarity*):

$$\frac{2|X \cap Y|}{|X \cup Y| + |X \cap Y|} = \frac{2|X \cap Y|}{|X| + |Y|}.$$

The **Czekanowsky–Dice distance** (or *Bray–Curtis non-metric coefficient*, *normalized symmetric difference distance*) is a **near-metric** on $\{0, 1\}^n$, defined by

$$1 - \frac{2|X \cap Y|}{|X| + |Y|} = \frac{|X \Delta Y|}{|X| + |Y|}.$$

- **Intersection distance**

The **intersection distance** is a distance on $\mathbb{R}^n$, defined by

$$1 - \frac{\sum \min\{x_i, y_i\}}{\min\{\sum x_i, \sum y_i\}}.$$

- **Motyka similarity**

The **Motyka similarity** is a similarity on $\mathbb{R}^n$, defined by

$$\frac{\sum \min\{x_i, y_i\}}{\sum (x_i + y_i)} = n \frac{\sum \min\{x_i, y_i\}}{\overline{x} + \overline{y}}.$$

- **Bray–Curtis similarity**

The **Bray–Curtis similarity** is a similarity on $\mathbb{R}^n$, defined by

$$\frac{2}{n(\overline{x} + \overline{y})} \sum \min\{x_i, y_j\}.$$

It is called *Renkonen %similarity* (or *percentage similarity*) if x, y are *frequency vectors*.

- **Bray–Curtis distance**

  The **Bray–Curtis distance** is a distance on $\mathbb{R}^n$, defined by

$$\frac{\sum |x_i - y_i|}{\sum (x_i + y_i)}.$$

- **Canberra distance**

  The **Canberra distance** is a distance on $\mathbb{R}^n$, defined by

$$\sum \frac{|x_i - y_i|}{|x_i| + |y_i|}.$$

- **Kulczynski similarity 1**

  The **Kulczynski similarity** 1 is a similarity on $\mathbb{R}^n$, defined by

$$\frac{\sum \min\{x_i, y_i\}}{\sum |x_i - y_i|}.$$

  The corresponding distance is

$$\frac{\sum |x_i - y_i|}{\sum \min\{x_i, y_i\}}.$$

- **Kulczynski similarity 2**

  The **Kulczynski similarity** 2 is a similarity on $\mathbb{R}^n$, defined by

$$\frac{n}{2}\left(\frac{1}{\overline{x}} + \frac{1}{\overline{y}}\right) \sum \min\{x_i, y_i\}.$$

  In binary case it takes form

$$\frac{|X \cap Y| \cdot (|X| + |Y|)}{2|X| \cdot |Y|}.$$

- **Baroni–Urbani–Buser similarity**

  The **Baroni–Urbani–Buser similarity** is a similarity on $\mathbb{R}^n$, defined by

$$\frac{\sum \min\{x_i, y_i\} + \sqrt{\sum \min\{x_i, y_i\} \sum (\max_{1 \leqslant j \leqslant n} x_j - \max\{x_i, y_i\})}}{\sum \max\{x_i, y_i\} + \sqrt{\sum \min\{x_i, y_i\} \sum (\max_{1 \leqslant j \leqslant n} x_j - \max\{x_i, y_i\})}}.$$

  In binary case it takes form

$$\frac{|X \cap Y| + \sqrt{|X \cap Y| \cdot |\overline{X \cup Y}|}}{|X \cup Y| + \sqrt{|X \cap Y| \cdot |\overline{X \cup Y}|}}.$$

17.2. RELATIVES OF EUCLIDEAN DISTANCE

- **Power (p, r)-distance**

The **power (p, r)-distance** is a distance on $\mathbb{R}^n$, defined by

$$\left(\sum (x_i - y_i)^p\right)^{\frac{1}{r}}.$$

For $p = r \geqslant 1$, it is the l_p**-metric**, including **Euclidean**, **Manhattan** (or *magnitude*) and **Chebyshev** (or *maximum-value*) **metrics** for $n = 2, 1$ and ∞, respectively.

The case $0 < p = r < 1$ is called **fractional distance** (not a metric); it is used for "dimensionality-coursed" data, i.e., when there are few observations and the number n of variables is large.

The weighted versions $(\sum w_i (x_i - y_i)^p)^{\frac{1}{p}}$ (with non-negative weights w_i) are also used, for $p = 1, 2$, in applications.

- **Penrose size distance**

The **Penrose size distance** is a distance on $\mathbb{R}^n$, defined by

$$\sqrt{n} \sum |x_i - y_i|.$$

It is proportional to the **Manhattan metric**. The *Czekanowsky mean character difference* is defined by $\frac{\sum |x_i - y_i|}{n}$.

- **Penrose shape distance**

The **Penrose shape distance** is a distance on $\mathbb{R}^n$, defined by

$$\sqrt{\sum ((x_i - \overline{x}) - (y_i - \overline{y}))^2}.$$

The sum of squares of two above **Penrose distances** is the **squared Euclidean distance**.

- **Lorentzian distance**

The **Lorentzian distance** is a distance on $\mathbb{R}^n$, defined by

$$\sum \ln\big(1 + |x_i - y_i|\big).$$

- **Binary Euclidean distance**

The **binary Euclidean distance** is a distance on $\mathbb{R}^n$, defined by

$$\sqrt{\sum (1_{x_i > 0} - 1_{y_i > 0})^2}.$$

• **Mean censored Euclidean distance**

The **mean censored Euclidean distance** is a distance on $\mathbb{R}^n$, defined by

$$\sqrt{\frac{\sum (x_i - y_i)^2}{\sum 1_{x_i^2+y_i^2 \neq 0}}}.$$

• **Normalized l_p-distance**

The **normalized l_p-distance**, $1 \leqslant p \leqslant \infty$, is a distance on $\mathbb{R}^n$, defined by

$$\frac{\|x-y\|_p}{\|x\|_p + \|y\|_p}.$$

The only integer value p, for which normalized l_p-distance is a metric, is $p = 2$. Moreover, in [Yian91] it is shown that, for any $a, b > 0$, the distance $\frac{\|x-y\|_2}{a+b(\|x\|_2+\|y\|_2)}$ is a metric.

• **Clark distance**

The **Clark distance** is a distance on $\mathbb{R}^n$, defined by

$$\left(\frac{1}{n}\sum\left(\frac{x_i - y_i}{|x_i| + |y_i|}\right)^2\right)^{\frac{1}{2}}.$$

• **Meehl distance**

The **Meehl distance** (or *Meehl index*) is a distance on $\mathbb{R}^n$, defined by

$$\sum_{1 \leqslant i \leqslant n-1} (x_i - y_i - x_{i+1} + y_{i+1})^2.$$

• **Hellinger distance**

The **Hellinger distance** is a distance on $\mathbb{R}^n_+$, defined by

$$\sqrt{2\sum\left(\sqrt{\frac{x_i}{\overline{x}}} - \sqrt{\frac{y_i}{\overline{y}}}\right)^2}.$$

(Cf. **Hellinger metric** in Probability Theory.)

The *Whittaker index of association* is defined by $\frac{1}{2}\sum |\frac{x_i}{\overline{x}} - \frac{y_i}{\overline{y}}|$.

• **Symmetric χ^2-measure**

The **symmetric χ^2-measure** is a distance on $\mathbb{R}^n$, defined by

$$\sum \frac{2}{\overline{x} \cdot \overline{y}} \cdot \frac{(x_i \overline{y} - y_i \overline{x})^2}{x_i + y_i}.$$

- **Symmetric χ^2-distance**

  The **symmetric χ^2-distance** (or *chi-distance*) is a distance on $\mathbb{R}^n$, defined by

$$\sqrt{\sum \frac{\overline{x}+\overline{y}}{n(x_i+y_i)}\left(\frac{x_i}{\overline{x}}-\frac{y_i}{\overline{y}}\right)^2}=\sqrt{\sum \frac{\overline{x}+\overline{y}}{n(\overline{x}\cdot\overline{y})^2}\cdot\frac{(x_i\overline{y}-y_i\overline{x})^2}{x_i+y_i}}.$$

- **Mahalanobis distance**

  The **Mahalanobis distance** (or *statistical distance*) is a distance on $\mathbb{R}^n$, defined by

$$\sqrt{(det\,A)^{\frac{1}{n}}(x-y)A^{-1}(x-y)^T},$$

  where A is a positive-definite matrix (usually, the *covariance matrix* of a finite subset of $\mathbb{R}^n$, consisting of *observation vectors*); cf. **Mahalanobis semi-metric**.

17.3. SIMILARITIES AND DISTANCES FOR BINARY DATA

Usually, such similarities s range from 0 to 1 or from -1 to 1; the corresponding distances are usually $1-s$ or $\frac{1-s}{2}$, respectively.

- **Hamann similarity**

  The **Hamann similarity** is a similarity on $\{0,1\}^n$, defined by

$$\frac{2|\overline{X\Delta Y}|}{n}-1=\frac{n-2|X\Delta Y|}{n}.$$

- **Rand similarity**

  The **Rand similarity** (or *Sokal–Michener similarity, simple matching*) is a similarity on $\{0,1\}^n$, defined by

$$\frac{|\overline{X\Delta Y}|}{n}.$$

  Corresponding metric $\frac{|X\Delta Y|}{n}$ is called *variance* (it is the binary case of *Czekanowsky mean character difference*), and $1-\frac{|X\Delta Y|}{n}$ is called *Gower similarity*.

- **Sokal–Sneath similarity 1**

  The **Sokal–Sneath similarity** 1 is a similarity on $\{0,1\}^n$, defined by

$$\frac{2|\overline{X\Delta Y}|}{n+|\overline{X\Delta Y}|}.$$

- **Sokal–Sneath similarity 2**

  The **Sokal–Sneath similarity** 2 is a similarity on $\{0, 1\}^n$, defined by

$$\frac{|X \cap Y|}{|X \cup Y| + |X \Delta Y|}.$$

- **Sokal–Sneath similarity 3**

  The **Sokal–Sneath similarity** 3 is a similarity on $\{0, 1\}^n$, defined by

$$\frac{|X \Delta Y|}{|\overline{X \Delta Y}|}.$$

- **Russel–Rao similarity**

  The **Russel–Rao similarity** is a similarity on $\{0, 1\}^n$, defined by

$$\frac{|X \cap Y|}{n}.$$

- **Simpson similarity**

  The **Simpson similarity** (*overlap similarity*) is a similarity on $\{0, 1\}^n$, defined by

$$\frac{|X \cap Y|}{\min\{|X|, |Y|\}}.$$

- **Braun–Blanquet similarity**

  The **Braun–Blanquet similarity** is a similarity on $\{0, 1\}^n$, defined by

$$\frac{|X \cap Y|}{\max\{|X|, |Y|\}}.$$

- **Roger–Tanimoto similarity**

  The **Roger–Tanimoto similarity** is a similarity on $\{0, 1\}^n$, defined by

$$\frac{|\overline{X \Delta Y}|}{n + |X \Delta Y|}.$$

- **Faith similarity**

  The **Faith similarity** is a similarity on $\{0, 1\}^n$, defined by

$$\frac{|X \cap Y| + |\overline{X \Delta Y}|}{2n}.$$

- **Tversky similarity**

The **Tversky similarity** is a similarity on $\{0, 1\}^n$, defined by

$$\frac{|X \cap Y|}{a|X \Delta Y| + b|X \cap Y|}.$$

It becomes **Tanimoto**, **Dice** and (the binary case of) **Kulczynsky** 1 **similarities** for $(a, b) = (1, 1)$, $(\frac{1}{2}, 1)$ and $(1, 0)$, respectively.

- **Gower–Legendre similarity**

The **Gower–Legendre similarity** is a similarity on $\{0, 1\}^n$, defined by

$$\frac{|\overline{X \Delta Y}|}{a|X \Delta Y| + |\overline{X \Delta Y}|} = \frac{|\overline{X \Delta Y}|}{n + (a-1)|X \Delta Y|}.$$

- **Anderberg similarity**

The **Anderberg similarity** (or *Sokal–Sneath* 4 *similarity*) is a similarity on $\{0, 1\}^n$, defined by

$$\frac{|X \cap Y|}{4}\left(\frac{1}{|X|} + \frac{1}{|Y|}\right) + \frac{|\overline{X \cup Y}|}{4}\left(\frac{1}{|\overline{X}|} + \frac{1}{|\overline{Y}|}\right).$$

- **Yule Q similarity**

The **Yule Q similarity** is a similarity on $\{0, 1\}^n$, defined by

$$\frac{|X \cap Y| \cdot |\overline{X \cup Y}| - |X \backslash Y| \cdot |Y \backslash X|}{|X \cap Y| \cdot |\overline{X \cup Y}| + |X \backslash Y| \cdot |Y \backslash X|}.$$

- **Yule Y similarity of colligation**

The **Yule Y similarity of colligation** is a similarity on $\{0, 1\}^n$, defined by

$$\frac{\sqrt{|X \cap Y| \cdot |\overline{X \cup Y}|} - \sqrt{|X \backslash Y| \cdot |Y \backslash X|}}{\sqrt{|X \cap Y| \cdot |\overline{X \cup Y}|} + \sqrt{|X \backslash Y| \cdot |Y \backslash X|}}.$$

- **Dispersion similarity**

The **dispersion similarity** is a similarity on $\{0, 1\}^n$, defined by

$$\frac{|X \cap Y| \cdot |\overline{X \cup Y}| - |X \backslash Y| \cdot |Y \backslash X|}{n^2}.$$

- **Pearson ϕ similarity**

  The **Pearson ϕ similarity** is a similarity on $\{0, 1\}^n$, defined by

$$\frac{|X \cap Y| \cdot |\overline{X \cup Y}| - |X \backslash Y| \cdot |Y \backslash X|}{\sqrt{|X| \cdot |\overline{X}| \cdot |Y| \cdot |\overline{Y}|}}.$$

- **Gower similarity 2**

  The **Gower similarity** 2 (or *Sokal–Sneath* 5 *similarity*) is a similarity on $\{0, 1\}^n$, defined by

$$\frac{|X \cap Y| \cdot |\overline{X \cup Y}|}{\sqrt{|X| \cdot |\overline{X}| \cdot |Y| \cdot |\overline{Y}|}}.$$

- **Pattern difference**

  The **pattern difference** is a distance on $\{0, 1\}^n$, defined by

$$\frac{4|X \backslash Y| \cdot |Y \backslash X|}{n^2}.$$

- **Q_0-difference**

  The Q_0**-difference** is a distance on $\{0, 1\}^n$, defined by

$$\frac{|X \backslash Y| \cdot |Y \backslash X|}{|X \cap Y| \cdot |\overline{X \cup Y}|}.$$

17.4. CORRELATION SIMILARITIES AND DISTANCES

- **Covariance similarity**

  The **covariance similarity** is a similarity on $\mathbb{R}^n$, defined by

$$\frac{\sum(x_i - \overline{x})(y_i - \overline{y})}{n} = \frac{\sum x_i y_i}{n} - \overline{x} \cdot \overline{y}.$$

- **Correlation similarity**

  The **correlation similarity** (or *Pearson correlation*, or, by its full name, *Pearson product-moment correlation linear coefficient*) s is a similarity on $\mathbb{R}^n$, defined by

$$\frac{\sum(x_i - \overline{x})(y_i - \overline{y})}{\sqrt{(\sum(x_j - \overline{x})^2)(\sum(y_j - \overline{y})^2)}}.$$

The dissimilarities $1-s$ and $1-s^2$ are called **correlation distance** (or *Pearson distance*) and *squared Pearson distance*, respectively. Moreover,

$$\sqrt{2(1-s)} = \sqrt{\sum\left(\frac{x_i - \overline{x}}{\sqrt{\sum(x_j - \overline{x})^2}} - \frac{y_i - \overline{y}}{\sqrt{\sum(y_j - \overline{y})^2}}\right)}$$

is a normalization of the Euclidean distance (cf. a different one, **normalized l_2-distance**).

In the case $\overline{x} = \overline{y} = 0$, the correlation similarity becomes $\frac{\langle x,y\rangle}{\|x\|_2 \cdot \|y\|_2}$.

- **Cosine similarity**

  The **cosine similarity** (or *Orchini similarity, angular similarity, normalized dot product*) is a similarity on $\mathbb{R}^n$, defined by

$$\frac{\langle x, y\rangle}{\|x\|_2 \cdot \|y\|_2} = \cos\phi,$$

  where ϕ is the angle between vectors x and y. In binary case, it becomes

$$\frac{|X \cap Y|}{\sqrt{|X| \cdot |Y|}}$$

  and called **Ochiai–Otsuka similarity**.

  In Record Linkage, cosine similarity is called **TF-IDF** (for term *Frequency – Inverse Document Frequency*).

  The **cosine distance** is defined by $1 - \cos\phi$.

- **Angular semi-metric**

  The **angular semi-metric** on $\mathbb{R}^n$ is the angle (measured in radians) between vectors x and y:

$$\arccos\frac{\langle x, y\rangle}{\|x\|_2 \cdot \|y\|_2}.$$

  (Not to be confused with **geodesic distance** in Probability Theory.)

- **Orloci distance**

  The **Orloci distance** (or *chord distance*) is a distance on $\mathbb{R}^n$, defined by

$$\sqrt{2\left(1 - \frac{\langle x, y\rangle}{\|x\|_2 \cdot \|y\|_2}\right)}.$$

  (Cf. **normalized Euclidean distance**.)

- **Similarity ratio**

The **similarity ratio** (or *Kohonen similarity*) is a similarity on $\mathbb{R}^n$, defined by

$$\frac{\langle x, y\rangle}{\langle x, y\rangle + \|x-y\|_2^2}.$$

Its binary case is the **Tanimoto similarity**.

- **Morisita–Horn similarity**

The **Morisita–Horn similarity** is a similarity on $\mathbb{R}^n$, defined by

$$\frac{2\langle x, y\rangle}{\|x\|_2^2 \cdot \frac{\overline{y}}{\overline{x}} + \|y\|_2^2 \cdot \frac{\overline{x}}{\overline{y}}}.$$

- **Spearman rank correlation**

In the case, when $x, y \in \mathbb{R}^n$ are *rankings* (or *permutations*), i.e., the components of each of them are different numbers $1, \dots, n$, one has $\overline{x} = \overline{y} = \frac{n+1}{2}$. For such *ordinal* data, the **correlation similarity** becomes

$$1 - \frac{6}{n(n^2-1)} \sum (x_i - y_i)^2.$$

It is the **Spearman ρ rank correlation**, called also *Spearman rho metric*, but it is not a distance. **Spearman ρ distance** is the Euclidean metric on permutations.

The **Spearman footrule** is defined by

$$1 - \frac{3}{n^2-1} \sum |x_i - y_i|.$$

It is l_1-version of the **Spearman rank correlation**. **Spearman footrule distance** is the **l_1-metric** on permutations.

Another correlation similarity for rankings is **Kendall τ rank correlation**, called also *Kendall τ metric* (but it is not a distance), which is defined by

$$\frac{2\sum_{1\leqslant i<j\leqslant n} sign(x_i - x_j) sign(y_i - y_j)}{n(n-1)}.$$

Kendall τ distance on permutations is defined by

$$\left|\left\{(i, j)\colon\ 1 \leqslant i < j \leqslant n, (x_i - x_j)(y_i - y_j) < 0\right\}\right|.$$

- **Cook distance**

The **Cook distance** is a distance on $\mathbb{R}^n$, giving a statistical measure of deciding if some i-th observation alone affects much regression estimates. It is a normalized **squared**

Euclidean distance between estimated parameters from regression models constructed from all data and from data without i-th observation.

Main similar distances, used in Regression Analysis for detecting influential observations, are *DFITS distance*, *Welsch distance*, and *Hadi distance*.

- **Distance-based machine learning**

The following setting is used for many real-world applications (neural networks, etc.), where data are incomplete and have both, continuous and nominal, attributes. Given an $m \times (n+1)$ matrix $((x_{ij}))$, its row $(x_{i0}, x_{i1}, \dots, x_{in})$ means *instance input vector* $x_i = (x_{i1}, \dots, x_{in})$ with output class x_{i0}; the set of m instances represents a training set during learning. For any new input vector $y = (y_1, \dots, y_n)$, the closest (in terms of selected distance d) instance x_i is sought, in order to *classify* y, i.e., predict its output class as x_{i0}.

The distance ([WiMa97]) $d(x_i, y)$ is defined by

$$\sqrt{\sum_{j=1}^{n} d_j^2(x_{ij}, y_j)}$$

with $d_j(x_{ij}, y_j) = 1$ if x_{ij} or y_j is unknown. If the *attribute* j (i.e., the range of values x_{ij} for $1 \leqslant i \leqslant m$) is nominal, then $d_j(x_{ij}, y_j)$ is defined, for example, as $1_{x_{ij} \neq y_j}$, or as

$$\sum_a \left| \frac{|\{1 \leqslant t \leqslant m\colon\ x_{t0} = a, x_{ij} = x_{ij}\}|}{|\{1 \leqslant t \leqslant m\colon\ x_{tj} = x_{ij}\}|} - \frac{|\{1 \leqslant t \leqslant m\colon\ x_{t0} = a, x_{ij} = y_j\}|}{|\{1 \leqslant t \leqslant m\colon\ x_{tj} = y_j\}|} \right|^q$$

for $q = 1$ or 2; the sum is taken by all output classes, i.e., values a from $\{x_{t0}\colon\ 1 \leqslant t \leqslant m\}$. For continuous attributes j, the number d_j is taken to be $|x_{ij} - y_j|$ divided by $\max_t x_{tj} - \min_t x_{tj}$, or by $\frac{1}{4}$ of the standard deviation of values x_{tj}, $1 \leqslant t \leqslant m$.

Chapter 18

Distances in Mathematical Engineering

In this chapter we group main distances used in *Robot Motion*, *Cellular Automata*, *Feedback Systems* and *Multi-objective Optimization*.

18.1. MOTION PLANNING DISTANCES

Automatic motion planning methods are applied in *Robotics*, *Virtual Reality Systems* and *Computer Aided Design*. A **motion planning metric** is a metric used in automatic motion planning methods.

Let a *robot* be a finite collection of rigid links organized in a kinematic hierarchy. If the robot has n degrees of freedom, this leads to an n-dimensional *manifold* C, called *configuration space* (or *C-space*) of the robot. The *workspace* W of the robot is the space in which the robot moves. Usually, it is modeled as the Euclidean space $\mathbb{E}^3$. The *obstacle region* CB is the set of all configurations $q \in C$, that either cause the robot to collide with obstacles B, or cause different links of the robot to collide among them. The closure $cl(C_{free})$ of $C_{free} = C\backslash\{CB\}$ is called *space of collision-free configurations*. A *motion planning algorithm* must find a collision-free path from an initial configuration to a goal configuration.

A **configuration metric** is a motion planning metric on the configuration space C of a robot.

Usually, the configuration space C consists of six-tuples $(x, y, z, \alpha, \beta, \gamma)$, where the first three coordinates define the position, and the last three the orientation. The orientation coordinates are the degrees in radians. Intuitively, a good measure of the distance between two configurations is a measure of the workspace region swept by the robot as it moves between them (the **swept volume**). However, the computation of such a metric is prohibitively expensive.

The simplest approach has been to consider C-space as a Cartesian space and to use Euclidean distance or its generalizations. For such configuration metrics, one normalizes the orientation coordinates so that they get the same magnitude as the position coordinates. Roughly, one multiplies the orientation coordinates by the maximum x, y or z range of the workspace bounding box. Examples of such configuration metrics are given below.

More generally, the configuration space of three-dimensional rigid body can be identified with the Lie group $ISO(3)$: $C \cong \mathbb{R}^3 \times \mathbb{R}P^3$. The general form of a matrix in $ISO(3)$ is given by

$$\begin{pmatrix} RX \\ 0\,1 \end{pmatrix},$$

where $R \in SO(3) \cong \mathbb{R}P^3$, and $X \in \mathbb{R}^3$. If X_q and R_q represent the translation and rotation components of the configuration $q = (X_q, R_q) \in ISO(3)$, then a configuration metric between configurations q and r is given by $w_{tr}\|X_q - X_r\| + w_{rot} f(R_q, R_r)$, where the **translation distance** $\|X_q - X_r\|$ is obtained using some norm $\|.\|$ on $\mathbb{R}^3$, and the **rotation distance** $f(R_q, R_r)$ is a positive scalar function which gives the distance between the rotations $R_q, R_r \in SO(3)$. The rotation distance is scaled relative to the translation distance via the weights w_{tr} and w_{rot}.

A **workspace metric** is a motion planning metric in the workspace $\mathbb{R}^3$.

There are many other types of metrics used in motion planning methods, in particular, the **Riemannian metrics**, the **Hausdorff metric**, the **growth distance**, etc.

- **Weighted Euclidean distance**

  The **weighted Euclidean distance** is a **configuration metric** on $\mathbb{R}^6$, defined by

$$\left(\sum_{i=1}^{3} |x_i - y_i|^2 + \sum_{i=4}^{6} \left(w_i |x_i - y_i|\right)^2\right)^{\frac{1}{2}}$$

  for any $x, y \in \mathbb{R}^6$, where $x = (x_1, \dots, x_6)$, x_1, x_2, x_3 are the position coordinates, x_4, x_5, x_6 are the orientation coordinates, and w_i is the normalization factor. The weighted Euclidean distance in $\mathbb{R}^6$ gives to position and orientation the same importance.

- **Scaled Euclidean distance**

  The **scaled Euclidean distance** is a **configuration metric** on $\mathbb{R}^6$, defined by

$$\left(s \sum_{i=1}^{3} |x_i - y_i|^2 + (1-s) \sum_{i=4}^{6} \left(w_i |x_i - y_i|\right)^2\right)^{\frac{1}{2}}$$

  for any $x, y \in \mathbb{R}^6$. The scaled Euclidean distance changes the relative importance of the position and orientation components through the scale parameter s.

- **Weighted Minkowskian distance**

  The **weighted Minkowskian distance** is a **configuration metric** on $\mathbb{R}^6$, defined by

$$\left(\sum_{i=1}^{3} |x_i - y_i|^p + \sum_{i=4}^{6} \left(w_i |x_i - y_i|\right)^p\right)^{\frac{1}{p}}$$

  for any $x, y \in \mathbb{R}^6$. It uses a parameter $p \geqslant 1$; as with Euclidean, both position and orientation have the same importance.

- **Modified Minkowskian distance**

The **modified Minkowskian distance** is a **configuration metric** on $\mathbb{R}^6$, defined by

$$\left(\sum_{i=1}^{3}|x_i - y_i|^{p_1} + \sum_{i=4}^{6}\left(w_i|x_i - y_i|\right)^{p_2}\right)^{\frac{1}{p_3}}$$

for all $x, y \in \mathbb{R}^6$. It distinguishes between position and orientation coordinates using the parameters $p_1 \geqslant 1$ (for the position) and $p_2 \geqslant 1$ (for the orientation).

- **Weighted Manhattan distance**

The **weighted Manhattan distance** is a **configuration metric** on $\mathbb{R}^6$, defined by

$$\sum_{i=1}^{3}|x_i - y_i| + \sum_{i=4}^{6} w_i|x_i - y_i|$$

for any $x, y \in \mathbb{R}^6$. It coincides, up to normalization factor, with the usual l_1**-metric** on $\mathbb{R}^6$.

- **Robot displacement metric**

The **robot displacement metric** is a **configuration metric** on a configuration space C of a robot, defined by

$$\max_{a \in A}\|a(q) - a(p)\|$$

for any configurations $q, r \in C$, where $a(q)$ is the position of the point a in the workspace $\mathbb{R}^3$, when the robot is at configuration q, and $\|.\|$ is one of the norms on $\mathbb{R}^3$, usually the Euclidean norm. Intuitively, the metric yields the maximum amount in workspace that any part of the robot is displaced when moving from one configuration to another (cf. **bounded box metric**).

- **Euler angle metric**

The **Euler angle metric** is a **rotation metric** on the group $SO(3)$ (for the case of using *roll-pitch-yaw Euler angles* for rotation), defined by

$$w_{rot}\sqrt{\Delta(\theta_1, \theta_2)^2 + \Delta(\phi_1, \phi_2)^2 + \Delta(\eta_1, \eta_2)^2}$$

for all $R_1, R_2 \in SO(3)$, given by *Euler angles* $(\theta_1, \phi_1, \eta_1)$, $(\theta_2, \phi_2, \eta_2)$, respectively, where $\Delta(\theta_1, \theta_2) = \min\{|\theta_1 - \theta_2|, 2\pi - |\theta_1 - \theta_2|\}$, $\theta_i \in [0, 2\pi]$, is the **metric between angles**, and w_{rot} is a scaling factor.

- **Unit quaternions metric**

The **unit quaternions metric** is a **rotation metric** on the *unit quaternion representation* of $SO(3)$, i.e., a representation of $SO(3)$ as the set of points (*unit quaternions*) on the *unit* *sphere* S^3 in $\mathbb{R}^4$ with identified antipodal points ($q \sim -q$). This representation of $SO(3)$ suggests a number of possible metrics on it, for example, the following ones:

1. $\| \ln(q^{-1}r) \|$,
2. $w_{rot}(1 - \|\lambda\|)$, $\lambda = \sum_{i=1}^{4} q_i r_i$,
3. $\min\{\|q - r\|, \|q + r\|\}$,
4. $\arccos \lambda$, $\lambda = \sum_{i=1}^{4} q_i r_i$,

where $q = q_1 + q_2 i + q_3 j + q_4 k$, $\sum_{i=1}^{4} q_i = 1$, $\|.\|$ is a norm on $\mathbb{R}^4$, and w_{rot} is a scaling factor.

- **Center of mass metric**

  The **center of mass metric** is a **workspace metric**, defined as the Euclidean distance between the *center of mass* of the robot in the two configurations. The center of mass is approximated by averaging all object vertices.

- **Bounded box metric**

  The **bounded box metric** is a **workspace metric**, defined as the maximum Euclidean distance between any vertex of the *bounding box* of the robot in one configuration and its corresponding vertex in the other configuration.

- **Pose distance**

  The **pose distance** provides a measure of dissimilarity between actions of *agents* (including robots and humans) for Learning by Imitation in Robotics.

  In this context, agents are considered as *kinematic chains*, and are represented in the form of a *kinematic tree*, such that every link in the kinematic chain is represented by an unique edge in the corresponding tree. The configuration of the chain is represented by *pose* of the corresponding tree which is obtained by an assignment of the pair (n_i, l_i) to every edge e_i. Here n_i is the unit normal, representing the orientation of the corresponding link in the chain, and l_i is the length of the link. The *pose class* consists of all poses of a given kinematic tree.

  The **pose distance** is a distance on a given pose class which is the sum of measures of dissimilarity for every pair of compatible segments in the given two poses.

- **Millibot metrics**

  Millibot is a team of heterogeneous, resource-limited robots. Robot teams can collectively share information. They are able to fuse range information from a variety of different platforms to build a global occupancy map that represent a single collective view of the environment. In the motion planning of the millibots for the construction of a **motion planning metric**, one casts a series of random points about a robot and pose each point as a candidate position for movement. The point with the highest overall utility is then selected, and the robot is directed to that point. Thus, the **free space metric**, determined by free space contour, only allows candidate points that do not drive robot through obstructions; **obstacle avoidance metric** penalizes for moves that get too close to obstacles; **frontier metric** rewards for moves that take robot towards open space; **formation**

metric rewards for moves that maintain formation; **localization metric**, based on separation angle between one or more localization pairs, rewards for moves that maximize localization (see [GKC04]); cf. **collision avoidance distance**, **piano movers distance**.

18.2. CELLULAR AUTOMATA DISTANCES

Let S, $2 \leqslant |S| < \infty$, denote a finite set (*alphabet*), and let S^∞ denote the set of bi-infinite sequences $\{x_i\}_{i=-\infty}^{\infty}$ (*configurations*) of elements (*letters*) of S. An (one-dimensional) *cellular automaton* is a continuous mapping $f : S^\infty \to S^\infty$ that commutes with the *translation map* $g : S^\infty \to S^\infty$, defined by $g(x_i) = x_{i+1}$. Once a metric on S^∞ is defined, the resulting metric space together with the self-mapping f form a **discrete dynamic system**. Cellular automata (generally, bi-infinite arrays instead of sequences) are used in Symbolic Dynamics, Computer Science and, as models, in Physics and Biology. The main distances between configurations $\{x_i\}_i$ and $\{y_i\}_i$ from S^∞ (see [BFK99]) follow.

- **Cantor metric**

  The **Cantor metric** is a metric on S^∞, defined by

$$2^{-\min\{i \geqslant 0:\, |x_i - y_i| + |x_{-i} - y_{-i}| \neq 0\}}.$$

  The corresponding metric space is compact.

- **Besicovitch semi-metric**

  The **Besicovitch semi-metric** is a semi-metric on S^∞, defined by

$$\overline{\lim}_{l \to \infty} \frac{|-l \leqslant i \leqslant l:\ x_i \neq y_i|}{2l+1}.$$

  The corresponding semi-metric space is **complete**. (Cf. **Besicovitch distance** on measurable functions.)

- **Weyl semi-metric**

  The **Weyl semi-metric** is a semi-metric on S^∞, defined by

$$\overline{\lim}_{l \to \infty} \max_{k \in \mathbb{Z}} \frac{|k+1 \leqslant i \leqslant k+l:\ x_i \neq y_i|}{l}.$$

  This and above semi-metric are **translation invariant**, but neither separable, nor locally compact. (Cf. **Weyl distance** on measurable functions.)

18.3. DISTANCES IN CONTROL THEORY

Control Theory consider feedback loop of a *plant* P (a function representing the object to be controlled, a system) and a *controller* C (a function to design). The output y, measured by a sensor, is fed back to the reference value r. Then controller takes the *error* $e = r - y$ to make inputs $u = Ce$. Subject to zero initial conditions, the input and output signals to the plant are related by $y = Pu$, where r, u, v and P, C are functions of the frequency variable s. So, $y = \frac{PC}{1+PC}r$ and $y \approx r$ (i.e., one controls the output by simply setting the reference) if PC is large for any value of s. If the system is modeled by a system of linear differential equations, then its *transfer function* $\frac{PC}{1+PC}$ is a rational function. The plant P is *stable* if it has no poles in the closed right half-plane $\mathbb{C}_+ = \{s \in \mathbb{C} : \Re s \geqslant 0\}$.

The *robust stabilization problem* is: given a *nominal* plant (a model) P_0 and some metric d on plants, find the centered in P_0 open ball of maximal radius, such that some controller (rational function) C stabilizes every element of this ball.

The *graph* $G(P)$ of the plant P is the set of all bounded input-output pairs $(u, y = Pu)$. Both, u and y, belong to the *Hardy space* $H^2(\mathbb{C}_+)$ of the right half-plane; the graph is a closed subspace of $H^2(\mathbb{C}_+) + H^2(\mathbb{C}_+)$. In fact, $G(P) = f(P)H^2(\mathbb{C}^2)$ for some function $f(P)$, called *graph symbol*, and $G(P)$ is a closed subspace of $H^2(\mathbb{C}^2)$.

All metrics below are *gap-like metrics*; they are topologically equivalent, and the stabilization is a robust property with respect of each of them.

- **Gap metric**

  The **gap metric** between plants P_1 and P_2 (introduced in Control Theory by Zames and El-Sakkary) is defined by

$$gap(P_1, P_2) = \|\Pi(P_1) - \Pi(P_2)\|_2,$$

  where $\Pi(P_i)$, $i = 1, 2$, is the orthogonal projection of the graph $G(P_i)$ of P_i seen as a closed subspace of $H^2(\mathbb{C}^2)$.

  We have

$$gap(P_1, P_2) = \max\{\delta_1(P_1, P_2), \delta_1(P_2, P_1)\},$$

  where $\delta_1(P_1, P_2) = \inf_{Q \in H^\infty} \|f(P_1) - f(P_2)Q\|_{H^\infty}$, and $f(P)$ is a graph symbol.

  If A is an $m \times n$ matrix with $m < n$, then its n columns span an n-dimensional subspace, and the matrix B of the orthogonal projection onto the column space of A is $A(A^T A)^{-1}A^T$. If the basis is orthonormal, then $B = AA^T$. In general, the **gap metric** between two subspaces of the same dimension is l_2-*norm* of the difference of their orthogonal projections; see also the definition of this distance as an **angle distance between subspaces**.

  In some applications (for example, when subspaces correspond to autoregressive models) the *Frobenius norm* is used instead of l_2-norm; cf. **Frobenius distance**.

- **Vidyasagar metric**

  The **Vidyasagar metric** (or *graph metric*) between plants P_1 and P_2 is defined by

$$\max\{\delta_2(P_1, P_2), \delta_2(P_2, P_1)\},$$

  where $\delta_2(P_1, P_2) = \inf_{\|Q\| \leqslant 1} \|f(P_1) - f(P_2)Q\|_{H^\infty}$.

The **behavioral distance** is the gap between *extended* graphs of P_1 and P_2; a term is added to the graph $G(P)$, in order to reflect all possible initial conditions (instead of usual setup with the initial conditions being zero).

- **Vinnicombe metric**

The **Vinnicombe metric** (ν*-gap metric*) between plants P_1 and P_2 is defined by

$$\delta_\nu(P_1, P_2) = \left\|(1 + P_2 P_2^*)^{-\frac{1}{2}}(P_2 - P_1)(1 + P_1^* P_1)^{-\frac{1}{2}}\right\|_\infty$$

if $wno(f^*(P_2)f(P_1)) = 0$, and it is equal to 1, otherwise. Here $f(P)$ is graph symbol function of plant P. See [Youn98] for the definition of the *winding number* $wno(f)$ of a rational function f and for good introduction in Feedback Stabilization.

18.4. MOEA DISTANCES

Most optimization problems have several objectives, but, for simplicity, only one of them is optimized, and others are handled as constraints. *Multi-objective optimization* consider (besides some inequality constraints) an objective vector function $f : X \subset \mathbb{R}^n \to \mathbb{R}^k$ from the *search* (or *genotype, decision variables*) space X to the *objective* (or *phenotype, decision vectors*) space $f(X) = \{f(x) : x \in X\} \subset \mathbb{R}^k$. A point $x^* \in X$ is *Pareto optimal* if, for every other $x \in X$, the decision vector $f(x)$ does not *Pareto dominate* $f(x^*)$, i.e., $f(x) \leqslant f(x^*)$. *Pareto optimal front* is the set $PF^* = \{f(x) : x \in X^*\}$, where X^* is the set of all Pareto optimal points.

Multi-objective evolutionary algorithms (MOEA, for short) produce, at each generation, an *approximation set* (found Pareto front PF_{known} approximating wished Pareto front PF^*) in objective space in which no element Pareto dominates another element. Examples of **MOEA metrics**, i.e., measures evaluating how close PF_{known} is to PF^*, follow.

- **Generational distance**

The **generational distance** is defined by

$$\frac{(\sum_{j=1}^m d_j^2)^{\frac{1}{2}}}{m},$$

where $m = |PF_{known}|$, and d_j is the Euclidean distance (in the objective space) between $f^j(x)$ (i.e., j-th member of PF_{known}) and the nearest member of PF^*. This distance is zero if and only if $PF_{known} = PF^*$.

The term **generational distance** (or *rate of turnover*) is used also for the minimal number of branches between two positions in any system of ranked descent represented by an hierarchical tree. Examples are: **phylogenetic distance** on a phylogenetic tree, the number of generations separating a photocopy from original block print, the number of generations separating audience of a memorial from the commemorated event.

- **Spacing**

  The **spacing** is defined by

$$\left(\frac{\sum_{j=1}^{m}(\overline{d}-d_j)^2}{m-1}\right)^{\frac{1}{2}},$$

  where $m = |PF_{known}|$, d_j is the l_1-distance (in the objective space) between $f^j(x)$ (i.e., j-th member of PF_{known}) and the nearest other member of PF_{known}, while $\overline{d}$ is the mean of all d_j.

- **Overall non-dominated vector ratio**

  The **overall non-dominated vector ratio** is defined by $\frac{|PF_{known}|}{|PF^*|}$.

Part V

Computer-Related Distances

Chapter 19. Distances on Real and Digital Planes
Chapter 20. Voronoi Diagram Distances
Chapter 21. Image and Audio Distances
Chapter 22. Distances in Internet and Similar Networks

Chapter 19

Distances on Real and Digital Planes

19.1. METRICS ON REAL PLANE

In the plane $\mathbb{R}^2$ we can use many various metrics. In particular, any l_p**-metric** (as well as any **norm metric** for a given norm $\|.\|$ on $\mathbb{R}^2$) can be used on the plane, and the most natural is the l_2-metric, i.e., the Euclidean metric $d_E(x, y) = \sqrt{(x_1 - y_1)^2 + (x_2 - y_2)^2}$ which gives the length of the straight line segment $[x, y]$, and is the **intrinsic metric** of the plane. However, there are other, often "exotic", metrics on $\mathbb{R}^2$. Many of them are used for the construction of *generalized Voronoi diagrams* on $\mathbb{R}^2$ (see, for example, **Moscow metric**, **network metric**, **nice metric**). Some of them are used in Digital Geometry.

Erdős-type distance problems (given, usually, for Euclidean metric on $\mathbb{R}^2$) are of interest for $\mathbb{R}^d$ and for other metrics on $\mathbb{R}^2$. Examples of such problems are to find out the following:

- the fewest number of different distances (or largest occurence of given distance) in an n-subset of $\mathbb{R}^2$; the largest size of a subset of $\mathbb{R}^2$ determining at most m distances;
- the minimum diameter of an n-subset of $\mathbb{R}^2$ with only integral distances (or, say, without a pair (d_1, d_2) of distances with $0 < |d_1 - d_2| < 1$);
- existence of an n-subset of $\mathbb{R}^2$ with, for each $1 \leqslant i \leqslant n$, a distance occuring exactly i times (examples are known for $n \leqslant 8$);
- *forbidden distances* of a partition of $\mathbb{R}^2$, i.e., distances not occuring within each part.

- **City-block metric**

  The **city-block metric** is the l_1**-metric** on $\mathbb{R}^2$, defined by

$$\|x - y\|_1 = |x_1 - y_1| + |x_2 - y_2|.$$

  This metric has many different names, for example, it is called **taxicab metric**, **Manhattan metric**, **rectilinear metric**, **right-angle metric**; on $\mathbb{Z}^2$ it is called **greed metric**, and 4**-metric**.

- **Chebyshev metric**

  The **Chebyshev metric** is the l_∞**-metric** on $\mathbb{R}^2$, defined by

$$\|x - y\|_\infty = \max\{|x_1 - y_1|, |x_2 - y_2|\}.$$

This metric is called also **uniform metric**, **sup metric**, and **box metric**; on $\mathbb{Z}^2$ it is called **lattice metric**, **chessboard metric**, **king-move metric**, and 8-**metric**.

- **(p, q)-relative metric**

  Let $0 < g \leqslant 1$, $p \geqslant \max\{1 - q, \frac{2-q}{3}\}$, and let $\|.\|_2$ be the Euclidean norm on $\mathbb{R}^2$ (in general, on $\mathbb{R}^n$).

  The (p, q)-**relative metric** is a metric on $\mathbb{R}^2$ (in general, on $\mathbb{R}^n$), defined by

$$\frac{\|x - y\|_2}{(\frac{1}{2}(\|x\|_2^p + \|y\|_2^p))^{\frac{q}{p}}}$$

  for x or $y \neq 0$ (and is equal to 0, otherwise). In the case of $p = \infty$ it has the form

$$\frac{\|x - y\|_2}{(\max\{\|x\|_2, \|y\|_2\})^q}.$$

  For $q = 1$ and any $1 \leqslant p < \infty$ one obtains the p-**relative metric**; for $q = 1$ and $p = \infty$ one obtains the **relative metric**.

  The construction above can be used for any *Ptolemaic* space $(V, \|.\|)$.

- ***M*-relative metric**

  Let $f : [0, \infty) \to (0, \infty)$ be a convex increasing function such that $\frac{f(x)}{x}$ is decreasing for $x > 0$. Let $\|.\|_2$ be the Euclidean norm on $\mathbb{R}^2$ (in general, on $\mathbb{R}^n$).

  The M-**relative metric** is a metric on $\mathbb{R}^2$ (in general, on $\mathbb{R}^n$) defined by

$$\frac{\|x - y\|_2}{f(\|x\|_2) \cdot f(\|y\|_2)}.$$

  In particular, the distance

$$\frac{\|x - y\|_2}{\sqrt[p]{1 + \|x\|_2^p}\,\sqrt[p]{1 + \|y\|_2^p}}$$

  is a metric on $\mathbb{R}^2$ (on $\mathbb{R}^n$) if and only if $p \geqslant 1$. A similar metric on $\mathbb{R}^2\backslash\{0\}$ (on $\mathbb{R}^n\backslash\{0\}$) can be defined by

$$\frac{\|x - y\|_2}{\|x\|_2 \cdot \|y\|_2}.$$

  The constructions above can be used for any *Ptolemaic* space $(V, \|.\|)$.

- **French Metro metric**

  Given a *norm* $\|.\|$ on $\mathbb{R}^2$, the **French metro metric** is a metric on $\mathbb{R}^2$, defined by

$$\|x - y\|$$

if $x = cy$ for some $c \in \mathbb{R}$, and by

$$\|x\| + \|y\|,$$

otherwise. For the Euclidean norm $\|.\|_2$, it is called **hedgehog metric**, **Paris metric**, or **radial metric**. In this case it can be defined as the minimum Euclidean length of all *admissible* connecting curves between two given points x and y, where a curve is called *admissible* if it consists of only segments of straight lines passing through the origin.

In graph terms, this metric is similar to the **path metric** of the tree consisting of a point from which radiate several disjoint paths.

- **Moscow metric**

  The **Moscow metric** (or **Karlsruhe metric**) is a metric on $\mathbb{R}^2$, defined as the minimum Euclidean length of all *admissible* connecting curves between x and $y \in \mathbb{R}^2$, where a curve is called *admissible* if it consists of only segments of straight lines passing through the origin, and of segments of circles centered at the origin (see, for example, [Klei88]).

  If the polar coordinates for points $x, y \in \mathbb{R}^2$ are (r_x, θ_x), (r_y, θ_y), respectively, then the distance between them is equal to $\min\{r_x, r_y\}\Delta(\theta_x - \theta_y) + |r_x - r_y|$ if $0 \leqslant \Delta(\theta_x, \theta_y) < 2$, and is equal to $r_x + r_y$ if $2 \leqslant \Delta(\theta_x, \theta_y) < \pi$, where $\Delta(\theta_x, \theta_y) = \min\{|\theta_x - \theta_y|, 2\pi - |\theta_x - \theta_y|\}$, $\theta_x, \theta_y \in [0, 2\pi)$, is the **metric between angles**.

- **Lift metric**

  The **lift metric** (or **raspberry picker metric**) is a metric on $\mathbb{R}^2$, defined by

$$|x_1 - y_1|$$

  if $x_2 = y_2$, and by

$$|x_1| + |x_2 - y_2| + |y_1|$$

  if $x_2 \neq y_2$ (see, for example, [Brya85]). It can be defined as the minimum Euclidean length of all *admissible* connecting curves between two given points x and y, where a curve is called *admissible* if it consists of only segments of straight lines parallel to x_1-axis, and of segments of x_2-axis.

- **British Rail metric**

  Given a *norm* $\|.\|$ on $\mathbb{R}^2$ (in general, on $\mathbb{R}^n$), the **British Rail metric** is a metric on $\mathbb{R}^2$ (in general, on $\mathbb{R}^n$), defined by

$$\|x\| + \|y\|$$

  for $x \neq y$ (and it is equal to 0, otherwise).

  It is also called **caterpillar metric**, and **shuttle metric**. For the Euclidean norm $\|.\|_2$ it is called **post-office metric**.

- **Radar screen metric**

  Given a *norm* $\|.\|$ on $\mathbb{R}^2$ (in general, on $\mathbb{R}^n$), the **radar screen metric** is a metric on $\mathbb{R}^2$

(in general, on $\mathbb{R}^n$), defined by

$$\min\{1, \|x - y\|\}.$$

- **Burago–Ivanov metric**

The **Burago–Ivanov metric** ([BuIv01]) is a metric on $\mathbb{R}^2$, defined by

$$\big|\|x\|_2 - \|y\|_2\big| + \min\{\|x\|_2, \|y\|_2\} \cdot \sqrt{\angle(x, y)},$$

where $\angle(x, y)$ is the angle between vectors x and y, and $\|.\|_2$ is the Euclidean norm on $\mathbb{R}^2$. The corresponding **internal metric** on $\mathbb{R}^2$ is equal to $|\,\|x\|_2 - \|y\|_2|$ if $\angle(x, y) = 0$, and is equal to $\|x\|_2 + \|y\|_2$, otherwise.

- **Flower-shop metric**

Let d be a metric on $\mathbb{R}^2$, and let f be a fixed point (a *flower-shop*) in the plane.

The **flower-shop metric** is a metric on $\mathbb{R}^2$, defined by

$$d(x, f) + d(f, y)$$

for $x \neq y$ (and is equal to 0, otherwise). So, a person living at point x, who wants to visit someone else living at point y, first goes to f, to buy some flowers. In the case $d(x, y) = \|x - y\|$ and $f = (0, 0)$, it is the **British rail metric**.

If $k > 1$ flower-shops $f_1, \dots, f_k$ are available, one buys the flowers, where the detour is a minimum, i.e., the distance between distinct points x, y is equal to $\min_{1 \leqslant i \leqslant k}(d(x, f_i) + d(f_i, y))$.

- **$2n$-gon metric**

Given a centrally symmetric regular $2n$-gon K on the plane, the **$2n$-gon metric** is a metric on $\mathbb{R}^2$, defined, for any $x, y \in \mathbb{R}^2$, as the shortest Euclidean length of a polygonal line from x to y with each of its sides parallel to some edge of K. The plane $\mathbb{R}^2$ equipped with the $2n$-gon metric is called *$2n$-gonal plane*.

If K is a square with the vertices $\{(\pm 1, \pm 1)\}$, one obtains the **Manhattan metric**.

- **Central Park metric**

The **Central Park metric** is a metric on $\mathbb{R}^2$, defined as the length of a shortest l_1-path (*Manhattan path*) between two points $x, y \in \mathbb{R}^2$ at the presence of a given set of areas which are traversed by a shortest Euclidean path (for example, Central Park in Manhattan).

- **Collision avoidance distance**

Let $\mathcal{O} = \{O_1, \dots, O_m\}$ be a collection of pairwise disjoint polygons on the Euclidean plane, represents a set of obstacles which are neither transparent, nor traversable.

The **collision avoidance distance** (or **piano movers distance**, **shortest path metric with obstacles**) is a metric on the set $\mathbb{R}^2 \backslash \{\mathcal{O}\}$, defined, for any $x, y \in \mathbb{R}^2 \backslash \{\mathcal{O}\}$, as the length of the shortest path among all possible continuous paths, connecting x and y, that do not intersect obstacles $O_i \backslash \partial O_i$ (a path can pass through points on the boundary ∂O_i of O_i), $i = 1, \dots m$.

- **Rectilinear distance with barriers**

Let $\mathcal{O} = \{O_1, \dots, O_m\}$ be a set of pairwise disjoint open polygonal barriers on $\mathbb{R}^2$. An *rectilinear path* (or *Manhattan path*) P_{xy} from x to y is a collection of horizontal and vertical segments on the plane, joining x and y. The path P_{xy} is called *feasible* if $P_{xy} \cap (\bigcup_{i=1}^{m} B_i) = \emptyset$.

The **rectilinear distance with barriers** (or *rectilinear distance in the presence of barriers*) is a metric on $\mathbb{R}^2 \backslash \{\mathcal{O}\}$, defined, for any $x, y \in \mathbb{R}^2 \backslash \{\mathcal{O}\}$, as the length of the shortest *feasible rectilinear path* from x to y.

The rectilinear distance in the presence of barriers is a restriction of the **Manhattan metric**, and usually it is considered on the set $\{q_1, \dots, q_n\} \subset \mathbb{R}^2$ of n *origin-destination points*: the problem to find a path of such kind arises, for example, in Urban Transportation, or in Plant and Facility Layout (see, for example, [LaLi81]).

- **Link distance**

Let $P \subset \mathbb{R}^2$ be a *polygonal domain* (on n vertices and h holes), i.e., a closed multiply-connected region whose boundary is a union of n line segments, forming $h + 1$ closed polygonal cycles. The **link distance** is a metric on P, defined, for any $x, y \in P$, as the minimum number of edges in a polygonal path from x to y within the polynomial domain P.

If the path is restricted to be rectilinear, one obtains the *rectilinear link distance*. If the path is C*-oriented* (i.e., each its edge is parallel to one of a set C fixed orientation), one obtains the C*-oriented link distance*.

- **Facility layout distances**

A *layout* is a partition of a rectangular plane region into smaller rectangles, called *departments*, by lines parallel to the sides of original rectangle. All interior vertices should be three-valent, and some of them, at least one on the boundary of each department, are *doors*, i.e., input-output locations. The problem is to design convenient notion of distance $d(x, y)$ between departments x and y which minimizes the *cost function* $\sum_{x,y} F(x, y)d(x, y)$, where $F(x, y)$ is some *material flow* between x and y. Main distances used are:

The **centroid distance**, i.e., the shortest Euclidean or **Manhattan** distance between *centroids* (the intersections of the diagonals) of x and y;

The **perimeter distance**, i.e., the shortest rectilinear distance between doors of x and y, but going only along the *walls*, i.e., department perimeters.

- **Quickest path metric**

A **quickest path metric** (or **network metric**) is a metric on $\mathbb{R}^2$ (or on a subset of $\mathbb{R}^2$) in the presence of a given *network*, i.e., a planar weighted graph $G = (V, E)$. For any $x, y \in \mathbb{R}^2$, it is the time needed for a *quickest path* between x and y in the presence of the network G, i.e., a path minimizing the travel time between x and y. After having accessed to G one can travel at some speed $v > 1$ along its edges. Movement off the network take place with unit speed with respect to a given metric d on the plane (for example, the Euclidean metric, or the **Manhattan metric**).

The **airlift metric** is a quickest path metric on $\mathbb{R}^2$ in the presence of an *airports network*, i.e., a planar graph $G = (V, E)$ on n vertices (*airports*) with positive edge weights $(w_e)_{e \in E}$ (*flight durations*). The graph may be entered and exited only at the airports. Movement off the network takes place with unit speed with respect to the Euclidean metric. We assume that going by car takes time equal to the Euclidean distance d_E, whereas the flight along an edge $e = uv$ of G takes time $w_e < d_E(u, v)$. In the simplest case, when there is an airlift between two points $a, b \in \mathbb{R}^2$, the distance between x and y is equal to

$$\min\{d_E(x, y), d_E(x, a) + w + d_E(b, y), d_E(x, b) + w + d_E(a, y)\},$$

where $w < d_2(a, b)$ is the flight duration from a to b.

The **city metric** is a quickest path metric on $\mathbb{R}^2$ in the presence of a *city public transportation network*, i.e., a planar straight line graph G with horizontal or vertical edges. G may be composed of many connected components, and may contain cycles. One is free to enter G at any point, be it at a vertex or on a edge (it is possible to postulate fixed entry points, too). Once having accessed G, one travels at fixed speed $v > 1$ in one of the available directions. Movement off the network takes place with unit speed with respect to the **Manhattan metric** (we imagine a large modern-style city with streets arranged in north-south and east-west directions).

The **subway metric** is a quickest path metric on $\mathbb{R}^2$ which is a variant of the city metric: a subway (in the form of a line in the plane) is used to alter walking distance within a city grid.

- **Periodic metric**

A metric d on $\mathbb{R}^2$ is called **periodic**, if there exists two linearly independent vectors v and u such that the *translation* by any vector $w = mv + nu$, $m, n \in \mathbb{Z}$, preserves distances, i.e., $d(x, y) = d(x + w, y + w)$ for any $x, y \in \mathbb{R}^2$ (cf. **translation invariant metric**).

- **Nice metric**

A metric d on $\mathbb{R}^2$ is called **nice** if it enjoys the following properties:

1. d induces the Euclidean topology;
2. The d-circles are bounded with respect to the Euclidean metric;
3. If $x, y \in \mathbb{R}^2$ and $x \neq y$, then there exists a point z, $z \neq x$, $z \neq y$, such that $d(x, y) = d(x, z) + d(z, y)$ holds;
4. If $x, y \in \mathbb{R}^2$, $x \prec y$ (where $\prec$ is a fixed order on $\mathbb{R}^2$, the lexicographic order, for example), $C(x, y) = \{z \in \mathbb{R}^2 : d(x, z) \leqslant d(y, z)\}$, $D(x, y) = \{z \in \mathbb{R}^2 : d(x, z) < d(y, z)$, and $\overline{D(x, y)}$ is the closure of $D(x, y)$, then $J(x, y) = C(x, y) \cap \overline{D(x, y)}$ is a curve homeomorphic to $(0, 1)$. The intersection of two such curves consists of finitely many connected components.

Every **norm metric** fulfills 1., 2., and 3. Property 2. means that the metric d is continuous at infinity with respect to the Euclidean metric. Property 4. is to ensure that the boundaries of the correspondent *Voronoi diagrams* are curves, and that not too many intersections exist in a neighborhood of a point, or at infinity. A nice metric d has a nice *Voronoi diagram*: in the Voronoi diagram $V(P, d, \mathbb{R}^2)$ (where $P = \{p_1, \dots, p_k\}, k \geqslant 2$, is the set of *generator points*) each *Voronoi region* $V(p_i)$ is a path-connected set with a non-empty interior, and the system $\{V(p_1), \dots V(p_k)\}$ forms a *partition* of the plane.

- **Radar discrimination distance**

The **radar discrimination distance** is a distance on $\mathbb{R}^2$, defined by

$$|\rho_x - \rho_y + \theta_{xy}|$$

if $x, y \in \mathbb{R}^2 \backslash \{0\}$, and by

$$|\rho_x - \rho_y|$$

if $x = 0$ or $y = 0$, where, for each "location" $x \in \mathbb{R}^2$, ρ_x denote the radial distance of x from the origin, and, for any $x, y \in R^2 \backslash \{0\}$, θ_{xy} denote the radian angle between them.

- **Ehrenfeucht–Haussler semi-metric**

Let S be a subset of $\mathbb{R}^2$ such that $x_1 \geqslant x_2 - 1 \geqslant 0$ for any $x \in S$.

The **Ehrenfeucht–Haussler semi-metric** (see [EhHa88]) on S is defined by

$$\log_2\left(\left(\frac{x_1}{y_2} + 1\right)\left(\frac{y_1}{x_2} + 1\right)\right).$$

- **Circle metric**

The **circle metric** is the **intrinsic metric** on the *unit circle* S^1 in the plane. As $S^1 = \{(x, y)\colon\ x^2 + y^2 = 1\} = \{e^{i\theta}\colon\ 0 \leqslant \theta < 2\pi\}$, it is the length of the shorter of two arcs, joining the points $e^{i\theta}, e^{i\vartheta} \in S^1$, and can be written by

$$\min\{|\theta - \vartheta|, 2\pi - |\theta - \vartheta|\} = \begin{cases} |\vartheta - \theta|, & \text{if } 0 \leqslant |\vartheta - \theta| \leqslant \pi, \\ 2\pi - |\vartheta - \theta|, & \text{if } |\vartheta - \theta| > \pi. \end{cases}$$

(Cf. **metric between angles**.)

- **Toroidal metric**

The *torus* $T \subset \mathbb{R}^2$ is the set $[0, 1) \times [0, 1) = \{x \in \mathbb{R}^2\colon\ 0 \leqslant x_1, x_2 < 1\}$.

The **toroidal metric** is a metric on T defined by

$$\sqrt{t_1^2 + t_2^2},$$

for any $x, y \in \mathbb{R}^2$, where $t_i = \min\{|x_i - y_i|, |x_i - y_i + 1|\}$ for $i = 1, 2$. (Cf. **torus metric**.)

- **Angular distance**

The **angular distance** traveled around a circle is the number of radians the path subtends, i.e.,

$$\theta = \frac{l}{r},$$

where l is the length of the path, and r is the radius of the circle.

- **Metric between angles**

The **metric between angles** Δ is a metric on the set of all angles in the plane, defined by

$$\min\{|\theta - \vartheta|, 2\pi - |\theta - \vartheta|\} = \begin{cases} |\vartheta - \theta|, & \text{if } 0 \leqslant |\vartheta - \theta| \leqslant \pi, \\ 2\pi - |\vartheta - \theta|, & \text{if } |\vartheta - \theta| > \pi \end{cases}$$

for any $\theta, \vartheta \in [0, 2\pi)$ (cf. **circle metric**).

- **Metric between directions**

On $\mathbb{R}^2$, the direction $\hat{l}$ is the class of all straight lines which are parallel to a given strait line $l \subset \mathbb{R}^2$. The **metric between directions** is a metric on the set $\mathcal{L}$ of all directions on the plane, defined, for any directions $\hat{l}, \hat{m} \in \mathcal{L}$, as the angle between any two representatives.

- **Circular-railroad quasi-metric**

The **circular-railroad quasi-metric** is a quasi-metric on the *unit circle* $S^1 \subset \mathbb{R}^2$, defined, for any $x, y \in S^1$, as the length of counterclockwise circular arc from x to y in S^1.

- **Inversive distance**

The **inversive distance** between two non-intersecting circles in the plane is defined as the natural logarithm of the ratio of the radii (the larger to the smaller) of two concentric circles into which the given circles can be inverted.

Let c be the distance between the centers of two non-intersecting circles of radii a and $b < a$. Then their inversive distance is given by

$$\cosh^{-1}\left|\frac{a^2 + b^2 - c^2}{2ab}\right|.$$

The *circumcircle* and *incircle* of a triangle with *circumradius* R and *inradius* r are at the inversive distance $2\sinh^{-1}(\frac{1}{2}\sqrt{\frac{r}{R}})$.

Given three non-collinear points, construct three tangent circles such that one is centered at each point and the circles are pairwise tangent to one another. Then there exist exactly two non-intersecting circles that are tangent to all three circles. These are called the inner and outer *Soddy circles*. The inversive distance between the Soddy circles is $2\cosh^{-1} 2$.

19.2. DIGITAL METRICS

Here we list special metrics which are used in *Computer Vision* (or *Pattern Recognition, Robot Vision, Digital Geometry*).

A *computer picture* (or *computer image*) is a subset of $\mathbb{Z}^n$ which is called *digital nD space*. Usually, pictures are represented in the *digital plane* (or *image plane*) $\mathbb{Z}^2$, or in the *digital space* (or *image space*) $\mathbb{Z}^3$. The points of $\mathbb{Z}^n$ are called *pixels*. An *nD m-quantized space* is a scaling $\frac{1}{m}\mathbb{Z}^n$.

A **digital metric** (see, for example, [RoPf68]) is any metric on a digital nD space. Usually, it should take integer values.

The metrics on $\mathbb{Z}^n$ that are mainly used are the l_1**-** and l_∞**-metrics**, as well as the l_2**-metric** after rounding to the nearest upper (or lower) integer. In general, given a list of *neighbors* of a pixel, it can be seen as a list of permitted *one-step moves* on $\mathbb{Z}^2$. Let associate a **prime distance**, i.e., a positive weight, to each type of such move. Many digital metrics can be obtained now as the minimum, over all admissible paths (i.e., sequences of permitted moves), of the sum of corresponding prime distances.

In practice, the subset $(\mathbb{Z}_m)^n = \{0, 1, \dots, m-1\}^n$ is considered instead of the full space $\mathbb{Z}^n$. $(\mathbb{Z}_m)^2$ and $(\mathbb{Z}_m)^3$ are called *m-grill* and *m-framework*, respectively. The most used metrics on $(\mathbb{Z}_m)^n$ are the **Hamming metric**, and the **Lee metric**.

- **Grid metric**

  The **grid metric** is the l_1**-metric** on $\mathbb{Z}^n$. The l_1-metric on $\mathbb{Z}^n$ can be seen as the **path metric** of an infinite graph: two points of $\mathbb{Z}^n$ are adjacent if their l_1-distance is equal to one. For $\mathbb{Z}^2$ this graph is the usual *grid*. Since each point has exactly four closest neighbors in $\mathbb{Z}^2$ for the l_1-metric, it is called also 4**-metric**.

  For $n = 2$, this metric is the restriction on $\mathbb{Z}^2$ of the **city-block metric** which is called also **taxicab metric**, **rectilinear metric**, or **Manhattan metric**.

- **Lattice metric**

  The **lattice metric** is the l_∞**-metric** on $\mathbb{Z}^n$. The l_∞-metric on $\mathbb{Z}^n$ can be seen as the **path metric** of an infinite graph: two points of $\mathbb{Z}^n$ are adjacent if their l_∞-distance is equal to one. For $\mathbb{Z}^2$, the adjacency corresponds to the king move in chessboard terms, and this graph is called l_∞*-grid*, while this metric is called also **chessboard metric**, **king-move metric**, or **king metric**. Since each point has exactly eight closest neighbors in $\mathbb{Z}^2$ for the l_∞-metric, it is called also 8**-metric**.

  This metric is the restriction on $\mathbb{Z}^n$ of the **Chebyshev metric** which is called also **sup metric**, or **uniform metric**.

- **Hexagonal metric**

  The **hexagonal metric** is a metric on $\mathbb{Z}^2$ with an *unit sphere* $S^1(x)$ (centered at $x \in \mathbb{Z}^2$), defined by $S^1(x) = S^1_{l_1}(x) \cup \{(x_1 - 1, x_2 - 1), (x_1 - 1, x_2 + 1)\}$ for x *even* (i.e., with even x_2), and by $S^1(x) = S^1_{l_1}(x) \cup \{(x_1 + 1, x_2 - 1), (x_1 + 1, x_2 + 1)\}$ for x *odd* (i.e., with odd x_2). Since any unit sphere $S^1(x)$ contains exactly six integral points, the hexagonal metric is called also 6**-metric** (see [LuRo76]).

For any $x, y \in \mathbb{Z}^2$, it can be written as

$$\max\left\{|u_2|, \frac{1}{2}(|u_2| + u_2) + \left\lfloor \frac{x_2+1}{2} \right\rfloor - \left\lfloor \frac{y_2+1}{2} \right\rfloor - u_1,\right.$$

$$\left.\frac{1}{2}(|u_2| - u_2) - \left\lfloor \frac{x_2+1}{2} \right\rfloor + \left\lfloor \frac{y_2+1}{2} \right\rfloor + u_1\right\},$$

where $u_1 = x_1 - y_1$, and $u_2 = x_2 - y_2$.

The hexagonal metric can be defined as the **path metric** on the *hexagonal grid* of the plane. In *hexagonal coordinates* (h_1, h_2) (in which h_1- and h_2-axes are parallel to the grid's edges) hexagonal distance between points (h_1, h_2) and (i_1, i_2) can be written as $|h_1 - i_1| + |h_2 - i_2|$ if $(h_1 - i_1)(h_2 - i_2) \geqslant 0$, and as $\max\{|h_1 - i_1|, |h_2 - i_2|\}$ if $(h_1 - i_1)(h_2 - i_2) \leqslant 0$. Here hexagonal coordinates (h_1, h_2) of a point x are related to its Cartesian coordinates (x_1, x_2) by $h_1 = x_1 - \lfloor \frac{x_2}{2} \rfloor$, $h_2 = x_2$ for x even, and by $h_1 = x_1 - \lfloor \frac{x_2+1}{2} \rfloor$, $h_2 = x_2$ for x odd.

The hexagonal metric is a better approximation to the Euclidean metric than either l_1-**metric**, or l_∞-**metric**.

- **Neighborhood sequence metric**

On the digital plane $\mathbb{Z}^2$, consider two types of motions: the *city-block motion*, restricting movements only to the horizontal or vertical directions, and the *chessboard motion*, also allowing diagonal movements. The use of both these motions is determined by a *neighborhood sequence* $B = \{b(1), b(2), \dots, b(l)\}$, where $b(i) \in \{1, 2\}$ is a particular type of neighborhood, with $b(i) = 1$ signifying unit change in 1 coordinate (*city-block neighborhood*), and $b(i) = 2$ meaning unit change also in 2 coordinates (*chessboard neighborhood*). The sequence B defines the type of motion to be used at every step (see [Das90]).

The **neighborhood sequence metric** is a metric on $\mathbb{Z}^2$, defined as the length of a shortest path between x and $y \in \mathbb{Z}^2$, determined by a given neighborhood sequence B. It can be written as

$$\max\{d_B^1(u), d_B^2(u)\},$$

where $u_1 = x_1 - y_1$, $u_2 = x_2 - y_2$,

$$d_B^1(u) = \max\{|u_1|, |u_2|\}, \qquad d_B^2(u) = \sum_{j=1}^{l} \left\lfloor \frac{|u_1| + |u_2| + g(j)}{f(l)} \right\rfloor,$$

$f(0) = 0$, $f(i) = \sum_{j=1}^{i} b(j)$, $1 \leqslant i \leqslant l$, $g(j) = f(l) - f(j-1) - 1$, $1 \leqslant j \leqslant l$.

For $B = \{1\}$ one obtains the **city-block metric**, for $B = \{2\}$ one obtains the **chessboard metric**. The case $B = \{1, 2\}$, i.e., the alternative use of these motions, results in **octagonal metric**, introduced in [RoPf68].

A proper selection of the B-sequence can make the corresponding metric very close to the Euclidean metric. It is always greater than the chessboard distance, but smaller than the city-block distance.

- **nD-neighborhood sequence metric**

The **nD-neighborhood sequence metric** is a metric on $\mathbb{Z}^n$, defined as the length of a shortest path between x and $y \in \mathbb{Z}^n$, determined by a given *nD-neighborhood sequence* B (see [Faze99]).

Formally, two points $x, y \in \mathbb{Z}^n$ are called *m-neighbors*, $0 \leqslant m \leqslant n$, if $0 \leqslant |x_i - y_i| \leqslant 1$, $1 \leqslant i \leqslant n$, and $\sum_{i=1}^{n} |x_i - y_i| \leqslant m$. A finite sequence $B = \{b(1), \dots, b(l)\}$, $b(i) \in \{1, 2, \dots, n\}$, is called *$nD$-neighborhood sequence with period l*. For any $x, y \in \mathbb{Z}^n$ the point sequence $x = x^0, x^1, \dots, x^k = y$, where x^i and x^{i+1}, $0 \leqslant i \leqslant k-1$, are r-neighbors, $r = b((i \ mod \ l) + 1)$, is called *path from x to y determined by B* with length k. The distance between x and y can be written as

$$\max_{1 \leqslant i \leqslant n} d_i(u) \quad \text{with } d_i(x, y) = \sum_{j=1}^{l} \left\lfloor \frac{a_i + g_i(j)}{f_i(l)} \right\rfloor,$$

where $u = (|u_1|, |u_2|, \dots, |u_n|)$ is the non-increasing ordering of $|u_m|$, $u_m = x_m - y_m$, $m = 1, \dots, n$, that is $|u_i| \leqslant |u_j|$ if $i < j$; $a_i = \sum_{j=1}^{n-i+1} u_j$; $b_i(j) = b(j)$ if $b(j) < n - i + 2$, and is $n - i + 1$, otherwise; $f_i(j) = \sum_{k=1}^{j} b_i(k)$ if $1 \leqslant j \leqslant l$, and is 0 if $j = 0$; $g_i(j) = f_i(l) - f_i(j-1) - 1$, $1 \leqslant j \leqslant l$.

The set of $3D$-neighborhood sequence metrics forms a complete distributive lattice under the natural comparison relation. This lattice has an important role in the approximation of the Euclidean metric by digital metrics.

- **Path-generated metric**

Consider *l_∞-grid*, i.e., the graph with the vertex-set $\mathbb{Z}^2$, and two vertices being *neighbors* if their l_∞-distance is equal to one. Let $\mathcal{P}$ be a collection of paths in l_∞-grid such that, for any $x, y \in \mathbb{Z}^2$, there exists at least one path from $\mathcal{P}$ between x and y, and if $\mathcal{P}$ contains a path Q, then it also contains every path contained in Q. Let $d_{\mathcal{P}}(x, y)$ be the length of the shortest path from $\mathcal{P}$ between x and $y \in \mathbb{Z}^2$. If $d_{\mathcal{P}}$ is a metric on $\mathbb{Z}^2$, then it is called **path-generated metric** (see, for example, [Melt91]).

Let G be one of the sets: $G_1 = \{\uparrow, \rightarrow\}$, $G_{2A} = \{\uparrow, \nearrow\}$, $G_{2B} = \{\uparrow, \nwarrow\}$, $G_{2C} = \{\nearrow, \nwarrow\}$, $G_{2D} = \{\rightarrow, \nwarrow\}$, $G_{3A} = \{\rightarrow, \uparrow, \nearrow\}$, $G_{3B}\{\rightarrow, \uparrow, \nwarrow\}$, $G_{4A} = \{\rightarrow, \nearrow, \nwarrow\}$, $G_{4B}\{\uparrow, \nearrow, \nwarrow, \}$, $G_5 = \{\rightarrow, \uparrow, \nearrow, \nwarrow\}$. Let $\mathcal{P}(G)$ be the set of paths which are obtained by concatenation of paths in G and the corresponding paths in the opposite directions. Any path-generated metric coincides with one of the metrics $d_{\mathcal{P}(G)}$. Moreover, one can obtain the following formulas:

1. $d_{\mathcal{P}(G_1)}(x, y) = |u_1| + |u_2|$;
2. $d_{\mathcal{P}(G_{2A})}(x, y) = \max\{|2u_1 - u_2|, |u_2|\}$;
3. $d_{\mathcal{P}(G_{2B})}(x, y) = \max\{|2u_1 + u_2|, |u_2|\}$;
4. $d_{\mathcal{P}(G_{2C})}(x, y) = \max\{|2u_2 + u_1|, |u_1|\}$;

5. $d_{\mathcal{P}(G_{2D})}(x, y) = \max\{|2u_2 - u_1|, |u_1|\}$;
6. $d_{\mathcal{P}(G_{3A})}(x, y) = \max\{|u_1|, u_2|, |u_1 - u_2|\}$;
7. $d_{\mathcal{P}(G_{3B})}(x, y) = \max\{|u_1|, |u_2|, |u_1 + u_2|\}$;
8. $d_{\mathcal{P}(G_{4A})}(x, y) = \max\{2\lceil(|u_1| - |u_2|)/2\rceil, 0\} + |u_2|$;
9. $d_{\mathcal{P}(G_{4B})}(x, y) = \max\{2\lceil(|u_2| - |u_1|)/2\rceil, 0\} + |u_1|$;
10. $d_{\mathcal{P}(G_5)}(x, y) = \max\{|u_1|, |u_2|\}$,

where $u_1 = x_1 - y_1$, $u_2 = x_2 - y_2$, and $\lceil . \rceil$ is the *ceiling function*: for any real x the number $\lceil x \rceil$ is the least integer greater than or equal to x.

The metric spaces obtained from G-sets which have the same numerical index are isometric. $d_{\mathcal{P}(G_1)}$ is the **city-block metric**, and $d_{\mathcal{P}(G_5)}$ is the **chessboard metric**.

- **Knight metric**

  The **knight metric** is a metric on $\mathbb{Z}^2$, defined as the minimum number of moves a chess knight would take to travel from x to $y \in \mathbb{Z}^2$. Its *unit sphere* S^1_{knight}, centered at the origin, contains exactly 8 integral points $\{(\pm 2, \pm 1), (\pm 1, \pm 2)\}$, and can be written as $S^1_{knight} = S^3_{l_1} \cap S^2_{l_\infty}$, where $S^3_{l_1}$ denotes the l_1-sphere of radius 3, and $S^2_{l_\infty}$ denotes the l_∞-sphere of radius 2, centered at the origin (see [DaCh88]).

  The distance between x and y is equal to 3 if $(M, m) = (1, 0)$, is equal to 4 if $(M, m) = (2, 2)$, and is equal to $\max\{\lceil \frac{M}{2} \rceil, \lceil \frac{M+m}{3} \rceil\} + (M + m) - \max\{\lceil \frac{M}{2} \rceil, \lceil \frac{M+m}{3} \rceil\} \pmod 2$, otherwise, where $M = \max\{|u_1|, |u_2|\}$, $m = \min\{|u_1|, |u_2|\}$, $u_1 = x_1 - y_1$, $u_2 = x_2 - y_2$.

- **Super-knight metric**

  Let $p, q \in \mathbb{N}$ such that $p + q$ is odd, and $(p, q) = 1$.

  An *(p, q)-super-knight* (or *(p, q)-leaper*) is a (variant) chess piece a move of which consists of a leap p squares in one orthogonal direction followed by a 90 degree direction change, and q squares leap to the destination square. Chess-variant terms exist for an $(p, 1)$-leaper with $p = 0, 1, 2, 3, 4$ (*Wazir*, *Ferz*, usual *Knight*, *Camel*, *Giraffe*), and for an $(p, 2)$-leaper with $p = 0, 1, 2, 3$ (*Dabbaba*, usual *Knight*, *Alfil*, *Zebra*).

  An **(p, q)-super-knight metric** (or *(p, q)-leaper metric*) is a metric on $\mathbb{Z}^2$, defined as the minimum number of moves a chess (p, q)-super-knight would take to travel from x to $y \in \mathbb{Z}^2$. Thus, its *unit sphere* $S^1_{p,q}$, centered at the origin, contains exactly 8 integral points $\{(\pm p, \pm q), (\pm q, \pm p)\}$. (See [DaMu90].)

  The **knight metric** is the $(1, 2)$-super-knight metric. The **city-block metric** can be considered as the *Wazir metric*, i.e., $(0, 1)$-super-knight metric.

- **Rook metric**

  The **rook metric** is a metric on $\mathbb{Z}^2$, defined as the minimum number of moves a chess rook would take to travel from x to $y \in \mathbb{Z}^2$. This metric can take only the values $\{0, 1, 2\}$, and coincides with the **Hamming metric** on $\mathbb{Z}^2$.

- **Chamfer metric**

  Given two positive numbers α, β with $\alpha \leqslant \beta < 2\alpha$, consider *$(\alpha, \beta)$-weighted l_∞-grid*, i.e., the infinite graph with the vertex-set $\mathbb{Z}^2$, two vertices being adjacent if their l_∞-distance is one, while horizontal/vertical and diagonal edges having *weights* α and β, respectively.

A **chamfer metric** (or (α, β)*-chamfer metric*, [Borg86]) is the **weighted path metric** in this graph. For any $x, y \in \mathbb{Z}^2$ it can be written as

$$\beta m + \alpha(M - m),$$

where $M = \max\{|u_1|, |u_2|\}$, $m = \min\{|u_1|, |u_2|\}$, $u_1 = x_1 - y_1$, $u_2 = x_2 - y_2$.

If the weights α and β are equal to the Euclidean lengths 1, $\sqrt{2}$ of horizontal/vertical and diagonal edges, respectively, then one obtains the Euclidean length of the shortest chessboard path between x and y. If $\alpha = \beta = 1$, one obtains the **chessboard metric**. The (3, 4)-chamfer metric is the most used one for digital images; it is called simply $(3, 4)$**-metric**.

An $3D$**-chamfer metric** is the **weighted path metric** of the graph with the vertex-set $\mathbb{Z}^3$ of *voxels*, two voxels being adjacent if their l_∞-distance is one, while weights α, β, and γ are associated, respectively, to the distance from 6 face neighbors, 12 edge neighbors, and 8 corner neighbors.

- **Weighted cut metric**

Consider *weighted* l_∞*-grid*, i.e., the graph with the vertex-set $\mathbb{Z}^2$, two vertices being adjacent if their l_∞-distance is one, and each edge having some positive *weight* (or *cost*). Usual **weighted path metric** between two pixels is the minimal cost of a path connecting them. The **weighted cut metric** between two pixels is the minimal cost (defined now as the sum of costs of crossed edges) of a *cut*, i.e., a plane curve connecting them while avoiding pixels.

- **Digital volume metric**

The **digital volume metric** is a metric on the set K of all bounded subsets (*pictures*, or *images*) of $\mathbb{Z}^2$ (in general, of $\mathbb{Z}^n$), defined by

$$vol(A \triangle B),$$

where $vol(A) = |A|$, i.e., the number of pixels contained in A, and $A \triangle B$ is the *symmetric difference* between sets A and B.

This metric is a digital analog of the **Nikodym metric**.

- **Hexagonal Hausdorff metric**

The **hexagonal Hausdorff metric** is a metric on the set of all bounded subsets (*pictures*, or *images*) of the *hexagonal greed* on the plane, defined by

$$\inf\{p, q : \ A \subset B + qH, \ B \subset A + pH\}$$

for any pictures A and B, where pH is the *regular hexagon of size* p (i.e., with $p+1$ pixels on each edge), centered at the origin and including its interior, and $+$ is the *Minkowski addition*: $A+B = \{x+y : \ x \in A, y \in B\}$ (cf. **Pompeiu–Hausdorff–Blaschke metric**). If A is a pixel x, then the distance between x and B is equal to $\sup_{y \in B} d_6(x, y)$, where d_6 is the **hexagonal metric**, i.e., the **path metric** on the hexagonal grid.

Chapter 20

Voronoi Diagram Distances

Given a finite set A of objects A_i in a space S, computing *Voronoi diagram* of A means partitioning the space S into *Voronoi regions* $V(A_i)$ in such a way that $V(A_i)$ contains all points of S that are "closer" to A_i than to any other object A_j in A.

Given a *generator set* $P = \{p_1, \dots p_k\}$, $k \geqslant 2$, of distinct points (*generators*) from $\mathbb{R}^n$, $n \geqslant 2$, the *ordinary Voronoi polygon* $V(p_i)$ associated with a generator p_i is defined by

$$V(p_i) = \{x \in \mathbb{R}^n : d_E(x, p_i) \leqslant d_E(x, p_j) \text{ for any } j \neq i\},$$

where d_E is the ordinary Euclidean distance on $\mathbb{R}^n$. The set

$$V(P, d_E, \mathbb{R}^n) = \{V(p_1), \dots, V(p_k)\}$$

is called *n-dimensional ordinary Voronoi diagram, generated by P*. The boundaries of (n-dimensional) Voronoi polygons are called ($(n-1)$-dimensional) *Voronoi facets*, the boundaries of Voronoi facets are called *$(n-2)$-dimensional Voronoi faces*, ..., the boundaries of two-dimensional Voronoi faces are called *Voronoi edges*, the boundaries of Voronoi edges are called *Voronoi vertices*.

A generalization of the ordinary Voronoi diagram is possible in three following ways:

1. The generalization with respect to the generator set $A = \{A_1, \dots, A_k\}$ which can be a set of lines, a set of areas, etc.;
2. The generalization with respect to the space S which can be a sphere (*spherical Voronoi diagram*), a cylinder (*cylindrical Voronoi diagram*), a cone (*conic Voronoi diagram*), a polyhedral surface (*polyhedral Voronoi diagram*), etc.;
3. The generalization with respect to the function d, where $d(x, A_i)$ measures the "distance" from a point $x \in S$ to a generator $A_i \in A$.

This generalized distance function d is called **Voronoi generation distance** (or *Voronoi distance*, *V-distance*), and allows many more functions than an ordinary metric on S. If F is a strictly increasing function of an V-distance d, i.e., $F(d(x, A_i)) \leqslant F(d(x, A_j))$ if and only if $d(x, A_i) \leqslant d(x, A_j)$, then the generalized Voronoi diagrams $V(A, F(d), S)$ and $V(A, d, S)$ coincide, and one says that the V-distance $F(d)$ is *transformable* to the V-distance d, and that the generalized Voronoi diagram $V(A, F(d), S)$ is a *trivial generalization* of the generalized Voronoi diagram $V(A, d, S)$. In applications, one often uses for trivial generalization of ordinary Voronoi diagram $V(P, d, \mathbb{R}^n)$ the **exponential distance**, the **logarithmic distance**, and the **power distance**. There are generalized Voronoi diagrams $V(P, D, \mathbb{R}^n)$, defined by V-distances, that are not transformable to the Euclidean distance d_E: the **multiplicatively weighted Voronoi distance**, the **additively weighted Voronoi distance**, etc.

For an additional information see, for example, [OBS92], [Klei89].

20.1. CLASSICAL VORONOI GENERATION DISTANCES

- **Exponential distance**

  The **exponential distance** is the Voronoi generation distance

$$D_{exp}(x, p_i) = e^{d_E(x, p_i)}$$

  for the trivial generalization $V(P, D_{exp}, \mathbb{R}^n)$ of the ordinary Voronoi diagram $V(P, d_E, \mathbb{R}^n)$, where d_E is the Euclidean distance.

- **Logarithmic distance**

  The **logarithmic distance** is the Voronoi generation distance

$$D_{\ln}(x, p_i) = \ln d_E(x, p_i)$$

  for the trivial generalization $V(P, D_{\ln}, \mathbb{R}^n)$ of the ordinary Voronoi diagram $V(P, d_E, \mathbb{R}^n)$, where d_E is the Euclidean distance.

- **Power distance**

  The **power distance** is the Voronoi generation distance

$$D_\alpha(x, p_i) = d_E(x, p_i)^\alpha, \quad \alpha > 0,$$

  for the trivial generalization $V(P, D_\alpha, \mathbb{R}^n)$ of the ordinary Voronoi diagram $V(P, d_E, \mathbb{R}^n)$, where d_E is the Euclidean distance.

- **Multiplicatively weighted distance**

  The **multiplicatively weighted distance** d_{MW} is the Voronoi generation distance of the generalized Voronoi diagram $V(P, d_{MW}, \mathbb{R}^n)$ (*multiplicatively weighted Voronoi diagram*), defined by

$$d_{MW}(x, p_i) = \frac{1}{w_i} d_E(x, p_i)$$

  for any point $x \in \mathbb{R}^n$ and any *generator point* $p_i \in P = \{p_1, \dots, p_k\}$, $k \geqslant 2$, where $w_i \in w = \{w_i, \dots, w_k\}$ is a given positive *multiplicative weight* of the generator p_i, and d_E is the ordinary Euclidean distance.

  For $\mathbb{R}^2$, the multiplicatively weighted Voronoi diagram is called *circular Dirichlet tessellation*. An edge in this diagram is a circular arc or a straight line.

  In the plane $\mathbb{R}^2$, there exists a generalization of the multiplicatively weighted Voronoi diagram, the *crystal Voronoi diagram*, with the same definition of the distance (where

w_i is the speed of growth of the crystal p_i), but a different partition of the plane, as the crystals can grow only in an empty area.

- **Additively weighted distance**

The **additively weighted distance** d_{AW} is the Voronoi generation distance of the generalized Voronoi diagram $V(P, d_{AW}, \mathbb{R}^n)$ (*additively weighted Voronoi diagram*), defined by

$$d_{AW}(x, p_i) = d_E(x, p_i) - w_i$$

for any point $x \in \mathbb{R}^n$ and any generator point $p_i \in P = \{p_1, \dots, p_k\}$, $k \geqslant 2$, where $w_i \in w = \{w_i, \dots, w_k\}$ is a given *additive weight* of the generator p_i, and d_E is the ordinary Euclidean distance.

For $\mathbb{R}^2$, the additively weighted Voronoi diagram is called *hyperbolic Dirichlet tessellation*. An edge in this Voronoi diagram is a hyperbolic arc or a straight line segment.

- **Additively weighted power distance**

The **additively weighted power distance** d_{PW} is the Voronoi generation distance of the generalized Voronoi diagram $V(P, d_{PW}, \mathbb{R}^n)$ (*additively weighted power Voronoi diagram*), defined by

$$d_{PW}(x, p_i) = d_E^2(x, p_i) - w_i$$

for any point $x \in \mathbb{R}^n$ and any *generator point* $p_i \in P = \{p_1, \dots, p_k\}$, $k \geqslant 2$, where $w_i \in w = \{w_i, \dots, w_k\}$ is a given *additive weight* of the generator p_i, and d_E is the ordinary Euclidean distance.

This diagram can be regarded as a Voronoi diagram of circles or as a Voronoi diagram with the *Laguerre geometry*.

The **multiplicatively weighted power distance** $d_{MPW}(x, p_i) = \frac{1}{w_i} d_E^2(x, p_i)$, $w_i > 0$, is transformable to the **multiplicatively weighted distance**, and gives a trivial extension of the multiplicatively weighted Voronoi diagram.

- **Compoundly weighted distance**

The **compoundly weighted distance** d_{CW} is the Voronoi generation distance of the generalized Voronoi diagram $V(P, d_{CW}, \mathbb{R}^n)$ (*compoundly weighted Voronoi diagram*), defined by

$$d_{CW}(x, p_i) = \frac{1}{w_i} d_E(x, p_i) - v_i$$

for any point $x \in \mathbb{R}^n$ and any generator point $p_i \in P = \{p_1, \dots, p_k\}$, $k \geqslant 2$, where $w_i \in w = \{w_i, \dots, w_k\}$ is a given positive *multiplicative weight* of the generator p_i, $v_i \in v = \{v_1, \dots, v_k\}$ is a given *additive weight* of the generator p_i, and d_E is the ordinary Euclidean distance.

An edge in the two-dimensional compoundly weighted Voronoi diagram is a part of a fourth-order polynomial curve, a hyperbolic arc, a circular arc, or a straight line.

20.2. PLANE VORONOI GENERATION DISTANCES

- **Shortest path distance with obstacles**

  Let $\mathcal{O} = \{O_1, \ldots, O_m\}$ be a collection of pairwise disjoint polygons on the Euclidean plane, representing a set of obstacles which are neither transparent nor traversable.

  The **shortest path distance with obstacles** d_{sp} is the Voronoi generation distance of the generalized Voronoi diagram $V(P, d_{sp}, \mathbb{R}^2 \backslash \{\mathcal{O}\})$ (*shortest path Voronoi diagram with obstacles*), defined, for any $x, y \in \mathbb{R}^2 \backslash \{\mathcal{O}\}$, as the length of the shortest path among all possible continuous paths, connecting x and y, that do not intersect obstacles $O_i \backslash \partial O_i$ (a path can pass through points on the boundary ∂O_i of O_i), $i = 1, \ldots m$.

  The shortest path is constructed with the aid of the *visibility polygon* and the *visibility graph* of $V(P, d_{sp}, \mathbb{R}^2 \backslash \{\mathcal{O}\})$.

- **Visibility shortest path distance**

  Let $\mathcal{O} = \{O_1, \ldots, O_m\}$ be a collection of pairwise disjoint line segments $O_l = [a_l, b_l]$ in the Euclidean plane, $P = \{p_1, \ldots, p_k\}$, $k \geqslant 2$, is the set of generator points,

$$VIS(p_i) = \left\{x \in \mathbb{R}^2 \colon\ [x, p_i] \cap \,]a_l, b_l[\, = \emptyset \text{ for all } l = 1, \ldots, m\right\}$$

  is the *visibility polygon* of the generator p_i, and d_E is the ordinary Euclidean distance.

  The **visibility shortest path distance** d_{vsp} is the Voronoi generation distance of the generalized Voronoi diagram $V(P, d_{vsp}, \mathbb{R}^2 \backslash \{\mathcal{O}\})$ (*visibility shortest path Voronoi diagram with line obstacles*), defined by

$$d_{vsp}(x, p_i) = \begin{cases} d_E(x, p_i), & \text{if } x \in VIS(p_i), \\ \infty, & \text{otherwise.} \end{cases}$$

- **Network distances**

  A *network* on $\mathbb{R}^2$ is a connected planar geometrical graph $G = (V, E)$ with the set V of vertices and the set E of edges (links).

  Let the generator set $P = \{p_1, \ldots, p_k\}$ be a subset of the set $V = \{p_1, \ldots, p_l\}$ of vertices of G, and the set L be given by points of links of G.

  The **network distance** d_{netv} on the set V is the Voronoi generation distance of the *network Voronoi node diagram* $V(P, d_{netv}, V)$, defined as the shortest path along the links of G from $p_i \in V$ to $p_j \in V$. It is the **weighted path metric** of the graph G, where w_e is the Euclidean length of the link $e \in E$.

  The **network distance** d_{netl} on the set L is the Voronoi generation distance of the *network Voronoi link diagram* $V(P, d_{netl}, L)$, defined as the shortest path along the links from $x \in L$ to $y \in L$.

  The **access network distance** d_{accnet} on $\mathbb{R}^2$ is the Voronoi generation distance of the *network Voronoi area diagram* $V(P, d_{accnet}, \mathbb{R}^2)$, defined by

$$d_{accnet}(x, y) = d_{netl}\big(l(x), l(y)\big) + d_{acc}(x) + d_{acc}(y),$$

where $d_{acc}(x) = \min_{l \in L} d(x, l) = d_E(x, l(x))$ is the *access distance* of a point x. In fact, $d_{acc}(x)$ is the Euclidean distance from x to the *access point* $l(x) \in L$ of x which is the nearest to x point on the links of G.

- **Airlift distance**

An *airports network* is an arbitrary planar graph G on n vertices (*airports*) with positive edge weights (*flight durations*). This graph may be entered and exited only at the airports. Once having accessed G, one travels at fixed speed $v > 1$ within the network. Movement off the network takes place with the unit speed with respect to the ordinary Euclidean distance.

The **airlift distance** d_{al} is the Voronoi generation distance of the *airlift Voronoi diagram* $V(P, d_{al}, \mathbb{R}^2)$, defined as the time needed for a *quickest path* between x and y in the presence of the airports network G, i.e., a path minimizing the travel time between x and y.

- **City distance**

A *city public transportation network*, like a subway or a bus transportation system, is a planar straight line graph G with horizontal or vertical edges. G may be composed of many connected components, and may contain cycles. One is free to enter G at any point, be it at a vertex or on an edge (it is possible to postulate fixed entry points, too). Once having accessed G, one travels at a fixed speed $v > 1$ in one of the available directions. Movement off the network takes place with the unit speed with respect to the **Manhattan metric** (we imagine a large modern-style city with streets arranged in north-south and east-west directions).

The **city distance** d_{city} is the Voronoi generation distance of the *city Voronoi diagram* $V(P, d_{city}, \mathbb{R}^2)$, defined as the time needed for the *quickest path* between x and y in the presence of the network G, i.e., a path minimizing the travel time between x and y.

The set $P = \{p_1, \dots, p_k\}$, $k \geqslant 2$, can be seen as a set of some city facilities (for example, post offices or hospitals): for some people several facilities of the same kind are equally attractive, and they want to find out which facility is reachable first.

- **Distance in a river**

The **distance in a river** d_{riv} is the Voronoi generation distance of the generalized Voronoi diagram $V(P, d_{riv}, \mathbb{R}^2)$ (*Voronoi diagram in a river*), defined by

$$d_{riv}(x, y) = \frac{-\alpha(x_1 - y_1) + \sqrt{(x_1 - y_1)^2 + (1 - \alpha^2)(x_2 - y_2)^2}}{v(1 - \alpha^2)},$$

where v is the speed of the boat on the still water, $w > 0$ is the speed of constant flow in the positive direction of the x_1-axis, and $\alpha = \frac{w}{v}$ $(0 < \alpha < 1)$ is the *relative flow speed*.

- **Boat-sail distance**

Let $\Omega \subset \mathbb{R}^2$ be a *domain* in the plane (*water surface*), let $f : \Omega \to \mathbb{R}^2$ be a continuous vector field on Ω, representing the velocity of the water flow (*flow field*); let $P = \{p_1, \dots, p_k\} \subset \Omega$, $k \geqslant 2$, be a set of k points in Ω (*harbors*).

The **boat-sail distance** ([NiSu03]) d_{bs} is the Voronoi generation distance of the generalized Voronoi diagram $V(P, d_{bs}, \Omega)$ (*boat-sail Voronoi diagram*), defined by

$$d_{bs}(x, y) = \inf_{\gamma} \delta(\gamma, x, y)$$

for all $x, y \in \Omega$, where

$$\delta(\gamma, x, y) = \int_0^1 \left| F \frac{\gamma'(s)}{|\gamma'(s)|} + f(\gamma(s)) \right|^{-1} ds$$

is the time necessary for the boat with the maximum speed F on the still water to move from x to y along the curve $\gamma : [0, 1] \to \Omega$, $\gamma(0) = x$, $\gamma(1) = y$, and the infimum is taken over all possible curves γ.

- **Peeper distance**

Let $S = \{(x_1, x_2) \in \mathbb{R}^2 \colon\ x_1 > 0\}$ be the half-plane in $\mathbb{R}^2$, let $P = \{p_1, \dots, p_k\}$, $k \geqslant 2$, be a set of points contained in the half-plane $\{(x_1, x_2) \in \mathbb{R}^2 \colon\ x_1 < 0\}$, and let the *window* be the open line segment $]a, b[$ with $a = (0, 1)$ and $b = (0, -1)$.

The **peeper distance** d_{pee} is the Voronoi generation distance of the generalized Voronoi diagram $V(P, d_{pee}, S)$ (*peeper's Voronoi diagram*), defined by

$$d_{pee}(x, p_i) = \begin{cases} d_E(x, p_i), & \text{if } [x, p] \cap\,]a, b[\ \neq \emptyset, \\ \infty, & \text{otherwise,} \end{cases}$$

where d_E is the ordinary Euclidean distance.

- **Snowmobile distance**

Let $\Omega \subset \mathbb{R}^2$ be a *domain* in the x_1x_2-plane of the space $\mathbb{R}^3$ (a *two-dimensional mapping*), and $\Omega^* = \{(q, h(q)) \colon\ q = (x_1(q), x_2(q)) \in \Omega, h(q) \in \mathbb{R}\}$ be the three-dimensional *land surface*, associated with the mapping Ω. Let $P = \{p_1, \dots, p_k\} \subset \Omega$, $k \geqslant 2$, be a set of k points in Ω (*snowmobile stations*).

The **snowmobile distance** d_{sm} is the Voronoi generation distance of the generalized Voronoi diagram $V(P, d_{sm}, \Omega)$ (*snowmobile Voronoi diagram*), defined by

$$d_{sm}(q, r) = \inf_{\gamma} \int_{\gamma} \frac{1}{F\left(1 - \alpha \frac{dh(\gamma(s))}{ds}\right)} ds$$

for any $q, r \in \Omega$, and calculating the minimum time necessary for the snowmobile with the speed F on a flat land to move from $(q, h(q))$ to $(r, h(r))$ along the *land path* γ^*: $\gamma^*(s) = (\gamma(s), h(\gamma(s)))$, associated with the *domain path* $\gamma : [0, 1] \to \Omega$, $\gamma(0) = q$, $\gamma(1) = r$ (the infimum is taken over all possible paths γ, and α is a positive constant).

A snowmobile goes uphill more slowly than goes downhill. The situation is opposite for a forest fire: the frontier of the fire goes uphill faster than goes downhill. This situation can be modeled using a negative value of α. The resulting distance is called **forest-fire distance**, and the resulting Voronoi diagram is called *forest-fire Voronoi diagram.*

- **Skew distance**

Let T be a *tilted plane* in $\mathbb{R}^3$, obtained by rotation the x_1x_2-plane around the x_1-axis through the angle α, $0 < \alpha < \frac{\pi}{2}$, with the coordinate system obtained by taking the coordinate system of the x_1x_2-plane, accordingly rotated. For a point $q \in T$, $q = (x_1(q), x_2(q))$, define *height* $h(q)$ as its x_3-coordinate in $\mathbb{R}^3$. Thus, $h(q) = x_2(q) \cdot \sin\alpha$. Let $P = \{p_1, \dots, p_k\} \subset T$, $k \geqslant 2$.

The **skew distance** ([AACL98]) d_{skew} is the Voronoi generation distance of the generalized Voronoi diagram $V(P, d_{skew}, T)$ (*skew Voronoi diagram*), defined by

$$d_{skew}(q,r) = d_E(q,r) + \big(h(r) - h(q)\big) = d_E(q,r) + \sin\alpha\big(x_2(r) - x_2(q)\big)$$

or, more generally, by

$$d_{skew}(q,r) = d_E(q,r) + k\big(x_2(r) - x_2(q)\big)$$

for all $q, r \in T$, where d_E is the ordinary Euclidean distance, and $k \geqslant 0$ is a constant.

20.3. OTHER VORONOI GENERATION DISTANCES

- **Voronoi distance for line segments**

The **Voronoi distance for** (a set of) **line segments** d_{sl} is the Voronoi generation distance of the generalized Voronoi diagram $V(A, d_{ls}, \mathbb{R}^2)$ (*line Voronoi diagram, generated by straight line segments*), defined by

$$d_{sl}(x, A_i) = \inf_{y \in A_i} d_E(x, y),$$

where the *generator set* $A = \{A_1, \dots, A_k\}$, $k \geqslant 2$, is a set of pairwise disjoint straight line segments $A_i = [a_i, b_i]$, and d_E is the ordinary Euclidean distance. In fact,

$$d_{ls}(x, A_i) = \begin{cases} d_E(x, a_i), & \text{if } x \in R_{a_i}, \\ d_E(x, b_i), & \text{if } x \in R_{b_i}, \\ d_E\Big(x - a_i, \frac{(x-a_i)^T(b_i-a_i)}{d_E^2(a_i,b_i)}(b_i - a_i)\Big), & \text{if } x \in \mathbb{R}^2 \backslash \{R_{a_i} \cup R_{b_i}\}, \end{cases}$$

where $R_{a_i} = \{x \in \mathbb{R}^2 \colon (b_i - a_i)^T(x - a_i) < 0\}$, $R_{b_i} = \{x \in \mathbb{R}^2 \colon (a_i - b_i)^T(x - b_i) < 0\}$.

- **Voronoi distance for arcs**

The **Voronoi distance for** (a set of circle) **arcs** d_{ca} is the Voronoi generation distance of the generalized Voronoi diagram $V(A, d_{ca}, \mathbb{R}^2)$ (*line Voronoi diagram, generated by circle arcs*), defined by

$$d_{ca}(x, A_i) = \inf_{y \in A_i} d_E(x, y),$$

where the *generator set* $A = \{A_1, \dots, A_k\}$, $k \geqslant 2$, is a set of pairwise disjoint circle arcs A_i (less than or equal to a semicircles) with radius r_i centered at x_{c_i}, and d_E is the ordinary Euclidean distance. In fact,

$$d_{ca}(x, A_i) = \min\{d_E(x, a_i), d_E(x, b_i), |d_E(x, x_{c_i}) - r_i|\},$$

where a_i and b_i are end points of A_i.

- **Voronoi distance for circles**

The **Voronoi distance for** (a set of) **circles** d_{ca} is the Voronoi generation distance of a generalized Voronoi diagram $V(A, d_{cl}, \mathbb{R}^2)$ (*line Voronoi diagram, generated by circles*), defined by

$$d_{cl}(x, A_i) = \inf_{y \in A_i} d_E(x, y),$$

where the *generator set* $A = \{A_1, \dots, A_k\}$, $k \geqslant 2$, is a set of pairwise disjoint circles A_i with radius r_i centered at x_{c_i}, and d_E is the ordinary Euclidean distance. In fact,

$$d_{ca}(x, A_i) = |d_E(x, x_{c_i}) - r_i|.$$

There exist different distances for the line Voronoi diagram, generated by circles. For example, $d^{\star}_{cl}(x, A_i) = d_E(x, x_{c_i}) - r_i$, or $d^{*}_{cl}(x, A_i) = d^2_E(x, x_{c_i}) - r^2_i$ (the *Laguerre Voronoi diagram*).

- **Voronoi distance for areas**

The **Voronoi distance for areas** d_{ar} is the Voronoi generation distance of the generalized Voronoi diagram $V(A, d_{ar}, \mathbb{R}^2)$ (*area Voronoi diagram*), defined by

$$d_{ar}(x, A_i) = \inf_{y \in A_i} d_E(x, y),$$

where $A = \{A_1, \dots, A_k\}$, $k \geqslant 2$, is a collection of pairwise disjoint connected closed sets (*areas*), and d_E is the ordinary Euclidean distance.

Note, that for any generalized generator set $A = \{A_1, \dots, A_k\}$, $k \geqslant 2$, one can use as the Voronoi generation distance the **Hausdorff distance** from a point x to a set A_i: $d_{Haus}(x, A_i) = \sup_{y \in A_i} d_E(x, y)$, where d_E is the ordinary Euclidean distance.

- **Cylindrical distance**

The **cylindrical distance** d_{cyl} is the **intrinsic metric** on the surface of a cylinder S which is used as the Voronoi generation distance in the *cylindrical Voronoi diagram* $V(P, d_{cyl}, S)$. If the axis of a cylinder with unit radius is placed at the x_3-axis in $\mathbb{R}^3$, the cylindrical distance between any points $x, y \in S$ with the cylindrical coordinates $(1, \theta_x, z_x)$ and $(1, \theta_y, z_y)$ is given by

$$d_{cyl}(x, y) = \begin{cases} \sqrt{(\theta_x - \theta_y)^2 + (z_x - z_y)^2}, & \text{if } \theta_y - \theta_x \leqslant \pi, \\ \sqrt{(\theta_x + 2\pi - \theta_y)^2 + (z_x - z_y)^2}, & \text{if } \theta_y - \theta_x > \pi. \end{cases}$$

- **Cone distance**

The **cone distance** d_{con} is the **intrinsic metric** on the surface of a cone S which is used as the Voronoi generation distance in the *conic Voronoi diagram* $V(P, d_{con}, S)$. If the axis of the cone S is placed at the x_3-axis in $\mathbb{R}^3$, and the radius of the circle made by the intersection of the cone S with the x_1x_2-plane is equal to one, then the cone distance between any points $x, y \in S$ is given by

$$d_{con}(x, y) = \begin{cases} \sqrt{r_x^2 + r_y^2 - 2r_x r_y \cos(\theta'_y - \theta'_x)}, \\ \quad \text{if } \theta'_y \leqslant \theta'_x + \pi \sin(\alpha/2), \\ \sqrt{r_x^2 + r_y^2 - 2r_x r_y \cos(\theta'_x + 2\pi \sin(\alpha/2) - \theta'_y)}, \\ \quad \text{if } \theta'_y > \theta'_x + \pi \sin(\alpha/2), \end{cases}$$

where (x_1, x_2, x_3) are the Cartesian coordinates of a point x on S, α is the top angle of the cone, θ_x is the counterclockwise angle from the x_1-axis to the ray from the origin to the point $(x_1, x_2, 0)$, $\theta'_x = \theta_x \sin(\alpha/2)$, $r_x = \sqrt{x_1^2 + x_2^2 + (x_3 - \coth(\alpha/2))^2}$ is the straight line distance from the top of the cone to the point (x_1, x_2, x_3).

- **Voronoi distances of order *m***

Given a finite set A of objects in a metric space (S, d), and an integer $m \geqslant 1$, consider the set of all *m-subsets* M_i of A (i.e., $M_i \subset A$, and $|M_i| = m$). The *Voronoi diagram of order m* of A is a partition of S into *Voronoi regions* $V(M_i)$ *of m-subsets* of A in such a way that $V(M_i)$ contains all points $s \in S$ which are "closer" to M_i than to any other m-set M_j: $d(s, x) < d(s, y)$ for any $x \in M_i$ and $y \in S \backslash M_i$. This diagram provides first, second, ..., m-th closest neighbors of a point in S.

Such diagrams can be defined in terms of some "distance function" $D(s, M_i)$, in particular, some m-**hemi-metric** on S. For $M_i = \{a_i, b_i\}$, there were considered the functions $|d(s, a_i) - d(s, b_i)|$, $d(s, a_i) + d(s, b_i)$, $d(s, a_i) \cdot d(s, b_i)$, as well as 2-**metrics** $d(s, a_i) + d(s, b_i) + d(a_i, b_i)$ and the area of triangle (s, a_i, b_i).

Chapter 21

Image and Audio Distances

21.1. IMAGE DISTANCES

Image Processing treat signals such as photographs, video, or tomographic output. In particular, *Computer Graphics* consists of image synthesis from some abstract models, while *Computer Vision* extracts some abstract information: say, the 3D (i.e., 3-dimensional) description of a scene from video footage of it. From about 2000, the analog image processing (by optical devices) gave way to the digital processing, and, in particular, digital image editing (for example, processing of images taken by popular digital cameras).

Computer graphics (and our brains) deals with *vector graphics images*, i.e., those represented geometrically by curves, polygons, etc. A *raster graphics image* (or *digital image, bitmap*) in 2D is a representation of 2D image as a finite set of digital values, called *pixels* (short for picture elements) placed on square grid $\mathbb{Z}^2$ or hexagonal grid. Typically, the image raster is a square $2^k \times 2^k$ grid with $k = 8, 9$ or 10. Video images and *tomographic* (i.e., obtained by sections) images are 3D (2D plus time); their digital values are called *voxels* (volume elements).

A *digital binary image* corresponds to only two values 0, 1 with 1 being interpreted as logical "true" and displayed as black; so, such image is identified with the set of black pixels. The elements of binary 2D image can be seen as complex numbers $x + iy$, where (x, y) are coordinates of a point on the real plane $\mathbb{R}^2$. A *continuous binary image* is a (usually, compact) subset of a **locally compact** metric space (usually, Euclidean space $\mathbb{E}^n$ with $n = 2, 3$).

The *gray-scale images* can be seen as point-weighted binary images. In general, a *fuzzy set* is a point-weighted set with weights (*membership values*). For the gray-scale images, xyi-representation is used, where plane coordinates (x, y) indicate shape, while the weight i (short for intensity, i.e., brightness) indicate *texture* (intensity pattern). Sometimes, the matrix $((i_{xy}))$ of gray-levels is used. *Brightness histogram* of a gray-scale image provides the frequency of each brightness value found in that image. If image has m brightness levels (bins of gray-scale), then there are 2^m different possible intensities. Usually, $m = 8$ and numbers $0, 1, \dots, 255$ represent intensity range from black to white; other typical values are $m = 10, 12, 14, 16$. Humans can differ between around 350000 different colors but between only 30 different gray-levels; so, color has much higher discriminatory power.

For color images, (RGB)-representation is most known, where space coordinates R, G, B indicate red, green and blue level; 3D histogram provides brightness at each point. Among many other 3D color models (spaces) are: (CMY) cube (Cyan, Magenta, Yellow colors), (HSL) cone (Hue-color type given as angle, Saturation in %, Luminosity in %), and (YUV), (YIQ) used, respectively, in PAL, NTSC television. CIE-approved conversion of (RGB) into luminance (luminosity) of gray-level is $0.299R + 0.587G + 0.114B$. *Color histogram* is a feature vector of length n (usually, $n = 64, 256$) with components representing ether the total number of pixels, or the percentage of pixels of given color in the image.

Images are often represented by *feature vectors*, including color histograms, color moments, textures, shape descriptors, etc. Examples of feature spaces are: *raw intensity* (pixel values), *edges* (boundaries, contours, surfaces), *salient features* (corners, line intersections, points of high curvature), and *statistical features* (moment invariants, centroids). Typical video features are in terms of overlapping frames and motions. *Image Retrieval* (similarity search) consists of (as for other data: audio recordings, DNA sequences, text documents, time-series, etc.) finding images whose features have values ether similar between them, or similar to given query or in given range.

There are two methods to compare images directly: intensity-based (color and texture histograms), and geometry-based (shape representations by *medial axis*, *skeletons*, etc.). Unprecise term *shape* is used for the extent (silhouette) of the object, for its local geometry or geometrical pattern (conspicuous geometric details, points, curves, etc.), or for that pattern modulo a similarity transformation group (translations, rotations, and scalings). Unprecise term *texture* means all what is left after color and shape have been considered, or it is defined via structure and randomness.

The similarity between vector representations of images is measured by usual practical distances: l_p**-metrics**, **weighted editing metrics**, **Tanimoto distance**, **cosine distance**, **Mahalanobis distance** and its extension, **Earth Mover distance**. Among probability distances, the following ones are most used: **Bhattacharya** 2, **Hellinger**, **Kullback–Leibler**, **Jeffrey** and (especially, for histograms) χ^2-, **Kolmogorov–Smirnov**, **Kuiper distances**.

The main distances applied for compact subsets X and Y of $\mathbb{R}^n$ (usually, $n = 2, 3$) or their digital versions are: **Asplund metric**, **Shephard metric**, **symmetric difference semi-metric** $Vol(X \Delta Y)$ (see **Nikodym metric**, **area deviation**, **digital volume metric** and their normalizations) and variations of the **Hausdorff distance** (see below).

For Image Processing, the distances below are between "true" and approximated digital images, in order to assess the performance of algorithms. For Image Retrieval, distances are between feature vectors of a query and reference.

- **Color distances**

  A *color space* is a 3-parameter description of colors. The need for exactly 3 parameters comes from the existence of 3 kinds of receptors in the human eye: for short, middle and long wavelengths, corresponding to blue, green, and red.

  The CIE (International Commission on Illumination) derived (XYZ) color space in 1931 from (RGB)-model and measurements of the human eye. In the CIE (XYZ) color space, the values X, Y and Z are also roughly red, green and blue, respectively.

  The basic assumption of Colorimetry, supported experimentally (Indow, 1991), is that the perceptual color space admits a metric, the true **color distance**. This metric is expected to be locally Euclidean, i.e., a **Riemannian metric**. Another assumption is that there is a continuous mapping from the metric space of *photic* (light) stimuli to this metric space. Cf. **probability-distance hypothesis** in Psychophysics that the probability with which one stimulus is discriminated from another is a (continuously increasing) function of some subjective quasi-metric between these stimuli.

Such *uniform color scale*, where equal distances in the color space correspond to equal differences in color, is not obtained yet and existing **color distances** are various approximations of it. First step in this direction was given by *MacAdam ellipses*, i.e., regions on a *chromaticity* (x, y) diagram which contains all colors looking indistinguishable to the average human eye. Those 25 ellipses define a metric in a color space. Here $x = \frac{X}{X+Y+Z}$ and $y = \frac{Y}{X+Y+Z}$ are projective coordinates, and the colors of the chromaticity diagram occupy a region of the real projective plane. The CIE $(L^*a^*b^*)$ (CIELAB) is an adaptation of CIE 1931 (XYZ) color space; it gives a partial linearization of the metric indicated by MacAdam ellipses. The parameters L^*, a^*, b^* of the most complete model are derived from L, a, b which are: the luminance L of the color from black $L = 0$ to white $L = 100$, its position a between green $a < 0$ and red $a > 0$, and its position b between green $b < 0$ and yellow $b > 0$.

- **Average color distance**

  For a given 3D color space and a list of n colors, let (c_{i1}, c_{i2}, c_{i3}) be the representation of the i-th color of the list in this space. For a color histogram $x = (x_1, \dots, x_n)$, its *average color* is the vector $(x_{(1)}, x_{(2)}, x_{(3)})$, where $x_{(j)} = \sum_{i=1}^{n} x_i c_{ij}$ (for example, the average red, blue and green values in (RGB)) of the pixels in the image.

  The **average color distance** between two color histograms ([HSEFN95]) is the Euclidean distance of their average colors.

- **Color component distances**

  Given an image (as a subset of $\mathbb{R}^2$), let p_i denote the area percentage of this image occupied by the color c_i. A *color component* of the image is a pair (c_i, p_i).

  The **Ma–Deng–Manjunath distance** between color components (c_i, p_i) and (c_j, p_j) is defined by

$$|p_i - p_j| \cdot d(c_i, c_j),$$

  where $d(c_i, c_j)$ is the distance between colors c_i and c_j in a given color space. Mojsilović et al. developed **Earth Mover distance**-like modification of this distance.

- **Histogram intersection quasi-distance**

  Given two color histograms $x = (x_1, \dots, x_n)$ and $y = (y_1, \dots, y_n)$ (with x_i, y_i representing number of pixels in the bin i), the Swain-Ballard's **histogram intersection quasi-distance** between them (cf. **intersection distance**) is defined by

$$1 - \frac{\sum_{i=1}^{n} \min\{x_i, y_i\}}{\sum_{i=1}^{n} x_i}.$$

For normalized histograms (total sum is 1) above quasi-distance became the usual l_1-**metric** $\sum_{i=1}^{n} |x_i - y_i|$. The Rosenfeld–Kak's *normalized cross correlation* between x and y is a similarity, defined by $\frac{\sum_{i=1}^{n} x_i, y_i}{\sum_{i=1}^{n} x_i^2}$.

- **Histogram quadratic distance**

  Given two color histograms $x = (x_1, \dots, x_n)$ and $y = (y_1, \dots, y_n)$ (usually, $n = 256$ or $n = 64$) representing the color percentages of two images, their **histogram quadratic distance** (used in IBM's Query By Image Content system) is **Mahalanobis distance**, defined by

$$\sqrt{(x-y)^T A(x-y)},$$

  where $A = ((a_{ij}))$ is a symmetric positive-definite matrix, and weight a_{ij} is some, perceptually justified, similarity between colors i and j. For example (see [HSEFN95]), $a_{ij} = 1 - \frac{d_{ij}}{\max_{1 \leqslant p.q \leqslant n} d_{pq}}$, where d_{ij} is the Euclidean distance between 3-vectors representing i and j in some color space. Another definition is given by $a_{ij} = 1 - \frac{1}{\sqrt{5}}((v_i - v_j)^2 + (s_i \cos h_i - s_j \cos h_j)^2 + (s_i \sin h_i - s_j \sin h_j)^2)^{\frac{1}{2}}$, where (h_i, s_i, v_i) and (h_j, s_j, v_j) are the representations of the colors i and j in the color space (HSV).

- **Gray-scale image distances**

  Let $f(x)$ and $g(x)$ denote brightness values of two digital gray-scale images f and g at the pixel $x \in X$, where X is a raster of pixels. Any distance between point-weighted sets (X, f) and (X, g) (for example, the **Earth Mover distance**) can be applied for measuring distances between f and g. However, the main used distances (called also *errors*) between images f and g are:

  1. The *root mean-square error* $RMS(f, g) = \left(\frac{1}{|X|}\sum_{x \in X}(f(x) - g(x))^2\right)^{\frac{1}{2}}$ (a variant is to use l_1-*norm* $|f(x) - g(x)|$ instead of l_2-*norm*);
  2. The *signal-to-noise ratio* $SNR(f, g) = \left(\frac{\sum_{x \in X} g(x)^2}{\sum_{x \in X}(f(x)-g(x))^2}\right)^{\frac{1}{2}}$;
  3. The *pixel misclassification error rate* $\frac{1}{|X|}|\{x \in X : f(x) \neq g(x)\}|$ (normalized **Hamming distance**);
  4. The *frequency root mean-square error* $\left(\frac{1}{|U|^2}\sum_{u \in U}(F(u) - G(u))^2\right)^{\frac{1}{2}}$, where F and G are the discrete Fourier transforms of f and g, respectively, and U is the frequency domain;
  5. The *Sobolev norm of order δ error* $\left(\frac{1}{|U|^2}\sum_{u \in U}(1+|\eta_u|^2)^{\delta}(F(u) - G(u))^2\right)^{\frac{1}{2}}$, where $0 < \delta < 1$ is fixed (usually, $\frac{1}{2}$), and η_u is the 2D frequency vector associated with position u in the frequency domain U.

- **Image compression L_p-metric**

  Given a number r, $0 \leqslant r < 1$, the **image compression L_p-metric** is the usual L_p-**metric** on $\mathbb{R}_+^{n^2}$ (the set of gray-scale images seen as $n \times n$ matrices) with p being a solution of the equation $r = \frac{p-1}{2p-1} \cdot e^{\frac{p}{2p-1}}$. So, $p = 1, 2$, or ∞ for, respectively, $r = 0$,

$r = \frac{1}{3}e^{\frac{2}{3}} \approx 0.65$, or $r \geqslant \frac{\sqrt{e}}{2} \approx 0.82$. Here r estimates *informative* (i.e., filled with non-zeros) part of the image. According to [KKN02], it is the best quality metric to select a lossy compression scheme.

- **Chamfering distances**

  The **chamfering distances** are distances approximating Euclidean distance by a weighted path distance on the graph $G = (\mathbb{Z}^2, E)$, where two pixels are neighbors if one can be obtained from another by an *one-step move* on $\mathbb{Z}^2$. The list of permitted moves is given, and a **prime distance**, i.e., a positive weight, is associated to each type of such move.

  An (α, β)**-chamfer metric** corresponds to two permitted moves – with l_1-distance 1 and with l_∞-distance 1 (diagonal moves only) – weighted α and β, respectively. The main applied cases are $(\alpha, \beta) = (1, 0)$ (the **city-block metric**, or **4-metric**), (1,1) (the **chessboard metric**, or **8-metric**), $(1, \sqrt{2})$ (the **Montanari metric**), $(3, 4)$ (the **$(3, 4)$-metric**), $(2, 3)$ (the **Hilditch–Rutovitz metric**), $(5, 7)$ (the **Verwer metric**).

  The **Borgefors metric** corresponds to three permitted moves – with l_1-distance 1, with l_∞-distance 1 (diagonal moves only), and knight moves – weighted 5, 7 and 11, respectively.

  An 3D**-chamfer metric** (or (α, β, γ)*-chamfer metric*) is the **weighted path metric** of the infinite graph with the vertex-set $\mathbb{Z}^3$ of voxels, two vertices being adjacent if their l_∞-distance is one, while weights α, β and γ are associated to 6 face, 12 edge and 8 corner neighbors, respectively. If $\alpha = \beta = \gamma = 1$, we obtain the l_∞-metric. The $(3, 4, 5)$- and $(1, 2, 3)$-chamfer metrics are the most used ones for digital 3D images.

- **Earth Mover distance**

  The **Earth Mover distance** is a discrete form of the **Monge–Kantorovich distance**. Roughly, it is minimal amount of work needed to transform earth or mass from one position (properly spread in space) to the other (a collection of holes). For any two finite sequences $(x_1, \dots, x_m)$ and $(y_1, \dots, y_n)$ over a metric space (X, d), consider *signatures*, i.e., point-weighted sets $P_1 = (p_1(x_1), \dots, p_1(x_m))$ and $P_2 = (p_2(y_1), \dots, p_2(y_n))$. For example ([RTG00]), signatures can represent clustered color or texture content of images: elements of X are centroids of clusters, and $p_1(x_i)$, $p_2(y_j)$ are sizes of corresponding clusters. The ground distance d is a **color distance**, say, the Euclidean distance in 3D CIE $(L^*a^*b^*)$ color space.

  Let $W_1 = \sum_i p_1(x_i)$ and $W_2 = \sum_j p_2(y_j)$ are the *total weights* of P_1 and P_2, respectively. Then the **Earth Mover distance** (or *transport distance*) between signatures P_1 and P_2 is defined as the function

$$\frac{\sum_{i,j} f_{ij}^* d(x_i, y_j)}{\sum_{i,j} f_{ij}^*},$$

  where the $m \times n$ matrix $S^* = ((f_{ij}^*))$ is an *optimal*, i.e., minimizing $\sum_{i,j} f_{ij} d(x_i, y_j)$, *flow*. A *flow* (of the weight of the earth) is an $m \times n$ matrix $S = ((f_{ij}))$ with following constraints:

1. All $f_{ij} \geqslant 0$;
2. $\sum_{i,j} f_{ij} = \min\{W_1, W_2\}$;
3. $\sum_i f_{ij} \leqslant p_2(y_j)$, and $\sum_j f_{ij} \leqslant p_1(x_i)$.

So, this distance is the average ground distance d that weights travel during an optimal flow.

In the case $W_1 = W_2$, above two inequalities 3. became equalities. Normalizing signatures to $W_1 = W_2 = 1$ (which not changes the distance) allow us to see P_1 and P_2 as probability distributions of random variables, say, X and Y. Then $\sum_{i,j} f_{ij} d(x_i, y_j)$ is just $\mathbb{E}_S[d(X, Y)]$, i.e., the Earth Mover distance coincides, in this case, with the **Kantorovich–Mallows–Monge–Wasserstein metric**. For, say, $W_1 < W_2$, it is not a metric in general. However, replacing, in above definition, the inequalities 3. by equalities:

3′. $\sum_i f_{ij} = p_2(y_j)$, and $\sum_j f_{ij} = \frac{p_1(x_i)W_1}{W_2}$,

produces Giannopoulos–Veltkamp's **proportional transport semi-metric**.

- **Parameterized curves distance**

The shape can be represented by a parameterized curve on the plane. Usually, such curve is *simple*, i.e., it has no self-intersections. Let $X = X(x(t))$ and $Y = Y(y(t))$ be two parameterized curves, where (continuous) parametrization functions $x(t)$ and $y(t)$ on $[0, 1]$ satisfy $x(0) = y(0) = 0$ and $x(1) = y(1) = 1$.

The most used **parameterized curves distance** is the minimum, over all monotone increasing parameterizations $x(t)$ and $y(t)$, of the maximal Euclidean distance $d_E(X(x(t)), Y(y(t)))$. It is Euclidean special case of the **dogkeeper distance** which is, in turn, the **Fréchet metric** for the case of curves. Among variations of this distance are dropping the monotonicity condition of the parametrization, or finding the part of one curve to which the other has the smallest such distance ([VeHa01]).

- **Non-linear elastic matching distances**

Consider a digital representation of curves. Let $r \geqslant 1$ be a constant, and let $A = \{a_1, \dots, a_m\}$, $B = \{b_1, \dots, b_n\}$ be finite ordered sets of consecutive points on two closed curves. For any order-preserving correspondence f between all points of A and all points of B, the *stretch* $s(a_i, b_j)$ *of* $(a_i, f(a_i) = b_j)$ is r if either $f(a_{i-1}) = b_j$ or $f(a_i) = b_{j-1}$, or zero, otherwise.

The **relaxed non-linear elastic matching distance** is the minimum, over all such f, of $\sum(s(a_i, b_j) + d(a_i, b_j))$, where $d(a_i, b_j)$ is the difference between the tangent angles of a_i and b_j. It is a **near-metric** for some r. For $r = 1$, it is called **non-linear elastic matching distance**.

- **Turning function distance**

For a plane polygon P, its *turning function* $T_P(s)$ is the angle between the counterclockwise tangent and the x-axis as the function of the arc length s. This function increases with each left hand turn and decreases with right hand turns.

Given two polygons of equal perimeters, their **turning function distance** is the L_p-**metric** between their turning functions.

- **Size function distance**

For a plane graph $G = (V, E)$ and a *measuring function* f on its vertex-set V (for example, the distance from $v \in V$ to the center of mass of V), the *size function* $S_G(x, y)$ is defined, on the points $(x, y) \in \mathbb{R}^2$, as the number of connected components of the restriction of G on vertices $\{v \in V : f(v) \leqslant y\}$ which contain a point v' with $f(v') \leqslant x$.

Given two plane graphs with vertex-sets belonging to a raster $R \subset \mathbb{Z}^2$, their Uras–Verri's **size function distance** is the normalized l_1-distance between their size functions over raster pixels.

- **Reflection distance**

For a finite union A of plane curves and each point $x \in \mathbb{R}^2$, let V_A^x denote the union of open line segments $]x, a[$, $a \in A$, which are *visible from* x, i.e., $]x, a[\cap A = \emptyset$. Denote by ρ_A^x the area of the set $\{x + v \in V_A^x : x - v \in V_A^x\}$.

The Hagedoorn–Veltkamp's **reflection distance** between finite unions A and B of plane curves is the normalized l_1-distance between the corresponding functions ρ_A^x and ρ_B^x, defined by

$$\frac{\int_{\mathbb{R}^2} |\rho_A^x - \rho_B^x| dx}{\int_{\mathbb{R}^2} \max\{\rho_A^x, \rho_B^x\} dx}.$$

- **Distance transform**

Given a metric space $(X = \mathbb{Z}^2, d)$ and a binary digital image $M \subset X$, the **distance transform** is a function $f_M : X \to \mathbb{R}_{\geqslant 0}$, where $f_M(x) = \inf_{u \in M} d(x, u)$ is the **point-set distance** $d(x, M)$. Therefore, a distance transform can be seen as a gray-scale digital image where each pixel is given a label (a gray-level) which corresponds to the distance to the nearest pixel of the background. Distance transforms, in Image Processing, are also called *distance fields* and, especially, **distance maps**; but we reserve the last term only for this notion in any metric space. A *distance transform of a shape* is the distance transform with M being the boundary of the image. For $X = \mathbb{R}^2$, the graph $\{(x, f(x)) : x \in X\}$ of $d(x, M)$ is called *Voronoi surface* of M.

- **Medial axis and skeleton**

Let (X, d) be a metric space, and let M be a subset of X. The **medial axis** of X is the set $MA(X) = \{x \in X : |\{m \in M : d(x, m) = d(x, M)\}| \geqslant 2\}$, i.e., all points of X which have in M at least two **elements of best approximation**. $MA(X)$ consists of all points of boundaries of *Voronoi regions* of points of M. The **skeleton** $Skel(X)$ of X is the set of the centers of all balls, in terms of the distance d which are inscribed in X and *maximal*, i.e., not belong to any other such ball. The **cut locus** of X is the closure $\overline{MA(X)}$ of the medial axis. In general, $MA(X) \subset Skel(X) \subset \overline{MA(X)}$. The *medial axis, skeleton* and *cut locus transforms* are point-weighted sets $MA(X)$, $Skel(X)$ and $\overline{MA(X)}$ (the restriction of the **distance transform** on those sets) with $d(x, M)$ being the weight of $x \in X$.

Usually, $X \subset \mathbb{E}^n$, and M is the boundary of X. The medial axis with M being continuous boundary can be considered as a limit of *Voronoi diagram* as the number of the generating points becomes infinite. For 2D binary images X, the skeleton is a curve, a single-pixel thin one, in digital case. The *exoskeleton* of X is the skeleton of the complement of X, i.e., of the background of the image for which X is the foreground.

- **Procrustes distance**

  The *shape* of a *form* (configuration of points in $\mathbb{R}^2$), seen as expression of translation-, rotation- and scale-invariant properties of form, can be represented by a sequence of *landmarks*, i.e., specific points on the form, selected accordingly to some rule. Each landmark point a can be seen as an element $(a', a'') \in \mathbb{R}^2$ or an element $a' + a''i \in \mathbb{C}$.

  Consider two shapes x and y, represented by theirs landmark vectors $(x_1, \dots, x_n)$ and $(y_1, \dots, y_n)$ from $\mathbb{C}^n$. Suppose that x and y are corrected for translation by setting $\sum_t x_t = \sum_t y_t = 0$. Then their **Procrustes distance** is defined by

$$\sqrt{\sum_{t=1}^{n} |x_t - y_t|^2},$$

  where two forms are, first, optimally (by least squares criterion) aligned to correct for scale, and their **Kendall shape distance** is defined by

$$\arccos \sqrt{\frac{(\sum_t x_t \overline{y}_t)(\sum_t y_t \overline{x}_t)}{(\sum_t x_t \overline{x}_t)(\sum_t y_t \overline{y}_t)}},$$

  where $\overline{\alpha} = a' - a''i$ is the *complex conjugate* of $\alpha = a' + a''i$.

- **Tangent distance**

  For any $x \in \mathbb{R}^n$ and a family of *transformations* $t(x, \alpha)$, where $\alpha \in \mathbb{R}^k$ is the vector of k parameters (for example, the scaling factor and rotation angle), the set $M_x = \{t(x, \alpha) \colon \alpha \in \mathbb{R}^k\} \subset \mathbb{R}^n$ is a manifold of dimension at most k. It is a curve if $k = 1$. The minimum Euclidean distance between manifolds M_x and M_y would be useful distance since it is invariant with respect to transformations $t(x, \alpha)$. However, the computation of such distance is too difficult in general; so, M_x is approximated by its *tangent subspace at point* x: $\{x + \sum_{i=1}^{k} \alpha_k x^i : \alpha \in \mathbb{R}^k\} \subset \mathbb{R}^n$, where *tangent vectors* x^i, $1 \leqslant i \leqslant k$, spanning it, are partial derivatives of $t(x, \alpha)$ with respect of α. The **one-sided** (or *directed*) **tangent distance** between elements x and y of $\mathbb{R}^n$ is a quasi-distance d, defined by

$$\sqrt{\min_{\alpha} \left\| x + \sum_{i=1}^{k} \alpha_k x^i - y \right\|^2}.$$

  The Simard–Le Cun–Denker's **tangent distance** is defined by $\min\{d(x, y), d(y, x)\}$.

In general, the *tangent set of a metric space X at a point x* is defined (by Gromov) as any limit point of the family of its dilations, for the dilation parameter going to infinity, taken in the pointed Gromov–Hausdorff topology (cf. **Gromov–Hausdorff distance**).

- **Figure of merit quasi-distance**

  Given two binary images, seen as non-empty finite subsets A and B of a finite metric space (X, d), their Pratt's **figure of merit quasi-distance** is defined by

$$\left(\max\{|A|, |B|\} \sum_{x\in B} \frac{1}{1+\alpha d(x, A)^2}\right)^{-1},$$

  where α is a scaling constant (usually, $\frac{1}{9}$), and $d(x, A) = \min_{y\in A} d(x, y)$ is the **point-set distance**.

  Similar quasi-distances are Peli–Malah's *mean error distance* $\frac{1}{|B|}\sum_{x\in B} d(x, A)$, and *mean square error distance* $\frac{1}{|B|}\sum_{x\in B} d(x, A)^2$.

- **p-th order mean Hausdorff distance**

  Given two binary images, seen as non-empty subsets A and B of a finite metric space (say, a raster of pixels) (X, d), their **p-th order mean Hausdorff distance** is ([Badd92]) a normalized **L_p-Hausdorff distance**, defined by

$$\left(\frac{1}{|X|}\sum_{x\in X} |d(x, A) - d(x, B)|^p\right)^{\frac{1}{p}},$$

  where $d(x, A) = \min_{y\in A} d(x, y)$ is the **point-set distance**. Usual Hausdorff metric is proportional to ∞ order mean Hausdorff distance.

  Venkatasubraminian's **Σ-Hausdorff distance** $d_{dHaus}(A, B) + d_{dHaus}(B, A)$ is equal to $\sum_{x\in A\cup B} |d(x, A) - d(x, B)|$, i.e., it is a version of L_1-Hausdorff distance.

  Another version of 1-st order mean Hausdorff distance is Lindstrom–Turk's *mean geometric error* between two images, seen as surfaces A and B. It is defined by

$$\frac{1}{Area(A) + Area(B)}\left(\int_{x\in A} d(x, B)\, dS + \int_{x\in B} d(x, A)\, dS\right),$$

  where $Area(A)$ denotes the area of surface A. If the images are seen as finite sets A and B, their *mean geometric error* is defined by

$$\frac{1}{|A| + |B|}\left(\sum_{x\in A} d(x, B) + \sum_{x\in B} d(x, A)\right).$$

- **Modified Hausdorff distance**

  Given two binary images, seen as non-empty finite subsets A and B of a finite metric space (X, d), their Dubuisson-Jain's **modified Hausdorff distance** is defined as the maximum of **point-set distances** averaged over A and B:

$$\max\left\{\frac{1}{|A|}\sum_{x\in A} d(x, B), \frac{1}{|B|}\sum_{x\in B} d(x, A)\right\}.$$

- **Partial Hausdorff quasi-distance**

  Given two binary images, seen as non-empty subsets A, B of a finite metric space (X, d), and integers k, l with $1 \leqslant k \leqslant |A|$, $1 \leqslant l \leqslant |B|$, their Huttenlocher–Rucklidge's **partial (k, l)-Hausdorff quasi-distance** is defined by

$$\max\{k^{th}_{x\in A} d(x, B), l^{th}_{x\in B} d(x, A)\},$$

  where $k^{th}_{x\in A} d(x, B)$ means k-th (rather than the largest $|A|$-th ranked one) among $|A|$ distances $d(x, B)$ ranked in increasing order. The case $k = \lfloor\frac{|A|}{2}\rfloor$, $l = \lfloor\frac{|B|}{2}\rfloor$ corresponds to the *modified median Hausdorff quasi-distance*.

- **Bottleneck distance**

  Given two binary images, seen as non-empty subsets A, B with $|A| = |B| = m$, of a metric space (X, d), their **bottleneck distance** is defined by

$$\min_{f}\max_{x\in A} d(x, f(x)),$$

  where f is any bijective mapping between A and B.

  Variations of above distance are:

  1. The **minimum weight matching**: $\min_f \sum_{x\in A} d(x, f(x))$;
  2. The **uniform matching**: $\min_f(\max_{x\in A} d(x, f(x)) - \min_{x\in A} d(x, f(x))$;
  3. The **minimum deviation matching**: $\min_f(\max_{x\in A} d(x, f(x)) - \frac{1}{|A|}\sum_{x\in A} d(x, f(x))$.

  Given an integer t with $1 \leqslant t \leqslant |A|$, the **$t$-bottleneck distance** between A and B ([InVe00]) is above minimum but with f being any mapping from A to B such that $|\{x \in A: f(x) = y\}| \leqslant t$. The cases $t = 1$ and $t = |A|$ correspond, respectively, to the bottleneck distance, and the **directed Hausdorff distance** $d_{dHaus}(A, B) = \max_{x\in A}\min_{y\in B} d(x, y)$.

- **Hausdorff distance up to G**

  Given a group $(G, \cdot, id)$ acting on the Euclidean space $\mathbb{E}^n$, the **Hausdorff distance up to G** between two compact subsets A and B (used in Image Processing) is their **generalized G-Hausdorff distance**, i.e., the minimum of $d_{Haus}(A, g(B))$ over all $g \in G$. Usually, G is the group of all isometries or all translations of $\mathbb{E}^n$.

- **Hyperbolic Hausdorff distance**

  For any compact subset A of $\mathbb{R}^n$, denote by $MAT(A)$ its *Blum's medial axis transform*, i.e., the subset of $X = \mathbb{R}^n \times \mathbb{R}_{\geqslant 0}$, whose elements are all pairs $x = (x', r_x)$ of the centers x' and the radii r_x of the maximal inscribed balls, in terms of the Euclidean distance d_E in $\mathbb{R}^n$. (Cf. **medial axis and skeleton** transforms for the general case.)

  The **hyperbolic Hausdorff distance** ([ChSe00]) is the **Hausdorff metric** on non-empty compact subsets $MAT(A)$ of the metric space (X, d), where the *hyperbolic distance d* on X is defined, for its elements $x = (x', r_x)$ and $y = (y', r_y)$, by

$$\max\{0, d_E(x', y') - (r_y - r_x)\}.$$

- **Non-linear Hausdorff metric**

  Given two compact subsets A and B of a metric space (X, d), their **non-linear Hausdorff metric** (or *Szatmári–Rekeczky–Roska wave distance*) is the **Hausdorff distance** $d_{Haus}(A \cap B, (A \cup B)^*)$, where $(A \cup B)^*$ is the subset of $A \cup B$ which forms a closed contiguous region with $A \cap B$, and the distances between points are allowed to be measured only along paths wholly in $A \cup B$.

- **Video quality metrics**

  Those metrics are between test and reference color video sequences, usually aimed at optimization of encoding/compression/decoding algorithms. Each of them is based on some perceptual model of human vision system, the simplest one being RMSE (root-mean-square error) and PSNR (peak signal-to-noise ratio) error measures. Among others, *threshold metrics* estimate the probability of detecting in video an *artifact* (i.e., a visible distortion that get added to a video signal during digital encoding). Examples are: Sarnoff's JND (just-noticeable differences) metric, Winkler's PDM (perceptual distortion metric), and Watson's DVQ (digital video quality) metric. DVQ is l_p**-metric** between feature vectors representing two video sequences. Some metrics measure special artifacts in the video: the appearance of block structure, blurriness, added "mosquito" noise (ambiguity in the edge direction), texture distortion, etc.

- **Time series video distances**

  The **time series video distances** are objective wavelet-based spatial-temporal **video quality metrics**. A video stream x is processed into time series $x(t)$ (seen as a curve on coordinate plane) which then (piecewise linearly) approximated by a set of n contiguous line segments that can be defined by $n + 1$ endpoints (x_i, x_i'), $0 \leqslant i \leqslant n$, on coordinate plane. In [WoPi99] are given following (cf. **Meehl distance**) distances between video streams x and y:

  1. *Shape*$(x, y) = \sum_{i=0}^{n-1} |(x'_{i+1} - x'_i) - (y'_{i+1} - y'_i)|$;
  2. *Offset*$(x, y) = \sum_{i=0}^{n-1} |\frac{x'_{i+1}+x'_i}{2} - \frac{y'_{i+1}+y'_i}{2}|$.

21.2. AUDIO DISTANCES

Audio (speech, music, etc.) *Signal Processing* is the processing of analog (continuous) or, mainly, digital representation of the air pressure waveform of the sound. A *sound spectrogram* (or *sonogram*) is a visual three-dimensional representation of an acoustic signal. It is obtained either by series of bandpass filters (an analog processing), or by application of the *short-time Fourier transform* to the electronic analog of an acoustic wave. Three axes represent time, frequency and *intensity* (acoustic energy). Often this three-dimensional curve is reduced to two dimensions by indicating the intensity with more thick lines or more intense gray or color values.

Sound is called *tone* if it is periodic (the lowest *fundamental* frequency plus its multiples, *harmonics* or *overtones*) and *noise*, otherwise. The frequency is measured in *cps* (the number of complete cycles per second) or Hz (hertz). The range of audible sound frequencies to humans is typically 20 Hz – 20 kHz.

Signal's *power* $P(f)$ is energy per unit of time; it is proportional to the square of signal's amplitude $A(f)$. *Decibel* dB is the unit used to express relative strength of two signals. One tenth of 1 dB is *bel*, the original outdated unit. Audio signal's amplitude in dB is $20\log_{10}\frac{A(f)}{A(f')} = 10\log_{10}\frac{P(f)}{P(f')}$, where f' ia a reference signal selected to correspond 0 dB (usually, the threshold of human hearing). The threshold of pain is about 120–140 dB.

Pitch and *loudness* are auditory subjective terms for frequency and amplitude.

Mel scale is a perceptual frequency scale, corresponding to the auditory sensation of tone height and based on *mel*, a unit of perceived frequency (pitch). It is connected to the acoustic frequency f hertz scale by $Mel(f) = 1127\ln(1+\frac{f}{700})$ (or, simply, $Mel(f) = 1000\log_2(1+\frac{f}{1000})$) so that 1000 Hz correspond to 1000 mels.

Bark scale (named after Barkhausen) is a psycho-acoustic scale of perceived intensity (loudness): it range from 1 to 24 corresponding to the first 24 critical bands of hearing (0, 100, 200, …, 1270, 1480, 1720, …, 950, 12000, 15500 Hz). Those bands correspond to spatial regions of the basilar membrane (of the inner ear), where oscillations, produced by the sound of given frequency, activate the hair cells and neurons. Bark scale is connected to the acoustic frequency f kilohertz scale by $Bark(f) = 13\arctan(0.76f) + 3.5\arctan(\frac{f}{0.75})^2$.

The main way humans control their *phonation* (speech, song, laughter) is by control over the *vocal tract* (the throat and mouth) shape. This shape, i.e., the cross-sectional profile of the tube from the closure in the *glottis* (the space between the vocal cords) to the opening (lips), is represented by the cross-sectional area function $Area(x)$, where x is the distance to glottis. The vocal tract acts as a resonator during vowel phonation, because it is kept relatively open. Those resonances reinforce the source sound (ongoing flow of lung air) at particular *resonant frequencies* (or *formants*) of the vocal tract, producing peaks in the *spectrum* of the sound. Each vowel has two characteristic formants, depending of the vertical and horizontal position of the tongue in the mouth. Source sound function is modified by frequency response function for a given area function. If the vocal tract is approximated as a sequence of concatenated tubes of constant cross-sectional area (of equal length, or epilarynx-pharynx-oral cavity, etc.), then *area ratio coefficients* are the ratios $\frac{Area(x_{i+1})}{Area(x_i)}$ for consecutive tubes; those coefficients can be computed by LPC (see below).

The *spectrum* of a sound is the distribution of magnitude (dB) (and sometimes the phases) in frequency (kHz) of the components of the wave. The *spectral envelope* is a smooth contour that connects the spectral peaks. The estimation of the spectral envelopes is based on either LPC (linear predictive coding), or FTT (fast Fourier transform using real cepstrum, i.e., the log amplitude spectrum of the sound).

FT (Fourier transform) maps time-domain functions into frequency-domain representations. The *cepstrum* of the signal $f(t)$ is $FT(\ln(FT(f(t) + 2\pi mi)))$, where m is the integer needed to unwrap the angle or imaginary part of the complex logarithm function. The complex and real cepstrum use, respectively, complex and real log function. The real cepstrum uses only the magnitude of the original signal $f(t)$, while the complex cepstrum uses also phase of $f(t)$. FFT method is based on linear spectral analysis. FFT performs Fourier transform on the signal and sample the discrete transform output at the desired frequencies usually in the *mel* scale.

Parameter-based distances used in recognition and processing of speech data are usually derived by LPC, modeling speech spectrum as a linear combination of the previous samples (as in autoregressive process). Roughly, LPC process each word of the speech signal in the following 6 steps: filtering, energy normalization, partition into frames, *windowing* (to minimize discontinuities at the borders of frames), obtaining LPC parameters by the autocorrelation method and conversion to the *LPC-derived cepstral coefficients*. LPC assumes that speech is produced by a buzzer at the glottis (with occasionally added hissing and popping sounds), and it removes the formants by filtering.

Majority of distortion measures between sonograms are variations of **squared Euclidean distance** (including covariance-weighted one, i.e., **Mahalanobis**, distance) and probabilistic distances belonging to following general types: generalized **total variation metric**, **f-divergence of Csizar** and **Chernoff distance**.

The distances for sound processing below are between vectors x and y representing two signals to compare. For recognition, they are a template reference and input signal, while for noise reduction, they are original (reference) and distorted signal (see, for example, [OASM03]). Often distances are calculated for small segments, between vectors representing short-time spectra, and then averaged.

- **Segmented signal-to-noise ratio**

The **segmented signal-to-noise ratio** $SNR_{seg}(x, y)$ between signals $x = (x_i)$ and $y = (y_i)$ is defined by

$$\frac{10}{m}\sum_{m=0}^{M-1}\left(\log_{10}\sum_{i=nm+1}^{nm+n}\frac{x_i^2}{(x_i - y_i)^2}\right),$$

where n is the number of frames, and M is the number of segments.

Usual *signal-to-noise ratio* $SNR(x, y)$ between x and y is given by

$$10\log_{10}\frac{\sum_{i=1}^{n} x_i^2}{\sum_{i=1}^{n}(x_i - y_i)^2}.$$

Another measure, used to compare two waveforms x and y in time-domain, is their **Czekanovski–Dice distance**, defined by

$$\frac{1}{n}\sum_{i=1}^{n}\left(1-\frac{2\min\{x_i, y_i\}}{x_i+y_i}\right).$$

- **Spectral magnitude-phase distortion**

  The **spectral magnitude-phase distortion** between signals $x = x(\omega)$ and $y = y(\omega)$ is defined by

$$\frac{1}{n}\left(\lambda\sum_{i=1}^{n}\big(|x(w)|-|y(w)|\big)^2+(1-\lambda)\sum_{i=1}^{n}\big(\angle x(w)-\angle y(w)\big)^2\right),$$

  where $|x(w)|$, $|y(w)|$ are magnitude spectra, and $\angle x(w)$, $\angle y(w)$ are phase spectra of x and y, respectively, while parameter λ, $0 \leqslant \lambda \leqslant 1$, is chosen in order to attach commensurate weights to the magnitude and phase terms. The case $\lambda = 0$ corresponds to the **spectral phase distance**.

  Given a signal $f(t) = ae^{-bt}u(t)$, $a, b > 0$, which has Fourier transform $x(w) = \frac{a}{b+iw}$, its *magnitude* (or *amplitude*) spectrum is $|x| = \frac{a}{\sqrt{b^2+w^2}}$, and its *phase* spectrum (in radians) is $\alpha(x) = \tan^{-1}\frac{w}{b}$, i.e., $x(w) = |x|e^{i\alpha} = |x|(\cos\alpha + i\sin\alpha)$.

- **RMS log spectral distance**

  The **RMS log spectral distance** (or *root-mean-square distance*) $LSD(x, y)$ between discrete spectra $x = (x_i)$ and $y = (y_i)$ is the following Euclidean distance:

$$\sqrt{\frac{1}{n}\sum_{i=1}^{n}(\ln x_i - \ln y_i)^2}.$$

  The square of RMS log spectral distance, via cepstrum representation $\ln x(\omega) = \sum_{j=-\infty}^{\infty} c_j e^{-ij\omega}$ (where $x(\omega)$ is the power spectrum, i.e., magnitude-squared Fourier transform) became, in complex cepstral space, the **cepstral distance**.

  The **log area ratio distance** $LAR(x, y)$ between x and y is defined by

$$\sqrt{\frac{1}{n}\sum_{i=1}^{n}10\big(\log_{10}Area(x_i)-\log_{10}Area(y_i)\big)^2},$$

  where $Area(z_i)$ means cross-sectional area of the segment of the vocal tract tube corresponding to z_i.

- **Bark spectral distance**

The **Bark spectral distance** is a perceptual distance, defined by

$$BSD(x, y) = \sum_{i=1}^{n} (x_i - y_i)^2,$$

i.e., is the **squared Euclidean distance** between *Bark spectra* (x_i) and (y_i) of x and y, where i-th component corresponds to i-th auditory critical band in Bark scale.

A modification of Bark spectral distance excludes critical bands i on which the loudness distortion $|x_i - y_i|$ is less than the noise masking threshold.

- **Itakura–Saito quasi-distance**

The **Itakura–Saito quasi-distance** (or *maximum likelihood distance*) $IS(x, y)$ between LPC-derived spectral envelopes $x = x(\omega)$ and $y = y(\omega)$ is defined by

$$\frac{1}{2\pi}\int_{-\pi}^{\pi}\left(\ln\frac{x(w)}{y(w)} + \frac{y(w)}{x(w)} - 1\right)dw.$$

The **cosh distance** is defined by $IS(x, y) + IS(y, x)$, i.e., is equal to

$$\frac{1}{2\pi}\int_{-\pi}^{\pi}\left(\frac{x(w)}{y(w)} + \frac{y(w)}{x(w)} - 2\right)dw = \frac{1}{2\pi}\int_{-\pi}^{\pi} 2\cosh\left(\ln\frac{x(w)}{y(w)} - 1\right)dw,$$

where $\cosh(t) = \frac{e^t + e^{-t}}{2}$ is the hyperbolic cosine function.

- **Log likelihood ratio quasi-distance**

The **log likelihood ratio quasi-distance** (or **Kullback–Leibler distance**) $KL(x, y)$ between LPC-derived spectral envelopes $x = x(\omega)$ and $y = y(\omega)$ is defined by

$$\frac{1}{2\pi}\int_{-\pi}^{\pi} x(w)\ln\frac{x(w)}{y(w)}dw.$$

The **Jeffrey divergence** $KL(x, y) + KL(y, x)$ is also used.

The **weighted likelihood ratio distance** between spectral envelopes $x = x(\omega)$ and $y = y(\omega)$ is defined by

$$\frac{1}{2\pi}\int_{-\pi}^{\pi}\left(\frac{\left(\ln\left(\frac{x(w)}{y(w)}\right) + \frac{y(w)}{x(w)} - 1\right)x(w)}{p_x} + \frac{\left(\ln\left(\frac{y(w)}{x(w)}\right) + \frac{x(w)}{y(w)} - 1\right)y(w)}{p_y}\right)dw,$$

where $P(x)$ and $P(y)$ denote the power of the spectra $x(w)$ and $y(w)$, respectively.

- **Cepstral distance**

The **cepstral distance** (or *squared Euclidean cepstrum metric*) $CEP(x, y)$ between

LPC-derived spectral envelopes $x = x(\omega)$ and $y = y(\omega)$ is defined by

$$\frac{1}{2\pi}\int_{-\pi}^{\pi}\left(\ln\frac{x(w)}{y(w)}\right)^2 dw = \frac{1}{2\pi}\int_{-\pi}^{\pi}\left(\ln x(w) - \ln y(w)\right)^2 dw = \sum_{j=-\infty}^{\infty}\left(c_j(x) - c_j(y)\right)^2,$$

where $c_j(z) = \frac{1}{2\pi}\int_{-\pi}^{\pi} e^{iwj}\ln|z(w)|\,dw$ is j-th cepstral (real) coefficient of z derived by Fourier transform or LPC.

- **Quefrency-weighted cepstral distance**

  The **quefrency-weighted cepstral distance** (or *weighted slope distance*) between x and y is defined by

  $$\sum_{i=-\infty}^{\infty} i^2\left(c_i(x) - c_i(y)\right)^2.$$

  "Quefrency" and "cepstrum" are anagrams of "frequency" and "spectrum", respectively.

  The **Martin cepstrum distance** between two AR (autoregressive) models is defined, in terms of their cepstrums, by

  $$\sqrt{\sum_{i=0}^{\infty} i\left(c_i(x) - c_i(y)\right)^2}.$$

  (Cf. general **Martin distance**, defined as an **angle distance between subspaces**, and **Martin metric** between strings which is an l_∞-analog of it.)

  The **Klatt slope metric** between discrete spectra $x = (x_i)$ and $y = (y_i)$ with n channel filters is defined by

  $$\sqrt{\sum_{i=1}^{n}\left((x_{i+1} - x_i) - (y_{i+1} - y_i)\right)^2}.$$

- **Phone distances**

  A *phone* is a sound segment that possess distinct acoustic properties, the basis sound unit. Cf. *phoneme*, i.e., a family of phones that speakers usually hear as a single sound; the number of phonemes range, among about 6000 languages spoken now, from 11 in Rotokas to 112 in !Xóõ (languages spoken by about 4000 people in Papua New Guinea and Botswana, respectively).

  Two main classes of **phone distance** (distances between two phones x and y) are:

  1. *Spectrogram-based distances* which are physical-acoustic distortion measures between the sound spectrograms of x and y;
  2. *Feature-based phone distances* which are usually **Manhattan distance** $\sum_i |x_i - y_i|$ between vectors (x_i) and (y_i) representing phones x and y with respect to given inventory of phonetic features (for example, nasality, stricture, palatalization, rounding, sillability).

- **Phonetic word distance**

The **phonetic word distance** between two words x and y is the following cost-based **editing metric** (i.e., the minimal cost of transforming x into y by substitution, deletion and insertion of phones). A word is seen as a string of phones. Given a phone distance function $r(u, v)$ on the International Phonetic Alphabet with additional phone 0 (the silence), the cost of substitution of phone u by v is $r(u, v)$, while $r(u, 0)$ is the cost of insertion or deletion of u. (Cf. distances for protein data based on **Dayhoff distance** on the set of 20 amino acids.)

- **Linguistic distance**

In Computational Linguistics, the **linguistic distance** (or **dialectology distance**) between language varieties X and Y is the mean, for fixed sample S of notions, **phonetic word distance** between *cognate* (i.e., having the same meaning) words s_X and s_Y, representing the same notion $s \in S$ in X and Y, respectively.

Stover distance (see http://sakla.net/concordances/index.html) between phrases with the same key word is the sum $\sum_{-n \leqslant i \leqslant +n} a_i x_i$, where $0 < a_i < 1$ and x_i is the proportion of non-mathched words between the phrases within a moving window. Phrases are first aligned, by the common key word, to compare the uses of it in context; also, the rarest words are replaced with a common pseudo-token.

- **Acoustics distances**

The **wavelength** is the distance the sound wave travels to complete one cycle. This distance is measured perpendicular to the wavefront in the direction of propagation between one peak of a sine wave and the next corresponding peak. The wavelength of any frequency may be found by dividing the speed of sound (331.4 m/s at sea level) in the medium by the fundamental frequency.

The **far field** is the part of a sound field in which sound waves can be considered planar and sound pressure decreases inversely with distance from the source. It corresponds to a reduction of about 6 dB in sound level for each doubling of distance.

The **near field** is the part of a sound field (usually within about two wavelengths from the source) where there is no simple relationship between sound level and distance.

The **proximity effect** is the anomaly of low frequencies being enhanced when a directional microphone is very close to the source.

The **critical distance** is the distance from the sound source at which the direct sound (produced by the sound source) and reverberant sound (produced by the direct sound bouncing off the walls, floor, etc.) are equal in intensity level.

The **blanking distance** is the minimum sensing range of an ultrasonic **proximity sensor**.

The **acoustic metric** is the term used occasionally for some distances between vowels; for example, Euclidean distance between vectors of formant frequencies of pronounced and intended vowel. (Not to be confused with **acoustic metrics** in General Relativity and Quantum Gravity.)

Chapter 22

Distances in Internet and Similar Networks

22.1. SCALE-FREE NETWORKS

A **network** is a graph, directed or undirected, with a positive number (weight) assigned to each of its arcs or edges. Real-world complex networks usually have a gigantic number N of vertices and are sparse, i.e., with relatively few edges.

They tend to be **small-world** ([Watt99]), i.e., interpolate between regular geometric lattices and random graphs in the following sense: they have large clustering coefficient (as lattices in local neighborhood), while average path distance between two vertices is small, about $\ln N$, as in a random graph.

The main subcase of a small-world network is a **scale-free network** ([Bara01]) in which the probability for a vertex to have degree k is similar to $k^{-\gamma}$ for some positive constant γ which usually belongs to the segment [2, 3]. This *power law* implies that very few vertices, called *hubs* (connectors, super-spreaders), are far more connected than other vertices. The power law (or **long range dependent**, *heavy-tail*) distributions, in space or time, were observed in many natural phenomena (both, physical and sociological).

- **Collaboration distance**

  The **collaboration distance** is the **path metric** (see http://www.ams.org/msnmain/cgd/) of the *Collaboration graph*, having about 0.4 million vertices (authors in Mathematical Reviews database) with xy being an edge if authors x and y have a joint publication among about 2 million papers itemized in this database. The vertex of largest degree, 1416, corresponds to Paul Erdős; the *Erdős number* of a mathematician is his collaboration distance to Paul Erdős.

  The **Barr's collaboration metric** (http://www.oakland.edu/enp/barr.pdf) is the **resistance distance** in the following extension of the Collaboration graph. First, put 1-ohm resistor between any two authors for every joint 2-authors paper. Then, for each n-authors paper, $n > 2$, add new vertex and connect it by $\frac{n}{4}$-ohm resistor to each of its co-authors.

- **Co-starring distance**

  The **co-starring distance** is the **path metric** of the *Hollywood graph*, having about 250000 vertices (actors in the Internet Movie database) with xy being an edge if the actors x and y appeared in a feature film together. The vertices of largest degree are Christofer Lee and Kevin Bacon; the trivia game *Six degrees of Kevin Bacon* uses the *Bacon number*, i.e., the co-starring distance to this actor.

Similar popular examples of such social scale-free networks are graphs of musicians (who played in the same rock band), baseball players (as team-mates), scientific publications (who cite each other), chess-players (who played each other), acquaintances among classmates in a college, business board membership, sexual contacts among members of a given group. The path metric of the last network is called **sexual distance**. Among other studied scale-free networks are air travel connections, word co-occurrences in human language, power grid of Western US, network of neurons of a worm, protein interaction networks and metabolic networks (with two substrates forming an edge if a reaction occurs between them via enzymes).

- **Forward quasi-distance**

  In a directed network, where edge-weights correspond to a point in time, the **forward quasi-distance** (**backward quasi-distance**) is the length of shortest directed path, but only among paths on which consecutive edge-weights are increasing (decreasing, respectively). The forward quasi-distance is useful in epidemiological networks (disease spreading by contact, or, say, heresy spreading within a church), while backward quasi-distance is appropriated in peer-to-peer file-sharing networks.

- **Betweenness centrality**

  For a **geodesic** metric space (X, d) (in particular, for the **path metric** of a graph), the **betweenness centrality** of a point $x \in X$ is defined by

$$g(x) = \sum_{y,z \in X} \frac{\text{Number of shortest } (y - z) \text{ paths through } x}{\text{Number of shortest } (y - z) \text{ paths}},$$

  and the **distance-mass function** is a function $M : \mathbb{R}_+ \to \mathbb{Q}$, defined by

$$M(a) = \frac{|\{y \in X:\ d(x, y) + d(y, z) = a \text{ for some } x, y \in X\}|}{|\{(x, z) \in X \times X:\ d(x, z) = a\}|}.$$

  It was conjectured in [GOJKK02] that many scale-free networks satisfy to power law $g^{-\gamma}$ (for the probability, for a vertex, to have betweenness centrality g), where γ is either 2, or ≈ 2.2 with distance-mass function $M(a)$ being either linear, or non-linear, respectively. In the linear case, for example, $\frac{M(a)}{a} \approx 4.5$ for the **Internet AS metric**, and ≈ 1 for the **Web hyperlink quasi-metric**.

- **Drift distance**

  The **drift distance** is the absolute value of the difference between observed and actual coordinates of a node in a NVE (Networked Virtual Environment). In models of such large-scale peer-to-peer NVE (for example, Massively Multiplayer Online Games), the users are represented as coordinate points on the plane (*nodes*) which can move at discrete *time-steps*, and each have a visibility range called *Area of Interest*. NVE creates a synthetic 3D world where each user assumes *avatar* (a virtual identity) to interact with other users or computer AI.

The term **drift distance** is also used for the current going through a material, in tire production, etc.

- **Semantic proximity**

  For the words in a document, there are short range syntactic relations and long range semantic correlations. The main document networks are Web and bibliographic databases (digital libraries, scientific databases, etc.); the documents in them are related by, respectively, hyperlinks and citation or collaboration.

  Also, some semantic tags (keywords) can be attached to the documents in order to index (classify) them: terms selected by author, title words, journal titles, etc.

  The **semantic proximity** between two keywords x and y is their **Tanimoto similarity** $\frac{|X \cap Y|}{|X \cup Y|}$, where X and Y are the sets of documents indexed by x and y, respectively. Their **keyword distance** is defined by $\frac{|X \Delta Y|}{|X \cap Y|}$; it is not a metric.

22.2. NETWORK-BASED SEMANTIC DISTANCES

Among main lexical networks (such as WordNet, Medical Search Headings, Roget's Thesaurus, Longman's Dictionary of Contemporary English) WordNet is the most popular lexical resource used in Natural Language Processing and Computational Linguistics. WordNet (see http://wordnet.princeton.edu) is an on-line lexical database in which English nouns, verbs, adjectives and adverbs are organized into *synsets* (synonym sets), each representing one underlying lexical concept. Two synsets can be linked semantically by one of following links: upwards x (*hyponym*) *IS-A* y (*hypernym*) link, downwards x (*meronym*) *CONTAINS* y (*holonym*) link, or a horizontal link expressing frequent co-occurrence (*antonymy*, etc.). *IS-A* links induce a partial order, called *IS-A* taxonomy. The version 2.0 of WordNet has 80000 noun concepts and 13500 verb concepts, organized in 9 and 554 separate *IS-A* hierarchies, respectively. In the resulting directed acyclic graph of concepts, for any two synsets (or concepts) x and y, let $l(x, y)$ denotes the length of shortest path between them, using only *IS-A* links, and let $LPS(x, y)$ denotes their *least common subsumer* (ancestor) by *IS-A* taxonomy. Let $d(x)$ denote the *depth* of x (i.e., its distance from the root in *IS-A* taxonomy) and let $D = \max_x d(x)$. The list of main related semantic similarities and distances follows.

- **Path similarity**

  The **path similarity** between synsets x and y is defined by

$$path(x, y) = \big(l(x, y)\big)^{-1}.$$

- **Leacock–Chodorow similarity**

  The **Leacock–Chodorow similarity** between synsets x and y is defined by

$$lch(x, y) = -\ln \frac{l(x, y)}{2D},$$

and the **conceptual distance** between them is defined by $\frac{l(x,y)}{D}$.

- **Wu–Palmer similarity**

  The **Wu–Palmer similarity** between synsets x and y is defined by

$$wup(x, y) = \frac{2d(LPS(x, y))}{d(x) + d(y)}.$$

- **Resnik similarity**

  The **Resnik similarity** between synsets x and y is defined by

$$res(x, y) = -\ln p\big(LPS(x, y)\big),$$

  where $p(z)$ is the probability of encountering an instance of concept z in a large corpus, and $-\ln p(z)$ is called *information content* of z.

- **Lin similarity**

  The **Lin similarity** between synsets x and y is defined by

$$lin(x, y) = \frac{2\ln p\big(LPS(x, y)\big)}{\ln p(x) + \ln p(y)}.$$

- **Jiang–Conrath distance**

  The **Jiang–Conrath distance** between synsets x and y is defined by

$$jcn(x, y) = 2\ln p\big(LPS(x, y)\big) - \big(\ln p(x) + \ln p(y)\big).$$

- **Lesk similarities**

  A *gloss* of a synonym set z is the member of this set giving a definition or explanation of underlying concept. The **Lesk similarities** are those defined by a function of overlap of glosses of corresponding concepts; for example, the **gloss overlap** is

$$\frac{2t(x, y)}{t(x) + t(y)},$$

  where $t(z)$ is the number of words in the synset z, and $t(x, y)$ is the number of common words in x and y.

- **Hirst–St-Onge similarity**

  The **Hirst–St-Onge similarity** between synsets x and y is defined by

$$hso(x, y) = C - L(x, y) - ck,$$

  where $L(x, y)$ is the length of a shortest path between x and y using all links, k is the number of changes of direction in that path, and C, c are constants.

The **Hirst–St-Onge distance** is defined by $\frac{L(x,y)}{k}$.

22.3. DISTANCES IN INTERNET AND WEB

Let us consider in detail the graphs of Web and of its hardware substrate, Internet, which are small-world and scale-free.

The *Internet* is a publicly available worldwide computer network which came from ARPANET (started in 1969 by US Department of Defense), NSFNet, Usenet, Bitnet, and other networks. In 1995, the National Science Foundation in the US gave up the stewardship of the Internet.

Its nodes are *routers*, i.e., devices that forward packets of data along networks from one computer to another, using IP (Internet Protocol relating names and numbers), TCP and UDP (for sending data), and (build on top of them) HTTP, Telnet, FTP and many other *protocols* (i.e., technical specifications of data transfer). Routers are located at *gateways*, i.e., at the places where at least two networks connect. The links that join the nodes together are various physical connectors, such as telephone wires, optical cables and satellite networks. Internet use *packet switching*, i.e., data (fragmented if needed) are forwarded not along a previously established path, but so as to optimize the use of available *bandwidth* (bit rate, in million bits per second) and minimize the *latency* (the time, in milliseconds, needed for a request to arrive).

Each computer linked to the Internet is given usually an unique "address", called its *IP address*. The number of possible IP addresses is $2^{32} \approx 4.3$ billion only. The most popular applications supported by the Internet are e-mail, file transfer, Web, and some multimedia.

The *Internet IP graph* has, as the vertex-set, the IP addresses of all computers linked to Internet; two vertices are adjacent if a router connects them directly, i.e., the passing datagram makes only one *hop*.

Internet also can be partitioned into ASs (administratively Autonomous Systems or domains). Within each AS the intra-domain routing is done by IGP (Interior Gateway Protocol), while inter-domain routing is done by BGP (Border Gateway Protocol) which assigns an ASN (16-bit number) to each AS. The *Internet AS graph* has ASs as vertices and edges represent the existence of a BGP peer connection between corresponding ASs.

The *World Wide Web* (*WWW* or *Web*, for short) is a major part of Internet content consisting of interconnected documents (resources). It corresponds to HTTP (Hyper Text Transfer Protocol) between browser and server, HTML (Hyper Text Markup Language) of encoding information for a display, and URLs (Uniform Resource Locators), giving unique "address" to web pages. The Web was started in 1989 in CERN which gave it for public use in 1993.

The *Web digraph* is a virtual network, the nodes of which are *documents* (i.e., static HTML pages or their URLs) which are connected by incoming or outcoming *hyperlinks*, i.e., hypertext links.

The number of nodes in the Web digraph was about 10 billion at 2005, and new pages are created at the rate of 7.3 million per day. Moreover, besides it lies the *Deep* or *Invisible* Web, i.e., searchable databases with number of pages (if not actual content) being about 500 times more than on static web pages. Those pages are not indexed by search engines;

they have dynamic URL and so, can be retrieved only by a direct query in real time. About 56%, 8%, 6% and 5% of web pages are in English, German, French and Japanese, respectively. In 2005, about 0.82 billion, i.e., 13% of the global population, were online. On the other hand, in the first 6 months of 2005 spam accounted for 61% of all e-mail traffic and 10.866 new Windows viruses and worms were detected.

There are several hundred thousand *cyber-communities*, i.e., clusters of nodes of the Web digraph, where link density is greater among members than between members and the rest. The cyber-communities (a customer group, a social network, a concept in a technical paper, etc.) are usually focused around a definite topic and contain a bipartite *hubs-authorities* subgraph, where all hubs (guides and resource lists) point to all authorities (useful and relevant pages on the topic). Examples of new media, created by Web: *(we)blogs* (digital diaries posted on Web), Wikipedia (the collaborative encyclopedia) and (in the project Semantic Web by WWW Consortium) linking to metadata.

In the average, nodes of the Web digraph are of size 10 Kilobytes, out-degree 7.2, and probability k^{-2} to have out-degree or in-degree k. A study in [BKMR00] of over 200 million web pages gave, approximatively, the largest connected component "core" of 56 million pages, with other 44 million of pages, connected to the core (newcomers?), 44 million to which the giant core is connected (corporations?) and 44 million connected to the core only by directed path. For randomly chosen nodes x and y, the probability of existence of directed path from x to y was 0.25 and the average length of such shortest path (if it exists) was 16, while maximal length of shortest path was over 28 in the core and over 500 in the whole digraph.

Distances below are examples of host-to-host **routing metrics**, i.e., values used by routing algorithms in the Internet, in order to compare possible routes. Examples of other such measures are: bandwidth consumption, communication cost, reliability (probability of packet loss).

- **Internet IP metric**

  The **Internet IP metric** (or *hop count*, *RIP metric*, *IP path length*) is the **path metric** in the *Internet IP graph*, i.e., the minimal number of hops (or, equivalently, routers, represented by their IP addresses) needed to forward a packet of data. RIP imposes a maximum distance of 15 and advertises by 16 non-reachable routes.

- **Internet AS metric**

  The **Internet AS metric** (or *BGP-metric*) is the **path metric** in the *Internet AS graph*, i.e., the minimal number of ISPs (Independent Service Providers), represented by their ASs, needed to forward a packet of data.

- **Geographic distance**

  The **geographic distance** is the **great circle distance** on the Earth from client x (destination) to the server y (source). However, for economical reasons, the data often do not follow such geodesics; for example, most data from Japan to Europe transit via US.

- **RTT-distance**

The **RTT-distance** is the RTT (Round Trip Time) of transmission between x and y, measured (in milliseconds) during the previous day; see [HFPMC02] for variations of this metric and connections with above three metrics.

- **Administrative cost distance**

The **administrative cost distance** is the nominal number (rating the trustworthiness of a routing information), assigned by the network to the route between x and y. For example, Cisco assigns values 0, 1, ..., 200, 255 for Connected Interface, Static Route, ..., Internal BGP, Unknown, respectively.

- **DRP-metrics**

DD (Distributed Director) system of Cisco use (with priorities and weights) the **administrative cost distance**, the **random metric** (selecting a random number for each IP address) and the **DRP** (Direct Response Protocol) metrics. DRP-metrics ask from all DRP-associated routers one of the following distances:

1. The **DRP-external metric**, i.e., the number of BGP (Border Gateway Protocol) hops between the client requesting service and the DRP server agent;
2. The **DRP-internal metric**, i.e., the number of IGP hops between the DRP server agent and the closest border router at the edge of the autonomous system;
3. The **DRP-server metric**, i.e., the number of IGP hops between the DRP server agent and the associated server.

- **Web hyperlink quasi-metric**

The **Web hyperlink quasi-metric** (or *click count*) is the length of the shortest directed path (if it exists) between two web pages (vertices in the Web digraph), i.e., the minimal number of needed mouse-clicks in this digraph.

- **Average-clicks Web quasi-distance**

The **average-clicks Web quasi-distance** between two web pages x and y in the Web digraph ([YOI03]) is the minimum $\sum_{i=1}^{m} \ln p\frac{z_i^+}{\alpha}$ over all directed paths $x = z_0, z_1, \dots, z_m = y$ connecting x and y, where z_i^+ is the out-degree of the page z_i. The parameter α is 1 or 0.85, while p (the average out-degree) is 7 or 6.

- **Dodge–Shiode WebX quasi-distance**

The **Dodge–Shiode WebX quasi-distance** between two web pages x and y of the Web digraph is the number $\frac{1}{h(x,y)}$, where $h(x, y)$ is the number of shortest directed paths connecting x and y.

- **Web similarity metrics**

Web similarity metrics form a family of indicators used to quantify the extent of relatedness (in content, links or/and usage) between two web pages x and y. For example: topical resemblance in overlap terms, *co-citation* (the number of pages, where both are given as hyperlinks), *bibliographical coupling* (the number of hyperlinks in common) and *co-occurrence frequency* $\min\{P(x|y), P(y|x)\}$, where $P(x|y)$ is the probability that a visitor of the page y will visit the page x.

In particular, **search-centric change metrics** are metrics used by search engines on Web, in order to measure the degree of change between two versions x and y of a web page. If X and Y are the set of all words (excluding HTML markup) in x and y, respectively, then the **word page distance** is the **Dice distance**

$$\frac{|X \Delta Y|}{|X| + |Y|} = 1 - \frac{2|X \cup Y|}{|X| + |Y|}.$$

If v_x and v_y are **TF-IDF** (Frequency – Inverse Document Frequency) weighted vector representations of x and y, then their **cosine page distance** is given by

$$1 - \frac{\langle v_x, v_y \rangle}{\|v_x\|_2 \cdot \|v_y\|_2}.$$

- **Network tomography metrics**

  Consider a network with fixed routing protocol, i.e., a *strongly connected* digraph $D = (V, E)$ with unique directed path $T(u, v)$ selected for any pair (u, v) of vertices. The routing protocol is described by binary *routing matrix* $A = ((a_{ij}))$, where $a_{ij} = 1$ if the arc $e \in E$, indexed i, belongs to the directed path $T(u, v)$, indexed j. The **Hamming distance** between two rows (columns) of A is called **distance between corresponding arcs** (directed paths) of the network.

  Consider two networks with the same digraph, but different routing protocols with routing matrices A and A', respectively. Then a **routing protocol semi-metric** ([Var04]) is the smallest Hamming distance between A and a matrix B, obtained from A' by permutations of rows and columns (both matrices are seen as strings).

Part VI

Distances in Natural Sciences

Chapter 23. **Distances in Biology**
Chapter 24. **Distances in Physics and Chemistry**
Chapter 25. **Distances in Geography, Geophysics, and Astronomy**
Chapter 26. **Distances in Cosmology and Theory of Relativity**

Chapter 23

Distances in Biology

The distances are mainly used in *Biology* in order to reconstruct the evolutionary history of organisms in the form of phylogenetic trees. In the classical approach those distances were based on the comparative morphology and physiology. The modern *Molecular Biology* compares DNA/protein sequences between organisms.

DNA is a sequence of *nucleotides* (or *nuclei acids*) A, T, G and C, and it can be seen as a word over this alphabet of 4 letters. The nucleotides A, G (short for adenine a guanine) are called *purines*, while T, C (short for thymine and cytosine) are called *pyrimidines* (in RNA, it is uracil U instead of T). Two strands of DNA are held together (in the form of a double helix) by weak hydrogen bonds between corresponding nucleotides (necessarily, a purine and a pyrimidine) in the strands alignment. Those pairs are called *base pairs*.

A *transition mutation* is a substitution of a base pair, so that a purine/pyrimidine is replaced by another purine/pyrimidine; for example, GC is replaced by AT. A *transversion mutation* is a substitution of a base pair, so that a purine/pyrimidine is replaced by a pyrimidine/purine base pair, or vice versa; for example, GC is replaced by TA.

DNA molecules occur (in the nuclei of eukaryote cells) in the form of long strings, called *chromosomes*. Most human cells contain 23 pairs of chromosomes, one set of 23 from each parent; human *gamete* (sperm or egg) is a *haploid*, i.e., contains only one set of 23 chromosomes. The (normal) males and females differ only in 23rd pair of chromosomes: XY for males, and XX for female.

A *gene* is a contiguous stretch of DNA which encodes (via transcription to RNA and then, translation) a protein or an RNA molecule. The location of a gene on its specific chromosome is called *gene locus*. Different versions (states) of a gene are called its *alleles*. Only less than 2% of human DNA are in genes; the functions, if any, of the remainder are unknown.

A *protein* is a large molecule which is a chain of *amino acids*; among them are hormones, catalysts (enzymes), antibodies, etc. There are twenty amino acids; the three-dimensional shape of a protein is defined by the (linear) sequence of amino acids, i.e., by a word in this alphabet in 20 letters.

The *genetic code* is universal to (almost) all organisms correspondence between some *codons* (i.e., ordered triples of nucleotides) and 20 amino acids. It express the *genotype* (information contained in genes, i.e., in DNA) as the *phenotype* (proteins). Three *stop codons* (UAA, UAG, and UGA) signify the end of a protein; any two, among 61 remaining codons, are called *synonymous* if they correspond to the same amino acids.

A *genome* is entire genetic constitution of a species or of a living organism. For example, the human genome is the set of 23 chromosomes consisting of about 3000 million base pairs of DNA and organized into about 20000–25000 genes.

IAM (for infinite-alleles model of evolution) assumes that an allele can change from any given state into any other given state. It corresponds to primary role for *genetic drift* (i.e., random variation in gene frequencies from one generation to another); especially, in small populations over *natural selection* (stepwise mutations). IAM is convenient for allozyme data (*allozyme* is a form of a protein which is encoded by one allele at a specific gene locus). *SMM* (for step-wise mutation model of evolution) is more convenient for (recently, most popular) micro-satellite data. *Micro-satellites* are highly variable repeating short sequences of DNA; their mutation rate is 1 per 1000–10000 replication events, while it is $1/1000000$ for allozymes. It turns out that micro-satellites alone contain enough information to plot the lineage tree of a organism. Micro-satellite data (for example, for DNA fingerprinting) consists of numbers of repeats of micro-satellites for each allele.

Evolutionary distance between two populations (or taxa) is a measure of genetic divergence estimating the *divergence time*, i.e., the time that has past since those populations existed as a single population.

Phylogenetic distance (or **genealogical distance**) between two taxa is the *branch length*, i.e., a minimum number of edges, separating them on a phylogenetic tree.

Immunological distance between two populations is a measure of the strength of antigen-antibody reactions, indicating the evolutionary distance separating them.

23.1. GENETIC DISTANCES FOR GENE-FREQUENCY DATA

In this section, a **genetic distance** between populations is a way of measuring the amount of evolutionary divergence by counting the number of allelic substitutions by loci.

A *population* is represented by a double-indexed vector $x = (x_{ij})$ with $\sum_{j=1}^{n} m_j$ components, where x_{ij} is the frequency of i-th *allele* (the label for a state of a gene) at the j-th gene locus (the position of a gene on a chromosome), m_j is the number of alleles at the j-th locus, and n is the number of considered loci.

Denote by $\sum$ summation over all i and j. Since x_{ij} is the frequency, it holds $x_{ij} \geqslant 0$, and $\sum_{i=1}^{m_j} x_{ij} = 1$.

- **Stephens et al. shared allele distance**

The **Stephens et al. shared allele distance** between populations is defined by

$$1 - \frac{\overline{SA(x, y)}}{\overline{SA(x)} + \overline{SA(y)}},$$

where, for two individuals a and b, $SA(a, b)$ denotes the number of shared alleles summed over all n loci and divided by $2n$, while $\overline{SA(x)}$, $\overline{SA(y)}$, and $\overline{SA(x, y)}$ are $SA(a, b)$ averaged over all pairs (a, b) with individuals a, b being in populations, represented by x, by y and, respectively, between them.

- **Dps distance**

  The **Dps distance** between populations is defined by

$$-\ln \frac{\sum \min\{x_{ij}, y_{ij}\}}{\sum_{j=1}^{n} m_j}.$$

- **Prevosti–Ocana–Alonso distance**

  The **Prevosti–Ocana–Alonso distance** between populations is defined (cf. L^1**-metric**) by

$$\frac{\sum |x_{ij} - y_{ij}|}{2n}.$$

- **Roger distance**

  The **Roger distance** is a metric between populations, defined by

$$\frac{1}{\sqrt{2}n} \sum_{j=1}^{n} \sqrt{\sum_{i=1}^{m_j} (x_{ij} - y_{ij})^2}.$$

- **Cavalli–Sforza–Edvards chord distance**

  The **Cavalli–Sforza–Edvards chord distance** between populations is defined by

$$\frac{2\sqrt{2}}{\pi} \sum_{j=1}^{n} \sqrt{1 - \sum_{i=1}^{m_j} \sqrt{x_{ij} y_{ij}}}.$$

  It is a metric. (Cf. **Hellinger distance**.)

- **Cavalli–Sforza arc distance**

  The **Cavalli–Sforza arc distance** between populations is defined by

$$\frac{2}{\pi} \arccos\left(\sum \sqrt{x_{ij} y_{ij}}\right).$$

  (Cf. **Fisher distance** in Probability.)

- **Nei–Tajima–Tateno distance**

  The **Nei–Tajima–Tateno distance** between populations is defined by

$$1 - \frac{1}{n} \sum \sqrt{x_{ij} y_{ij}}.$$

- **Nei minimum genetic distance**

  The **Nei minimum genetic distance** between populations is defined by

$$\frac{1}{2n}\sum(x_{ij}-y_{ij})^2.$$

  (Cf. **squared Euclidean distance**.)

- **Nei standard genetic distance**

  The **Nei standard genetic distance** between populations is defined by

$$-\ln I,$$

  where I is Nei *normalized identity of genes*, defined by $\frac{\langle x,y\rangle}{\|x\|_2\cdot\|y\|_2}$ (cf. **Bhattacharya distances** in Probability and **angular semi-metric**).

- **Sangvi χ^2 distance**

  The **Sangvi χ^2 distance** between populations is defined by

$$\frac{2}{n}\sum\frac{(x_{ij}-y_{ij})^2}{x_{ij}+y_{ij}}.$$

- ***F*-statistics distance**

  The ***F*-statistics distance** between populations is defined by

$$\frac{\sum(x_{ij}-y_{ij})^2}{2(n-\sum x_{ij}y_{ij})}.$$

- **Fuzzy set distance**

  The Dubois–Prade's **fuzzy set distance** between populations is defined by

$$\frac{\sum 1_{x_{ij}\neq y_{ij}}}{\sum_{j=1}^{n} m_j}.$$

- **Kinship distance**

  The **kinship distance** between populations is defined by

$$-\ln\langle x,y\rangle,$$

  and $\langle x,y\rangle$ is called *kinship coefficient*.

- **Reynolds–Weir–Cockerham distance**

  The **Reynolds–Weir–Cockerham distance** (or *co-ancestry distance*) between populations is defined by

$$-\ln(1-\theta),$$

where **co-ancestry coefficient** θ of two individuals (or two populations) is the probability that a randomly picked allele from one individual (or from genetic pool of one population) is *identical by descent* (i.e., corresponding genes are physical copies of the same ancestral gene) to a randomly picked allele in another. Two genes can be identical by state (i.e., with the same allele label), but not identical by descent. The co-ancestry coefficient θ of two individuals is the **inbreeding coefficient** of their following generation.

- **Goldstein and al. distance**

  The **Goldstein and al. distance** between populations is defined by

$$\frac{1}{n}\sum(ix_{ij} - iy_{ij})^2.$$

- **Average square distance**

  The **average square distance** between populations is defined by

$$\frac{1}{n}\sum_{k=1}^{n}\left(\sum_{1\leqslant i<j\leqslant m_j}(i-j)^2 x_{ik}y_{jk}\right).$$

- **Shriver–Boerwinkle stepwise distance**

  The **Shriver–Boerwinkle stepwise distance** between populations is defined by

$$\frac{1}{n}\sum_{k=1}^{n}\sum_{1\leqslant i,j\leqslant m_k}|i-j|(2x_{ik}y_{jk} - x_{ik}x_{jk} - y_{ik}y_{jk}).$$

23.2. DISTANCES FOR DNA DATA

Distances between DNA or protein sequences are usually measured in terms of substitutions, i.e., mutations, between them. A *DNA sequence* will be seen as a sequence $x = (x_1, \dots, x_n)$ over 4-letter alphabet of four nucleotides A, T, C, G; $\sum$ denotes $\sum_{i=1}^{n}$.

- **No. of differences**

  The **No. of DNA differences** is just the **Hamming metric** between DNA sequences:

$$\sum 1_{x_i \neq y_i}.$$

- ***p*-distance**

  The ***p*-distance** d_p between DNA sequences is defined by

$$\frac{\sum 1_{x_i \neq y_i}}{n}.$$

- **Jukes–Cantor nucleotide distance**

The **Jukes–Cantor nucleotide distance** between DNA sequences is defined by

$$-\frac{3}{4}\ln\left(1-\frac{4}{3}d_p(x,y)\right),$$

where d_p is the p-**distance**. If the rate of substitution varies with the gamma distribution, and a is the parameter describing the shape of this distribution, then the **gamma distance for the Jukes–Cantor model** is defined by

$$\frac{3a}{4}\left(\left(1-\frac{4}{3}d_p(x,y)\right)^{-1/a}-1\right).$$

- **Tajima–Nei distance**

The **Tajima–Nei distance** between DNA sequences is defined by

$$-b\ln\left(1-\frac{d_p(x,y)}{b}\right),$$

where

$$b=\frac{1}{2}\left(1-\sum_{j=A,T,C,G}\left(\frac{1_{x_i=y_i=j}}{n}\right)^2+\frac{1}{c}\sum\left(\frac{1_{x_i\neq y_i}}{n}\right)^2\right),\quad \text{and}$$

$$c=\frac{1}{2}\sum_{i,k\in\{A,T,G,C\},\ j\neq k}\frac{(\sum 1_{(x_i,y_i)=(j,k)})^2}{(\sum 1_{x_i=y_i=j})(\sum 1_{x_i=y_i=k})}.$$

Let $P=\frac{1}{n}|\{1\leqslant i\leqslant n\colon\ \{x_i,y_i\}=\{A,G\}\text{ or }\{T,C\}\}|$, and $Q=\frac{1}{n}|\{1\leqslant i\leqslant n\colon \{x_i,y_i\}=\{A,T\}\text{ or }\{G,C\}\}|$, i.e., P and Q are the frequencies of, respectively, transition and transversion mutations between x and y. The following four distances are given in terms of P and Q.

- **Jin–Nei gamma distance**

The **Jin–Nei gamma distance** between DNA sequences is defined by

$$\frac{a}{2}\left((1-2P-Q)^{-1/a}+\frac{1}{2}(1-2Q)^{-1/a}-\frac{3}{2}\right),$$

where the rate of substitution varies with the gamma distribution, and a is the parameter describing the shape of this distribution.

- **Kimura 2-parameter distance**

The **Kimura 2-parameter distance** between DNA sequences is defined by

$$-\frac{1}{2}\ln(1-2P-Q)-\frac{1}{2}\ln\sqrt{1-2Q}.$$

- **Tamura 3-parameter distance**

  The **Tamura 3-parameter distance** between DNA sequences is defined by

$$-b\ln\left(1-\frac{P}{b}-Q\right)-\frac{1}{2}(1-b)\ln(1-2Q),$$

  where $f_x = \frac{1}{n}|\{1 \leqslant i \leqslant n\colon\ x_i = G \text{ or } C\}|$, $f_y = \frac{1}{n}|\{1 \leqslant i \leqslant n\colon\ y_i = G \text{ or } C\}|$, and $b = f_x + f_y - 2f_x f_y$.

  In the case $f_x = f_y = \frac{1}{2}$ (so, $b = \frac{1}{2}$), it is the **Kimura 2-parameter distance**.

- **Tamura–Nei distance**

  The **Tamura–Nei distance** between DNA sequences is defined by

$$-\frac{2f_Af_G}{f_R}\ln\left(1-\frac{f_R}{2f_Af_G}P_{AG}-\frac{1}{2f_R}P_{RY}\right)-\frac{2f_Tf_C}{f_Y}\ln\left(1-\frac{f_Y}{2f_Tf_C}P_{TC}-\frac{1}{2f_Y}P_{RY}\right)$$
$$-2\left(f_Rf_Y-\frac{f_Af_Gf_Y}{f_R}-\frac{f_Tf_Cf_R}{f_Y}\right)\ln\left(1-\frac{1}{2f_Rf_Y}P_{RY}\right),$$

  where $f_j = \frac{1}{2n}\sum(1_{x_i=j}+1_{y_i=j})$ for $j = A, G, T, C$, and $f_R = f_A+f_G$, $f_Y = f_T+f_C$, while $P_{RY} = \frac{1}{n}|\{1 \leqslant i \leqslant n\colon\ |\{x_i, y_i\} \cap \{A, G\}| = |\{x_i, y_i\} \cap \{T, C\}| = 1\}|$ (the proportion of transversion differences), $P_{AG} = \frac{1}{n}|\{1 \leqslant i \leqslant n\colon\ \{x_i, y_i\} = \{A, G\}\}|$ (the proportion of transitions within purines), and $P_{TC} = \frac{1}{n}|\{1 \leqslant i \leqslant n\colon\ \{x_i, y_i\} = \{T, C\}\}|$ (the proportion of transitions within pyrimidines).

- **Garson et al. hybridization metric**

  H-measure between two DNA n-sequences x and y is defined by

$$H(x, y) = \min_{-n\leqslant k\leqslant n}\sum 1_{x_i\neq y^*_{i+k}},$$

  where indexes $i + k$ are modulo n, and y^* is the reversal of y followed by *Watson–Crick complementation*, i.e., interchange of all A, T, G, C by T, A, C, G, respectively.

  An *DNA cube* is any maximal set of DNA n-sequences, such that $H(x, y) = 0$ for any two of them. The **Garson at al. hybridization metric** between DNA cubes A and B is defined by

$$\min_{x\in A, y\in B} H(x, y).$$

23.3. DISTANCES FOR PROTEIN DATA

A *protein sequence* will be seen as a sequence $x = (x_1, \dots, x_n)$ over 20-letter alphabet of 20 amino acids; $\sum$ denotes $\sum_{i=1}^{n}$.

There are several notions of similarity/distance on the set of 20 amino acids, based, for example, on their hydrophilicity, polarity, charge, shape, etc. Most important are 20×20 *Dayhoff PAM250* matrix which express relative mutability of 20 amino acids.

- **PAM distance**

  The **PAM distance** (or **Dayhoff–Eck distance**, *PAM value*) between protein sequences is defined as the minimal number of accepted (i.e., fixed) point mutations per 100 amino acids needed to transform one protein into another. 1 PAM is an unit of evolution: it corresponds to 1 point mutation per 100 amino acids. PAM values 80, 100, 200, 250 correspond to the distance (in %) 50, 60, 75, 92 between proteins.

- **No. of protein differences**

  The **No. of protein differences** is just the **Hamming metric** between protein sequences:

$$\sum 1_{x_i \neq y_i}.$$

- **Amino *p*-distance**

  The **amino *p*-distance** (or *uncorrected distance*) d_p between protein sequences is defined by

$$\frac{\sum 1_{x_i \neq y_i}}{n}.$$

- **Amino Poisson correction distance**

  The **amino Poisson correction distance** between protein sequences is defined, via **amino *p*-distance** d_p, by

$$-\ln\bigl(1 - d_p(x, y)\bigr).$$

- **Amino gamma distance**

  The **amino gamma distance** (or *Poisson correction gamma distance*) between protein sequences is defined, via **amino *p*-distance** d_p, by

$$a\bigl(\bigl(1 - d_p(x, y)\bigr)^{-1/a} - 1\bigr),$$

  where the substitution rate varies with $i = 1, \dots, n$ according to gamma distribution, and a is the parameter describing the shape of the distribution. For $a = 2.25$ and $a = 0.65$, it estimates **Dayhoff** and **Grishin distances**, respectively. In some applications, this distance with $a = 2.25$ is called simply **Dayhoff distance**.

- **Jukes–Cantor protein distance**

  The **Jukes–Cantor protein distance** between protein sequences is defined, via **amino *p*-distance** d_p, by

$$-\frac{19}{20}\ln\left(1 - \frac{20}{19}d_p(x, y)\right).$$

- **Kimura protein distance**

  The **Kimura protein distance** between protein sequences is defined, via **amino p-distance** d_p, by

$$-\ln\left(1 - d_p(x, y) - \frac{d_p^2(x, y)}{5}\right).$$

- **Grishin distance**

  The **Grishin distance** d between protein sequences can be obtained, via **amino p-distance** d_p, from the formula

$$\frac{\ln(1 + 2d(x, y))}{2d(x, y)} = 1 - d_p(x, y).$$

- **Edgar k-mer distance**

  The **Edgar k-mer distance** between sequences $x = (x_1, \dots, x_m)$ and $y = (y_1, \dots, y_n)$ over a compressed amino acid alphabet is defined by

$$\ln\left(\frac{1}{10} + \frac{\sum_a \min\{x(a), y(a)\}}{\min\{m, n\} - k + 1}\right),$$

  where a is any *k-mer* (a word of length k over the alphabet), while $x(a)$ and $y(a)$ are the number of times a occurs in x and y, respectively, as a block (contiguous subsequence). (Cf. **q-gram similarity**.)

23.4. OTHER BIOLOGICAL DISTANCES

- **RNA structural distance**

  An RNA (sequence) is a string over the alphabet $\{A, C, G, T\}$ of nucleotides (bases). Inside a cell, such string folds in $3D$ space, because of pairing of nucleotide bases (usually, by bonds A–U, G–C and G–U). The *secondary structure* of an RNA is, roughly, the set of helices (or the list of paired bases) making up the RNA. This structure can be represented as planar graph and further, as rooted tree. The *tertiary structure* is the geometric form the RNA takes in space.

  An **RNA structural distance** between two RNA sequences is a distance between their secondary structures. Examples of such RNA distances are: **tree edit distance** (and other distances on rooted trees given in Chapter 15), and the **base-pair distance**, i.e., the **symmetric difference metric** between secondary structures seen as sets of paired bases.

  In *in silico* (i.e., computer-simulated) RNA evolution, the **fitness** of an RNA sequence x is the **metric transform** $f(d(x, x_T))$, where $f : \mathbb{R}_{\geqslant 0} \to \mathbb{R}_{\geqslant 0}$ is a scaling function, and $d(x, x_T)$ is an RNA structural distance between the sequence x and the selected fixed target RNA sequence x_T.

- **Distances for genome permutations**

The *genomes* of two related species, given by the order of genes along chromosomes, for large (i.e., happening on large portion of the chromosome) mutations, can be seen as genome rearrangements and, so, as *permutations* (or *rankings*) of homologous genes. Therefore, many chromosomal rearrangements can be presented then as *indels* (insertions or deletions), inversions, transpositions, reversals and other *editing operations*. Some of those operations have biological meaning for DNA/protein sequences as well.

A **distance for genome permutations** is the **edit distance** with respect of given set of editing operations, i.e., the minimal number of those editing operations needed to transform one permutation into another. If one attach a positive number (cost or weight) to each permitted editing operation, then the distance is minimal sum of weights in a sequence of operations transforming one permutation into another. If one takes into account the directionality of the genes, a chromosome is described by a *signed permutation*, i.e., by a vector $x = (x_1, \dots, x_n)$, where $|x_i|$ are different numbers $1, \dots, n$, and any x_i can be positive or negative.

An example of distance measures between genomes (or species), seen as collections of sets of genes, is Ferretti–Nadeau–Sankoff **syntenic distance**. It is the minimal number of mutation moves – *translocations* (exchanges of genes between two chromosomes), *fusions* (of two chromosomes in one) and *fissions* (of one chromosome in two) – needed to transfer one genome into another.

- **Genome distance**

The **genome distance** between two loci on a chromosome is the number of base pairs separating them on the chromosome.

- **Map distance**

The **map distance** between two loci on a genetic map is the recombination frequency expressed as a percentage; it is measured in *centimorgans* cM (or *map units*), where 1 cM corresponds to their statistically corrected recombination frequency 1%.

Typically, a linkage map distance of 1 cM (*genetic scale*) corresponds to a **genome distance** (*physical scale*) of about one *megabase* (million base pairs).

- **Metabolic distance**

The **metabolic distance** (or *pathway distance*) between enzymes is the minimum number of metabolic steps separating two enzymes in the metabolic pathways.

- **Gendron et al. distance**

The **Gendron et al. distance** between two base-base interactions, represented by 4×4 *homogeneous transformation matrices* X and Y, is defined by

$$\frac{S(XY^{-1}) + S(X^{-1}Y)}{2},$$

where $S(M) = \sqrt{l^2 + (\theta/\alpha)^2}$, l is the length of translation, θ is the angle of rotation, and α represents a scaling factor between the translation and rotation contributions.

- **Biotope distance**

The *biotopes* here are represented as binary sequences $x = (x_1, \dots, x_n)$, where $x_i = 1$ means the presence of the species i. The **biotope distance** (or **Tanimoto distance**) between biotopes x and y is defined by

$$\frac{|\{1 \leqslant i \leqslant n\colon\ x_i \neq y_i\}|}{|\{1 \leqslant i \leqslant n\colon\ x_i + y_i > 0\}|}.$$

- **Taxonomic distance**

Given a finite metric space (X, d) (usually, an Euclidean space) and a selected, as typical by some criterion, vertex $x_0 \in X$, called *prototype* (or *centroid*), the **taxonomic distance** of every $x \in X$ is the number $d(x, x_0)$. Usually, the elements of X represent phenotypes or morphological traits. The average of $d(x, x_0)$ by $x \in X$ estimates corresponding *variability*.

The term **taxonomic distance** is also used in Phylogenetic Taxonomy for every dissimilarity between two *taxa*, i.e., entities or groups which are arranged into an hierarchy.

- **Victor–Purpura distance**

A *spike train* x is a time sequence $(x_1, \dots, x_n)$ of n events (for example, neuronal spikes, or hearth beats). The time sequence lists either absolute spike times, or inter-spike time intervals. A human brain has about 100 billion of *neurons* (nerve cells). A neuron reacts on a stimulus by producing a spike train which is a sequence of short electrical pulses called *spikes*.

The **Victor–Purpura distance** between two spike trains x and y is a cost-based **editing metric** (i.e., the minimal cost of transforming x into y) by the following operations with their associated costs: insert a spike (cost 1), delete a spike (cost 1), shift a spike by an amount of time t (cost qt, where $q > 0$ is a parameter).

In order to compare reactions of a population of neurons on two different stimuli, the **Chernoff distance** between corresponding distributions of spike counts is used.

- **Oliva et al. perception distance**

Let $\{s_1, \dots, s_n\}$ be the set of *stimuli*, and let q_{ij} be the conditional probability that a subject will perceive stimulus s_j, when the stimulus s_i was shown; so, $q_{ij} \geqslant 0$, and $\sum_{j=1}^{n} q_{ij} = 1$. Let q_i is the probability of presenting stimulus s_i.

The **Oliva at al. perception distance** ([OSLM04]) between stimuli s_i and s_j is defined by

$$\frac{1}{q_i + q_j} \sum_{k=1}^{n} \left| \frac{q_{ik}}{q_i} - \frac{q_{jk}}{q_j} \right|.$$

- **Probability-distance hypothesis**

In Psychophysics, the **probability-distance hypothesis** is a hypothesis that the probability with which one stimulus is discriminated from another is a (continuously increasing) function of some subjective quasi-metric between these stimuli (see [Dzha01]). Under this hypothesis, such subjective metric is a **Finsler metric** if and only if it coincides in the small with the **intrinsic metric** (i.e., the infimum of the lengths of all paths connecting two stimuli).

- **Marital distance**

  The **marital distance** is a distance between birthplaces of spouses (or zygotes).

- **Isolation-by-distance**

  Isolation-by-distance is a biological model predicting that the genetic distance between populations increases exponentially with respect to geographic distance. Therefore, emergence of regional differences (races) and new species is explained by restricted gene flow and adaptive variations. Isolation-by-distance was studied, for example, via surname structure (cf. **Lasker distance**).

- **Malecot's distance model**

  The **Malecot's distance model** is a migratory model of isolation by distance, expressed by the following *Malecot's equation* for dependency of alleles at two loci (*allelic association*, or *linkage disequilibrium*) ρ_d:

$$\rho_d = (1 - L)Me^{\varepsilon d} + L,$$

  where d is distance between two loci (either **genome distance** in base pairs, or **map distance** in centimorgans), ε is a constant for a specified region, $L = \lim_{d \to 0} \rho_d$, and $M \leqslant 1$ is a parameter expressing mutation rate.

- **Lasker distance**

  The **Lasker distance** (Rodrigues-Larralde et al., 1989) between two human populations x and y, characterized by surname frequency vectors (x_i) and (y_i), is the number $-\ln 2R_{x,y}$, where $R_{x,y} = \frac{1}{2}\sum_i x_i y_i$ is Lasker's *coefficient of relationship by isonymy*. Surname structure is related to inbreeding and (in patrilinear societies) to random genetic drift, mutation and migration. Surnames can be considered as alleles of one locus, and their distribution can be analyzed by the theory of neutral mutations; an isonymy points to a common ancestry.

- **Surname distance model**

  A **surname distance model** was used in [COR05], in order to estimate the preference transmission from parents to children by comparing, for 47 provinces of mainland Spain, the 47×47 distance matrices for **surname distance** with those of **consumption distance** and **cultural distance**. The distances were l_1-distances $\sum_i |x_i - y_i|$ between the frequency vectors (x_i), (y_i) of provinces x, y, where z_i is, for the province z, either the frequency of i-th surname (**surname distance**, or the budget share of i-th good **consumption distance**, or **cultural distance** the population rate for i-th cultural issue (rate of weddings, newspaper readership, etc.), respectively.

  Other considered there (matrices of) distances are:

- the *geographical distance* (in kilometers, between the capitals of two provinces);
- the *income distance* $|m(x) - m(y)|$, where $m(z)$ is mean income in the province z;
- the *climatic distance* $\sum_{1 \leqslant i \leqslant 12} |x_i - y_i|$, where z_i is the average temperature in the province z during i-th month;
- the *migration distance* $\sum_{1 \leqslant i \leqslant 47} |x_i - y_i|$, where z_i is the percentage of people (living in the province z) born in the province i.

Strong *vertical preference transmission*, i.e., correlation between surname and consumption distances, was detected only for food items.

- **Distance model of altruism**

  In Evolutionary Ecology, altruism is explained by kin selection and group selection, and it supposed to be a driving force of the transition from unicellular organisms to multicellularity. The **distance model of altruism** (see [Koel00]) suggests that altruists spread locally, i.e., with small interaction distance and offspring dispersal distance, while the evolutionary response of egoists is to invest in increasing of those distances. The intermediate behaviors are not maintained, and evolution will lead to a stable bimodal spatial pattern.

- **Distance running model**

  The **distance running model** is a model of antropogenesis proposed in [BrLi04]. Bipedality is a key derived behavior of hominids which appeared 4.5–6 million years ago. However, australopithecines were still animals. The genus Homo which emerged about 2 million years ago already could produce rudimentary tools. Bramble–Lieberman model attributes this transition to a suite of adaptations specific to running long distances in the savanna. They specify how endurance running, a derived capability of Homo, defined the human body form, producing balanced head, low/wide shoulders, narrow chest, short forearms, large hip, etc.

Chapter 24

Distances in Physics and Chemistry

24.1. DISTANCES IN PHYSICS

Physics studies the behavior and properties of matter in a wide variety of contexts, ranging from the sub-microscopic particles from which all ordinary matter is made (*Particle Physics*) to the behavior of the material Universe as a whole (*Cosmology*). Physics forces which act at a distance (i.e., a push or pull which acts without "physical contact") are nuclear and molecular attraction, and, beyond atomic level, gravity (completed, perhaps, by anti-gravity), static electricity, and magnetism. Last two forces can be both, push and pull. Distances on small scale are treated in this chapter, while large distances (in Astronomy and Cosmology) are object of chapters 25 and 26. In fact, the distances having physical meaning range from 1.6×10^{-35} m (*Planck length*) to 7.4×10^{26} m (the estimated size of observable Universe). At present, Theory of Relativity, Quantum Theory and Newton's laws permit to describe and predict the behavior of physical systems in range $10^{-15} - 10^{25}$ m.

- **Mechanic distance**

  The **mechanic distance** is the position of a particle as a function of time t. For a particle with initial position x_0 and initial speed v_0, which acted upon by a constant acceleration a, it is given by

  $$x(t) = x_0 + v_0 t + \frac{1}{2}at^2.$$

  The distance fallen under uniform acceleration a, in order to reach a speed v, is given by $x = \frac{v^2}{2a}$.

  A *free falling body* is a body which is falling subject only to acceleration by gravity g. The distance fallen by it, after a time t, is $\frac{1}{2}gt^2$; it is called the **free fall distance**.

- **Terminal distance**

  The **terminal distance** is a distance of an object, moving in a resistive medium, from an initial position to a stop.

  Given an object of mass m moving in a resistive medium (where the drag per unit mass is proportional to speed with constant of proportionality β, and there is no other force acting on a body), the position $x(t)$ of a body with initial position x_0 and initial velocity v_0 is given by $x(t) = x_0 + \frac{v_0}{\beta}(1 - e^{-\beta t})$. The speed of the body $v(t) = x'(t) = v_0 e^{-\beta t}$

decreases to zero over time, and the body reaches a **maximum terminal distance**

$$x_{terminal} = \lim_{t\to\infty} x(t) = x_0 + \frac{v_0}{\beta}.$$

For a projectile, moving with initial position (x_0, y_0) and initial velocity (v_{x0}, v_{y0}), the position $(x(t), y(t))$ is given by $x(t) = x_0 + \frac{v_{x0}}{\beta}(1 - e^{-\beta t})$, $y(t) = (y_0 + \frac{v_{y0}}{\beta} - \frac{g}{\beta^2}) + \frac{v_{y0}\beta - g}{\beta^2}e^{-\beta t}$. The horizontal motion gets stopped to reach a maximum terminal distance

$$x_{terminal} = x_0 + \frac{v_{x0}}{\beta}.$$

- **Ballistics distances**

  Ballistics is the study of the motion of *projectiles*, i.e., bodies which are propelled (or thrown) with some initial velocity, and then allowed to be acted upon by the forces of gravity and possible drag.

  The horizontal distance traveled by a projectile is called **range**, the maximum upward distance reached by it is called **height**, and the path of the object is called **trajectory**.

  For a projectile launched with a velocity v_0 at an angle θ to the horizontal, the **range** is given by

$$x(t) = v_0 t \cos\theta,$$

  where t is the time of motion. On a level plane, where the projectile lands at the same altitude as it was launched, the full range is

$$x_{max} = \frac{v_0^2 \sin 2\theta}{g},$$

  which is maximized when $\theta = \pi/4$. If the altitude of the landing point is Δh higher that of the launch point, then

$$x_{max} = \frac{v_0^2 \sin 2\theta}{2g}\left(1 + \left(1 - \frac{2\Delta hg}{v_0^2 \sin^2\theta}\right)^{1/2}\right).$$

  The **height** is given by

$$\frac{v_0 \sin^2\theta}{2g},$$

  and is maximized when $\theta = \pi/2$.

  The arc length of the **trajectory** is given by

$$\frac{v_0^2}{g}\left(\sin\theta + \cos^2\theta gd^{-1}(\theta)\right),$$

where $gd(x) = \int_0^x \frac{dt}{\cosh t}$ is the *Gudermannian function*. The arch length is maximized when $gd^{-1}(\theta)\sin\theta = (\int_0^\theta \frac{dt}{\cos t})\sin\theta = 1$, and approximate solution is given by $\theta \approx 0.9855$.

- **Acoustic metric**

In Acoustics, the **acoustic metric** (or *sonic metric*) is a characteristic of sound-carrying properties of a given medium: air, water, etc.

In General Relativity and Quantum Gravity, it is a characteristic of signal-carrying properties in a given *analog model* (with respect to Condense Matter Physics), where, for example, the propagation of scalar fields in curved *space-time* is modeled (see, for example, a survey [BLV05]) as the propagation of sound in a moving fluid, or slow light in moving fluid dielectric, or *superfluid* (quasi-particles in quantum fluid), etc. The passage of a signal through an acoustic metric itself modifies the metric; for example, the motion of sound in air moves air and modifies the local speed of the sound. Such "effective" (i.e., recognized by its "effects") Lorentzian metric governs, instead of the background metric, the propagation of fluctuations: the particles associated to the perturbations follow geodesics of that metric.

In fact, if a fluid is barotropic and inviscid, and the flow is irrotational, then the propagation of sound is described by an **acoustic metric** which depends on the density ρ of flow, velocity $\mathbf{v}$ of flow and local speed s of sound in the fluid. It can be given by the *acoustic tensor*

$$g = g(t, \mathbf{x}) = \frac{\rho}{s}\begin{pmatrix} -(s^2 - v^2) & \vdots & -\mathbf{v}^T \\ \cdots & & \cdots \\ -\mathbf{v} & \vdots & 1_3 \end{pmatrix},$$

where 1_3 is the 3×3 identity matrix, and $v = \|\mathbf{v}\|$. The *acoustic line element* can be written as

$$ds^2 = \frac{\rho}{s}\left(-(s^2 - v^2)\,dt^2 - 2\mathbf{v}\,d\mathbf{x}\,dt + (d\mathbf{x})^2\right) = \frac{\rho}{s}\left(-s^2\,dt^2 + (d\mathbf{x} - \mathbf{v}\,dt)^2\right).$$

The signature of this metric is (3, 1), i.e., it is a **Lorentzian metric**. If the speed of the fluid becomes supersonic, then the sound waves will be unable to come back, i.e., there exists a *mute hole*, the acoustic analog of a *black hole*.

- **Healing length**

For a superfluid, the **healing length** is a length over which the wave function can vary while still minimizing energy.

For *Bose–Einstein condensates*, the healing length is the width of the bounding region over which the probability density of the condensate drops to zero.

- **Optical distance**

In Optics and Telecommunications, the **optical distance** (or *optical path length*) is a distance traveled by light: the physical distance in a medium multiplied by the index of refraction of the medium. By *Fermat's principle* light always follows the shortest optical path.

For a series of continuous layers with index of refraction $n(s)$ as a function of distance s, it is given by

$$\int_C n(s)\,ds.$$

For a series of discrete layers with indices of refraction n_i and thicknesses s_i, it is equal to

$$\sum_{i=1}^{N} n_i s_i = \frac{\delta}{k_0},$$

where δ is the phase shift, and k_0 is the vacuum wave number.

- **Spatial coherence length**

The **spatial coherence length** is the propagation distance from a coherent source to a point where an electromagnetic wave maintains a specific degree of coherence. This notion is used in Telecommunication Engineering (usually, for optical regime) and in synchrotron X-ray Optics (the advanced characteristics of synchrotron sources provide highly coherent X-rays). The spatial coherence length is about 20 cm, 100 m, and 100 km for helium-neon, semiconductor and fiber lasers, respectively. Cf. *temporal coherence length* which describes the correlation between signals observed at different moments of time.

For vortex-loop phase transitions (superconductors, superfluid, etc.), **coherence length** is the diameter of the largest loop which is thermally exited.

- **Inverse-square distance laws**

Any law stating that some physical quantity is inversely proportional to the square of the distance from the source that quantity.

Law of universal gravitation (Newton–Bullialdus): the gravitational attraction between two massive point-like objects is directly proportional to the product of their masses and inversely proportional to the square of the distance between them. The existence of extra dimensions, thought by M-theory, will be experimentally checked in 2007 (the opening at CERN, near Geneva, of LHC, i.e., large hadron collider), basing on the inverse proportionality of the gravitational attraction in n-dimensional space to the $(n-1)$-degree of the distance between objects; if the Universe have 4-th dimension, LHC will find out the inverse proportionality to the cube of the small inter-particle distance.

Coulomb's law: the force of attraction or repulsion between two (stationary) charged point particles is directly proportional to the product of charges and inversely proportional to the square of the distance between them.

The *intensity* (power per unit area in the direction of propagation) of a spherical wavefront (light, sound, etc.) radiating from a point source decreases (assuming that there are no losses caused by absorption or scattering) is inversely proportional to the square r^2 of the distance from the source. However, for a radio wave, it decrease like $\frac{1}{r}$.

- **Range of fundamental forces**

The fundamental forces (or interactions) are gravity and electromagnetic, weak and strong forces. The **range** of a force is considered *short* if it decays (approach 0) exponentially as d increases. Both, electromagnetic force and gravity, are forces of infinite range which obey **inverse-square distance laws**. Both, weak and strong forces, are very short range (about 10^{-18} m and 10^{-15} m) which is limited by the uncertainty principle.

- **Long range order**

A physical system has **long range order** if remote portions of the same sample exhibit correlated behavior. For example, in crystals and some liquids, the positions of an atom and its neighbors define positions of all other atoms. Examples of long range order in solids are: magnetism, charge density waves, superconductivity. **Short range** is the first- or second-nearest neighbors of an atom. More precisely, the system has **long range order**, *quasi-long range order* or is *disordered* if corresponding correlation function decays, for large distances, to a constant, to zero as a polynomial, or to zero exponentially (cf. **long range dependency**).

- **Action at a distance (in Physics)**

An **action at a distance** is the interaction, without known mediator, of two objects separated in space. Einstein used term *spooky action at a distance* for quantum mechanical interaction (like *entanglement* and *quantum non-locality*) which is instantaneous, regardless of distance. Main conceptions of interaction at a distance are Newton instantaneous long range interaction and Faraday–Maxwell short range interaction. Already controversial (since speed of light is maximal) long range interaction reach status of marginality for "mental action at a distance": telepathy, clairvoyance, precognition, psychokinesis, etc.

The term *short range interaction* is also used for the transmission of action on distance by a material medium from a point to a point with certain velocity dependent on properties of this medium. Also, in Information Storage, the term *near-field interaction* is used for very short distance interaction using scanning probe techniques.

- **Interaction distance**

The **interaction distance** between two particles is the farthest distance of their approach at which it is discernible that they will not pass at the *impact parameter*, i.e. their distance of closest approach if they had continued to move in their original direction at their original speed.

- **Hopping distance**

The *hopping* is atomic-scale long range dynamics that control diffusivity and conductivity. For example, oxidation of DNA (loss of an electron) generates a radical cation which can migrate long (more that 20 nanometers) distance, called **hopping distance**, from site to site (to "hop" from one aggregate to another) before it is trapped by reaction with water.

- **Skin depth**

  The **skin depth** of a substance is the distance to which incident electromagnetic radiation penetrates. The skin depth is given by

$$\frac{c}{\sqrt{2\pi\sigma\mu\omega}},$$

  where c is the speed of light, σ is the electrical conductivity, μ is the permeability, and ω is the angular frequency. (Not to be confused with source-skin **medical distance**).

- **Gyroradius**

  The **gyroradius** (or *cyclotron radius*) is the radius of the circular orbit of a charged particle (for example, energetic electron that is ejected from Sun) gyrating around its gliding center.

24.2. DISTANCES IN CHEMISTRY

Main chemical substances are ionic (held together by ionic bonds), metallic (giant close packed structures held together by metallic bonds), giant covalent (as diamond and graphite), or molecular small covalent). Molecules are made of fixed number of atoms joined together by covalent bonds; they range from small (single-atom molecules in the noble gases) to very large ones (as in polymers, proteins or DNA). The **interatomic distance** of two atoms is the distance (in angstroms or picometers) between their nuclei.

- **Atomic radius**

  Quantum Mechanics implies that an atom is not a ball having exactly defined boundary. Hence, **atomic radius** is defined as the distance from the atomic nucleus to the outmost stable electron orbital in a atom that is at equilibrium. Atomic radii represent the sizes of isolated, electrically neutral atoms, unaffected by bonding.

  Atomic radii are estimated from **bond distances** if the atoms of the element form bonds; otherwise (like the noble gases), only **Van-der-Waals radii** are used.

  The atomic radii of elements increase as one moves dawn the column (or to the left the row) in the Periodic Table of Elements.

- **Bond distance**

  The **bond distance** (or *bond length*) is the distance between the nuclei of two bonded atoms. For example, typical bond distances for carbon-carbon bonds in an organic molecule are 1.53, 1.34 and 1.20 angstroms for single, double and triple bonds, respectively.

  Depending on the type of bonding of the element, its atomic radius is called *covalent* or *metallic*. The **metallic radius** is one half of the **metallic distance**, i.e., closest internuclear distance in a *metallic crystal* (a closely packed crystal lattice of metallic element).

  Covalent radii of atoms (of elements that form covalent bonds) are inferred from bond distances between pairs of covalently-bonded atoms: they are equal to the sum of the covalent radii of two atoms. If the two atoms are of the same kind, then their covalent radius is one half of their bond distance. Covalent radii for elements whose atoms cannot bond to each another is inferred by combining, in various molecules, radii of those that bond with bond distances between pairs of atoms of different kind.

- **Van-der-Waals contact distance**

  Intermolecular distance data are interpreted by viewing atoms as hard spheres. The spheres of two neighboring non-bonded atoms (in touching molecules or atoms) are supposed just touch. So, their interatomic distance, called **Van-der-Waals contact distance**, is the sum of radii, called **Van-der-Waals radii**, of their hard spheres. Van-der-Waals radius of carbon is 1.7 angstroms, while its covalent radius is 0.76. Van-der-Waals contact distance corresponds to a "weak bond", when repulsion forces of electronic shells exceed London (attractive electrostatic) forces.

- **Interionic distance**

  An *ion* is an atom that has an positive or negative electrical charge. The **interionic distance** is the distance between the centers of two adjacent (bonded) ions. **Ionic radii** are inferred from ionic bond distances in real molecules and crystals.

  The ion radii of *cations* (positive ions, for example, sodium Na^+) are smaller than the atomic radii of the atoms they come from, while *anions* (negative ions, for example, chlorine Cl^-) are larger than their atoms.

- **Hydrodynamic radius**

  The **hydrodynamic radius** of a molecule, undergoing diffusion in a solution, is the hypothetical radius of a hard sphere which diffuses with the same speed.

- **Range of molecular forces**

  Molecular forces (or interactions) are the following electromagnetic forces: ionic bonds (charges), hydrogen bonds (dipolar), dipole-dipole interactions, London forces (the attraction part of Van-der-Waals forces) and steric repulsion (the repulsion part of Van-der-Waals forces). If the distance (between two molecules or atoms) is d, then (experimental observation) potential energy function P inversely relate to d^n with $n = 1, 3, 3, 6, 12$ for five above forces, respectively. The **range** (or the *radius*) of an interaction is considered *short* if P approach 0 rapidly as d increases. It also called *short* if it is at most 3 angstroms; so, only the range of steric repulsion is short. (Cf. **inverse-square distance laws**.)

- **Chemical distance**

  Various chemical systems (single molecules, their fragments, crystals, polymers, clusters) are well represented by graphs where vertices (say, atoms, molecules acting as monomers, molecular fragments) are linked by, say, chemical bonding, Van-der-Waals interactions, hydrogen bonding, reactions path. In Organic Chemistry, a *molecular graph* $G(x) = (V(x), E(x))$ is a graph representing a molecule x, so that the vertices $v \in V(x)$ are atoms and the edges $e \in E(x)$ correspond to electron pair bonds. The **Wiener num-**

ber of a molecule is one half of the sum of all pairwise distances between vertices of its molecular graph.

The (bonds and electrons) *BE-matrix* of a molecule x is the $|V(x)| \times |V(x)|$ matrix $((e_{ij}(x)))$, where $e_{ii}(x)$ is the number of free unshared valence electrons of the atom A_i, and, for $i \neq j$, $e_{ij}(x) = e_{ji}(x) = 1$ if there is a bond between atoms A_i and A_j, and $= 0$, otherwise.

Given two *stoichiometric* (i.e., with the same number of atoms) molecules x and y, their **Dugundji–Ugi chemical distance** is the **Hamming metric**

$$\sum_{1 \leqslant i,j \leqslant |V|} \left| e_{ij}(x) - e_{ij}(y) \right|,$$

and their **Pospichal–Kvasnička chemical distance** is

$$\min_{P} \sum_{1 \leqslant i,j \leqslant |V|} \left| e_{ij}(x) - e_{P(i)P(j)}(y) \right|,$$

where P is any permutation of atoms.

Above distance is equal to $|E(x)| + |E(y)| - 2|E(x, y)|$, where $E(x, y)$ is the edge-set of the maximum common subgraph (not induced, in general) of the molecular graphs $G(x)$ and $G(y)$. (Cf. **Zelinka distance** and **Mahalanobis distance**.)

The **Pospichal–Kvasnička reaction distance**, assigned to a molecular transformation $x \to y$, is the minimum number of *elementary transformations* needed to transform $G(x)$ onto $G(y)$.

- **Molecular rms radius**

The **molecular rms radius** (or *radius of gyration*) is the root-mean-square distance of atoms in a molecule from their common center of gravity; it is defined by

$$\sqrt{\frac{\sum_{1 \leqslant i \leqslant n} d_{0i}^2}{n+1}} = \sqrt{\frac{\sum_i \sum_j d_{ij}^2}{(n+1)^2}},$$

where n is the number of atoms, d_{0i} is the Euclidean distance of i-th atom from the center of gravity of the molecule (in a specified conformation), and d_{ij} is the Euclidean distance between i-th and j-th atoms.

- **Mean molecular radius**

The **mean molecular radius** is the number $\frac{r_i}{n}$, where n is the number of atoms in the molecule, and r_i is the Euclidean distance of i-th atom from the geometric center $\frac{\sum_j x_{ij}}{n}$ of the molecule (here x_{ij} is i-th Cartesian coordinate of the j-th atom).

Chapter 25

Distances in Geography, Geophysics, and Astronomy

25.1. DISTANCES IN GEOGRAPHY AND GEOPHYSICS

- **Great circle distance**

  The **great circle distance** (or **spherical distance**, **orthodromic distance**) is the shortest distance between points x and y on the surface of the Earth measured along a path on the Earth's surface. It is the length of the *great circle* arc, passing through x and y, in the spherical model of the planet.

  Let δ_1 and ϕ_1 be, respectively, the latitude and the longitude of x, and δ_2 and ϕ_2 those of y; let r be the Earth's radius. Then the great circle distance is equal to

  $$r \arccos\big(\sin\delta_1 \sin\delta_2 + \cos\delta_1 \cos\delta_2 \cos(\phi_1 - \phi_2)\big).$$

  In the spherical coordinates (θ, ϕ), where ϕ is the azimuthal angle, and θ is the colatitude, the great circle distance between $x = (\theta_1, \phi_1)$ and $y = (\theta_2, \phi_2)$ is equal to

  $$r \arccos\big(\cos\theta_1 \cos\theta_2 + \sin\theta_1 \sin\theta_2 \cos(\phi_1 - \phi_2)\big).$$

  For $\phi_1 = \phi_2$, the formula above reduces to $r|\theta_1 - \theta_2|$.

  The **spheroidal distance** is the distance between two points on the Earth's surface in the spheroidal model of the planet. The shape of the Earth more closely resembles a flattened spheroid with extreme values for the radius of curvature of 6336 km at the equator and 6399 km at the poles.

- **Loxodromic distance**

  The **loxodromic distance** is a distance between two points on the Earth's surface on a path with a constant direction on the compass. It is never shorter than the great circle distance.

- **Nautical distance**

  The **nautical distance** is the length in nautical miles of the *rhumb line* (a curve that crosses each meridian at the same angle) joining any two places on the Earth's surface. One nautical mile is equal to 1852 m.

- **Horizon distance**

  In Television, the **horizon distance** is the distance of the farthest point on the Earth's surface visible from a transmitting antenna.

In Radio, the **horizon distance** is the distance on the Earth's surface reached by a direct wave; due to atmospheric refraction, it is sometimes greater than the distance to the visible horizon.

- **Skip distance**

  The **skip distance** is the shortest distance that permits a radio signal (of given frequency) to travel from the transmitter to the receiver by reflection (hop) in the ionosphere.

- **Tolerance distance**

  In GIS (computer-based Geographic Information System), the **tolerance distance** is the maximal distance between points which must be established so that gaps and overshoots can be corrected (lines snapped together) as long as they fall within tolerance distance.

  For a sensor, the **tolerance distance** is a **range distance** within which a localization error is acceptable to the application.

- **Map's distance**

  The **map's distance** is the distance between two points on the map; cf. **map distance** between two loci on a genetic map.

  The **horizontal distance** is determined by multiplying the map's distance by the numerical scale of the map.

- **Horizontal distance**

  The **horizontal distance** (or **ground distance**) is the distance on a true level plane between two points, as scaled off of the map (it does not take into account the relief between two points). There are two types of horizontal distance: **straight-line distance** (the length of the straight-line segment between two points as scaled off of the map), and **distance of travel** (the length of the shortest path between two points as scaled off of the map, in the presence of roads, rivers, etc.).

- **Slope distance**

  The **slope distance** (or **slant distance**) is the inclined distance (as opposed to true horizontal or vertical distance) between two points.

- **Road travel distance**

  The **road travel distance** (or **actual distance**, **wheel distance**, *road distance*) between any two points (for example, two cities) of a region is the length of the shortest road connecting them. Since it is often not feasible to measure the actual distances for all pairs of points, it is a common practice to use *distance estimators*.

  Empirical observation shows that road travel distances are often simply a linear function of **great circle distances**; in Swedish towns one can let road distance $= 1.21 \cdot d$, where d is the **great circle distance**. In USA the multiplier is about 1.15 in an east-west direction, and about 1.21 in the north-south direction.

  Cf. **official distance**: the driving distance used for payment of travel.

- **Moho distance**

The **Moho distance** is the distance from a point on the Earth's surface to the *Moho interface* (or *Mohorovicic seismic discontinuity*) beneath it. The *Moho interface* is the boundary between the Earth's brittle outer crust and the hotter softer mantle; Moho distance ranges between 5–10 km beneath the ocean floor to 35–65 km beneath the continents. Cf. the world deepest cave (Krubera, Caucasus: 2.1 km), deepest mine (Western Deep Levels gold mine, South Africa: about 4 km) and deepest drill (Kola Superdeep Borehole: 12.3 km). The temperature rises usually by 1° every 33 m. Japanese research vessel *Chikyu* is scheduled to drill (from September 2007, 200 km off Nagoya cost) till Moho interface.

The Earth's mantle extends from the Moho discontinuity to the mantle-core boundary at a depth of approximately 2890 km. The mantle is divided into the upper and the lower mantle by a discontinuity at about 660 km. Other seismic discontinuities are at about 60–90 km (Hales discontinuity), 50–150 km (Gutenberg discontinuity), 220 km (Lehmann discontinuity), 410 km, 520 km, and 710 km.

- **Distances in Seismology**

The Earth's crust is broken into tectonic plates that move around (at some centimeters per year) driven by the thermal convection of the deeper mantle and by gravity. At their boundaries, plates stick most of the time and slip suddenly. An *earthquake*, i.e., a sudden (several seconds) motion or trembling in the Earth, caused by abrupt release of slowly accumulated strain, was, from 1906, seen mainly as a rupture (sudden appearance, nucleation and propagation of new crack or fault) due to elastic rebound. However, from 1966, it is seen within the framework of slippage along pre-existing fault or plate interface, as the result of stick-slip frictional instability. So, an earthquake happens when dynamic friction becomes less than static friction. The advancing boundary of the slip region is called *rupture front*. The standard approach assumes that the fault is a definite surface of tangential displacement discontinuity, embedded in a liner elastic crust.

90% of earthquakes are of tectonic origin, but they can be caused also by volcanic eruption, nuclear explosion and work in a large dam, well or mine. Earthquake can be measured by **focal depth**, speed of slip, intensity (modified Mercali scale of earthquake effects), magnitude, acceleration (main destruction factor), etc. The Richter logarithmic scale of magnitude is computed from the amplitude and frequency of shock waves received by seismograph, adjusted to account for **epicentral distance**. An increase of 0.1 of the Richter magnitude corresponds to an increase of 10 times in amplitude of the waves; the largest recorded value is 9.5 (Chile, 1960).

Distance attenuation models, used in earthquake engineering for buildings and bridges, derive usually acceleration decay with increase of some **site-source distance**, i.e., the distance between seismological stations and the crucial (for the given model) "central" point of the earthquake.

The simplest model is the *hypocenter* (or focus), i.e., the point inside the Earth from which an earthquake originates (the waves first emanate, the seismic rupture or slip begins). The *epicenter* is the point of the Earth's surface directly above the hypocenter.

The terminology below is also used for other catastrophes, such as an impact or explosion of a nuclear weapon, meteorite or comet, but for an explosion in the air, the term *hypocenter* refers to the point on the Earth surface directly below the burst. The list of main Seismology distances follows.

The **focal depth**: the distance between hypocenter and epicenter; the average focal depth is 100–300 km.

The **hypocentral distance**: the distance from the station to the hypocenter.

The **epicentral distance** (or **earthquake distance**): the **great circle distance** from the station to epicenter.

The **Joyner–Boore distance**: the distance from the station to the closest point of the Earth's surface, located over the *rupture surface*, i.e., the rupturing portion of the fault plane.

The **rupture distance**: the distance from the station to the closest point on the rupture surface.

The **seismogenic depth distance**: the distance from the station to the closest point of the rupture surface within the *seismogenic zone*, i.e., the depth range where the earthquake may occur; usually, at depth 8–12 km.

Also used are the distances from the station to:

- the center of static energy release and the center of static deformation of the fault plane;
- the surface point of maximal macroseismic intensity, i.e., of maximal ground acceleration (it can be different from epicenter);
- the epicenter such that the reflection of body waves from the *Moho interface* (the crust-mantle boundary) contribute more to ground motion than directly arriving shear waves (it called *critical Moho distance*);
- sources of noise and disturbances: oceans, lakes, rivers, railroads, buildings.

The **space-time link distance** between two earthquakes x and y is defined by

$$\sqrt{d^2(x, y) + C|t_x - t_y|^2},$$

where $d(x, y)$ is the distance between their epicenters or hypocenters, $|t_x - t_y|$ is the time lag, and C is a scaling constant needed to connect distance $d(x, y)$ and time.

25.2. DISTANCES IN ASTRONOMY

A *celestial object* (or *celestial body*) is a term describing astronomical objects such as stars and planets. The *celestial sphere* is the projection of celestial objects into their apparent positions in the sky as viewed from the Earth. The *celestial equator* is the projection of the Earth's equator onto the celestial sphere. The *celestial poles* are the projections of Earth's North and South poles onto the celestial sphere. The *hour circle* of a celestial object is the great circle of the celestial sphere, passing through the object and the celestial poles. The *ecliptic* is the intersection of the plane that contains the orbit of the Earth with the celestial sphere: seen from the Earth, it is the path that the Sun appears to follow across the sky over the course of a year. The *vernal equinox point* (or the *First point in Aries*) is one of the two points on the celestial sphere, where the celestial equator intersects the ecliptic: it is the position of the Sun on the celestial sphere at the time of the vernal equinox.

The *horizon* is the line that separates Earth from sky. It divides the sky into the upper hemisphere that the observer can see, and the lower hemisphere that he can not. The pole of the upper hemisphere (the point of the sky directly overhead) is called *zenith*, the pole of the lower hemisphere is called *nadir*.

In general, an **astronomical distance** is a distance from one celestial body to another (measured in light-years, parsecs, or Astronomical Units). The average distance between stars (in a galaxy like our own) is several light-years. The average distance between galaxies (in a cluster) is only about 20 times their diameter, i.e., several megaparsecs.

- **Latitude**

In spherical coordinates (r, θ, ϕ), the **latitude** is the **angular distance** δ from the xy-plane (*fundamental plane*) to a point, measured from the origin; $\delta = 90^\circ - \theta$, where θ is the **colatitude**.

In *geographic coordinate system* (or *earth-mapping coordinate system*), the **latitude** is the angular distance from the Earth's equator to an object, measured from the center of the Earth. Latitude is measured in degrees, from -90° (South pole) to $+90^\circ$ (North pole). *Parallels* are the lines of constant latitude.

In Astronomy, the **celestial latitude** is the latitude of a celestial object on the celestial sphere from the intersection of the fundamental plane with the celestial sphere in given *celestial coordinate system*. In the *equatorial coordinate system* the fundamental plane is the plane of the Earth's equator; in the *ecliptic coordinate system* the fundamental plane is the plane of ecliptic; in the *galactic coordinate system* the fundamental plane is the plane of Milky Way; in the *horizontal coordinate system* the fundamental plane is the observer's horizon. Celestial latitude is measured in degrees.

- **Longitude**

In spherical coordinates (r, θ, ϕ), the **longitude** is the **angular distance** ϕ in the xy-plane from x-axis to the intersection of a great circle, that passes through a point, with xy-plane.

In *geographic coordinate system* (or *earth-mapping coordinate system*), the **longitude** is the angular distance measured eastward along the Earth's equator from the *Greenwich meridian* (or *Prime meridian*) to the intersection of the meridian that passes through the object. Longitude is measured in degrees, from 0° to 360°. A *meridian* is a great circle, passing through Earth's North and South poles; the meridians are the lines of constant longitude.

In Astronomy, the **celestial longitude** is the longitude of a celestial object on the celestial sphere measured eastward, along the intersection of the fundamental plane with the celestial sphere in given *celestial coordinate system*, from the chosen point. In the *equatorial coordinate system* the fundamental plane is the plane of the Earth's equator; in the *ecliptic coordinate system* – the plane of ecliptic; in the *galactic coordinate system* – the plane of Milky Way; in the *horizontal coordinate system* – the observer's horizon. Celestial longitude is measured in units of time.

- **Colatitude**

  In spherical coordinates (r, θ, ϕ), the **colatitude** is the **angular distance** θ from the z-axis to a point, measured from the origin; $\theta = 90° - \delta$, where δ is the **latitude**.

  In *geographic coordinate system* (or *earth-mapping coordinate system*), the **colatitude** is the angular distance from the Earth's North pole to an object, measured from the center of the Earth. Colatitude is measured in degrees.

- **Declination**

  In the *equatorial coordinate system* (or *geocentric coordinate system*), the **declination** δ is the **celestial latitude** of a celestial object on the celestial sphere, measured from the celestial equator. Declination is measured in degrees, from $-90°$ to $+90°$.

- **Right ascension**

  In the *equatorial coordinate system* (or *geocentric coordinate system*), fixed to the stars, the **right ascension** RA is the **celestial longitude** of a celestial object on the celestial sphere, measured eastward along the celestial equator from the First point in Aries to the intersection of the hour circle of the celestial object. Right ascension is measured in units of time (hours, minutes and seconds) with one hour of time approximately equal to $15°$. The time needed for one complete cycle of the precession of the equinoxes is called *Platonic year* (or *Great year*); it is about 257 centures and slightly decreases. This cycle is important in Astrology and Maya calendar.

- **Hour angle**

  In the *equatorial coordinate system* (or *geocentric coordinate system*), fixed to the Earth, the **hour angle** is the **celestial longitude** of a celestial object on the celestial sphere, measured along the celestial equator from the observer's meridian to the intersection of the circle of the celestial object. Hour angle is measured in units of time (hours, minutes and seconds). It gives the time elapsed since the celestial object's last transit at the observer's meridian (for a positive hour angle), or the time unit the next transit (for a negative hour angle).

- **Polar distance**

  In the *equatorial coordinate system* (or *geocentric coordinate system*), the **polar distance** (or *codeclination*) *PD* is the **colatitude** of a celestial object, i.e., the **angular distance** from the celestial pole to a celestial object on the celestial sphere, similar as **declination** δ is measured from the celestial equator: $PD = 90° \pm \delta$. Polar distance is expressed in degrees, and cannot exceed $90°$ in magnitude. An object on the celestial equator has $PD = 90°$.

- **Ecliptic latitude**

  In the *ecliptic coordinate system*, the **ecliptic latitude** is the **celestial latitude** of a celestial object on the celestial sphere from the ecliptic. Ecliptic latitude is measured in degrees.

- **Ecliptic longitude**

  In *ecliptic coordinate system*, the **ecliptic longitude** is the **celestial longitude** of a celestial object on the celestial sphere measured eastward along the ecliptic from the First point in Aries. Ecliptic longitude is measure in units of time.

- **Altitude**

  In the *horizontal coordinate system* (or *Alt/Az coordinate system*) , the **altitude** ALT is the **celestial latitude** of an object from the horizon. It is the complement of the **zenith angle** ZA: $ALT = 90° - ZA$. Altitude is measured in degrees.

- **Azimuth**

  In the *horizontal coordinate system* (or *Alt/Az coordinate system*), the **azimuth** is the **celestial longitude** of an object, measured eastward along the horizon from the North point. Azimuth is measured in degrees, from 0 to 360°.

- **Zenith angle**

  In the *horizontal coordinate system* (or *Alt/Az coordinate system*), the **zenith angle** ZA is the **colatitude** of an object, measured from the zenith.

- **Lunar distance**

  The **lunar distance** is the **angular distance** between the Moon and another celestial object.

- **Elliptic orbit distance**

  The **elliptic orbit distance** is a distance from a mass M which a body has in an elliptic orbit about the mass M at the focus. This distance is given by

$$\frac{a(1-e^2)}{1+e\cos\theta},$$

  where a is the *semi-major axis*, e is the *eccentricity*, and θ is the orbital angle.

  The *semi-major axis* a of an ellipse (or an elliptic orbit) is half of its major axis; it is the average (over the eccentric anomaly) elliptic orbit distance. Such average distance over the **true anomaly** is the *semi-minor axis*, i.e., the half of its minor axis.

  The *eccentricity* e of an ellipse (or an elliptic orbit) is the ratio of half the distance between the foci c and the semi-major axis a: $e = \frac{c}{a}$. For an elliptic orbit, $e = \frac{r_+ - r_-}{r_+ + r_-}$, where r_+ is the **apoapsis distance**, and r_- is the **periapsis distance**.

- **Periapsis distance**

  The **periapsis distance** is the closest distance r_- a body reaches in an elliptic orbit about a mass M. $r_- = a(1-e)$, where a is the *semi-major axis*, and e is the *eccentricity*.

The **perigee** is the periapsis of an elliptical orbit around the Earth. The **perihelion** is the periapsis of an elliptical orbit around the Sun. The **periastron** is the point of closest approach of two stars which are in orbit around each other.

- **Apoapsis distance**

The **apoapsis distance** is the farthest distance r_+ a body reaches in an elliptic orbit about a mass M. $r_+ = a(1 + e)$, where a is the *semi-major axis*, and e is the *eccentricity*.

The **apogee** is the apoapsis of an elliptical orbit around the Earth. The **aphelion** is the apoapsis of an elliptical orbit around the Sun. The **apastron** is the point of greatest separation between two stars which are in orbit around each other.

- **True anomaly**

The **true anomaly** is the **angular distance** of a point in an orbit past the point of **periapsis** measured in degrees.

- **Titius–Bode law**

Titius–Bode law is an empirical (not explained well yet) law approximating the mean planetary distance from Sun (i.e. its orbital *semi-major axis*) by $\frac{3k+4}{10}$ AU. Here 1 AU denotes such mean distance for Earth (i.e., about 1.5×108 km ≈ 8.3 light-minutes) and $k = 0, 2^0, 2^1, 2^2, 2^3, 2^4, 2^5, 2^6, 2^7$ for Mercury, Venus, Earth, Mars, Ceres (the largest one in Asteroid Belt), Jupiter, Saturn, Uranus, Pluto. However, Neptune not fits in the law while Pluto fits Neptune's spot $k = 2^7$.

- **Primary-satellite distances**

Consider two celestial bodies: a *primary* M and a smaller one m (a satellite, orbiting around M, or a secondary star, or a comet passing by).

The **mean distance** is the arithmetic mean of the maximum and minimum distances of a body m from its primary M.

Let ρ_M, ρ_m and R_M, R_m denote densities and radii of M and m. Then the **Roche limit** of the pair (M, m) is the maximal distance between them within which m will disintegrate due to tidal forces of M exceeding the gravitational self-attraction of m. This distance is $R_M \sqrt[3]{2\frac{\rho_M}{\rho_m}} \approx 1.26 R_M \sqrt[3]{\frac{\rho_M}{\rho_m}}$ if m is a rigid spherical body, and it is about $2.423 R_M \sqrt[3]{\frac{\rho_M}{\rho_m}}$ if body m is fluid. The Roche limit is relevant only if it exceeds R_M. It is $0.80 R_M$, $1.49 R_M$ and $2.80 R_M$ for pairs (the Sun, the Earth), (the Earth, the Moon) and (the Earth, a comet), respectively. A possible origin of the rings of Saturn is a moon which came closer to Saturn than its Roche limit.

Let $d(m, M)$ denote the distance between m and M; let S_m and S_M denote masses of m and M. Then the **Hill sphere of m in presence of M** is an approximation of the gravitational *sphere of influence* of m in the face of perturbation from M. Its radius is about $d(m, M)\sqrt[3]{\frac{S_m}{3S_M}}$. For example, the radius of Hill sphere of the Earth is 0.01 AU; the Moon, at distance 0.0025 AU, is well within the Hill sphere of the Earth.

The pair (M, m) can be characterized by five **Lagrange points** L_i, $1 \leqslant i \leqslant 5$, where a third, much smaller body (say, a spacecraft), will be relatively stable because its centrifugal force is equal to combined gravitational attraction of M and m. Those points are:

L_1, L_2, L_3 lying on the line trough centers of M and m so that $d(L_3, m) = 2d(M, m)$, $d(M, L_2) = d(M, L_1) + d(L_1, m) + d(m, L_2)$, and $d(L_1, m) = d(m, L_2)$;

L_4 and L_5 lying on the orbit of m around M and forming equilateral triangles with the centers of M and m. Those two points are more stable; each of them form with M and m a partial solution of (unsolved) gravitational *three-body problem*.

Chapter 26

Distances in Cosmology and Theory of Relativity

26.1. DISTANCES IN COSMOLOGY

The *Universe* is defined as the whole space-time continuum in which we exist, together with all the energy and matter within it.

Cosmology is the study of the large-scale structure of the Universe. Specific cosmological questions of interest include the *isotropy* of the Universe (on the largest scales, the Universe looks the same in all directions, i.e., is invariant to rotations), the *homogeneousness* of the Universe (any measurable property of the Universe is the same everywhere, i.e., it is invariant to translations), the density of the Universe, the equality of matter and anti-matter, and the origin of density fluctuations in galaxies.

In the 1929, E. Hubble discovered that all galaxies have a positive *redshift*, i.e., all galaxies, except for a few nearby galaxies like Andromeda, are receding from the Milky Way. By the Copernican principle (that we are not at a special place in the Universe), we deduce that all galaxies are receding from each other, i.e., we live in a dynamic, expanding Universe, and the further a galaxy is away from us, the faster it is moving away (this is now called *Hubble law*). The *Hubble flow* is the general outward movement of galaxies and clusters of galaxies resulting from the expansion of the Universe. It occurs radially away from the observer, and obeys the Hubble law. Galaxies can overcome this expansion on scales smaller than that of clusters of galaxies; the clusters, however, are being forever driven apart by the Hubble flow.

In Cosmology, the prevailing scientific theory about the early development and shape of the Universe is the *Big Bang Theory*. The observation that galaxies appear to be receding from each other can be combined with the General Theory of Relativity to extrapolate the condition of the Universe back in time. This leads to the construction that as one goes back in time, the Universe becomes increasingly hot and dense, then leads to a gravitational singularity, at which all distances become zero, and temperatures and pressures become infinite. The term *Big Bang* is used to refer to a hypothesized point in time when the observed expansion of the Universe began. Based on measurements of the expansion of the Universe, it is currently believed that the Universe has an age of 13.7 ± 0.2 billion years. It should be longer if the expansion accelerates, as was supposed recently. N. Dauphas, basing on the abundance ratio of uranium/thorium chondritic meteorites, estimated ([Dau05]) this age as 14.5 ± 2 billion years.

In Cosmology (or, more exactly, *Cosmography*, the measurement of the Universe) there are many ways to specify the distance between two points, because in the expanding Universe, the distances between comoving objects are constantly changing, and Earth-bound observers look back in time as they look out in distance. The unifying aspect is that all distance measures somehow measure the separation between events on *radial null trajectories*, i.e., trajectories of photons with terminate at the observer. In general, the **cosmological distance** is a distance far beyond the boundaries of our Galaxy.

The geometry of the Universe is determined by several *cosmological parameters*: the *expansion parameter* (or the *scale factor*) a, the *Hubble constant* H, the *density* ρ and the *critical density* ρ_{crit} (the density required for the Universe to stop expansion and, eventually, collapse back onto itself), the *cosmological constant* Λ, the *curvature of the Universe* k. Many of these quantities are related under the assumptions of a given *cosmological model*. The most common cosmological models are closed and open *Friedmann–Lemaître cosmological models* and *Einstein–de Sitter cosmological model* (cf. also *Einstein cosmological model, de Sitter cosmological model, Eddington–Lemaître cosmological model*). The Einstein–de Sitter cosmological model assumes a homogeneous, isotropic, constant curvature Universe with zero cosmological constant Λ and pressure P. For constant mass M of the Universe, $H^2 = \frac{8}{3}\pi G\rho$, $t = \frac{2}{3}H^{-1}$, $a = \frac{1}{R_C}(\frac{9GM}{2})^{1/3}t^{2/3}$, where $G = 6.67 \times 10^{-11}$ m^3 kg^{-1} s^{-2} is the *gravitational constant*, $R_C = |k|^{-\frac{1}{2}}$ is the absolute value of the *radius of curvature*, and t is the age of the Universe.

Expansion parameter $a = a(t)$ is a *scale factor*, related the size of the Universe $R = R(t)$ at time t to the size of the Universe $R_0 = R(t_0)$ at time t_0 by $R = aR_0$. Most commonly in modern usage it is chosen to be dimensionless, with $a(t_{obser}) = 1$, where t_{obser} is the present age of the Universe.

The *Hubble constant* H is the constant of proportionality between the speed of expansion v and the size of the Universe R, i.e., $v = HR$. This equality is just the *Hubble law* with the Hubble constant $H = \frac{a'(t)}{a(t)}$. The current value of the Hubble constant, as estimated recently, $H_0 = 71 \pm 4$ km s^{-1} Mpc^{-1}, where the subscript 0 refers to the present epoch because, in general, H changes with time. The *Hubble time* and the **Hubble distance** are defined by $t_H = \frac{1}{H_0}$ and $D_H = \frac{c}{H_0}$ (here c is the speed of light), respectively.

The mass density $\rho = \rho_0$ in the present epoch and the value of the cosmological constant Λ are dynamical properties of the Universe. They can be made into dimensionless density parameters Ω_M and Ω_Λ by $\Omega_M = \frac{8\pi G\rho_0}{3H_0^3}$, $\Omega_\Lambda = \frac{\Lambda}{3H_0^3}$. A third density parameter Ω_R measures the "curvature of space", and can be defined by the relation $\Omega_M + \Omega_\Lambda + \Omega_R = 1$.

These parameters totally determine the geometry of the Universe if it is homogeneous, isotropic, and matter-dominated.

The velocity of a galaxy is measured by the *Doppler effect*, i.e., the fact that light emitted from a source is shifted in wavelength by the motion of the source. A relativistic form of the Doppler shift exists for objects traveling very fast, and is given by $\frac{\lambda_{obser}}{\lambda_{emit}} = \sqrt{\frac{c+v}{c-v}}$, where λ_{emit} is the emitted wavelength, and λ_{obser} is the shifted (observed) wavelength. The change in wavelength with respect to the source at rest is called *redshift* (if moving away), and is denoted by the letter z. Relativistic redshift z for a particle is given by $z = \frac{\Delta\lambda_{obser}}{\lambda_{emit}} = \frac{\lambda_{obser}}{\lambda_{emit}} - 1 = \sqrt{\frac{c+v}{c-v}} - 1$.

The cosmological redshift is directly related to the scale factor $a = a(t)$: $z+1 = \frac{a(t_{obser})}{a(t_{emit})}$. Here $a(t_{obser})$ is the value of the scale factor at the time the light from the object is observed, and $a(t_{emit})$ is the value of the scale factor at the time it was emitted.

- **Hubble distance**

The **Hubble distance** is a constant

$$D_H = \frac{c}{H_0} = 4220 \text{ Mpc} \approx 1.3 \times 10^{26} \text{ m} \approx 1.377 \times 10^{10} \text{ light-years},$$

where c is the speed of light, and $H_0 = 71 \pm 4 \text{ km s}^{-1} \text{ Mpc}^{-1}$ is the *Hubble constant.*

It is the distance from us to the *cosmic light horizon* which marks the edge of the visible Universe, i.e., the radius of a sphere, centered upon the Earth which is approximately 13.7 billion light-years. It is often referred as **lookback distance** because astronomers, who view distant objects, are "looking back" into the history of the Universe.

For small v/c or small distance d in the expanding Universe the velocity is proportional to the distance, and all distance measures, for example, **angular diameter distance**, **luminosity distance**, etc., converge. Taking the linear approximation, this reduces to $d \approx zD_H$, where z is the *redshift*. But this is true only for small redshifts.

- **Comoving distance**

In the standard Big Bang model are used *comoving coordinates*, where the spatial reference frame is attached to the average positions of galaxies. With this set of coordinates, both time and expansion of the Universe can be ignored and the shape of space is seen as a spatial hypersurface at constant cosmological time.

The **comoving distance** (or *coordinate distance*, *cosmological distance*, χ) is a distance in comoving coordinates between two points in space at a single cosmological time, i.e., the distance between two nearby objects in the Universe which remains constant with epoch if the two objects are moving with the Hubble flow. It is the distance between them which would be measured with rulers at the time they are being observed (the **proper distance**) divided by the ratio of the scale factor of the Universe then to now. In other words, it is the proper distance multiplied by $(1 + z)$, where z is the *redshift*:

$$d_{comov}(x, y) = d_{proper}(x, y) \cdot \frac{a(t_{obser})}{a(t_{emit})} = d_{proper}(x, y) \cdot (1 + z).$$

At the time t_{obser}, i.e., in the present epoch, $a = a(t_{obser}) = 1$, and $d_{comov} = d_{proper}$, i.e., the comoving distance between two nearby events (close in redshift or distance) is the proper distance between them. In general, for a cosmological time t, it holds $d_{comov} = \frac{d_{proper}}{a(t)}$.

The total **line-of sight comoving distance** D_C from us to a distant object is computed by integrating the infinitesimal $d_{comov}(x, y)$ contributions between nearby events along the time ray from the time t_{emit}, when the light from the object was emitted, to the time t_{obser}, when the object is observed:

$$D_C = \int_{t_{emit}}^{t_{obser}} \frac{c\,dt}{a(t)}.$$

In terms of redshift, D_C from us to a distant object is computed by integrating the infinitesimal $d_{comov}(x, y)$ contributions between nearby events along the radial ray from $z = 0$ to the object: $D_C = D_H \int_0^z \frac{dz}{E(z)}$, where D_H is the **Hubble distance**, and $E(z) = (\Omega_M(1+z)^3 + \Omega_R(1+z)^2 + \Omega_\Lambda)^{\frac{1}{2}}$.

In a sense, the comoving distance is the fundamental distance measure in Cosmology since all other distances can simply derived in terms of it.

- **Proper distance**

The **proper distance** (or **physical distance**, *ordinary distance*) is a distance between two nearby events in the frame in which they happen at the same time. It is the distance measured by a ruler at the time of observation. So, for a cosmological time t, it holds

$$d_{proper}(x, y) = d_{comov} \cdot a(t),$$

where d_{comov} is the **comoving distance**, and $a(t)$ is the *scale factor*.

In the present epoch (i.e., at the time t_{obser}) it holds $a = a(t_{obser}) = 1$, and $d_{proper} = d_{comov}$. So, the proper distance between two nearby events (i.e., close in redshift or distance) is the distance which we would measure locally between the events today if those two points were locked into the Hubble flow.

- **Proper motion distance**

The **proper motion distance** (or **transverse comoving distance**, *contemporary angular diameter distance*) D_M is a distance from us to a distant object, defined as the ratio of the actual transverse velocity (in distance over time) of the object to its *proper motion* (in radians per unit time). It is given by

$$D_M = \begin{cases} D_H \frac{1}{\sqrt{\Omega_R}} \sinh(\sqrt{\Omega_R} D_C / D_H), & \text{for} \Omega_R > 0, \\ D_C, & \text{for } \Omega_R = 0, \\ D_H \frac{1}{\sqrt{|\Omega_R|}} \sin(\sqrt{|\Omega_R|} D_C / D_H), & \text{for} \Omega_R < 0, \end{cases}$$

where D_H is the **Hubble distance**, and D_C is the **line-of-sight comoving distance**. For $\Omega_\Lambda = 0$, there is an analytic solution (z is the *redshift*):

$$D_M = D_H \frac{2(2 - \Omega_M(1-z) - (2 - \Omega_M)\sqrt{1 + \Omega_M z})}{\Omega_M^2 (1+z)}.$$

The proper motion distance D_M coincides with the line-of-sight comoving distance D_C if and only if the curvature of the Universe is equal to zero. The **comoving distance** between two events at the same redshift or distance but separated on the sky by some angle $\delta\theta$ is equal to $D_M \delta\theta$.

The distance D_M is related to the **luminosity distance** D_L by $D_M = \frac{D_L}{1+z}$, and to the **angular diameter distance** D_A by $D_M = (1+z) D_A$.

- **Luminosity distance**

The **luminosity distance** D_L is a distance from us to a distant object, defined by the relationship between observed flux S and emitted luminosity L:

$$D_L = \sqrt{\frac{L}{4\pi S}}.$$

This distance is related to the **proper motion distance** D_M by $D_L = (1+z)D_M$, and to the **angular diameter distance** D_A by $D_L = (1+z)^2 D_A$, where z is the *redshift*.

The luminosity distance does take into account the fact that the observed luminosity is attenuated by two factors, the relativistic redshift and the Doppler shift of emission, each of which contributes an $(1+z)$ attenuation: $L_{obser} = \frac{L_{emiss}}{(1+z)^2}$.

The *corrected luminosity distance* D'_L is defined by $D'_L = \frac{D_L}{1+z}$.

- **Distance modulus**

The **distance modulus** DM is defined by $DM = 5\ln(\frac{D_L}{10pc})$, where D_L is the **luminosity distance**. The distance modulus is the difference between the absolute magnitude and apparent magnitude of an astronomical object. Distance moduli are most commonly used when expressing the distances to other galaxies. For example, the Large Magellanic Cloud is at a distance modulus 18.5, the Andromeda Galaxy's distance modulus is 24.5, and the Virgo Cluster has the DM equal to 31.7.

- **Angular diameter distance**

The **angular diameter distance** (or *angular size distance*) D_A is a distance from us to a distant object, defined as the ratio of an object's physical transverse size to its angular size (in radians). It is used to convert angular separations in telescope images into proper separations at the source. It is special for not increasing indefinitely as $z \to \infty$; it turns over at $z \sim 1$, and thereafter more distant objects actually appear larger in angular size. Angular diameter distance is related to the **proper motion distance** D_M by $D_A = \frac{D_M}{1+z}$, and to the **luminosity distance** D_L by $D_A = \frac{D_L}{(1+z)^2}$, where z is the *redshift*.

- **Light-travel distance**

The **light-travel distance** (or *light-travel time distance*) D_{lt} is a distance from us to a distant object, defined by $D_{lt} = c(t_{obser} - t_{emit})$, where t_{obser} is the time, when the object was observed, and t_{emit} is the time, when the light from the object was emitted.

It is not a very useful distance, because it is very hard to determine t_{emit}, the age of the Universe at the time of emission of the light which we see.

- **Parallax distance**

The **parallax distance** D_P is a distance from us to a distant object, defined from measuring of *parallaxes*, i.e., its apparent changes of position in the sky caused by the motion of the observer on the Earth around the Sun.

The *cosmological parallax* is measured as the difference in the angles of line of sight to the object from two endpoints of the diameter of the orbit of the Earth which is used as a *baseline*. Given a baseline, the parallax $\alpha - \beta$ depends on the distance, and knowing this and the length of the baseline (two astronomical units AU, where AU $\approx$ 150 million kilometers is the distance from the Earth to the Sun) one can compute the distance to the star by the formula

$$D_P = \frac{2}{\alpha - \beta},$$

where D_P is in parsecs, α and β are in arc-seconds.

In Astronomy, "parallax" usually means the *annual parallax* p which is the difference in the angles of a star seen from the Earth and from the Sun. Therefore the distance of a star (in parsecs) is given by $D_P = \frac{1}{p}$.

- **Radar distance**

The **radar distance** D_R is a distance from us to a distant object, measured by a *radar*.

Radar typically consists of a high frequency radio pulse sent out for a short interval of time. When it encounters a conducting object, sufficient energy is reflected back to allow the radar system to detect it. Since radio waves travel in air at close to their speed in vacuum, one can calculate the distance D_R of the detected object from the round-trip time t between the transmitted and received pulses as

$$D_R = \frac{1}{2}ct,$$

where c is the speed of light.

- **Cosmological distance ladder**

For measuring distances to astronomical objects, one uses a kind of "ladder" of different methods; each method goes only to a limited distance, and each method which goes to a larger distance builds on the data of the preceding methods.

The starting point is knowing the distance from the Earth to the Sun; this distance is called one *astronomical unit* (AU), and is roughly 150 million kilometers. Copernicus made the first, roughly accurate, solar system model, using data taken in ancient times, in his famous *De Revolutionibus* (1543). Distances in inner solar system are measured by bouncing radar signals off planets or asteroids, and measuring the time until the echo is received. Modern models are very accurate.

The next step in the ladder consists of simple geometrical methods; with them, one can go to a few hundred light-years. The distance to nearby stars can be determined by their *parallaxes*: using Earth's orbit as a baseline, the distances to stars are measured by triangulation. This is accurate to about 1% at 50 light-years, 10% at 500 light-years.

Using data acquired by the geometrical methods, and adding *photometry* (i.e., measurements of the brightness) and spectroscopy, one gets the next step in the ladder for stars so far away that their parallaxes are not measurable yet. As the brightness decreases proportionally to the square of the distance, if we know the *absolute brightness* of a star (i.e., its in the standard reference distance 10 parsecs), and its *apparent brightness* (i.e., the actual brightness which we observe on the Earth) we can say how far away the star is. To define the absolute brightness, one can use a *Hertzsprung–Russel diagram*: stars of similar type have similar brightnesses; thus, if we know a star's type (from its color and/or spectrum), we can find its distance by comparing its apparent with its absolute magnitude; the latter derived from geometric parallaxes to nearby stars.

For even larger distances in the Universe, one need an additional element: *standard candles*, i.e., several types of cosmological objects, for which one can determine their absolute brightness without knowing their distances. *Primary standard candles* are *Cepheids* variable stars. They periodically change their size and temperature. There is a relationship between the brightness of these pulsating stars and the period of their oscillations, and this relationship can be used to determine their absolute brightness. Cepheids can be identified as far as in the Virgo Cluster (60 million light-years). Another type of standard candle (*secondary standard candles*) which is brighter than Cepheids and, hence, can be used to determine the distances to galaxies even hundreds of millions of light-years away, are supernovae and entire galaxies.

For really large distances (several hundreds of millions of light-years or even several billions of light-years), the cosmological redshift and the Hubble law are used. A complication is that it is not clear what is meant by "distance" here, and there are several types of distances used in Cosmology (**luminosity distance**, **proper motion distance**, **angular diameter distance**, etc.).

Depending of situation, there is a large variety of special techniques to measure distances in Cosmology, such as *secular parallax distance*, *statistical parallax distance*, *Bondi radar distance*, *kinematic distance*, *expansion parallax distance*, *light echo distance*, *spectroscopic parallax distance*, *RR Lyrae distance*, etc.

26.2. DISTANCES IN THEORY OF RELATIVITY

The *Minkowski space-time* (or *Minkowski space*, *Lorentz space-time*, *flat space-time*) is the usual geometric setting for the Einstein Special Theory of Relativity. In this setting the three ordinary dimensions of space are combined with a single dimension of time to form a four-dimensional *space-time* $\mathbb{R}^{1,3}$ in the absence of gravity.

Vectors in $\mathbb{R}^{1,3}$ are called *four-vectors* (or *events*). They can be written as (ct, x, y, z), where the first component is called *time-like component* (c is the speed of light, and t is the time) while the other three components are called *spatial components*. In *spherical coordinates*, they can be written as (ct, r, θ, ϕ). In the Theory of Relativity, the *spherical coordinates* are a system of curvilinear coordinates (ct, r, θ, ϕ), where c is the speed of light, t is the time, r is the *radius* from a point to the origin with $0 \leqslant r < \infty$, ϕ is the azimuthal angle in the xy-plane from x-axis with $0 \leqslant \phi < 2\pi$ (*longitude*), and θ is the polar angle from the z-axis with $0 \leqslant \theta \leqslant \pi$ (*colatitude*). Four-vectors are classified according to the sign of their squared *norm*:

$$\|v\|^2 = \langle v, v \rangle = c^2t^2 - x^2 - y^2 - z^2.$$

They are said to be *time-like*, *space-like*, and *light-like* (*isotropic*) if their squared norms are positive, negative, or equal to zero, respectively.

The set of all light-like vectors forms the *light cone*. If the coordinate origin is singled out, the space can be broken up into three domains: domains of *absolute future* and *absolute past*, falling into the light cone, whose points are joined to the origin by time-like vectors with positive or negative value of time coordinate, respectively, and domain of *absolute elsewhere*, falling out of the light cone, whose points are joined to the origin by space-like vectors.

A *world line* of an object is the sequence of events, that marks the time history of the object. A world line traces out the path of a single point in the Minkowski space. It is an one-dimensional *curve*, represented by the coordinates as a function of one parameter. World line is a *time-like* curve in space-time, i.e., at any point its *tangent vector* is a time-like four-vector. All world lines fall within the light cone, formed by *light-like* curves, i.e., the curves which tangent vectors are light-like four-vectors, corresponded to the motion of light and other particles of zero rest mass.

World lines of particles at constant speed (equivalently, of free falling particles) are called *geodesics*. In Minkowski space they are straight lines.

A geodesic in the Minkowski space, which joins two given events x and y, is the longest curve among all world lines which join these two events. This follows from the **inverse triangle inequality**

$$\|x + y\| \geqslant \|x\| + \|y\|,$$

according to which a time-like broken line joining two events is shorter than the single time-like geodesic joining them, i.e., the proper time of the particle moving freely from x to y is greater than the proper time of any other particle whose world line joins these events. This fact is usually called *twin paradox*.

The *space-time* is a four-dimensional *manifold* which is the usual mathematical setting for the Einstein General Theory of Relativity. Here the three spatial components with a single time-like component form a four-dimensional space-time in the presence of gravity. Gravity is equivalent to the geometric properties of space-time, and in the presence of gravity the geometry of space-time is curved. Thus, the space-time is a four-dimensional curved manifold for which the tangent space to any point is the Minkowski space, i.e., it is a *pseudo-Riemannian manifold* of signature $(1, 3)$.

In the General Theory of Relativity, gravity is described by the properties of the local geometry of space-time. In particular, the gravitational field can be built out of a **metric tensor**, a quantity describing geometrical properties space-time such as distance, area, and angle. Matter is described by its *stress-energy tensor*, a quantity which contains the density and pressure of matter. The strength of coupling between matter and gravity is determined by the *gravitational constant*.

The *Einstein field equation* is an equation in the General Theory of Relativity, that describes how matter creates gravity and, conversely, how gravity affects matter. A solution of the Einstein field equation is a certain **Einstein metric** appropriate for the given mass and pressure distribution of the matter.

Black hole is a massive astrophysical object that is theorized to be created from the collapse of a neutron star. The gravitational forces are so strong in a black hole that they overcome neutron degeneracy pressure and, roughly speaking, collapse to a point (known as a *singularity*). Even light cannot escape the gravitational pull of a black hole within the black hole's so-called *Schwarzschild radius* (or *gravitational radius*). Uncharged, zero angular momentum black holes are called *Schwarzschild black holes*. Uncharged non-zero angular momentum black holes are called *Kerr black holes*. Non-spinning charged black holes are called *Reissner–Nordström black holes*. Charged, spinning black holes are called *Kerr–Newman black holes*. Corresponding metrics describe how space-time is curved by matter in the presence of these black holes.

For an additional information see, for example, [Wein72].

- **Minkowski metric**

The **Minkowski metric** is a **pseudo-Riemannian metric**, defined on the *Minkowski space* $\mathbb{R}^{1,3}$, i.e., a four-dimensional real vector space which is considered as the *pseudo-Euclidean space* of signature $(1, 3)$. It is defined by its **metric tensor**

$$((g_{ij})) = \begin{pmatrix} 1 & 0 & 0 & 0 \\ 0 & -1 & 0 & 0 \\ 0 & 0 & -1 & 0 \\ 0 & 0 & 0 & -1 \end{pmatrix}.$$

The *line element* ds^2, and the *space-time interval element* ds of this metric are given by

$$ds^2 = c^2\,dt^2 - dx^2 - dy^2 - dz^2.$$

In *spherical coordinates* (ct, r, θ, ϕ), one has $ds^2 = c^2\,dt^2 - dr^2 - r^2\,d\theta^2 - r^2 \sin^2\theta\,d\phi^2$.

The pseudo-Euclidean space $\mathbb{R}^{3,1}$ of signature $(3, 1)$ with the *line element*

$$ds^2 = -c^2\,dt^2 + dx^2 + dy^2 + dz^2$$

can also be used as a space-time model of the Einstein Special Theory of Relativity. The metric of signature $(1, 3)$ is commonly used by people from a Particle Physics background, whereas the metric of signature $(3, 1)$ is typically used by people from a Relativity background.

- **Lorentz metric**

A **Lorentz metric** (or **Lorentzian metric**) is a **pseudo-Riemannian metric** of signature $(1, p)$.

A *Lorentzian manifold* is a manifold equipped with the Lorentz metric. The curved space-time of the General Theory of Relativity can be modeled as a Lorentzian manifold M of signature $(1, 3)$. The *Minkowski space* $\mathbb{R}^{1,3}$ with the flat **Minkowski metric** is a model of a flat Lorentzian manifold.

In the *Lorentzian Geometry* the following definition of distance is commonly used. Given a rectifiable non-space-like curve $\gamma : [0, 1] \to M$ in the space-time M, the *length* of the curve is defined as $l(\gamma) = \int_0^1 \sqrt{-\langle \frac{d\gamma}{dt}, \frac{d\gamma}{dt} \rangle} dt$. For a space-like curve we set $l(\gamma) = 0$. Then the **Lorentz distance** between two points $p, q \in M$ is defined as

$$\sup_{\gamma \in \Gamma} l(\gamma)$$

if $p \prec q$, i.e., if the set Γ of *future directed* non-space-like curves from p to q is non-empty. Otherwise, the Lorentz distance is equal to 0.

- **Lorentz–Minkowski distance**

The **Lorentz–Minkowski distance** is a distance on $\mathbb{R}^n$ (or on $\mathbb{C}^n$), defined by

$$\sqrt{|x_1 - y_1|^2 - \sum_{i=2}^{n} |x_i - y_i|^2}.$$

- **Galilean distance**

The **Galilean distance** is a distance on $\mathbb{R}^n$, defined by

$$|x_1 - y_1|$$

if $x_1 \neq y_1$, and by

$$\sqrt{(x_2 - y_2)^2 + \cdots + (x_n - y_n)^2}$$

if $x_1 = y_1$. The space $\mathbb{R}^n$ equipped with the Galilean distance is called *Galilean space*. For $n = 4$, it is a mathematical setting for the space-time of classical mechanics according to Galilei–Newton in which the distance between two events taking place at the points p and q at the moments of time t_1 and t_2 is defined as the time interval $|t_1 - t_2|$, while if these events take place at the same time, it is defined as the distance between the points p and q.

- **Einstein metric**

In the General Theory of Relativity, described how space-time is curved by matter, the **Einstein metric** is a solution to the *Einstein field equation*

$$R_{ij} - \frac{g_{ij} R}{2} + \Lambda g_{ij} = \frac{8\pi G}{c^4} T_{ij},$$

i.e., a **metric tensor** $((g_{ij}))$ of signature $(1, 3)$, appropriate for the given mass and pressure distribution of the matter. Here $E_{ij} = R_{ij} - \frac{g_{ij} R}{2} + \Lambda g_{ij}$ is the *Einstein curvature tensor*, R_{ij} is the *Ricci curvature tensor*, R is the *Ricci scalar*, Λ is the *cosmological constant*, G is the *gravitational constant*, and T_{ij} is a *stress-energy tensor*. *Empty space* (*vacuum*) corresponds to the case of vanished Ricci tensor: $R_{ij} = 0$.

The static Einstein metric for a homogeneous and isotropic Universe is given by the *line element*

$$ds^2 = -dt^2 + \frac{dr^2}{(1-kr^2)} + r^2(d\theta^2 + \sin^2\theta\, d\phi^2),$$

where k is the curvature of the space-time, and the *scale factor* is equal to 1.

- **de Sitter metric**

The **de Sitter metric** is a maximally symmetric vacuum solution to the *Einstein field equation* with a positive cosmological constant Λ, given by the *line element*

$$ds^2 = dt^2 + e^{2\sqrt{\frac{\Lambda}{3}}t}(dr^2 + r^2\, d\theta^2 + r^2 \sin^2\theta\, d\phi^2).$$

Without a cosmological constant (i.e., with $\Lambda = 0$), the most symmetric solution to the Einstein field equation in the vacuum is the flat **Minkowski metric**.

The **anti-de Sitter metric** corresponds to the negative value of Λ.

- **Schwarzschild metric**

The **Schwarzschild metric** is a solution to the *Einstein field equation* for empty space (vacuum) around a spherically symmetric mass distribution; this metric gives a representation of an Universe around a black hole of a given mass, from which no energy can be extracted. It was found by K. Schwarzschild in 1916, only a few months after the publication of the Einstein field equation, and was the first exact solution of this equation.

The *line element* of this metric is given by

$$ds^2 = \left(1 - \frac{r_g}{r}\right)c^2\, dt^2 - \frac{1}{(1-\frac{r_g}{r})}dr^2 - r^2(d\theta^2 + \sin^2\theta\, d\phi^2),$$

where $r_g = \frac{2Gm}{c^2}$ is the *Schwarzschild radius*, m is the mass of the black hole, and G is the *gravitational constant*.

This solution is only valid for radii larger than r_g, as at $r = r_g$ there is a coordinate singularity. This problem can be removed by a transformation to a different choice of space-time coordinates, called *Kruskal–Szekeres coordinates*. As $r \to +\infty$, the Schwarzschild metric approaches the **Minkowski metric**.

- **Kruskal–Szekeres metric**

The **Kruskal–Szekeres metric** is a solution to the *Einstein field equation* for empty space (vacuum) around a static spherically symmetric mass distribution, given by the *line element*

$$ds^2 = 4\frac{r_g}{r}\left(\frac{r_g}{R}\right)^2 e^{-\frac{r}{r_g}}(c^2\, dt'^2 - dr'^2) - r^2(d\theta^2 + \sin^2\theta\, d\phi^2),$$

where $r_g = \frac{2Gm}{c^2}$ is the *Schwarzschild radius*, m is the mass of the black hole, G is the *gravitational constant*, R is a constant, and *Kruskal–Szekeres coordinates* (t', r', θ, ϕ) are obtained from the *spherical coordinates* (ct, r, θ, ϕ) by *Kruskal–Szekeres transformation* $r'^2 - ct'^2 = R^2(\frac{r}{r_g} - 1)e^{\frac{r}{r_g}}$, $\frac{ct'}{r'} = \tanh(\frac{ct}{2r_g})$.

In fact, the Kruskal–Szekeres metric is the **Schwarzschild metric**, written in Kruskal–Szekeres coordinates. It shows, that the singularity of the space-time in the Schwarzschild metric at the Schwarzschild radius r_g is not a real physical singularity.

- **Kottler metric**

The **Kottler metric** is the unique spherically symmetric vacuum solution to the *Einstein field equation* with a cosmological constant Λ. It is given by the *line element*

$$ds^2 = -\left(1 - \frac{2m}{r} - \frac{\Lambda r^2}{3}\right)dt^2 + \left(1 - \frac{2m}{r} - \frac{\Lambda r^2}{3}\right)^{-1} dr^2 + r^2(d\theta^2 + \sin^2\theta\, d\phi^2).$$

It is called also **Schwarzschild–de Sitter metric** for $\Lambda > 0$, and **Schwarzschild–anti-de Sitter metric** for $\Lambda < 0$.

- **Reissner–Nordström metric**

The **Reissner–Nordström metric** is a solution to the *Einstein field equation* for empty space (vacuum) around a spherically symmetric mass distribution in the presence of a charge; this metric gives a representation of an Universe around a charged black hole.

The *line element* of this metric is given by

$$ds^2 = \left(1 - \frac{2m}{r} + \frac{e^2}{r^2}\right)dt^2 - \left(1 - \frac{2m}{r} + \frac{e^2}{r^2}\right)^{-1} dr^2 - r^2(d\theta^2 + \sin^2\theta\, d\phi^2),$$

where m is the mass of the hole, e is the charge ($e < m$), and we have used units with the speed of light c and the *gravitational constant* G equal to one.

- **Kerr metric**

The **Kerr metric** (or **Kerr–Schild metric**) is an exact solution to the *Einstein field equation* for empty space (vacuum) around an axial symmetric, rotating mass distribution; this metric gives a representation of an Universe around a rotating black hole.

Its *line element* is given (in *Boyer–Lindquist form*) by

$$ds^2 = \rho^2\left(\frac{dr^2}{\Delta} + d\theta^2\right) + (r^2 + a^2)\sin^2\theta\, d\phi^2 - dt^2 + \frac{2mr}{\rho^2}(a\sin^2\theta\, d\phi - dt)^2,$$

where $\rho^2 = r^2 + a^2\cos^2\theta$, and $\Delta = r^2 - 2mr + a^2$. Here m is the mass of the black hole, and a is the angular velocity as measured by a distant observer.

The generalization of the Kerr metric for charged black hole is known as the **Kerr–Newman metric**. When $a = 0$, the Kerr metric becomes the **Schwarzschild metric**.

- **Kerr–Newman metric**

The **Kerr–Newman metric** is an exact, unique, and complete solution to the *Einstein field equation* for empty space (vacuum) around an axial symmetric, rotating mass distribution in the presence of a charge; this metric gives a representation of an Universe around a rotating charged black hole.

The *line element* of the exterior metric is given by

$$ds^2 = -\frac{\Delta}{\rho^2}\left(dt - a\sin^2\theta\, d\phi\right)^2 + \frac{\sin^2\theta}{\rho^2}\left(\left(r^2 + a^2\right) d\phi - a\, dt\right)^2 + \frac{\rho^2}{\Delta}\, dr^2 + \rho^2\, d\theta^2,$$

where $\rho^2 = r^2 + a^2\cos^2\theta$, and $\Delta = r^2 - 2mr + a^2 + e^2$. Here m is the mass of the black hole, e is the charge, and a is the angular velocity. When $e = 0$, the Kerr-Newman metric becomes the **Kerr metric**.

- **Static isotropic metric**

The **static isotropic metric** is a most general solution to the *Einstein field equation* for empty space (vacuum); this metric can represent a static isotropic gravitational field. The *line element* of this metric is given by

$$ds^2 = B(r)\, dt^2 - A(r)\, dr^2 - r^2\left(d\theta^2 + \sin^2\theta\, d\phi^2\right),$$

where $B(r)$ and $A(r)$ are arbitrary functions.

- **Eddington–Robertson metric**

The **Eddington–Robertson metric** is a generalization of the **Schwarzschild metric** to assume that mass m, the *gravitational constant* G, and the density ρ are altered by unknown dimensionless parameters α, β, and γ (all equal to 1 in the *Einstein field equation*).

The *line element* of this metric is given by

$$ds^2 = \left(1 - 2\alpha\frac{mG}{r} + 2(\beta - \alpha\gamma)\left(\frac{mG}{r}\right)^2 + \cdots\right)dt^2 - \left(1 + 2\gamma\frac{mG}{r} + \cdots\right)dr^2 - r^2\left(d\theta^2 + \sin^2\theta\, d\phi^2\right).$$

- **Janis–Newman–Wincour metric**

The **Janis–Newman–Wincour metric** is the most general spherically symmetric static and asymptotically flat solution to the *Einstein field equation* coupled to a massless scalar field. It is given by the *line element*

$$ds^2 = -\left(1 - \frac{2m}{\gamma r}\right)^{\gamma} dt^2 + \left(1 - \frac{2m}{\gamma r}\right)^{-\gamma} dr^2 + \left(1 - \frac{2m}{\gamma r}\right)^{1-\gamma} r^2\left(d\theta^2 + \sin^2\theta\, d\phi^2\right),$$

where m and γ are constants. For $\gamma = 1$ one obtains the **Schwarzschild metric**. In this case the scalar field vanishes.

- **Robertson–Walker metric**

The **Robertson–Walker metric** (or **Friedmann–Lemaître–Robertson–Walker metric**) is a solution to the *Einstein field equation* for an isotropic and homogeneous Universe filled with a constant density and negligible pressure; this metric gives a representation of an Universe filled with a pressureless dust. The *line element* of this metric is usually written in the *spherical coordinates* (ct, r, θ, ϕ):

$$ds^2 = c^2\,dt^2 - a(t)^2\left(\frac{dr^2}{1-kr^2} + r^2\left(d\theta^2 + \sin^2\theta\,d\phi^2\right)\right),$$

where $a(t)$ is the *scale factor*, and k is the *curvature* of the space-time.

There exists other form for the *line element*:

$$ds^2 = c^2\,dt^2 - a(t)^2\left(dr'^2 + \tilde{r}^2\left(d\theta^2 + \sin^2\theta\,d\phi^2\right)\right),$$

where r' gives the **comoving distance** from the observer, and $\tilde{r}$ gives the **proper motion distance**, i.e., $\tilde{r} = R_C \sinh(r'/R_C)$, or r', or $R_C \sin(r'/R_C)$ for negative, zero or positive curvature, respectively, where $R_C = 1/\sqrt{|k|}$ is the absolute value of the *radius of curvature*.

- **GCSS metric**

A **GCSS** (i.e., **general cylindrically symmetric stationary**) **metric** is a solution to the *Einstein field equation*, given by the *line element*

$$ds^2 = -f\,dt^2 + 2k\,dt\,d\phi + e^{\mu}\left(dr^2 + dz^2\right) + l\,d\phi^2,$$

where the space-time is divided into two regions: the interior, with $0 \leqslant r \leqslant R$, to a cylindrical surface of radius R centered along z, and the exterior, with $R \leqslant r < \infty$. Here f, k, μ and l are functions only of r, and $-\infty < t, z < \infty$, $0 \leqslant \phi \leqslant 2\pi$, the hypersurfaces $\phi = 0$ and $\phi = 2\pi$ are identical.

- **Lewis metric**

The **Lewis metric** is a **cylindrically symmetric stationary metric** which is a solution to the *Einstein field equation* for empty space (vacuum) in the exterior of a cylindrical surface. The *line element* of this metric has the form

$$ds^2 = -f\,dt^2 + 2k\,dt\,d\phi - e^{\mu}\left(dr^2 + dz^2\right) + l\,d\phi^2,$$

where $f = ar^{-n+1} - \frac{c^2}{n^2 a} r^{n+1}$, $k = -Af$, $l = \frac{r^2}{f} - A^2 f$, $e^{\mu} = r^{\frac{1}{2}(n^2-1)}$ with $A = \frac{cr^{n+1}}{naf} + b$. The constants n, a, b, and c can be either real, or complex, the corresponding solutions belong to the *Weyl class* or *Lewis class*, respectively. In the last case, the metric coefficients become

$$\begin{aligned} f &= r\left(a_1^2 - b_1^2\right)\cos(m\ln r) + 2ra_1b_1\sin(m\ln r), \\ k &= -r(a_1a_2 - b_1b_2)\cos(m\ln r) - r(a_1b_2 + a_2b_1)\sin(m\ln r), \\ l &= -r\left(a_2^2 - b_2^2\right)\cos(m\ln r) - 2ra_2b_2\sin(m\ln r), \\ e^{\mu} &= r^{-\frac{1}{2}(m^2+1)}, \end{aligned}$$

where m, a_1, a_2, b_1, and b_2 are real constants with $a_1b_2 - a_2b_1 = 1$. Such metrics form a subclass of *Kasner type metrics*.

- **Van Stockum metric**

The **van Stockum metric** is a stationary cylindrically symmetric solution to the *Einstein field equation* for empty space (vacuum) with a rigidly rotating infinitely long dust cylinder. The *line element* of this metric for the interior of the cylinder is given (in comoving, i.e., corotating coordinates) by

$$ds^2 = -dt^2 + 2ar^2\,dt\,d\phi + e^{-a^2r^2}\left(dr^2 + dz^2\right) + r^2\left(1 - a^2r^2\right)d\phi^2,$$

where $0 \leqslant r \leqslant R$, R is the radius of the cylinder, and a is the angular velocity of the dust particles. There are three vacuum exterior solutions (i.e., **Lewis metrics**) that can be matched to the interior solution, depending on the mass per unit length of the interior (the *low mass case*, the *null case*, and the *ultrarelativistic case*). Under some conditions (for example, if $ar > 1$), the existence of *closed time-like curves* (and, hence, time-travel) is allowed.

- **Levi-Civita metric**

The **Levi-Civita metric** is a static cylindrically symmetric vacuum solution to the *Einstein field equation*, with the *line element*, given (in the Weyl form) by

$$ds^2 = -r^{4\sigma}\,dt^2 + r^{4\sigma(2\sigma-1)}\left(dr^2 + dz^2\right) + C^{-2}r^{2-4\sigma}\,d\phi,$$

where the constant C refers to the deficit angle, and the parameter σ is mostly understood in accordance with the Newtonian analogy of the Levi-Civita solution – the gravitational field of an infinite uniform line-mass (*infinite wire*) with the linear mass density σ. In the case $\sigma = -\frac{1}{2}$, $C = 1$ this metric can be transformed either into the *Taub's plane symmetric metric*, or into the *Robinson–Trautman metric*.

- **Weyl–Papapetrou metric**

The **Weyl–Papapetrou metric** is a stationary axially symmetric solution to the *Einstein field equation*, given by the *line element*

$$ds^2 = F\,dt^2 - e^{\mu}\left(dz^2 + dr^2\right) - L\,d\phi^2 - 2K\,d\phi\,dt,$$

where F, K, L and μ are functions only of r and z, $LF + K^2 = r^2$, $\infty < t, z < \infty$, $0 \leqslant r < \infty$, and $0 \leqslant \phi \leqslant 2\pi$, the hypersurfaces $\phi = 0$ and $\phi - 2\pi$ are identical.

- **Bonnor dust metric**

The **Bonnor dust metric** is a solution to the *Einstein field equation*, which is an axially symmetric metric, describing a cloud of rigidly rotating dust particles moving along circular geodesics about the z-axis in hypersurfaces of $z = constant$. The *line element* of this metric is given by

$$ds^2 = dt^2 + \left(r^2 - n^2\right)d\phi^2 + 2n\,dt\,d\phi + e^{\mu}\left(dr^2 + dz^2\right),$$

where, in Bonnor's comoving (i.e., corotating) coordinates, $n = \frac{2hr^2}{R^3}$, $\mu = \frac{h^2r^2(r^2-8z^2)}{2R^8}$, $R^2 = r^2 + z^2$, and h is a rotation parameter. As $R \to \infty$, the metric coefficients tend to Minkowski values.

- **Weyl metric**

The **Weyl metric** is a general static axially symmetric vacuum solution to the *Einstein field equation*, given, in Weyl canonical coordinates, by the *line element*

$$ds^2 = e^{2\lambda}\,dt^2 - e^{-2\lambda}\left(e^{2\mu}\left(dr^2 + dz^2\right) + r^2\,d\phi^2\right),$$

where λ and μ are functions only of r and z such that $\frac{\partial^2\lambda}{\partial r^2} + \frac{1}{r}\cdot\frac{\partial\lambda}{\partial r} + \frac{\partial^2\lambda}{\partial z^2} = 0$, $\frac{\partial\mu}{\partial r} = r(\frac{\partial\lambda^2}{\partial r} - \frac{\partial\lambda^2}{\partial z})$, and $\frac{\partial\mu}{\partial z} = 2r\frac{\partial\lambda}{\partial r}\frac{\partial\lambda}{\partial z}$.

- **Zipoy–Voorhees metric**

The **Zipoy–Voorhees metric** (or γ**-metric**) is a **Weyl metric**, obtained for

$$e^{2\lambda} = \left(\frac{R_1 + R_2 - 2m}{R_1 + R_2 + 2m}\right)^{\gamma}, \qquad e^{2\mu} = \left(\frac{(R_1 + R_2 + 2m)(R_1 + R_2 - 2m)}{4R_1R_2}\right)^{\gamma^2},$$

where $R_1^2 = r^2 + (z - m)^2$, $R_2^2 = r^2 + (z + m)^2$. Here λ corresponds to the Newtonian potential of a line segment of mass density $\gamma/2$ and length $2m$, symmetrically distributed along the z-axis. The case $\gamma = 1$ corresponds to the **Schwarzschild metric**, the cases $\gamma > 1$ ($\gamma < 1$) correspond to an oblate (prolate) spheroid, and for $\gamma = 0$ one obtains the flat Minkowski space-time.

- **Straight spinning string metric**

The **straight spinning string metric** is given by the *line element*

$$ds^2 = -(dt - a\,d\phi)^2 + dz^2 + dr^2 + k^2r^2\,d\phi^2,$$

where a and $k > 0$ are constants. It describes the space-time around a straight spinning string. The constant k is related to the string's mass-per-length μ by $k = 1 - 4\mu$, and the constant a is a measure for the string's spin. For $a = 0$ and $k = 1$, one obtains the **Minkowski metric** in cylindrical coordinates.

- **Tomimatsu–Sato metric**

A **Tomimatsu–Sato metric** ([ToSa73]) is one of the metrics from an infinite family of spinning mass solutions to the *Einstein field equation*, each of which has the form $\xi = U/W$, where U and W are some polynomials. The simplest solution has $U = p^2(x^4-1)+q^2(y^4-1)-2ipqxy(x^2-y^2)$, $W = 2px(x^2-1)-2iqy(1-y^2)$, where $p^2+q^2=1$. The *line element* for this solution is given by

$$ds^2 = \Sigma^{-1}\big((\alpha\, dt + \beta\, d\phi)^2 - r^2(\gamma\, dt + \delta\, d\phi)^2\big) - \frac{\Sigma}{p^4(x^2-y^2)^4}\big(dz^2 + dr^2\big),$$

where $\alpha = p^2(x^2-1)^2+q^2(1-y^2)^2$, $\beta = -\frac{2q}{p}W(p^2(x^2-1)(x^2-y^2)+2(px+1)W)$, $\gamma = -2pq(x^2-y^2)$, $\delta = \alpha+4((x^2-1)+(x^2+1)(px+1))$, $\Sigma = \alpha\delta-\beta\gamma = |U+W|^2$.

- **Gödel metric**

The **Gödel metric** is an exact solution to the *Einstein field equation* with cosmological constant for a rotating Universe, given by the *line element*

$$ds^2 = -\big(dt^2 + C(r)\, d\phi\big)^2 + D^2(r)\, d\phi^2 + dr^2 + dz^2,$$

where (t, r, ϕ, z) are the usual *cylindrical coordinates*. The *Gödel Universe* is homogeneous if $C(r) = \frac{4\Omega}{m^2}\sinh^2(\frac{mr}{2})$, $D(r) = \frac{1}{m}\sinh(mr)$, where m and Ω are constants. The Gödel Universe allows the possibility of *closed time-like curves*, and hence, time-travel. The condition required to avoid such curves is $m^2 > 4\Omega^2$.

- **Plane wave metric**

The **plane wave metric** is a vacuum solution to the *Einstein field equation*, given by the *line element*

$$ds^2 = 2\, dw\, du + 2f(u)\big(x^2+y^2\big)\, du^2 - dx^2 - dy^2.$$

It is conformally flat, and describes a pure radiation field. The space-time with this metric is called *plane gravitational wave*.

- **Wils metric**

The **Wils metric** is a solution to the *Einstein field equation*, given by the *line element*

$$ds^2 = 2x\, dw\, du - 2w\, du\, dx + \big(2f(u)x\big(x^2+y^2\big) - w^2\big)du^2 - dx^2 - dy^2.$$

It is conformally flat, and describes a pure radiation field which is not a *plane wave*.

- **Koutras–McIntosh metric**

The **Koutras–McIntosh metric** is a solution to the *Einstein field equation*, given by the *line element*

$$ds^2 = 2(ax+b)\, dw\, du - 2aw\, du\, dx + \big(2f(u)(ax+b)\big(x^2+y^2\big) - a^2w^2\big)\, du^2 - dx^2 - dy^2.$$

It is conformally flat and describes a pure radiation field which, in general, is not a *plane wave*. It gives the **plane wave metric** for $a = 0$, $b = 1$, and the **Wils metric** for $a = 1$, $b = 0$.

- **Edgar–Ludwig metric**

The **Edgar–Ludwig metric** is a solution to the *Einstein field equation*, given by the *line element*

$$ds^2 = 2(ax + b)\,dw\,du - 2aw\,du\,dx \\ + \big(2f(u)(ax + b)\big(g(u)y + h(u) + x^2 + y^2\big) - a^2w^2\big)du^2 - dx^2 - dy^2.$$

This metric is a generalization of the **Koutras–McIntosh metric**. It is the most general metric, describing a conformally flat pure radiation (or null fluid) field which, in general, is not a *plane wave*. If plane waves are excluded, it has the form

$$ds^2 = 2x\,dw\,du - 2w\,du\,dx + \big(2f(u)x\big(g(u)y + h(u) + x^2 + y^2\big) - w^2\big)du^2 \\ - dx^2 - dy^2.$$

- **Bondi radiating metric**

The **Bondi radiating metric** describes the asymptotic form of a radiating solution to the *Einstein field equation*, given by the *line element*

$$ds^2 = -\left(\frac{V}{r}e^{2\beta} - U^2r^2e^{2\gamma}\right)du^2 - 2e^{2\beta}\,du\,dr - 2Ur^2e^{2\gamma}\,du\,d\theta \\ + r^2\big(e^{2\gamma}\,d\theta^2 + e^{-2\gamma}\sin^2\theta\,d\phi^2\big),$$

where u is the retarded time, r is the **luminosity distance**, $0 \leqslant \theta \leqslant \pi$, $0 \leqslant \phi \leqslant 2\pi$, and U, V, β, γ are functions of u, r, and θ. This metric is used in the theory of gravitation waves.

- **Taub–NUT de Sitter metric**

The **Taub–NUT de Sitter metric** is a positive-definite (i.e., Riemannian) solution to the *Einstein field equation* with a cosmological constant Λ, given by the *line element*

$$ds^2 = \frac{r^2 - L^2}{4\Delta}dr^2 + \frac{L^2\Delta}{r^2 - L^2}(d\psi + \cos\theta\,d\phi)^2 + \frac{r^2 - L^2}{4}\big(d\theta^2 + \sin^2\theta\,d\phi^2\big),$$

where $\Delta = r^2 - 2Mr + L^2 + \frac{\Lambda}{4}(L^4 + 2L^2r^2 - \frac{1}{3}r^4)$, L and M are parameters, and θ, ϕ, ψ are *Euler angles*. If $\Lambda = 0$, one obtains the **Taub–NUT metric**, using some regularity conditions.

- **Eguchi–Hanson de Sitter metric**

The **Eguchi–Hanson de Sitter metric** is a positive-definite (i.e., Riemannian) solution to the *Einstein field equation* with a cosmological constant Λ, given by the *line element*

$$ds^2 = \left(1 - \frac{a^4}{r^4} - \frac{\Lambda r^2}{6}\right)^{-1} dr^2 + \frac{r^2}{4}\left(1 - \frac{a^4}{r^4} - \frac{\Lambda r^2}{6}\right)(d\psi + \cos\theta\, d\phi)^2 + \frac{r^2}{4}\left(d\theta^2 + \sin^2\theta\, d\phi^2\right),$$

where a is a parameter, and θ, ϕ, ψ are *Euler angles*. If $\Lambda = 0$, one obtains the **Eguchi–Hanson metric**.

- **Barriola–Vilenkin monopole metric**

The **Barriola–Vilenkin monopole metric** is given by the *line element*

$$ds^2 = -dt^2 + dr^2 + k^2 r^2\left(d\theta^2 + \sin^2\theta\, d\phi^2\right),$$

with a constant $k < 1$. There is a deficit solid angle and a singularity at $r = 0$; the plane $t = constant$, $\theta = \frac{\pi}{2}$ has the geometry of a cone. This metric is an example of a conical singularity; it can be used as a model for *monopoles* that might exist in the Universe (cf. **monopole metric**).

A *magnetic monopole* is a hypothetical isolated magnetic pole, "a magnet with only one pole". It has been theorized that such things might exist in the form of tiny particles similar to electrons or protons, forming from topological defects in a similar manner to cosmic strings, but no such particle has ever been found.

- **Bertotti–Robinson metric**

The **Bertotti–Robinson metric** is a solution to the *Einstein field equation* in an Universe with an uniform magnetic field. The *line element* of this metric is given by

$$ds^2 = Q^2\left(-dt^2 + \sin^2 t\, dw^2 + d\theta^2 + \sin^2\theta\, d\phi^2\right),$$

where Q is a constant, $t \in [0, \pi]$, $w \in (-\infty, +\infty)$, $\theta \in [0, \pi]$, and $\phi \in [0, 2\pi]$.

- **Morris–Thorne metric**

The **Morris–Thorne metric** is a *wormhole* solution to the *Einstein field equation* with the *line element*

$$ds^2 = e^{\frac{2\Phi(w)}{c^2}} c^2\, dt^2 - dw^2 - r(w)^2\left(d\theta^2 + \sin^2\theta\, d\phi^2\right),$$

where $w \in [-\infty, +\infty]$, r is a function of w, that reaches some minimal value above zero at some finite value of w, and $\Phi(w)$ is a gravitational potential allowed by the space-time geometry.

A *wormhole* is a hypothetical "tube" in space connecting widely separated positions in an Universe. All wormholes require exotic material with negative energy density, in order to hold them open.

- **Misner metric**

The **Misner metric** is a metric, representing two black holes. Misner (1960) provided a prescription for writing a metric connecting a pair of black holes, instantaneously at rest, whose throats are connected by a *wormhole*. The *line element* of this metric has the form

$$ds^2 = -dt^2 + \psi^4\left(dx^2 + dy^2 + dz^2\right),$$

where the *conformal factor* ψ is given by

$$\psi = \sum_{n=-N}^{N} \frac{1}{\sinh(\mu_0 n)} \frac{1}{\sqrt{x^2 + y^2 + (z + \coth(\mu_0 n))^2}}.$$

The parameter μ_0 is a measure of the ratio of mass to separation of the throats (equivalently, a measure of the distance of a loop in the surface, passing through one throat and out the other). The summation limit N tends to infinity.

The topology of the *Minsler space-time* is that of a pair of asymptotically flat sheets connected by a number of Einstein–Rosen bridges. In the simplest case, the Misner space can be considered as a two-dimensional space with topology $\mathbb{R} \times S^1$ in which light progressively tilt as one moves forward in time, and has *closed time-like curves* after certain point.

- **Alcubierre metric**

The **Alcubierre metric** is a solution to the *Einstein field equation*, representing *warp drive space-time* in which the existence of *closed time-like curves* is allowed. What is violated in this case is only the relativistic principle that a space-going traveler may move with any velocity up to, but not including or overcoming, the speed of light. Alcubierre's construction corresponds to a wrap drive in that it causes space-time to contract in front of spaceship bubble and expand behind, thus providing the spaceship with a velocity that can be much greater than the speed of light relative to distant objects, while the spaceship never locally travels faster than light.

The *line element* of this metric has the form

$$ds^2 = -dt^2 + \left(dx - vf(r)\,dt\right)^2 + dy^2 + dz^2,$$

with $v = \frac{dx_s(t)}{dt}$ the apparent velocity of the warp drive spaceship, $x_s(t)$ the trajectory of the spaceship along coordinate x, the radial coordinate being defined by $r = ((x - x_s(t))^2 + y^2 + z^2)^{\frac{1}{2}}$, and $f(r)$ an arbitrary function subjected to the boundary conditions that $f = 1$ at $r = 0$ (the location of the spaceship), and $f = 0$ at infinity.

- **Rotating C-metric**

The **rotating C-metric** is a solution to the *Einstein–Maxwell equations*, describing two oppositely charged black holes, uniformly accelerating in opposite directions. The *line element* of this metric has the form

$$ds^2 = A^{-2}(x+y)^{-2}\left(\frac{dy^2}{F(y)} + \frac{dx^2}{G(x)} + k^{-2}G(X)\,d\phi^2 - k^2A^2F(y)\,dt^2\right),$$

where $F(y) = -1 + y^2 - 2mAy^3 + e^2A^2y^4$, $G(x) = 1 - x^2 - 2mAx^3 - e^2A^2x^4$, m, e, and A are parameters related to the mass, charge and acceleration of the black holes, and k is a constant fixed by regularity conditions.

- **Kaluza–Klein metric**

The **Kaluza–Klein metric** is a metric in the *Kaluza–Klein model* of 5-dimensional (in general, multi-dimensional) space-time which sought to unify classical gravity and electromagnetism.

T. Kaluza (1919) obtained that if the Einstein theory of pure gravitation is extended to a five-dimensional space-time, the *Einstein field equations* can be split into ordinary four-dimensional gravitation tensor field, plus an extra vector field which is equivalent to Maxwell's equation for the electromagnetic field, plus an extra scalar field (known as the "dilation") which is equivalent to the massless Klein–Gordon equation.

O.Klein (1926) assumed the fifth dimension to have circular topology, so that the fifth coordinate is periodic, and extra dimension is curled up to an unobservable size. An alternative proposal is that the extra dimension is (extra dimensions are) extended, and the matter is trapped in four-dimensional submanifold. This approach has properties similar to four-dimensional – all dimensions are extended and equal at the beginning, and the signature has the form $(p, 1)$.

In a model of large extra dimension, the fifth-dimensional metric of an Universe can be written in Gaussian normal coordinates in the form

$$ds^2 = -(dx_5)^2 + \lambda^2(x_5)\sum_{\alpha,\beta}\eta_{\alpha\beta}\,dx_\alpha\,dx_\beta,$$

where $\eta_{\alpha\beta}$ is the four-dimensional **metric tensor**, and $\lambda^2(x_5)$ is the arbitrary function of the fifth coordinate.

- **Quantum metrics**

A **quantum metric** is a general term used for a metric expected to describe the space-time at quantum scales (of order *Planck-length* l_P). Extrapolating the predictions of both, Quantum Mechanics and General Relativity, the metric structure of this space-time is determined by vacuum fluctuations of very high energy (10^{19} GeV corresponding to the *Planck-mass* m_P) creating black holes with radii of order l_P. The space-time became "quantum foam": violent warping and turbulence. It looses smooth continuous structure (apparent macroscopically) of a *Riemannian manifold*, to become discrete, fractal, non-differentiable: breakdown at l_P of the functional integral in the classical field equations.

Examples of quantum metric spaces are given by: Rieffel's **compact quantum metric space**, **Fubini–Study metric** on quantum states, statistical geometry of fuzzy lumps ([ReRo01]) and quantization of the **metric cone** in [IsKuPe90].

Part VII

Real-World Distances

Chapter 27. Length Measures and Scales
Chapter 28. Non-Mathematical and Figurative Meaning of Distance

Chapter 27

Length Measures and Scales

Here we give selected information on most important length units and present, in length terms, a list of interesting objects.

27.1. LENGTH SCALES

The main length measure systems are: Metric, Imperial (British and American), Japanese, Thai, Chinese Imperial, Old Russian, Ancient Roman, Ancient Greek, Biblical, Astronomical, Nautical, and Typographical.

There are many other specialized length scales; for example, to measure cloth, shoe size, gauges (as interior diameters of shotguns, wires, jewelry rings), sizes for abrasive grit, sheet metal thickness, etc. Also, many units express relative or reciprocal distances.

- **International Metric System**

  The **International Metric System** (or SI, short for *Système International*) is a modernized version of the metric system of units, established by an international treaty (the *Treaty of the Meter* from 20 May 1875), which provides a logical and interconnected framework for all measurements in science, industry and commerce. The system is built on a foundation consisting of following seven *SI base units*, assumed to be mutually independent:

  1. length: **meter** (m); it is equal to the distance traveled by light in a vacuum in 1/299792458 of a second; 2. time: *second* (s); 3. mass: *kilogram* (kg); 4. temperature: *kelvin* (K); 5. electric current: *ampere* (A); 6. luminous intensity: *candela* (cd); 7. amount of substance: *mole* (mol).

  Originally, on 26 March 1791, the *mètre* (French for meter) was defined as $\frac{1}{10000000}$ of the distance from the North pole to the equator along the meridian that passes through Paris. In 1799 the standard of *mètre* became a meter-long platinum–iridium bar kept in Sèvres, a town outside Paris, for people to come and compare their rulers with. (The metric system, introduced in 1793, was so unpopular that Napoleon was forced to abandon it and France returned to the *mètre* only in 1837.) In 1960, the meter was officially defined in terms of wavelength.

- **Metrication**

  The **metrication** is ongoing (especially, in US and UK) process of conversion to **International Metric System** SI. Officially, only US, Liberia and Muanmar do not switched to SI. For example, US uses only miles for road distance signs (milestones). The altitudes in aviation are usually described in feet; in shipping, nautical miles and knots are used.

Hard metric means designing in the metric measures from the start and conformation, where appropriate, to internationally recognized sizes and designs.

Soft metric means multiplying an inch-pound number by a metric conversion factor and rounding it to an appropriate level of precision; so, the soft converted products do not change size. *American Metric System* consists of converting traditional units to embrace the uniform base 10 method that the Metric System uses. Such SI-Imperial hybrid units, used in soft metrication, are, for example, *kyloyard* (914.4 m), *kylofoot* (304.8 m), *mil* (24.5 micron), and *microinch* (or *min*, 25.4 nm).

- **Meter-related terms**

We present this large family of terms by following typical examples.

Meter: besides the unit of length, this term is used in poetry, music and for any of various measuring instruments.

Metrometer: in Medicine, an instrument measuring the size of the womb; the same term is used for a computer tool analyzing French Verse.

Metering: an equivalent term for a measuring.

Metrology: scientific study of measurement.

Metrosophy: a Cosmology based on strict number correspondences.

Metronomy: measurement of time by an instrument.

Allometry: the study of the change of proportions of various parts of an organism as a consequence of growth; *archeometry*: science of exact measuring referring remote past, and so on.

- **Metric length measures**

kilometer (km) $= 1000$ meters $= 10^3$ m;

meter (m) $= 10$ decimeters $= 10^0$ m;

decimeter (dm) $= 10$ centimeters $= 10^{-1}$ m;

centimeter (cm) $= 10$ millimeters $= 10^{-2}$ m;

millimeter (mm) $= 1000$ micrometers $= 10^{-3}$ m;

micrometer (micron) $= 1000$ nanometers $= 10^{-6}$ m;

nanometer (nm) $= 10$ angströms $= 10^{-9}$ m.

The lengths 10^{3t} m, $t = -8, -7, \dots, -1, 1, \dots, 7, 8$, are given by prefixes: yocto-, zepto-, atto-, fempto-, pico-, nano-, micro-, milli-, kilo-, mega-, giga-, tera-, peta-, exa-, zetta-, yotta-, respectively.

- **Imperial length measures**

The **Imperial length measures** (as slightly adjusted by international agreement of July 1, 1959) are:

league = 3 miles;

(*US survey*) *mile* = 5280 feet ≈ 1609.347 m;

international mile = 1609.344 m;

yard = 3 feet = 0.9144 m;

foot = 12 inches = 0.3048 m;

inch = 2.54 cm (for firearms, *caliber*);

line = $\frac{1}{12}$ inch;

mickey = $\frac{1}{200}$ inch;

mil (British *thou*) = $\frac{1}{1000}$ inch (*mil* is also an angle measure $\frac{\pi}{3200} \approx 0.001$ radian).

The following are old measures: *barleycorn* = $\frac{1}{3}$ inch; *digit* = $\frac{3}{4}$ inches; *palm* = 3 inches; *hand* = 4 inches; *shaftment* = 6 inches; *span* = 9 inches; *cubit* = 18 inches.

In addition, *Surveyor's Chain measures* are: *furlong* = 10 chains = $\frac{1}{8}$ mile; *chain* = 100 links = 66 feet; *rope* = 20 feet; *rod* (or *pole*) = 16.5 feet; *link* = 7.92 inches. Mile, furlong and fathom (6 feet) come from the slightly shorter Greco-Roman milos (milliare), stadion and orguia, mentioned in the New Testament.

Prototypical Biblical measures were: *cubit* and its multiples by 4, $\frac{1}{2}$, $\frac{1}{6}$ and $\frac{1}{24}$ called *fathom*, *span*, *palm* and *digit*, respectively. But the basic length of the Biblical cubit is unknown; it is estimated now as about 17.6 inches for the common (used in commerce) cubit and 20–22 inches for the sacred one (used for building). The *Talmudic cubit* is 56.02 cm, i.e., slightly longer than 22 inches.

Accordingly to http://en.wikipedia.org/wiki/List_of_Strange_units_of_measurement, an old unit, called *distance* and equal to = 221763 inches (about 5633 m) has the following strange definition: it is equal to 3 miles + 3 furlongs + 9 chains + 3 rods + 9 feet + 9 shaftements + 9 hands + 9 barleycorns.

For measuring cloth, old measures are used: *bolt* = 40 yards; *ell* = $\frac{5}{4}$ yard; *goad* = $\frac{3}{2}$ yard; *quarter* (or *span*) = $\frac{1}{4}$ yard; *finger* = $\frac{1}{8}$ yard; *nail* = $\frac{1}{16}$ yard.

- **Nautical length units**

The **nautical length units** (also used in aerial navigation) are:

sea league = 3 sea (nautical) miles;

nautical mile = 1852 m;

geographical mile ≈ 1855 m (the average distance on the Earth's surface, represented by one minute of latitude);

cable = 120 fathoms = 720 feet = 219.456 m;

short cable = $\frac{1}{10}$ nautical mile ≈ 608 feet;

fathom = 6 feet.

- **ISO paper sizes**

In the widely used ISO paper size system, the height-to-width ratio of all pages is the *Lichtenberg ratio*, i.e., $\sqrt{2}$. The system consists of formats An, Bn and (used for envelopes) Cn with $0 \leqslant n \leqslant 10$, having widths $2^{-\frac{1}{4}-\frac{n}{2}}$, $2^{-\frac{n}{2}}$ and $2^{-\frac{1}{8}-\frac{n}{2}}$, respectively. Above measures are in meters; so, the area of An is 2^{-n} square meter. They are rounded and expressed usually in millimeters; for example, format A4 is 210×297 and format B7 (used also for EU and US passports) is 88×125.

- **Typographical length units**

  The **ATA system** (British and American) uses:

  line $= \frac{1}{12}$ inch $\approx 2.117 \times 10^{-3}$ m;

  agate line $= \frac{1}{14}$ inch;

  pica[PostScript] = 2 lines;

  point[PostScript] (or *agate[Adobe]*) $= \frac{1}{6}$ line = 100 gutenbergs;

  pica = 12 points $\approx$ 1.99925 lines $\approx 4.218 \times 10^{-3}$ m;

  point $\frac{1}{72.72}$ inch = 20 twips $\approx 3.515 \times 10^{-4}$ m;

  pixel = 15 twips;

  kyu (or *Q*) 2.5×10^{-3} m $\approx$ 14.173 twips;

  twip (short for twentieth of a points) $\approx 1.764 \times 10^{-5}$ m.

  The **Didot system** (European) uses:

  cicero = 12 Didot points $\approx$ 1.07 pica;

  Didot point $\approx$ 21.397 twips $\approx 3.761 \times 10^{-4}$ m.

- **Very small length units**

  Angström (*A*) $= 10^{-10}$ m;

  angström star (or *Bearden unit*): $A^* \approx 1.0000148$ angström (used, from 1965, to measure wavelengths of X-rays and distances between atoms in crystals);

  X unit (or *Siegbahn unit*) $\approx 1.0021 \times 10^{-13}$ m (used formerly to measure wavelength of X-rays and gamma-rays);

  bohr (the atomic unit of length): α_0, the mean radius, $\approx 5.291772 \times 10^{-11}$ m, of orbit of the electron of an hydrogen atom (in the Bohr model);

  reduced Compton wavelength (i.e., $\frac{\hbar}{mc}$) for electron mass m_e: $\overline{\lambda}_C = \alpha\alpha_0 \approx 3.862 \times 10^{-13}$ m, where $\hbar$ is the *reduced Planck's constant* (or *Dirac's constant*), c is the speed of light, and $\alpha \approx \frac{1}{137}$ is the *fine-structure constant*;

  classical electron radius: $r_e = \alpha\overline{\lambda}_C = \alpha^2\alpha_0 \approx 2.81794 \times 10^{-15}$ m;

  Planck length (the smallest physical length): $l_P = \sqrt{\frac{\hbar G}{c^3}} \approx 1.6162 \times 10^{-35}$ m, where G is the Newton universal *gravitational constant*. It is the reduced Compton wavelength and also half of the Schwarzschild radius, for the *Plank mass* $m_P = \sqrt{\frac{\hbar c}{G}} \approx 2.176 \times 10^{-8}$ kg. *Planck time* is $t_p = cl_P \approx 5.4 \times 10^{-44}$ s.

  In fact, $10^{35} l_P \approx 1$ US mile, $10^{43} T_P \approx 54$ s and $10^9 m_P \approx 21.76$ kg ≈ 1 (classical) talent. L. Cottrell (http://planck.com/humanscale.htm) proposed a "postmetric" human-scale adaptation of Planck units system based on above three units, calling them (Planck) *mile*, *minute*, and *talent*.

- **Astronomical length units**

The *Hubble distance* (the edge of the cosmic light horizon) is $D_H = \frac{c}{H_0} \approx 4.22$ gigaparsec ≈ 13.7 light-Gyr (used to measure, as percents of D_H, distances $d > \frac{1}{2}$ megaparsec in terms of redshift z: $d = zD_H$ if $z \leqslant 1$, and $d = \frac{(z+1)^2-1}{(z+1)^2+1} D_H$, otherwise);

gigaparsec $= 10^3$ megaparsec;

hubble (or light-gigayear, light-Gyr, light-Ga) $= 10^9$ (billion) light-years ≈ 306.595 megaparsec;

megaparsec $= 10^3$ kiloparsec ≈ 3.262 MLY;

MLY $= 10^6$ (million) light-years;

kiloparsec $= 10^3$ parsecs;

parsec $= \frac{648000}{\pi}$ AU ≈ 3.261634 light-years $= 3.08568 \times 10^{16}$ m (the distance from an imaginary star, when lines drawn from it to both, the Earth and the Sun, form the maximum angle, i.e., *parallax*, of one second);

light-year $\approx 9.46073 \times 10^{15}$ m $\approx 5.2595 \times 10^5$ light-minutes $\approx \pi \times 10^7$ light-seconds (the distance light travels in vacuum in a year; used to measure interstellar distances);

spat (used formerly) $= 10^{12}$ m ≈ 6.6846 AU;

astronomical unit (AU) $= 1.49597871 \times 10^{11}$ m ≈ 8.32 light-minutes (the average distance between the Earth and the Sun; used to measure distances within the solar system);

light-second $\approx 2.998 \times 10^8$ m;

picoparsec ≈ 30.86 km (cf. other funny units such as *microcentury* ≈ 52.5 minutes, usual length of lectures, and *nanocentury* $\approx \pi$ seconds).

27.2. ORDERS OF MAGNITUDE FOR LENGTH

In this section we present a selection of orders of length magnitudes, expressed in meters.

1.616×10^{-35} *Planck length* (smallest possible physical length): probably, the "quantum foam" (violent warping and turbulence of *space-time*, no smooth spatial geometry); the dominant structures are little (multiply-connected) *wormholes* and *bubbles* popping into existence and back out of it;

10^{-34}: length of a putative string: M-Theory suppose that all forces and all 25 elementary particles come by vibration of such strings (which smooth quantum foam on sub-Planck distances) and hope to unify Quantum Mechanics and General Relativity;

$10^{-24} = 1$ **yoctometer**;

$10^{-21} = 1$ **zeptometer**;

$10^{-18} = 1$ **attometer**: weak nuclear force range, size of a quark;

$10^{-15} = 1$ **femtometer** (formerly, *fermi*);

1.3×10^{-15}: strong nuclear force range, medium-sized nucleus;

$10^{-12} = 1$ **picometer** (formerly, *bicron* or *stigma*): distance between atomic nuclei in a White Dwarf star;

10^{-11}: wavelength of hardest (shortest) X-rays and largest wavelength of gamma rays;

5×10^{-11}: diameter of the smallest (hydrogen H) atom; 1.5×10^{-10}: diameter of the smallest (hydrogen H_2) molecule;

$10^{-10} = 1$ *angström*: diameter of a typical atom, limit of resolution of the electron microscope;

1.54×10^{-10}: length of a typical covalent bond (C-C);

$10^{-9} = 1$ **nanometer**: diameter of typical molecule;

2×10^{-9}: diameter of the DNA helix;

10^{-8}: wavelengths of softest X-rays and most extreme ultraviolet;

1.1×10^{-8}: diameter of prion (smallest self-replicating biological entity);

9×10^{-8}: the smallest feature of computer chip in 2005, human immunodeficiency virus, HIV; in general, known viruses range from 2×10^{-8} (parvovirus B-19) to 8×10^{-7} (Mimivirus);

10^{-7}: size of chromosomes, maximum size of a particle that can fit through a surgical mask;

2×10^{-7}: limit of resolution of the light microscope;

$3.8 - 7.4 \times 10^{-7}$: wavelength of visible (to humans) light, i.e., the color range of violet through red;

$10^{-6} = 1$ **micrometer** (formerly, *micron*);

$10^{-6} - 10^{-5}$: diameter of a typical bacterium; in general, known (in non-dormant state) bacteria range from 1.5×10^{-7} (Micoplasma genitalium: "minimal cell") to 7×10^{-4} (Thiomargarita of Namibia);

7×10^{-6}: diameter of the nucleus of a typical eukaryotic cell;

8×10^{-6}: mean width of human hair (ranges from 1.8×10^{-6} to 18×10^{-6});

10^{-5}: typical size of (a fog, mist, or cloud) water droplet;

10^{-5}, 1.5×10^{-5}, and 2×10^{-5}: widths of cotton, silk, and wool fibers;

5×10^{-4}: diameter of a human ovum, MEMS micro-engine;

$10^{-3} = 1$ **millimeter**: farthest infrared wavelength;

5×10^{-3}: length of average red ant; in general, insects range from 1.7×10^{-4} (Megaphragma caribea) to 3.6×10^{-1} (Pharnacia kirbyi);

8.9×10^{-3}: Schwarzschild radius ($\frac{2Gm}{c^2}$: the one below which mass m collapses into a black hole) of the Earth;

$10^{-2} = 1$ **centimeter**;

$10^{-1} = 1$ **decimeter**: wavelengths of lowest microwave and highest UHF radio frequency, 3 GHz;

1 **meter**: wavelength of lowest UHF and highest VHF radio frequency, 300 MHz;

1.435: standard gauge of a railway track;

2.77–3.44: wavelength of the broadcast radio FM band, 108–87 MHz;

5.5 and 30.1: height of the tallest animal, the giraffe, and length of a blue whale, the largest animal;

$10 = 1$ **decameter**: wavelength of the lowest VHF and highest shortwave radio frequency, 30 MHz;

26: highest measured ocean wave;

100 = 1 **hectometer**: wavelength of the lowest shortwave radio frequency and highest medium wave radio frequency, 3 MHz;

112.8: height of the world's tallest tree, a Coast Redwood;

137, 300, 508, and 541: heights of the Great Pyramid of Giza, of the Eiffel Tower, of Taipei 101 Tower (tallest in 2005), and of the planned Freedom Tower at the World Trade Center site;

187–555: wavelength of the broadcast radio AM band, 1600–540 kHz;

340: distance which sound travels in air in one second;

$10^3 = 1$ **kilometer**;

2.95×10^3: Schwarzschild radius of the Sun;

3.79×10^3: mean depth of oceans;

4×10^3: the radius of the asteroid that may have killed off the dinosaurs;

10^4: wavelength of the lowest medium wave radio frequency, 300 kHz;

8.8×10^3 and 10.9×10^3: height of the highest mountain, Mount Everest and depth of the Mindanao Trench;

$5 \times 10^4 = 50$ km: the maximal distance on which the light of a match can be seen (at least 10 photons arrive on the retina during 0.1 s);

$1.11 \times 10^5 = 111$ km: one degree of latitude on the Earth;

10^5–10^6: range of voice frequency;

1.69×10^5: length of Delaware Water Supply Tunnel (New York), the world's longest tunnel;

2×10^5: wavelength of tsunami;

$10^6 = 1$ **megameter**;

3.48×10^6: diameter of the Moon;

5×10^6: diameter of LHS 4033, the smallest known White Dwarf star;

6.4×10^6 and 6.65×10^6: length of the Great Wall of China and length of Nile river;

1.28×10^7 and 4.01×10^7: equatorial diameter of the Earth and length of the Earth's equator;

3.84×10^8: Moon's orbital distance from the Earth;

$10^9 = 1$ **gigameter**;

1.39×10^9: diameter of the Sun;

5.8×10^{10}: orbital distance of Mercury;

1.496×10^{11} (1 astronomical unit, AU): mean distance between the Earth and the Sun (orbital distance of the Earth);

5.7×10^{11}: length of longest observed comet tail (Hyakutake, 1996);

$10^{12} = 1$ **terameter** (formerly, *spat*);

$2.1 \times 10^{12} \approx 7$ AU: diameter of the largest known supergiant star, KY Cygni;

$4.5 \times 10^{12} \approx 30$ AU: orbital distance of Neptune;

30–50 AU: distance from the Sun to Kuiper asteroid belt; the diameter of NGC 4061, the largest known black hole, is within 30–270 AU;

$10^{15} = 1$ **petameter**;

50000–100000 AU: distance from the Sun to Oort cloud (supposed spherical cloud of comets);

$3.99 \times 10^{16} = 266715$ AU $= 4.22$ light-years $= 1.3$ parsec: distance to Proxima Centauri, the nearest star;

$10^{18} = 1$ **exameter**;

$1.57 \times 10^{18} \approx 50.9$ parsec: distance to supernova 1987A;

$9.46 \times 10^{18} \approx 306.6$ parsec $\approx 10^5$ light-years: diameter of the galactic disk of our Milky Way galaxy;

$2.62 \times 10^{20} \approx 8.5$ kiloparsec (2.77×10^4 light-years): the distance from the Sun to the Galactic Center (in Sagittarius A^*);

$3.98 \times 10^{20} \approx 12.9$ kiloparsec: distance to Canis Major Dwarf, the nearest galaxy;

$10^{21} = 1$ **zettameter**;

$2.23 \times 10^{22} = 725$ kiloparsec: distance to Andromeda Nebula, the closest large galaxy;

$5 \times 10^{22} = 1.6$ megaparsec: diameter of Local Group of galaxies;

$5.7 \times 10^{23} = 60$ MLY: distance to Virgo cluster, the nearest major cluster (which dominates the Local Supercluster and where was found the first dark matter galaxy and first extragalactic stars);

$10^{24} = 1$ **yottameter**;

$2 \times 10^{24} = 60$ megaparsec: diameter of the Local (or Virgo) Supercluster;

$2.36 \times 10^{24} = 250$ MLY: distance to the Great Attractor (a gravitational anomaly in the Local Supercluster);

200 MLY: width of the Great Wall and Lyman alpha blobs, largest observed superstructures in the Universe (the space looks uniform on larger scales);

12080 MLY $=$ 3704 megaparsec: distance to the farthest known quasar SDSS J1148+5251 (redshift 6.43, while 6.5 is supposed to be the "wall of invisibility" for visible light);

13230 MLY: distance to the farthest known galaxy Abell 1835 IR1916 (redshift 10);

$1.3 \times 10^{26} = 13.7$ light-Gyr $= 4.22$ gigaparsec: the distance (estimated by the Wilkinson Microwave Anisotropy Probe) that cosmic background radiation has traveled since the Big Bang (Hubble radius $D_H = \frac{c}{H_0}$, the cosmic light horizon, age of Universe);

$7.4 \times 10^{26} = 78000$ MLY: the present (comoving) distance to the edge of the observable Universe (the size of observable Universe is larger than Hubble radius, since Universe is expanding);

The hypothesis of parallel universes estimate that one can find another identical copy of our Universe within the distance $10^{10^{118}}$ m.

Chapter 28

Non-Mathematical and Figurative Meaning of Distance

In this chapter we present selected practical distances, used in daily life and work outside of science, as well as examples of distances used as a metaphor for remoteness (the fact of being apart, being unknown, coldness of manner, etc.).

28.1. REMOTENESS-RELATED DISTANCES

- **Approximative human-scale distances**

  The **arm's length** is a distance (about 0.7 m, i.e., within **personal distance**) discouraging familiarity or conflict (analogs: Italian braccio, Turkish pik, and Old Russian sazhen). The **reach distance** is the difference between maximum reach and arm's length distance.

  The **shouting distance** is short, easily reachable distance. The **spitting distance** is a very close distance.

  The **striking distance** is the distance through which an object can be reached by striking.

  The **stone's throw** is a distance about 25 fathoms (46 m).

  The **hailing distance** is the distance within which the human voice can be heard.

  The **walking distance** is the distance normally (depending on the context) reachable by walking. For example, some UK high schools define 2 and 3 miles as statutory walking distance for children before and after 11 years.

- **Distances between people**

  In [Hall69], four interpersonal bodily distances were introduced: the *intimate distance* for embracing or whispering (15–45 cm), the *personal-casual distance* for conversations among good friends (45–120 cm), the *social-consultive distance* for conversations among acquaintances (1.2–3.6 m), and the *public distance* used for public speaking (over 3.6 m). What distance is appropriate for a given social situation depends on culture and personal preference. For example, under Islamic law, proximity (being in the same room or secluded place) between a man and a woman is permitted only in the presence of their *mahram* (a spouse or anybody from the same sex or pre-puberty one from the opposite sex). For an average westerner, personal space is about 70 cm in front, 40 cm behind and 60 cm on either side.

Distancing behavior of people can be measured, for example, by *stop distance* (when the subject stops approach since she/he begins to feel uncomfortable), or by *quotient of approach*, i.e., the percentage of moves made that reduce the interpersonal distance to all moves made.

The **people angular distance in a posture** is the spatial orientation, measured in degrees, of an individual's shoulders relative to those of another; the position of a speaker's upper body in relation to a listener's (for example, facing or angled away); the degree of body alignment between a speaker and a listener as measured in the coronal (vertical) plane which divides the body into front and back. This distance reveals how one feels about people nearby: the upper body unwittingly angles away from disliked persons and during disagreement.

- **Emotional distance**

The **emotional distance** (or *psychic distance*) is the degree of emotional detachment toward a person, group of people or events; indifference by personal withdrawal, reserve.

Bogardus Social Distance Scale measures, in fact, not social but this distance; it offers following eight response items: would marry, would have as a guest in household, would have as next door neighbor, would have in neighborhood, would keep in the same town, would keep out of my town, would exile, would kill. Dodd and Nehnevasja attached, in 1954, increasing distances 10^t meters, $0 \leqslant t \leqslant 7$, to 8 levels of Bogardus scale.

The **propinquity effect** is the tendency for people to get emotionally involved, as to form friendships or romantic relationships, with those who have higher *propinquity* (physical/psychological proximity) with them, i.e., whom they encounter often. Walmsley proposed that emotional involvement decreases as $d^{-\frac{1}{2}}$ with increasing **subjective distance** d.

- **Social distance**

In Sociology, the **social distance** is the extent to which individuals or groups are removed from or excluded from participating in one another's lives; a degree of understanding and intimacy which characterize personal and social relations generally. This notion was originated by G. Simmel in 1903; in his view, forms are the stable outcomes of distances interposed between subject and object (which in turn is a division of self).

Bogardus Social Distance Scale (cf. **emotional distance** above) is scored so that the responses for each ethnic/racial group are averaged across all respondents which yields a RDQ (racial distance quotient) ranging from 1.00 to 8.00.

An example of relevant models: [Aker97] defines an *agent* x as a pair (x_1, x_2) of numbers, where x_1 represents the initial, i.e., inherited, social position, and the position expected to be acquired, x_2. The agent x chooses the value x_2 so as to maximize

$$f(x_1) + \sum_{y \neq x} \frac{e}{(h + |x_1 - y_1|)(g + |x_2 - y_1|)},$$

where e, h, g are parameters, $f(x_1)$ represents the intrinsic value of x, and $|x_1 - y_1|$, $|x_2 - y_1|$ are inherited and acquired *social distances* of x from any agent y (with the social position y_1) of the given society.

- **Rummel sociocultural distances**

R.J. Rummel defined ([Rumm76]) the main sociocultural distances between two persons as follows.

1. **Personal distance**: one at which people begin to encroach on each other's territory of personal space.
2. **Psychological distance**: perceived difference in motivation, temperaments, abilities, moods, and states (subsuming *intellectual distance*).
3. **Interests-distance**: perceived difference in wants, means, and goals (including **ideological distance** on socio-political programs).
4. **Affine distance**: degree of sympathy, liking or affection between two.
5. **Social attributes distance**: differences in income, education, race, sex, occupation, etc.
6. **Status-distance**: differences in wealth, power, and prestige (including **power distance**).
7. **Class-distance**: degree to which one person is in general authoritatively superordinate to the other.
8. **Cultural distance**: differences in meanings, values and norms reflected in differences in philosophy-religion, science, ethics-law, language, and fine arts.

- **Cultural distance**

In [KoSi88], the **cultural distance between countries** $x = (x_1, \dots, x_5)$ and $y = (y_1, \dots, y_5)$ (usually, US) is derived as the following composite index

$$\sum_{i=1}^{5} \frac{(x_i - y_i)^2}{5V_i},$$

where V_i is the variance of the index i, and indexes are ([Hofs80]):

1. Power distance;
2. Uncertainty avoidance (the extent to which the members of a culture feel threatened by uncertain or unknown situations);
3. Individualism versus collectivism;
4. Masculinity versus femininity;
5. Confucian dynamism (ranges from long-term to short-term orientation).

The **power distance** above, measures the extent to which the less powerful members of institutions and organizations within a country expect and accept that power is distributed unequally, i.e., how much a culture has respect for authority. For example, Latin Europe and Japan fall in the middle range.

- **Effective trade distance**

The **effective trade distance** between countries x and y with populations $x_1, \dots, x_m$

and $y_1, \ldots, y_n$ of their main agglomerations is defined in [HeMa02] as

$$\left(\sum_{1 \leqslant i \leqslant m} \frac{x_i}{\sum_{1 \leqslant t \leqslant m} x_t} \sum_{1 \leqslant j \leqslant n} \frac{y_j}{\sum_{1 \leqslant t \leqslant n} y_t} d_{ij}^r \right)^{\frac{1}{r}},$$

where d_{ij} is bilateral distance (in kilometers) of corresponding agglomerations, and r measures the sensitivity of trade flows to d_{ij}.

As an **internal distance of a country**, measuring the average distance between producers and consumers, [HeMa02] proposes $.67\sqrt{\frac{area}{\pi}}$.

- **Technology distances**

The **technological distance** between two firms is a distance (usually, χ^2- or **cosine distance**) between their *patent portfolio*, i.e., vectors of the number of patents granted in (usually, 36) technological sub-categories. Another measures are based on the number of patent citations, co-authorship networks etc.

Granstrand's **cognitive distance** between two firms is the **Steinhaus distance** $\frac{\mu(A \triangle B)}{\mu(A \cup B)} = 1 - \frac{\mu(A \cap B)}{\mu(A \cup B)}$ between their technological profiles (sets of ideas) A and B seen as subsets of a *measure space* $(\Omega, \mathcal{A}, \mu)$.

Economic model of O.Olsson defines the metric space (I, d) of all ideas (as in human thinking), $I \subset \mathbb{R}_+^n$, with some *intellectual distance* d. The closed, bounded and connected *knowledge set* $A_t \subset I$ extends with time t. New elements are, normally, convex combinations of previous ones: *innovations* within gradual technological progress. Exceptionally, breakthroughs (Kuhn's paradigm shifts) occur.

Patel's **economic distance** between two countries is the time (in years) for a lagging country to catch up to the same per capita income level as the present one of an advanced country. Fukuchi–Satoh's **technology distance** between countries is the time (in years) when a lagging country realizes a similar technical structure as advanced one has now. The basic assumption of the popular *Convergence Hypothesis* is that the technology distance between two countries is smaller than the economic one.

In Production Economics, a *technology* is modeled as a set of pairs (x, y), where $x \in \mathbb{R}_+^m$ is an *input* vector, $y \in \mathbb{R}_+^m$ is an *output* vector, and x can produce y. Such set T should satisfy standard economical regularity conditions. The **technology directional distance function** of input/output x, y toward (projected and evaluated) direction $(-d_x, d_y) \in \mathbb{R}_-^m \times \mathbb{R}_+^m$ is $\sup\{k \geqslant 0\colon ((x - kd_x), (y + kd_y)) \in T\}$. The **Shephard output distance function** is $\sup\{k \geqslant 0\colon (x, \frac{y}{k}) \in T\}$. The *frontier* $f_s(x)$ is the maximum feasible output given input x in a given system or year s. The **distance to frontier** of a production point $(y = g_s(x), x)$ is $\frac{g_s(x)}{f_s(x)}$. The *Malmquist index* measuring TFP (total factor productivity) change between periods s, s' (or comparing to another unit in the same period) is $\frac{g_s'(x)}{f_s(x)}$. The term *distance to frontier* is also used for the inverse of TFP in a given industry (or of GDP per worker in a given country) relative to the existing maximum (the frontier, usually, US).

- **Death of Distance**

  Death of Distance is the title of the influential book [Cair01] arguing that the telecommunication revolution (the Internet, mobile telephones, digital television, etc.) initiated the "death of distance" implying fundamental changes: three-shift work, lower taxes, prominence of English, outsourcing, new ways of government control and citizens communication, etc. The proportion of long-distance relationships in foreign relations increased. But the "death of distance" allows also both, management-at-a-distance and concentration of elites within the "latte belt".

  Similarly (see [Ferg03]), the steam-powered ships and the telegraph (as railroads before and cars later) led, via falling transportation costs, to the "annihilation of distance" in the 19th and 20th centuries. Further in the past, archaeological evidence points out the appearance of systematic long-distance object exchange ($\approx$140000 years ago), and the innovation of projectile weapons (40000 years ago) which allowed humans to kill large game (and other humans) from safe distance.

  However, modern technology eclipsed distance only in that the time to reach a destination has shrunk. In fact, the distance (cultural, political, geographic, and economic) "still matters" for, say, a company's strategy on the emerging markets, for political legitimacy, etc.

- **Moral distance**

  The **moral distance** is a measure of moral indifference or empathy, toward a person, group of people, or events.

  The **distancing** is a separation in time or space that reduces the empathy that a person may have for the suffering of others, i.e., that increases moral distance. The term *distancing* is also used (in books by M.D. Kantor) for APD (Avoidant Personality Disorder): fear of intimacy and commitment (confirmed bachelors, "*femmes fatales*", etc.)

- **Technology-related distancing**

  The *Moral Distancing Hypothesis* postulates that technology increases the propensity for unethical conduct by creating a **moral distance** between an act and the moral responsibility for it.

  Print technologies divided people into separate communication systems and distanced them from face-to-face response, sound and touch. Television involved audile-tactile senses and made the distance less inhibiting, but it exacerbated the *cognitive distancing*: story and image are biased against space/place and time/memory. This distancing has not diminished with computers; only interactivity increased. In terms of M. Hunter: technology only re-articulates *communication distance*, because it also must be regarded as the space between understanding and not. The collapsing of spatial barriers diminish economic but not social and cognitive distance.

  On the other hand, the *Psychological Distancing Model* in [Well86] relates the immediacy of communication to the number of information channels: sensory modalities decrease progressively as one moves from face-to-face to telephone, videophone, and e-mail. On-line settings tend to filter out social and relation cues. Also, the lack of instant feedback, because of e-mail communication, is asynchronous and can be isolating. For example, moral and cognitive effects of distancing in on-line education are not known at present.

- **Transactional distance**

  The **transactional distance** is a perceived degree of separation during interaction between students and teachers, and within each group. This distance decreases with *dialog* (a purposeful positive interaction meant to improve the understanding of the student), with larger autonomy of the learner, and with lesser predetermined structure of instructional program. This notion was introduced by M.G. Moore in 1993 as a paradigm for distance education.

- **Antinomy of distance**

  The **antinomy of distance**, as introduced in [Bull12] for aesthetic experiences by beholder and artist, is that both should find the right amount of **emotional distance** (neither too involved, nor too detached), in order to create or appreciate art. The fine line between objectivity and subjectivity can be crossed easily, and the amount of distance can fluctuate in time.

  The **aesthetic distance** is a degree of emotional involvement of the individual, who undergoes experiences and objective reality of the art, in a work of art. Some examples are: the perspective of a member of the audience in relation to the performance, the psychological and the emotional distance between the text and the reader, the **actor-character distance** in Stanislavsky system of acting.

  A variation of antinomy of distance appears in critical thinking: need to put some emotional and intellectual distance between oneself and ideas, in order to better evaluate their validity. Another variation is detailed in *Paradox of Dominance: Distance and Connection* (http://www.leatherpage.com/rscurrent.htm/)

  The **historical distance**, in terms of [Tail04], is the position the historian adopts *vis-à-vis* his objects – whether far-removed, up-close, or somewhere in between; it is the fantasy through which the living mind of the historian, encountering the inert and unrecoverable, positions itself to make the material look alive. The antinomy of distance appears again because historians engage the past not just intellectually but morally and emotionally. The formal properties of historical accounts are influenced by their affective, ideological and cognitive commitments.

  Related problem is how much distance people must put between themselves and their pasts in order to remain psychologically viable; S.Freud showed that often there is no such distance with childhoods.

- **Kristeva non-metric space**

  J. Kristeva's (1980) basic psychoanalytic distinction is between pre-Oedipal and Oedipal aspects of personality development. Narcissistic identification and maternal dependency, anarchic component drives, polymorphic erotogenicism, and primary processes characterize the pre-Oedipal. Paternal competition and identification, specific drives, phallic erotogenicism, and secondary processes characterize Oedipal aspects. Kristeva describes the pre-Oedipal feminine phase by an enveloping, amorphous, **non-metric space** (Plato's *chora*) that both nourishes and threatens; it also defines and limits self-identity. She characterizes the Oedipal male phase by a metric space (Aristotle's *topos*); the self and the self-to-space are more precise and well defined in topos. Kristeva posits also that the semiotic process is rooted in feminine libidinal, pre-Oedipal energy which needs channeling for social cohesion.

J. Deleuze and F. Guattari (1980) divided their *multiplicities* (networks, manifolds, spaces) into *striated* (metric, hierarchical, centered and numerical) and *smooth* ("non-metric, rhizomic and acentered, that occupy space without counting it and can be explored only by legwork").

Above French poststructuralists use metaphor *non-metric* in line with systematic use of topological terms by psychoanalyst J. Lacan. In particular, he sought space J (of *Jouissance*, i.e., sexual relations) as a bounded metric space.

Back to Mathematics, the **non-metricity tensor** is the *covariant derivative* of a **metric tensor**. It can be non-zero for **pseudo-Riemannian metrics** and vanishes for **Riemannian metrics**.

- **Simone Weil distance**

  "The Distance" is the title of a philosophico-theological essay by Simone Weil from her *Waiting for God*, Putnam, New York, 1951. She connects God love to the distance; so, his absence can be interpreted as a presence: "every separation is a link". Therefore, she posits, the crucifixion of Christ (the greatest love/distance) was necessary "in order that we should realize the distance between ourselves and God ... for we do not realize distance except in the downward direction". Cf. Lurian kabbalistic notions of *tzimzum* (God contraction, "withdrawal"), and shattering of the vessels (evil as the force of separation which lost its distancing function and become husks).

  Also, a song "From a Distance", written by Julie Gold, is about how God is watching us and how, despite the distance (physical and emotional) distorting perceptions, there is still a little peace and love in this world.

- **Swedenborg heaven distances**

  Famous scientist and visionary E. Swedenborg, in Section 22 (Nos. 191–199, *Space in Heaven*) of his main work *Heaven and Hell* (1952, first edition in Latin, London, 1758), posits: "distances and so, space, depend completely on interior state of angels". A move in heaven is just a change of such state, the length of a way corresponds to the will of a walker, approaching reflects similarity of states. In the spiritual realm and afterlife, for him, "instead of distances and space, exist only states and their changes".

- **Far Near Distance**

  Far Near Distance is the name of the program of the House of World Cultures in Berlin which present a panorama of contemporary positions of all artists of Iranian origin. Examples of similar use of distance terms in modern popular culture are: "Some near distance" is the title of art exhibition of Mark Lewis (Bilbao, 2003), "A Near Distance" is a paper collage by Perle Fine (New York, 1961), "Quiet Distance" is a fine art print by

Ed Mell, "Distance" is a Japanese movie by Hirokazu Koreeda (2001), "The Distance" is an album by American rock "The Silver Bullet Band", "Near Distance" is a musical composition by Chen Yi (New York, 1988), "Near Distance" is a lyrics by Manchester quartet "Puressence".

The terms *near distance* and *far distance* are also used in Ophthalmology and for settings in some sensor devices.

- **Quotes on "near-far" distances**

"Better is a nearby neighbor, than a far off brother." (Bible)

"It is when suffering seems near to them that men have pity; as for disasters that are ten thousand years off in the past or the future, men cannot anticipate them, and either feel no pity for them, or at all events feel it in no comparable measure." (Aristotle)

"The path of duty lies in what is near, and man seeks for it in what is remote." (Mencius)

"Sight not what is near through aiming at what is far." (Euripides)

"Good government occurs when those who are near are made happy, and those who are far off are attracted." (Confucius)

"By what road", I asked a little boy, sitting at a cross-road, "do we go to the town?" – "This one", he replied, "is short but long and that one is long but short". I proceeded along the "short but long road". When I approached the town, I discovered that it was hedged in by gardens and orchards. Turning back I said to him, "My son, did you not tell me that this road was short?" – "And", he replied, "Did I not also tell you: "But long?" I kissed him upon his head and said to him, "Happy are you, O Israel, all of you are wise, both young and old". (Erubin, Talmud)

"The Prophet Muhammad was heard saying: "The smallest reward for the people of paradise is an abode where there are 80000 servants and 72 wives, over which stands a dome decorated with pearls, aquamarine, and ruby, as wide as the distance from Al-Jabiyyah [a Damascus suburb] to Sana'a [Yemen]". (Hadith, Islamic Tradition)

"There is no object so large ... that at great distance from the eye it does not appear smaller than a smaller object near." (Leonardo da Vinci)

"Nothing makes Earth seems so spacious as to have friends at a distance; they make the latitudes and longitudes." (Henri David Thoreau)

Tobler's first law of Geography: everything is related to everything else, but near things are more related than distant things. **Nearness principle** (or *least effort principle*): given a distribution of equally desirable locations, the closest destination is most frequently chosen.

28.2. VISION-RELATED DISTANCES

- **Vision distances**

The **inter-pupillary distance** (or *inter-ocular distance*): in Ophthalmology, the distance between the centers of the pupils of the two eyes when the visual axes are parallel. Typically, it is 2.5 inches (6.35 cm).

The **near acuity** is the eye's ability to distinguish an object's shape and details at a near distance such as 40 cm; the **distance acuity** is the eye's ability to do it at a far distance such as 6 m.

The **optical near devices** are designed for magnifying close objects and print; the **optical distance devices** are for magnifying things in the distance (from about 3 m to far away).

The **near distance**: in Ophthalmology, the distance between the object plane and the *spectacle* (eyeglasses) plane.

The **infinite distance**: in Ophthalmology, the distance of 20 feet (6.1 m) or more; so called because rays entering the eye from an object at that distance are practically as parallel as if they came from a point at an infinite distance.

Distance vision is a vision for objects that at least 20 feets from the viewer.

The **angular eye distance** is the aperture of the angle made at the eye by lines drawn from the eye to two objects.

The **RPV-distance** (or *resting point of vergence*) is the distance at which the eyes are set to *converge* (turn inward toward the nose) when there is no close object to converge on. It averages about 45 inches (1.14 m) when looking straight ahead and comes in to about 35 inches (0.89 m) with 30-degree downward gaze angle. Ergonomists recommend RPV-distance as eye-screen distance in sustained viewing, in order to minimize eyestrain.

The **default accommodation distance** (or *resting point of accommodation*, *RPA-distance*) is the distance at which the eyes focus when there is nothing to focus on.

- **Size-distance paradox**

Emmert's law states that a retinal image is proportional in perceived size (apparent height) to the perceived distance of the surface it is projected upon. This law is based on the fact that the perceived size of an object doubles every time its perceived distance from the observer is cut in half and vice versa. Emmert's law accounts for *constancy scaling*, i.e., the fact that the size of an object is perceived to remain constant despite the changes in the retinal image (as objects become more distant they begin, because of visual perspective, appear smaller).

The *size-distance invariance hypothesis* posits that the ratio of perceived size and perceived distance is the tangent of the physical visual angle. In particular, the objects which appear closer should also appear smaller. But with *moon illusion* it comes to **size-distance paradox**. The Moon (and, similarly, the Sun) illusion is that, despite of constancy of its visual angle (roughly, 0.52 degree), the horizon moon may appear to be about twice the diameter of the zenith moon. This illusion is still not understood completely; it is supposed to be cognitive: the size of the zenith moon is underestimated since it is perceived as approaching.

The most common optical illusions distort size or length; for example, Mueller-Lyer, Sander, and Ponzo illusions.

- **Symbolic distance effect**

  In Psychology, the brain compares two concepts (or objects) with higher accuracy and faster reaction time if they differ more on the relevant dimension.

- **Subjective distance**

  The **subjective distance** (or *cognitive distance*) is a mental representation of actual distance molded by an individual's social, cultural and general life experiences. Cognitive distance errors occur either because information about two points is not coded/stored in the same branch of memory, or because of errors in retrieval of this information. For example, the length of a route with many turns and landmarks is usually overestimated.

- **Egocentric distance**

  In Psychophysiology, the **egocentric distance** is the perceived absolute distance from the self (observer or listener) to an object or a stimulus (such as a sound source). Usually, visual egocentric distance underestimates actual physical distance to far objects, and overestimates it for near objects. In Visual Perception, the *action space* of a subject is 1–30 m; the smaller and larger spaces are called *personal space*, and *vista space*, respectively.

  The exocentric distance is perceived relative distance between objects.

- **Distance cues**

  The **distance cues** are cues used to estimate the **egocentric distance**.

  For a listener from a fixed location, main auditory distance cues include: *intensity* (in open space it decreases of 5 dB for each doubling of the distance), *direct-to-reverberant energy ratio* (in the presence of sound reflecting surfaces), *spectrum*, and *binaural differences*.

  For an observer, main visual distance cues include:
  - *relative size*, *relative brightness*, *light and shade*;
  - *height in the visual field* (in the case of flat surfaces lying below the level of the eye, the more distant parts appear higher);
  - *interposition* (when one object partially occludes another from view);
  - *binocular disparities*, *convergence* (depending on the angle of the optical axes of the eyes), *accommodation* (the state of focus of the eyes);
  - *aerial perspective* (the objects in the distance became bluer and paler), *distance hazing* (the objects in the distance became decreased in contrast, more fuzzy);
  - *motion perspective* (the stationary objects appear, to moving observer, to glide past).

  Examples of the techniques, using above distance cues to create an optical illusion for the viewer, are:
  - *distance fog*: an 3D computer graphics technique so that objects further from the camera are progressively more blurred (obscured by haze);
  - *forced perspective*: a film-making technique to make objects appear either far away, or vice versa depending on their positions to the camera and each other.

- **Distance-related shots**

A film *shot* is what is recorded between the time the camera starts (the director's call for "action"), and the time it stops (the call to "cut").

Main **distance-related shots** (camera set-ups) are:
- *establishing shot*: a shot, at the beginning of a sequence which establish the location of the action and/or the time of day;
- *long shot*: a shot taken from at least 50 yards (45.72 m) from the action;
- *medium shot*: a shot from 5–15 yards (4.57–13.72 m) including a small group entirely, shows group/objects in relation to surroundings;
- *close-up*: a shot taking the actor from the neck upwards, or an object from a similarly close position;
- *two-shot*: a shot that features two persons in the foreground;
- *insert*: an inserted shot (usually a close up) used to reveal greater detail.

28.3. EQUIPMENT DISTANCES

- **Focus distances**

The **working distance**: the distance from the front lens of a microscope to the object when the instrument is correctly focused.

The **object distance**: the distance from the lens of camera to the object being photographed, i.e., being focused on.

The **image distance**: the distance from the lens to the image (picture on the screen); when a converging lens is placed between the object and the screen, the sum of inverse object and image distances is equal to inverse focal distance.

The **focal distance** (*focal length*): the distance from the optical center of a lens (or a curved mirror) to the focus (to the image).

The **depth of field**: the distance in front of and behind the subject which appear to be in focus, i.e., the region where the blurring is tolerated.

The **hyperfocal distance**: the distance from the lens to the nearest point (*hyperfocal point*) that is in focus when the lens is focused at infinity; beyond this point all objects are well defined and clear. It is the nearest distance at which the far end of the **depth of field** stretches to infinity. (Cf. **infinite distance**).

- **Distances in Stereoscopy**

A way of 3D imaging is creating a pair of 2D images by a two-camera system.

The **inter-camera distance** (or *base line length*, *inter-ocular lens spacing*) is the distance between the two cameras from which the left and right eye images are rendered.

The **convergence distance** is the distance between the center of camera *base line* to the *convergence point* where the two lenses should converge for good stereoscopy. This distance should be 15–30 times **inter-camera distance**.

The **picture plane distance** is the distance at which the object will appear on (but not behind or in front) the *picture plane* (the apparent surface of the image). The *window* is a masking border of the screen frame such that objects, appearing at (but not behind or outside) it, appear to be at the same distance from the viewer as this frame. In human viewing, the picture plane distance is about 30 times *inter-ocular distance*.

- **Miss distance**

  The **miss distance** is the distance between the lines of sight representing two estimates from two sensor sites to the target. (Cf. **line-line distance**.)

- **Offset distance**

  In nuclear warfare, the **offset distance** is the distance the desired (or actual) ground zero is offset from the center of the area (or point) target.

  In Computation, **offset** is the distance from the beginning of a string to the end of the segment on that string. For a vehicle, **offset** of a wheel is the distance from its hub mounting surface to the centerline of the wheel.

- **Standoff distance**

  The **standoff distance** is the distance of object from the source of the explosion (in warfare), or from the laser beam delivery point (in laser material processing). Also, in mechanics and electronics, it is the distance separating two parts from one another (for example, for insulating: cf. **clearance distance**).

- **Proximity fuse**

  The **proximity fuse** is a fuse that is designed to detonate an explosive automatically when close enough to the target.

- **Proximity sensors**

  Proximity sensors are variety of ultrasonic, laser, photoelectric and fiber optic sensors designed to measure distance from itself to a target.

  Compare with following simple distance estimation (for prey recognition) by some insects: the velocity of the mantid's head movement is kept constant during peering, and so, the distance to the target is inversely proportional to the velocity of the retinal image.

- **Precise distance measurement**

  The resolution of TEM (transmission electronic microscope) is about 0.2 nm (2×10^{-10} m), i.e., the typical separation between two atoms in a solid. This resolution is 1000 times greater than a light microscope and about 500000 times greater than that of a human eye. However, only nanoparticles can fit in the vision field of an electronic microscope.

  The methods, based on measuring the wavelength of laser light, are used to measure macroscopic distances non-treatable by electronic microscope. However, the uncertainty of such methods is at least the wavelength of light, say, 633 nm.

  The recent adaptation of *Fabry–Perot metrology* (measuring the frequency of light stored between two highly reflective mirrors) to laser light permit to measure relatively long (up to 5 cm) distances with uncertainty only 0.01 nm.

- **Radio distance measurement**

  Distance measuring equipment (DME) is an air navigation technology that measures distances by timing the propagation delay of UHF signals to a *transponder* (receiver-transmitter that will generate a reply signal upon proper interrogation) and back. DME is expected to be phased out by global satellite-based systems: GPS and, planned for 2010, Galileo (EU) and GLOSNASS (Russia/India).

  The GPS (Global Positioning System) is a radio navigation system which permits one to get her/his exact position on the globe (anywhere, anytime). It consists of 24 satellites and a monitoring system operated by the US Department of Defense. Non-military part of GPS can be used just by the purchase of an adequate receiver and the accuracy is 10 m.

  The **GPS pseudo-distance** (or *pseudo-range*) from a receiver to a satellite is the travel time of a satellite time signal to a receiver multiplied by propagation time of radio signal (about the speed of light). It is called *pseudo-distance* because of the error: the receiver clock is not so perfect as ultra-precise clock of satellite. The GPS receiver calculates its position (in latitude, longitude, altitude, etc.) by solving a system of equations using its pseudo-distances from at least four satellites and the knowledge of their positions.

- **Radio distances**

  Line-of-sight distance is the distance which radio signal travel, from one antenna to another, by a path where both antennas are visible to one another, and there are no obstructions. In fact, waves can travel below the horizon, since the signal can interact with the ground and/or the ionosphere.

  If two frequencies of radio are used (for instance, 12,5 kHz and 25 kHz in maritime communication), the **interoperability distance** and **adjacent channel separation distance** are the range within which all receivers work with all transmitters, and, respectively, the minimal distance which should separate adjacently tunes narrow-band transmitter and wide-band receiver, in order to avoid interference.

  DX is amateur radio slang (and Morse code) for distance; **DXing** is a distant radio exchange (amplifiers required).

- **Transmission distance**

  The **transmission distance** is a **range distance**: for a given signal transmission system (fiber optic cable, wireless, etc.), it is the maximal distance the system can support within acceptable path loss level.

  For a given network of contact that can transmit an infection (or, say, an idea with the belief system considered as the immune system), the **transmission distance** is the **graphic metric** (edges correspond to events of infection) via the most recent common ancestor, between (infectious agents isolated from) infected individuals.

- **Instrument distances**

  The **load distance**: the distance (on a lever) from the fulcrum to the load. The **effort distance** (or *resistance distance*): the distance (on a lever) from the fulcrum to the effort.

The **K-distance**: the distance from the outside fiber of a rolled steel beam to the web toe of the fillet of a rolled shape.

The **end distance**: the distance from a bolt, screw, or nail to the end of a (wood) structural member. The **edge distance**: the distance from a bolt, screw, or nail to the edge of a (wood) structural member.

- **Creepage distance**

  The **creepage distance** is the shortest path along the surface of the insulation material between two conductive parts. The **clearance distance** is the shortest (straight-line) distance between two conductive parts.

- **Solvent migration distance**

  In Chromatography, the **solvent migration distance** is the distance traveled by the front line of the liquid or gas entering chromatographic bed for *elution* (the process of using a solvent to extract an absorbed substance from a solid medium).

- **Spray distance**

  The **spray distance** is the distance maintained between the thermal spraying gun nozzle tip and the surface of the workpiece during spraying.

- **Vertical separation distance**

  The **vertical separation distance** is the distance between the bottom of a sewage septic system's drain field and the underlying water table. This separation distance allows pathogens (disease-causing bacteria, viruses, or protozoa) in the effluent to be removed by the soil before it comes in contact with the groundwater.

- **Protective action distance**

  The **protective action distance** is the distance downwind from the incident (a spill involving dangerous goods which are considered toxic by inhalation) in which persons may become incapacitated.

- **Sight distances**

  Sight distance (or *clear sight distance*) is the length of highway visible to a driver. A **safe sight distance** is the necessary sight distance needed by the driver in order to accomplish fixed task; the main safe distances, used in road design, are:

  stopping sight distance – to stop the vehicle before reaching an unexpected obstacle,

  maneuver sight distance – to drive around an unexpected small obstacle,

  passing sight distance – to overtake safely,

  road view sight distance – to anticipate on the alignment (eventually curved and horizontal/vertical) of the road (for instance, choosing a speed).

  Also, adequate sight distances are required locally: at intersections and in order to process information on traffic signs.

• **Vehicle distances**

The **braking distance**: the distance a motor vehicle travels from the moment the brakes are applied until the vehicle completely stops.

The **reaction distance**: the distance a motor vehicle travels from the moment the driver sees a hazard until he applies the brakes (corresponding to human perception time plus human reaction time). (Not to be confused with reaction **animal distance**.)

The **stopping distance**: the distance a motor vehicle travels from where the driver perceives the need to stop to the actual stopping point (corresponding to vehicle reaction time plus vehicle braking capability).

The **official distance**: the DoD (US Department of Defense) recognized driving distance between two locations that will be used for travel or payment of billing (not to be confused with **administrative cost distance** in Internet.)

The **distance-based exit number**: a number assigned to a road junction, usually an exit from a freeway, expressing in miles (or kilometers) the distance from the beginning of the highway to the exit. A *milestone* (or *kilometer sign*) is one of a series of numbered markers placed along a road at regular intervals. Zero Milestone in Washington, DC is attended as the reference point for all road distances in US.

The **accelerate-stop distance**: the runway plus stop-way length declared available and suitable for the acceleration and deceleration of an airplane aborting a takeoff.

The **endurance distance**: total distance that a ground vehicle or ship can be self-propelled at any specified endurance speed.

The **distance made good** is a nautical term: the distance traveled after correction for current, *leeway* (the sideways movement of the boat away from the wind) and other errors that may not have been included in the original distance measurement. *Log* is a device to measure the distance traveled through the water which further corrected to a distance made good.

The **distance line**: in Diving, a temporary marker (typically, 50 meters of thin polypropylene line) of shortest route between two points. It is used to navigate back to the start in poor visibility.

28.4. MISCELLANY

• **Range distances**

The **range distances** are practical distances emphasizing a maximum distance for effective operation such as vehicle travel without refueling, a bullet reach, visibility, movement limit, home range of an animal, etc.

For example, the **dispersal distance** in Biology can refer to seed dispersal by pollination, to natal dispersal, to breeding dispersal, to migration dispersal, etc.

The range of a risk factor (toxicity, blast etc.) indicates minimal **safe distancing**. The range of a device (for example, a remote control), which is specified by the manufacturer and used as a reference, is called **operating distance** (or *nominal sensing distance*). The maximal distance allowed for activation of a sensor-operated switch is called **activation distance**.

- **Spacing distances**

  The following examples illustrate this large family of practical distances emphasizing a minimum distance (cf. **minimum distance** in Coding, nearest-neighbor **animal distance** and **first-neighbor distance** for atoms in a solid).

  The **miles in trail**: a specified minimum distance, in nautical miles, required to be maintained between airplanes.

  The **isolation distance**: a specified minimum distance required (because of pollination) to be maintained between variations of the same species of crop for the purpose to keep seed pure (for example, 10 feet $\approx$ 3 m for rice).

  The **stop-spacing distance**: the interval between stops of a bus; the mean stop-spacing distance in the US (for light rail systems) ranges from 500 m (Philadelphia) to 1742 m (Los Angeles).

  The **character spacing**: the interval between characters in a given computer font.

  The **music distance**: the interval between notes.

  The **just noticeable difference** (JND): the smallest percent change in a dimension (for distance/position, etc.) that can be reliably perceived.

- **Quality metrics**

  This vast family of measures (or standards of measure) concern different attributes of objects (usually, equipment). In such terms, our distances and similarities are "similarity metrics", i.e., metrics (measures) quantifying the extent of relatedness between two objects. Examples of more abstract quality metrics are given below.

  The **software metric** is a measure of software quality which indicate the complexity, understandability, description, testability and intricacy of code.

  The **trust metric** is: in Computer Security, a measure to evaluate a set of peer certificates resulting in a set of accounts accepted, and, in Sociology, a measure of how a member of the group is trusted by the others in the group. For example, UNIX access metric is a combination of only *read*, *write* and *execute* kinds of access to a resource. Much finer *Advogato* trust metric (used in the community of open source developers to rank them) is based on bonds of trust formed when a person issues a certificate about someone else.

  The **risk metric** is used in Insurance and (to evaluate a portfolio) in Finance.

- **Action at a distance (in Computing)**

  The **action at a distance** (in Computing) is a class of programming problems in which the state in one part of a program's data structure varies wildly because of difficult-to-identify operations in another part of the program. The *Law of Demeter* is a guideline for developing software: "only talk to your immediate friends" (units closely related to it), and each unit should have only limited knowledge about other units.

- **Action distance**

The **action distance** is the distance between the set of information generated by the Active Business Intelligence system and the set of actions appropriate to a specific business situation. Action distance is the measure of the effort required to understand information and to affect action based on that information. It could be the physical distance between information displayed and action controlled.

- **Distance decay**

The **distance decay** (or *distance lapse rate*) is the attenuation of a pattern or process with distance. In spatial interaction, it is the mathematical representation of inverse ratio between quantity of obtained substance and the distance from its source. This decay measures the effect of distance on accessibility: it can reflect a reduction in demand due to the increasing travel cost. Examples of distance-decay curves: Pareto model $\ln I_{ij} = a - b \ln d_{ij}$, and the model $\ln I_{ij} = a - bd_{ij}^p$ with $p = \frac{1}{2}$, 1, or 2 (here I_{ij} and d_{ij} are interaction and distance between points i, j, while a and b are parameters).

- **Distance curve**

A **distance curve** is a plot (or a graph) of a given parameter against corresponding distance. Examples of distance curves, in terms of a process under consideration, are: **time-distance curve** (for travel time of wave-train, seismic signals, etc.), *drawdown-distance curve*, *melting-distance curve* and *wear volume versus distance curve*.

Force-distance curve is, in SPM (Scanning Probe Microscopy), a plot of the vertical force that the tip of the probe applies to the sample surface, while a contact-AFM (Atomic Force Microscopy) image is being taken. Also, *frequency-distance* and *amplitude-distance* curves are used in SPM.

The term *distance curve* is also used for charting growth, for instance, a child's height or weight at each birthday. A plot of the rate of growth against age is called **velocity-distance curve**. The last term is also used for speed of aircraft.

- **Mass-distance function**

A **mass-distance function** is a function proportional to $\frac{xy}{d(x,y)}$. It also called *gravity function* because it express the gravitational attraction between masses x and y at (Euclidean) distance $d(x, y)$; cf. **inverse-square laws**. Such functions are often used in social sciences; for example, it can express the communication with x and y being the population of the sender and the receiver location, where $d(x, y)$ is the physical distance between them.

A **mass-distance decay curve** is a plot of "mass" decay when the distance to the center of "gravity" increases. Such curves are used to find out *offender's heaven* (the point of origin; cf. **distances in Criminology**), the galactic mass within a given radius from its center (using *rotation-distance curves*), etc.

- **Long range dependence**

A (second order stationary) stochastic process X_k, $k \in \mathbb{Z}$, is called **long range dependent** (or *long memory*) if there exist numbers $\alpha, 0 < \alpha < 1$, and $c_\rho > 0$ such that

$\lim_{k \to \infty} c_\rho k^\alpha \rho_k = 1$ holds, where $\rho(k)$ is the autocorrelation function. So, correlations decay very slowly (asymptotically hyperbolic) to zero implying that $\sum_{k \in \mathbb{Z}} |\rho_k| = \infty$, and that events that far apart are correlated (long memory). If above sum is finite and decay is exponential, then process is *short range*. Examples of such processes are the exponential, normal and Poisson processes, which are memoryless, and, in physical terms, the systems in thermodynamic equilibrium. Above power law decay for correlations as a function of time, translates into a power law decay of the Fourier spectrum as a function of frequency f and called $\frac{1}{f}$ *noise*.

A process has *self-similarity exponent* (or *Hust parameter*) H if X_k and $t^{-H} X_{tk}$ have the same finite-dimensional distributions for any positive t. The cases $H = \frac{1}{2}$ and $H = 1$ correspond, respectively, to purely random process and to exact self-similarity (or *scale-invariance*): the same behavior on all scales (cf. **fractal** and **scale-free network**). The processes with $\frac{1}{2} < H < 1$ are long range dependent with $\alpha = 2(1 - H)$.

Long range dependence corresponds to *heavy-tailed* (or *power low*) distributions. The *distribution function* and *tail* of a non-negative random variable X is $F(x) = P(X \leqslant x)$ and $\overline{F(x)} = P(X > x)$. A distribution $F(X)$ is *heavy-tailed* if there exist a number $\alpha, 0 < \alpha < 1$, such that $\lim_{x \to \infty} x^\alpha \overline{F(x)} = 1$. Many such distributions occur in real world (for example, in Physics, Economics, Internet) in both, space (distances) and time (durations). A standard example is the Pareto distribution $\overline{F(x)} = x^{-\alpha}, x \geqslant 1$, where $\alpha > 0$ is a parameter. (Cf. **distance decay** above.)

- **Distances in Medicine**

The **inter-occlusal distance**: in Dentistry, the distance between the occluding surfaces of the maxillary and mandibular teeth when the mandible is in physiologic rest position.

The **inter-arch distance**: in Dentistry, the vertical distance between the maxillary and mandibular arches. The **inter-ridge distance**: the vertical distance between the maxillary and mandibular ridges.

The **inter-proximal distance**: the **spacing distance** between adjacent teeth; *mesial drift* is the movement of the teeth slowly toward the front of the mouth with decreasing of the inter-proximal distance by wear.

The **inter-pediculate distance**: the distance between the vertebral pedicles as measured on the radiograph.

The **source-skin distance**: the distance from the focal spot on the target of the x-ray tube to the skin of the subject as measured along the central ray.

The **inter-aural distance**: the distance between the ears. The **inter-ocular distance**: the distance between the eyes.

The **anogenital distance**: the length of the *perineum*, i.e., the region between anus and genital area (the anterior base of the penis for a male). For a male it is normally twice what it is for a female; so, this distance is a measure of physical masculinity. Other such measures are second-to-forth digit (index to ring finger) ratio which is lower for men in the same population, and mental rotation ability, higher for men.

The **sedimentation distance** (or ESR, *erythrocyte sedimentation rate*): the distance red blood cells travel in one hour in a sample of blood as they settle to the bottom of a test tube. ESR indicates inflammation and increases in many diseases.

Examples of distances considered, in brain MRI imaging, for *cortical maps* (i.e., outer layer regions of cerebral hemispheres representing sensory inputs or motor outputs) are: MRI **distance map** from gray/white matter interface, **cortical distance** (say, between activation locations of spatially adjacent stimuli), *cortical thickness* and *lateralization metrics*.

- **Distances in Criminology**

The **geographic profiling** (or *geoforensic analysis*) aims to identify the spatial behavior (target selection and, especially, likely *point of origin*, i.e., the residence or workplace) of a serial criminal offender as it relates to the spatial distribution of linked crime sites.

The **offender's buffer zone** (or *coal-sack effect*) is an area surrounding *offender's heaven* (point of origin) from which little or no criminal activity will be observed; usually, such zone occurs for premeditated personal offenses. The primary streets and network arterials, that lead into the buffer zone, tend to intersect near the estimated offender's heaven. An 1 km buffer zone was found for UK serial rapists. Most personal offenses occur within about 2 km from offender's heaven, while property theft occur further away.

The **journey-to-crime decay function** is a graphical **distance curve** used to represent how the number of offenses committed by an offender decreases as the distance from his/her residence increases. Such functions are variations of center of gravity functions based on Newton's law of attraction between two bodies.

Given n crime sites (x_i, y_i), $1 \leqslant i \leqslant n$ (where x_i and y_i are latitude and longitude of i-th site), the *Newton–Swoope Model* predicts offender's heaven to be within the circle around the point $(\frac{\sum_i x_i}{n}, \frac{\sum_i y_i}{n})$ with search radius being

$$\sqrt{\frac{\max |x_{i_1} - x_{i_2}| \cdot \max |y_{i_1} - y_{i_2}|}{\pi (n-1)^2}},$$

where maxima are by (i_1, i_2), $1 \leqslant i_1 < i_2 \leqslant n$. The *Ganter–Gregory Circle Model* predicts offender's heaven to be within a circle around first offense crime site with diameter being the maximum distance between crime sites.

The *centrographic models* estimate offender's haven as a *center*, i.e., a point from which a given function of travel distances to all crime sites is minimized; the distances are the Euclidean distance, the Manhattan distance, the **wheel distance** (i.e., the actual travel path), perceived travel time, etc. Many of those models are reverse of Location Theory models aiming to maximize the placement of distribution facilities in order to minimize travel costs. Those models (*Voronoi polygons*, etc.) are based on the **nearness principle** (*least effort principle*).

- **Animal distances**

The **individual distance**: the distance which an animal attempts to maintain between itself and other animals.

The **group distance**: the distance which a group of animals attempts to maintain between it and other groups.

The **reaction distance**: the distance on which the animal reacts to the appearance of prey; *catching distance*: the distance on which the predator can strike a prey.

The **escape distance**: the distance on which the animal reacts on the appearance of a predator or dominating animal of the same species. This *flight initiation distance* is related to (shorter) corresponding *alert distance*.

The **nearest-neighbor distance**: more or less constant distance which an animal maintain, in directional movement of large groups (such as schools of fish or flocks of birds), from its immediate neighbors. The mechanism of *allelomimesis* ("do what your neighbor does") prevents the structural breakdown of a group and can generate seemingly intelligent evasive maneuvers in the presence of predators.

The **distance-to-shore**: the distance to the coastline used, for example, to study clustering of whale strandings by distorted echolocations, anomalies of magnetic field etc.

A **distance pheromone** is a soluble (for example, in the urine) substance emitted by an animal, as an olfactory chemosensory cue, in order to obtain mates. In contrast, a *contact pheromone* is such unsoluble substance; it coats the animal's body and is a contact cue.

- **Horse-racing distances**

The **horse-racing distances** are measured in the approximate length of a horse, i.e., about 8 feet (2.44 m). Winning margins are measured in **lengths**, ranging from half the length to the **distance**, i.e., more than 20 lengths. Smaller margins are: *short-head*, *head*, or *neck*. Also, the *hand*, i.e., 4 inches (10.2 cm), is used for measuring the height of horses.

- **Triathlon race distances**

The **Ironman distance** (started in Hawaii, 1978): 3.5 km swim followed by 180 km bike followed by 42.2 km (*marathon distance*) run.

The international **Olympic distance** (started in Sydney, 2000) is 1.5 km (*metric mile*), 40 km and 10 km of swim, bike and run, respectively.

Also used: the *sprint distance* (0.75 km, 20 km, 5 km), and the *long distance* (2 km, 80 km, 20 km).

- **Sabbath distance**

The **Sabbath distance** (or *rabbinical mile*) is a **range distance**: 2000 Talmudic cubits (1120.4 m) which an observant Jew should not exceed in a public thoroughfare from any given private place on the Sabbath day.

Other Talmudic length units are: day's march, parsa, stadium (40, 4, $\frac{4}{5}$ of rabbinical mile, respectively), and span, hasit, hand-breath, thumb, middle finger, little finger ($\frac{1}{2}$, $\frac{1}{3}$, $\frac{1}{6}$, $\frac{1}{24}$, $\frac{1}{30}$, $\frac{1}{36}$ of Talmudic cubit, respectively).

- **Galactocentric distance**

  The star's **Galactocentric distance** is its distance from the Galactic Center. The Sun's Galactocentric distance is about 8.5 kiloparsec, i.e., 27700 light-years.

- **Cosmic light horizon**

  The **cosmic light horizon** (or **Hubble distance**, *age of Universe*) is an increasing **range distance**: the maximum distance that a light signal could have traveled since Big Bang, the beginning of the Universe. At present, 13–14 $\times$ 10^9 light-years, i.e., about 46 $\times$ 10^{60} Planck lengths.

References

[Abel91] Abels H. *The Gallery Distance of Flags*, Order, Vol. 8, pp. 77–92, 1991.

[AAH00] Aichholzer O., Aurenhammer F. and Hurtado F. *Edge Operations on Non-crossing Spanning Trees*, Proc. 16th European Workshop on Computational Geometry CG'2000, pp. 121–125, 2000.

[AACL98] Aichholzer O., Aurenhammer F., Chen D.Z., Lee D.T., Mukhopadhyay A. and Papadopoulou E. *Voronoi Diagrams for Direction-Sensitive Distances*, Proc. 13th Symposium on Computational Geometry, ACM Press, New York, 1997.

[Aker97] Akerlof G.A. *Social Distance and Social Decisions*, Econometrica, Vol. 65, No. 5, pp. 1005–1027, 1997.

[Amar85] Amari S. *Differential-Geometrical Methods in Statistics*, Lecture Notes in Statistics, Springer-Verlag, 1985.

[Amba76] Ambartzumian R. *A Note on Pseudo-Metrics on the Plane*, Z. Wahrsch. Verw. Gebiete, Vol. 37, pp. 145–155, 1976.

[ArWe92] Arnold R. and Wellerding A. *On the Sobolev Distance of Convex Bodies*, Aeq. Mathematicae, Vol. 44, pp. 72–83, 1992.

[Badd92] Baddeley A.J. *Errors in Binary Images and an L^p Version of the Hausdorff Metric*, Nieuw Archief voor Wiskunde, Vol. 10, pp. 157–183, 1992.

[Bara01] Barabási A.L. *The Physics of the Web*, Physics World, July 2001.

[Barb35] Barbilian D. *Einordnung von Lobayschewskys Massenbestimmung in either gewissen allgemeinen Metrik der Jordansche Bereiche*, Casopis Mathematiky a Fysiky, Vol. 64, pp. 182–183, 1935.

[BLV05] Barceló C., Liberati S. and Visser M. *Analogue Gravity*, arXiv: gr-qc/0505065, Vol. 2, 2005.

[BLMN05] Bartal Y., Linial N., Mendel M. and Naor A. *Some Low Distorsion Metric Ramsey Problems*, Discrete and Computational Geometry, Vol. 33, pp. 27–41, 2005.

[Bata95] Batagelj V. *Norms and Distances over Finite Groups*, J. of Combinatorics, Information and System Sci., Vol. 20, pp. 243–252, 1995.

[Beer99] Beer G. *On Metric Boundedness Structures*, Set-Valued Analysis, Vol. 7, pp. 195–208, 1999.

[BGLVZ98] Bennet C.H., Gács P., Li M., Vitánai P.M.B. and Zurek W. *Information Distance*, IEEE Transactions on Information Theory, Vol. 44, Nr. 4, pp. 1407–1423, 1998.

[BGT93] Berrou C., Glavieux A. and Thitimajshima P. *Near Shannon Limit Error-correcting Coding and Decoding: Turbo-Codes*, Proc. of IEEE Int. Conf. on Communication, pp. 1064–1070, 1993.

[BFK99] Blanchard F., Formenti E. and Kurka P. *Cellular Automata in the Cantor, Besicovitch and Weyl Topological Spaces*, Complex Systems, Vol. 11, pp. 107–123, 1999.

[BCFS97] Block H.W., Chhetry D., Fang Z. and Sampson A.R. *Metrics on Permutations Useful for Positive Dependence*, J. of Statistical Planning and Inference, Vol. 62, pp. 219–234, 1997.

[Blum70] Blumenthal L.M. *Theory and Applications of Distance Geometry*, Chelsea Publ., New York, 1970.

[Borg86] Borgefors G. *Distance Transformations in Digital Images*, Comp. Vision, Graphic and Image Processing, Vol. 34, pp. 344–371, 1986.

[BrLi04] Bramble D.M. and Lieberman D.E. *Endurance Running and the Evolution of Homo*, Nature, Vol. 432, pp. 345–352, 2004.

[BKMR00] Broder A.Z., Kumar S. R., Maaghoul F., Raghavan P., Rajagopalan S., Stata R., Tomkins A. and Wiener G. *Graph Structure in the Web: Experiments and Models*, Proc. 9th WWW Conf., Amsterdam, 2000.

[BGL95] Brualdi R.A., Graves J.S. and Lawrence K.M. *Codes with a Poset Metric*, Discrete Math., Vol. 147, pp. 57–72, 1995.

[Brya85] Bryant V. *Metric Spaces: Iteration and Application*, Cambridge Univ. Press, 1985.

[Bull12] Bullough E. *"Psychical Distance" as a Factor in Art and as an Aestetic Principle*, British J. of Psychology, Vol. 5, pp. 87–117, 1912.

[BuIv01] Burago D., Burago Y. and Ivanov S. *A Course in Metric Geometry*, Amer. Math. Soc., Graduate Studies in Math., Vol. 33, 2001.

[BuKe53] Busemann H. and Kelly P.J. *Projective Geometry and Projective Metrics*, Academic Press, New York, 1953.

[Buse55] Busemann H. *The Geometry of Geodesics*, Academic Press, New York, 1955.

[BuPh87] Busemann H. and Phadke B.B. *Spaces with Distinguished Geodesics*, Marcel Dekker, New York, 1987.

[Cair01] Cairncross F. *The Death of Distance 2.0: How the Communication Revolution will Change our Lives*, Harvard Business School Press, 2nd edition, 2001.

[CSY01] Calude C.S., Salomaa K. and Yu S. *Metric Lexical Analysis*, Springer-Verlag, 2001.

[CJT93] Chartrand G., Johns G.L. and Tian S. *Detour Distance in Graphs*, Ann. of Discrete Math., Vol. 55, pp. 127–136, 1993.

[ChLu85] Cheng Y.C. and Lu S.Y. *Waveform Correlation by Tree Matching*, IEEE Trans. Pattern Anal. Machine Intell., Vol. 7, pp. 299–305, 1985.

[Chen72] Chentsov N.N. *Statistical Decision Rules and Optimal Inferences*, Nauka, Moscow, 1972.

[ChFi98] Chepoi V. and Fichet B. *A Note on Circular Decomposable Metrics*, Geom. Dedicata, Vol. 69, pp. 237–240, 1998.

[ChSe00] Choi S.W. and Seidel H.-P. *Hyperbolic Hausdorff Distance for Medial Axis Transform*, Research Report MPI-I-2000-4-003 of Max-Planck-Institute für Informatik, 2000.

[COR05] Collado M.D., Ortuno-Ortin I. and Romeu A. *Vertical Transmission of Consumption Behavior and the Distribution of Surnames*, http://www.econ.upf.es/docs/seminars/collado.pdf

[Cops68] Copson E.T. *Metric Spaces*, Cambridge Univ. Press, 1968.

[Corm03] Cormode G. *Sequence Distance Embedding*, PhD Thesis, Univ. of Warwick, 2003.

[CPQ96] Critchlow D.E., Pearl D.K. and Qian C. *The Triples Distance for Rooted Bifurcating Phylogenetic Trees*, Syst. Biology, Vol. 45, pp. 323–334, 1996.

[CCL01] Croft W. B., Cronon-Townsend S. and Lavrenko V. *Relevance Feedback and Personalization: A Language Modeling Perspective*, in DELOS-NSF Workshop on Personalization and Recommender Systems in Digital Libraries, pp. 49–54, 2001.

[DaCh88] Das P.P. and Chatterji B.N. *Knight's Distance in Digital Geometry*, Pattern Recognition Letters, Vol. 7, pp. 215–226, 1988.

[Das90] Das P.P. *Lattice of Octagonal Distances in Digital Geometry*, Pattern Recognition Letters, Vol. 11, pp. 663–667, 1990.

[DaMu90] Das P.P. and Mukherjee J. *Metricity of Super-Knight's Distance in Digital Geometry*, Pattern Recognition Letters, Vol. 11, pp. 601–604, 1990.

[Dau05] Dauphas N. *The U/Th Production Ratio and the Age of the Milky Way from Meteorites and Galactic Halo Stars*, Nature, Vol. 435, pp. 1203–1205, 2005.

[Day81] Day W.H.E. *The Complexity of Computing Metric Distances between Partitions*, Math. Social Sci., Vol. 1, pp. 269–287, 1981.

[DeDu03] Deza M.M. and Dutour M. *Cones of Metrics, Hemi-Metrics and Super-Metrics*, Ann. of European Academy of Sci., pp. 141–162, 2003.

[DeHu98] Deza M. and Huang T. *Metrics on Permutations, a Survey*, J. of Combinatorics, Information and System Sci., Vol. 23, Nos. 1–4, pp. 173–185, 1998.

[DeLa97] Deza M.M. and Laurent M. *Geometry of Cuts and Metrics*, Springer-Verlag, 1997.

[Dzha01] Dzhafarov E.N. *Multidimensional Fechnerian Scaling: Probability-Distance Hypothesis*, J. of Math. Psychology, Vol. 46, pp. 352–374, 2001.

[EhHa88] Ehrenfeucht A. and Haussler D. *A New Distance Metric on Strings Computable in Linear Time*, Discrete Applied Math., Vol. 20, pp. 191–203, 1988.

[EM98] *Encyclopedia of Mathematics*, Hazewinkel M. (ed.), Kluwer Academic Publ., 1998.

[Ernv85] Ernvall S. *On the Modular Distance*, IEEE Trans. Inf. Theory, Vol. IT-31, Nr. 4, pp. 521–522, 1985.

[EMM85] Estabrook G.F., McMorris F.R. and Meacham C.A. *Comparison of Undirected Phylogenetic Trees Based on Subtrees of Four Evolutionary Units*, Syst. Zool., Vol. 34, pp. 193–200, 1985.

[FaMu03] Farrán J.N. and Munuera C. *Goppa-Like Bounds for the Generalized Feng–Rao Distances*, Discrete Applied Math., Vol. 128, pp. 145–156, 2003.

[Faze99] Fazekas A. *Lattice of Distances Based on 3D-neighborhood Sequences*, Acta Mathematica Academiae Paedagogicae Nyiregyháziensis, Vol. 15, pp. 55–60, 1999.

[Ferg03] Ferguson N. *Empire: The Rise and Demise of the British World Order and Lessons for Global Power*, Basic Books, 2003.

[Frie98] Frieden B.R. *Physics from Fisher Information*, Cambridge Univ. Press, 1998.

[Gabi85] Gabidulin E.M. *Theory of Codes with Maximum Rank Distance*, Probl. Peredachi Inform., Vol. 21, No. 1, pp. 1–12, 1985.

[GaSi98] Gabidulin E.M. and Simonis J. *Metrics Generated by Families of Subspaces*, IEEE Transactions on Information Theory, Vol. 44, No. 3, pp. 1136–1141, 1998.

[GiOn96] Gilbert E.G. and Ong C.J. *Growth distances: New Measures for Object Separation and Penetration*, IEEE Transactions in Robotics, Vol. 12, No. 6, 1996.

[Gile87] Giles J.R. *Introduction to the Analysis of Metric Spaces*, Australian Math. Soc. Lecture Series, Cambridge Univ. Press, 1987.

[GoMc80] Godsil C.D. and McKay B.D. *The Dimension of a Graph*, Quart. J. Math. Oxford Series (2), Vol. 31, No. 124, pp. 423–427, 1980.

[GOJKK02] Goh K.I., Oh E.S., Jeong H., Kahng B. and Kim D. *Classification of Scale Free Networks*, Proc. Nat. Acad. Sci. USA, Vol. 99, pp. 12583–12588, 2002.

[Gopp71] Goppa V.D. *Rational Representation of Codes and* (L,g)*-codes*, Probl. Peredachi Inform., Vol. 7, No. 3, pp. 41–49, 1971.

[Coto82] Gotoh O. *An Improved Algorithm for Matching Biological Sequences*, J. of Molecular Biology, Vol. 162, pp. 705–708, 1982.

[GKC04] Grabowski R., Khosa P. and Choset H. *Development and Deployment of a Line of Sight Virtual Sensor for Heterogeneous Teams*, Proc. IEEE Int. Conf. on Robotics and Automation, New Orleans, 2004.

[Grub93] Gruber P.M. *The Space of Convex Bodies*, in *Handbook of Convex Geometry*, Gruber P.M. and Wills J.M. (eds.), Elsevier Sci. Publ., 1993.

[HSEFN95] Hafner J., Sawhney H.S., Equitz W., Flickner M. and Niblack W. *Efficient Color Histogram Indexing for Quadratic Form Distance Functions*, IEEE Transactions on Pattern Analysis and Machine Intelligence, Vol. 17, No. 7, pp. 729–736, 1995.

[Hall69] Hall E.T. *The Hidden Dimension*, Anchor Books, New York, 1969.

[Hami66] Hamilton W.R. *Elements of Quaternions*, 2nd edition, 1899–1901, enlarged by C.J. Joly, reprinted by Chelsea Publ., New York, 1969.

[HeMa02] Head K. and Mayer T. *Illusory Border Effects: Distance mismeasurement inflates estimates of home bias in trade*, CEPII Working Paper No 2002-01, 2002.

[Hemm02] Hemmerling A. *Effective Metric Spaces and Representations of the Reals*, Theoretical Comp. Sci., Vol. 284, No. 2, pp. 347–372, 2002.

[Hofs80] Hofstede G. *Culture's Consequences: International Differences in Work-Related Values*, Sage Publ., California, 1980.

[Hube94] Huber K. *Codes over Gaussian Integers*, IEEE Trans. Inf. Theory, Vol. 40, No. 1, pp. 207–216, 1994.

[Hube93] Huber K. *Codes over Eisenstein–Jacobi Integers*, Contemporary Math., Vol. 168, pp. 165–179, 1994.

[HFPMC02] Huffaker B., Fomenkov M., Plummer D.J., Moore D. and Claffy K., *Distance Metrics in the Internet*, IEEE Int. Telecommunication Symposium (ITS-2002), September 2002, http://www.caida.org/outreach/papers/2002/Distance

[InVe00] Indyk P. and Venkatasubramanian S. *Approximate Congruence in Nearly Linear Time*, http://www.research.att.com/~suresh/papers/hallj/hallj.pdf

[Isbe64] Isbell J. *Six Theorems about Metric Spaces*, Comment. Math. Helv., Vol. 39, pp. 65–74, 1964.

[IsKuPe90] Isham C.J., Kubyshin Y. and Penteln P. *Quantum Norm Theory and the Quantization of Metric Topology*, Class. Quantum Gravity, Vol. 7, pp. 1053–1074, 1990.

[IvSt95] Ivanova R. and Stanilov G. *A Skew-Symmetric Curvature Operator in Riemannian Geometry*, in Symposia Gaussiana, Conf. A, Behara M., Fritsch R. and Lintz R. (eds.), pp. 391–395, 1995.

[JWZ94] Jiang T., Wang L. and Zhang K. *Alignment of Trees – an Alternative to Tree Edit*, in *Combinatorial Pattern Matching, Lecture Notes in Computer Science*, Vol. 807, Crochemore M. and Gusfield D. (eds.), Springer-Verlag, 1994.

[Klei88] Klein R. *Voronoi Diagrams in the Moscow Metric*, Graph Theoretic Concepts in Comp. Sci., Vol. 6, 1988.

[Klei89] Klein R. *Concrete and Abstract Voronoi Diagrams*, Lecture Notes in Comp. Sci., Springer-Verlag, 1989.

[KlRa93] Klein D.J. and Randic M. *Resistance Distance*, J. of Math. Chemistry, Vol. 12, pp. 81–95, 1993.

[Koel00] Koella J.C. *The Spatial Spread of Altruism Versus the Evolutionary Response of Egoists*, Proc. Royal Soc. London, Series B, Vol. 267, pp. 1979–1985, 2000.

[KoSi88] Kogut B. and Singh H. *The Effect of National Culture on the Choice of Entry Mode*, J. of Int. Business Studies, Vol. 19, No. 3, pp. 411–432, 1988.

[KKN02] Kosheleva O., Kreinovich V. and Nguyen H.T. *On the Optimal Choice of Quality Metric in Image Compression*, Fifth IEEE Southwest Symposium on Image Analysis and Interpretation, 7–9 April 2002, Santa Fe. IEEE Comp. Soc. Digital Library, Electronic Edition, pp. 116–120, 2002.

[LaLi81] Larson R.C. and Li V.O.K. *Finding Minimum Rectilinear Distance Paths in the Presence of Barriers*, Networks, Vol. 11, pp. 285–304, 1981.

[LCLM04] Li M., Chen X., Li X., Ma B. and Vitányi P. *The Similarity Metric*, IEEE Trans. Inf. Theory, Vol. 50–12, pp. 3250–3264, 2004.

[LuRo76] Luczak E. and Rosenfeld A. *Distance on a Hexagonal Grid*, IEEE Trans. on Computers, Vol. 25, No. 5, pp. 532–533, 1976.

[MaMo95] Mak King-Tim and Morton A.J. *Distances between Traveling Salesman Tours*, Discrete Applied Math., Vol. 58, pp. 281–291, 1995.

[MaSt99] Martin W.J. and Stinson D.R. *Association Schemes for Ordered Orthogonal Arrays and (T, M, S)-nets*, Canad. J. Math., Vol. 51, pp. 326–346, 1999.

[McCa97] McCanna J.E. *Multiply-Sure Distances in Graphs*, Congressus Numerantium, Vol. 97, pp. 71–81, 1997.

[Melt91] Melter R.A. *A Survey of Digital Metrics*, Contemporary Math., Vol. 119, 1991.

[Monj98] Monjardet B. *On the Comparison of the Spearman and Kendall Metrics between Linear Orders*, Discrete Math., Vol. 192, pp. 281–292, 1998.

[Mura85] Murakami H. *Some Metrics on Classical Knots*, Math. Ann., Vol. 270, pp. 35–45, 1985.

[NeWu70] Needleman S.B. and Wunsh S.D. *A General Method Applicable to the Search of the Similarities in the Amino Acids Sequences of Two Proteins*, J. of Molecular Biology, Vol. 48, pp. 443–453, 1970.

[NiSu03] Nishida T. and Sugihara K. *FEM-Like Fast Marching Method for the Computation of the Boat-Sail Distance and the Associated Voronoi Diagram*, http://www.keisu.t.u-tokyo.ac.jp/Research/METR/2003/METR03-45.pdf

[OBS92] Okabe A., Boots B. and Sugihara K. *Spatial Tessellation: Concepts and Applications of Voronoi Diagrams*, Wiley, 1992.

[OSLM04] Oliva D., Samengo I., Leutgeb S. and Mizumori S. *A Subgective Distance between Stimuli: Quantifying the Metric Structure of Representations*, Neural Computation, Vol.17, No. 4, pp. 969–990, 2005.

[Orli32] Orlicz W. *Über eine gewisse Klasse von Raumen vom Typus B'*, Bull. Int. Acad. Pol. Series A, Vol. 8–9, pp. 207–220, 1932.

[OASM03] Ozer H., Avcibas I., Sankur B. and Memon N.D. *Steganalysis of Audio Based on Audio Quality Metrics*, in *Security and Watermarking of Multimedia Contents V*, Proc. of SPIE, Vol. 5020, pp. 55–66, 2003.

[Page65] Page E.S. *On Monte-Carlo Methods in Congestion Problem. 1. Searching for an Optimum in Discrete Situations*, J. Oper. Res., Vol. 13, No. 2, pp. 291–299, 1965.

[Petz96] Petz D. *Monotone Metrics on Matrix Spaces*, Linear Algebra Appl., Vol. 244, 1996.

[PM] PlanetMath.org, http://planetmath.org/encyclopedia/

[Rach91] Rachev S.T. *Probability Metrics and the Stability of Stochastic Models*, Wiley, New York, 1991.

[ReRo01] Requardt M. and Roy S. *(Quantum) Spacetime as a Statistical Geometry of Fuzzy Lumps and the Connection with Random Metric Spaces*, Class. Quantum Gravity, Vol. 18, pp. 3039–3057, 2001.

[RoTs96] Rosenbloom M.Y. and Tsfasman M.A. *Codes for the m-Metric*, Problems of Information Transmission, Vol. 33, No. 1, pp. 45–52, 1997.

[RoPf68] Rosenfeld A. and Pfaltz J. *Distance Functions on Digital Pictures*, Pattern Recognition, Vol. 1, pp. 33–61, 1968.

[RTG00] Rubner Y., Tomasi C. and Guibas L.J. *The Earth Mover's Distance as a Metric for Image Retrieval*, Int. J. of Comp. Vision, Vol. 40, No. 2, pp. 99–121, 2000.

[Rumm76] Rummel R.J. *Understanding Conflict and War*, Sage Publ., California, 1976.

[ScSk83] Schweizer B. and Sklar A. *Probabilistic Metric Spaces*, North-Holland, 1983.

[Selk77] Selkow S.M. *The Tree-to-Tree Editing Problem*, Inform. Process. Lett., Vol. 6, No. 6, pp. 184–186, 1977.

[ShKa97] Sharma B.D. and Kaushik M.L. *Error-Correcting Codes through a New Metric*, 41st Annual Conf. Int. Stat. Inst., New Delhi, 1997.

[Tai79] Tai K.-C. *The Tree-to-Tree Correction Problem*, J. of the Association for Comp. Machinery, Vol. 26, pp. 422–433, 1979.

[Tail04] Tailor B. *Introduction: How Far, How Near: Distance and Proximity in the Historical Imagination*, History Workshop J., Vol. 57, pp. 117–122, 2004.

[ToSa73] Tomimatsu A. and Sato H. *New Exact Solution for the Gravitational Field of a Spinning Mass*, Phys. Rev. Letters, Vol. 29, pp. 1344–1345, 1972.

[Var04] Vardi Y. *Metrics Useful in Network Tomography Studies*, Signal Processing Letters, Vol. 11, No. 3, pp. 353–355, 2004.

[VeHa01] Veltkamp R.C. and Hagendoorn M. *State-of-the-Art in Shape Matching*, in *Principles of Visual Information Retrieval*, Lew M. (ed.), pp. 87–119, Springer-Verlag, 2001.

[Watt99] Watts D.J. *Small Worlds: The Dynamics of Networks between Order and Randomness*, Princeton Univ. Press, 1999.

[Wein72] Weinberg S. *Gravitation and Cosmology: Principles and Applications of the General Theory of Relativity*, Wiley, New York, 1972.

[Weis99] Weisstein E.W. *CRC Concise Encyclopedia of Mathematics*, CRC Press, 1999.

[Well86] Wellens R.A. *Use of a Psychological Model to Assess Differences in Telecommunication Media*, in *Teleconferencing and Electronic Communication*, Parker L.A. and Olgren O.H. (eds.), pp. 347–361, Univ. of Wisconsin Extension, 1986.

[WFE] Wikipedia, the Free Encyclopedia, http://en.wikipedia.org

[WiMa97] Wilson D.R. and Martinez T.R. *Improved Heterogeneous Distance Functions*, J. of Artificial Intelligence Research, Vol. 6, p. 134, 1997.

[WoPi99] Wolf S. and Pinson M.H. *Spatial-Temporal Distortion Metrics for In-Service Quality Monitoring of Any Digital Video System*, Proc. of SPIE Int. Symp. on Voice, Video, and Data Commun., September 1999.

[Yian91] Yianilos P.N. *Normalized Forms for Two Common Metrics*, NEC Research Institute, Report 91-082-9027-1, 1991.

[Youn98] Young N. *Some Function-Theoretic Issues in Feedback Stabilisation*, Holomorphic Spaces, MSRI Publication, Vol. 33, 1998.

[YOI03] Yutaka M., Ohsawa Y. and Ishizuka M. *Average-Clicks: A New Measure of Distance on the World Wide Web*, Journal of Intelligent Information Systems, Vol. 20, No. 1, pp. 51–62, 2003.

[Zeli75] Zelinka B. *On a Certain Distance between Isomorphism Classes of Graphs*, Casopis Pest. Mat., Vol. 100, pp. 371–373, 1975.

Subject Index

(1, 2)-B-metric, 21
1-degree edit distance, 203
1-sum distance, 46
2-metric, 36
$(2k+1)$-gonal distance, 5
$(2k+1)$-gonal inequality, 5
$2k$-gonal distance, 4
$2k$-gonal inequality, 4
$2n$-gon metric, 243
(3, 4)-metric, 266
3D-chamfer metric, 266
4-metric, 248
6-metric, 248
8-metric, 248
α-scaled metric, 44
χ^2-distance, 184
δ-hyperbolic metric, 6
γ-metric, 333
$\natural$-Gordian distance, 121
$\natural$-inversion distance, 121
Σ-Hausdorff distance, 270
τ-distance space, 38
Υ-metric, 192

absolute summation distance, 212
absolute value metric, 155
ACME distance, 214
acoustic metric, 303
acoustics distances, 278
action at a distance (in Computing), 365
action at a distance (in Physics), 305
action distance, 366
activation distance, 365
actual distance, 310
additively weighted distance, 255
additively weighted power distance, 255
administrative cost distance, 285
aesthetic distance, 355
affine distance, 72
affine metric, 72
affine pseudo-distance, 72
airlift distance, 257
Albert quasi-metric, 4
Alcubierre metric, 337
Ali–Silvey distance, 186
alignment distance, 203
altitude, 315
amino gamma distance, 295
amino p-distance, 295
amino Poisson correction distance, 295
analytic metric, 113
Anderberg similarity, 225
Anderson–Darling distance, 179
angle distances between subspaces, 163
angular diameter distance, 322
angular distance, 247
angular semi-metric, 227
animal distances, 369
anti-de Sitter metric, 328
antinomy of distance, 355
antipodal extension distance, 46
antipodal metric space, 8
aphelion, 316
apoapsis distance, 316
apogee, 316
Apollonian metric, 78
Appert partially ordered distance, 43
approach space, 39
approximative human-scale distances, 350
area deviation, 122
arithmetic codes distance, 211
arithmetic r-norm metric, 154
as-crow-flies distance, 54
Asplund metric, 125
Assouad pseudo-distance, 5
Assouad–Nagata dimension, 11
astronomical distance, 313
astronomical length units, 346
asymptotic dimension, 11
Atiyah–Hitchin metric, 104
atomic radius, 306
attributed tree metrics, 206
average color distance, 264
average distance, 13

average distance property, 12
average linkage, 217
average square distance, 292
average-clicks Web quasi-distance, 285
azimuth, 315
Azukawa metric, 106
Azukawa semi-distance, 107

Baddeley–Molchanov distance function, 40
Baire metric, 150
Baire space, 33
ballistics distances, 302
Banach space, 53
Banach–Mazur compactum, 57
Banach–Mazur distance, 28
Banach–Mazur metric, 126
bar product distance, 209
Barbilian metric, 78
Bark spectral distance, 276
Baroni–Urbani–Buser similarity, 220
Barriola–Vilenkin monopole metric, 336
base-pair distance, 296
Berger metric, 88
Bergman metric, 102
Bergman p-metric, 167
Bertotti–Robinson metric, 336
Berwald metric, 93
Besicovitch distance, 167
Besicovitch semi-metric, 234
Besov metric, 168
betweenness centrality, 280
Bhattacharya distance 1, 184
Bhattacharya distance 2, 184
bi-discrepancy semi-metric, 181
bi-invariant metric, 134
bi-Lipschitz equivalent metrics, 18
bi-Lipschitz mapping, 17
binary Euclidean distance, 221
binary relation distance, 200
biotope distance, 298
Birnbaum–Orlicz distance, 182
Blaschke distance, 26
Blaschke metric, 90
Bloch metric, 167
block graph, 196
boat-sail distance, 257
Bogolubov–Kubo–Mori metric, 98
Bohr metric, 166
Bombieri metric, 158
bond distance, 306
Bondi radiating metric, 335
Bonnor dust metric, 333
Boolean metric space, 41
Borgefors metric, 266
bottleneck distance, 271
bounded box metric, 233
bounded metric, 22
bounded metric space, 22
box metric, 241
Braun–Blanquet similarity, 224
Bray–Curtis distance, 220
Bray–Curtis similarity, 219
Bregman distance, 182
Bregman quasi-distance, 171
British Rail metric, 242
Bryant metric, 93
bundle metric, 100
Burago–Ivanov metric, 243
Burbea–Rao distance, 182
Bures distance, 97
Bures metric, 97
Busemann metric, 69
Busemann metric of sets, 49
bush metric, 6
Bushell metric, 129

(c, p)-norm metric, 161
c-isomorphism of metric spaces, 18
C-metric, 137
C-quasi-conformal metrical mapping, 17
c-uniformly perfect metric space, 17
Calabi metric, 103
Calabi–Yau metric, 101
Canberra distance, 220
Cantor metric, 234
Carathéodory metric, 106
Carathéodory semi-distance, 106
Carnot–Carathéodory metric, 88
Cartan metric, 110
Catalan surface metric, 120
category of metric spaces, 19
caterpillar metric, 242
CAT(k) space, 65
Cauchy completion, 22
Cavalli–Sforza arc distance, 290
Cavalli–Sforza–Edvards chord distance, 290
Cayley metric, 152
Cayley–Klein–Hilbert metric, 77
Cayley–Menger matrix, 15
center of mass metric, 233
Central Park metric, 243
centroid distance, 244
centroid linkage, 217
cepstral distance, 276
chain metric, 153
chamfer metric, 251
chamfering distances, 266

Chebyshev center, 14
Chebyshev metric, 240
Chebyshev set, 14
chemical distance, 307
Chernoff distance, 186
chessboard metric, 248
chordal distance, 164
chordal metric, 156
circle metric, 246
circular cut semi-metric, 193
circular decomposable semi-metric, 194
circular metric in digraphs, 192
circular-railroad quasi-metric, 247
city distance, 257
city-block metric, 240
clarity similarity, 187
Clark distance, 222
clearance distance, 363
closest partition metric, 204
co-starring distance, 279
coarse embedding, 19
coarse isometry, 18
coarse-path metric, 3
coarsely equivalent metrics, 19
colatitude, 314
collaboration distance, 279
collision avoidance distance, 243
color component distances, 264
color distances, 263
common subgraph distance, 197
common supergraph distance, 197
commutation distance, 138
comoving distance, 320
compact quantum metric space, 29
compact space, 34
complete linkage, 217
complete metric, 22
complete metric space, 22
complete Riemannian metric, 87
completely normal space, 33
completely regular space, 32
complex Finsler metric, 104
complex modulus metric, 156
compoundly weighted distance, 255
cone distance, 261
cone metric, 129
configuration metric, 230
conformal metric, 79
conformal space, 85
conformal structure, 85
conformally invariant metric, 115
connected space, 33
Connes metric, 98
constrained edit distance, 203
constructive metric space, 30
continuous dual space, 175
contraction, 17
contraction distance, 198
convex distance function, 24
convex metric, 114
convolution metric, 180
Cook distance, 228
correlation distance, 227
correlation similarity, 226
cosh distance, 276
cosine distance, 227
cosine page distance, 286
cosine similarity, 227
cosmic light horizon, 370
cosmological distance, 319
cosmological distance ladder, 323
countably-normed space, 35
covariance similarity, 226
covering radius, 12
Cramer–von Mises distance, 179
creepage distance, 363
Crnkovic–Drachma distance, 179
crossing-free edge move metric, 200
crossing-free edge slide metric, 200
crossing-free transformation metrics, 200
crossover metric, 204
cultural distance, 352
cut semi-metric, 193
cycloidal metric, 88
cylindrical distance, 260
Czekanowsky–Dice distance, 219

D-chromatic number, 12
D-distance graph, 196
Daniels–Guilbaud semi-metric, 152
Dayhoff–Eck distance, 295
de Sitter metric, 328
Death of Distance, 354
declination, 314
decomposable semi-metric, 194
degenerate metric, 84
degree-2 distance, 204
delta distance, 121
Desarguesian space, 64
design distance, 209
detour distance, 191
DeWitt supermetric, 93
diagonal metric, 84
dialectology distance, 278
digital metric, 248
digital volume metric, 252
dilated metric, 44

dilation, 18
directed Hausdorff distance, 26
Dirichlet metric, 168
discrepancy metric, 181
discrete dynamic system, 15
discrete metric, 21
discrete space, 35
dispersion similarity, 225
dissimilarity, 2
distance, 2
distance constant of operator algebra, 175
distance cues, 359
distance curve, 366
distance decay, 366
distance decoder, 197
distance degree regular graph, 196
distance distribution, 210
distance in a river, 257
distance labelling scheme, 197
distance map, 14
distance matrix, 15
distance measuring equipment, 362
distance model of altruism, 300
distance modulus, 322
distance monotone metric space, 8
distance of negative type, 4
distance of travel, 310
distance on building, 41
distance pheromone, 369
distance ratio metric, 79
distance running model, 300
distance space, 2
distance to frontier, 353
distance transform, 268
distance up to nearest integer, 49
distance vision, 358
distance-balanced graph, 196
distance-based machine learning, 229
distance-decreasing semi-metric, 105
distance-hereditary graph, 196
distance-polynomial graph, 195
distance-regular graph, 195
distance-related shots, 360
distance-transitive graph, 195
distances between people, 350
distances for genome permutations, 297
distances in Criminology, 368
distances in Medicine, 367
distances in Seismology, 311
distances in Stereoscopy, 360
distancing, 354
distortion, 18
Dodge–Shiode WebX quasi-distance, 285
dogkeeper distance, 166
dominating metric, 21
Dps distance, 290
Drapal–Kepka distance, 142
Dress dimension, 20
drift distance, 280
DRP-metrics, 285
dual distance, 209
dual metrics, 56
Dudley metric, 180
Dugundji–Ugi chemical distance, 308
Duncan metric, 150
DXing, 362

Earth Mover distance, 266
earthquake distance, 312
eccentricity, 13
ecliptic latitude, 315
ecliptic longitude, 315
Eddington–Robertson metric, 330
Edgar k-mer distance, 296
Edgar–Ludwig metric, 335
edge distance, 198
edge jump distance, 199
edge move distance, 198
edge rotation distance, 199
edge shift distance, 199
edge slide distance, 199
editing compression metric, 147
editing metric, 24
editing metric with moves, 147
effective free distance, 210
effective metric space, 30
effective trade distance, 352
Eggleston distance, 125
egocentric distance, 359
Eguchi–Hanson de Sitter metric, 336
Eguchi–Hanson metric, 104
Ehrenfeucht–Haussler semi-metric, 246
Einstein metric, 327
element of best approximation, 14
Ellenberg similarity, 218
ellipsoid metric, 118
elliptic metric, 73
elliptic orbit distance, 315
emotional distance, 351
energy, 13
engineer semi-metric, 178
Enomoto–Katona metric, 25
entropy metric, 187
epicentral distance, 312
equicut semi-metric, 193
equidistant metric, 21
equivalent metrics, 21

Erdős-type distance problems, 240
escape distance, 369
Euclidean metric, 54
Euclidean space, 54
Euler angle metric, 232
even cut semi-metric, 193
evolutionary distance, 289
exponential distance, 254
extended metric, 3
extended real line metric, 155
extension distances, 46
extremal metric, 115

f-divergence of Csizar, 183
F-norm metric, 50
F-rotation distance, 200
F-statistics distance, 291
F^*-metric, 51
facility layout distances, 244
factor distance, 147
Faith similarity, 224
Fano metric, 215
Far Near Distance, 356
Feng–Rao distance, 209
Ferrand metric, 79
Ferrand second metric, 80
fidelity similarity, 183
figure of merit quasi-distance, 270
finite l_p-semi-metric, 194
finite nuclear norm metric, 173
finite polynomial metric space, 196
finite subgroup metric, 144
Finsler metric, 91
first-countable space, 33
Fisher distance, 96
Fisher information metric, 95
Fisher–Rao metric, 95
flag metric, 69
flat metric, 114
flat space, 86
Florian metric, 124
flower-shop metric, 243
focus distances, 360
force-distance curve, 366
forest-fire distance, 259
formation metric, 234
forward quasi-distance, 280
four-point inequality, 6
four-point inequality metric, 6
fractal, 11
fractional distance, 221
framework, 190
Fréchet metric, 27
Fréchet permutation metric, 153
Fréchet product metric, 47
Fréchet surface metric, 116
Fréchet–Nikodym–Aronszayn distance, 25
free distance, 210
free fall distance, 301
free space metric, 233
French Metro metric, 241
Frobenius distance, 164
Frobenius norm metric, 161
frontier metric, 233
Fubini–Study metric, 102
functional transform metric, 45
Funk metric, 92
fuzzy set distance, 291

G-distance, 114
G-invariant metric, 87
G-norm metric, 139
g-transform metric, 45
Gabidulin–Simonis metrics, 213
galactocentric distance, 370
Galilean distance, 327
gallery distance of flags, 145
gallery distance on building, 41
gallery metric, 24
gap distance, 164
gap metric, 235
Garson et al. hybridization metric, 294
gate extension distance, 46
gauge metric, 140
GCSS metric, 331
Gehring metric, 79
Gendron et al. distance, 297
genealogical distance, 289
general linear group semi-metric, 139
generalized absolute value metric, 155
generalized chordal metric, 157
generalized Fano metric, 216
generalized G-Hausdorff metric, 27
generalized metric, 41
generalized Riemannian spaces, 86
generalized torus semi-metric, 140
generational distance, 236
genetic distance, 289
genome distance, 297
geodesic, 8
geodesic distance, 63
geodesic metric space, 63
geodetic graph, 195
geographic distance, 284
Gibbons–Manton metric, 108
Gleason distance, 77
Gleason similarity, 219

Gluskin–Khrabrov distance, 28
Gödel metric, 334
Godsil–McKay dimension, 11
Goldstein and al. distance, 292
Golmez partially ordered distance, 43
Goppa designed minimum distance, 209
Gordian distance, 121
Gower similarity 2, 226
Gower–Legendre similarity, 225
GPS pseudo-distance, 362
Gram matrix, 16
graphic metric, 191
gray-scale image distances, 265
great circle distance, 309
great circle metric, 73
greatest agreement subtree distance, 206
Grenander distance, 127
grid metric, 248
Grishin distance, 296
Gromov δ-hyperbolic inequality, 6
Gromov product similarity, 7
Gromov–Hausdorff metric, 27
Gromov-hyperbolic metric space, 65
ground distance, 310
group norm metric, 50
growth distance, 126
gyroradius, 306

(h, Δ)-metric, 114
half-Apollonian metric, 78
half-plane projective metric, 68
half-space parabolic distance, 80
Hamann similarity, 223
Hamiltonian metric, 191
Hamming metric, 25
Hamming metric on permutations, 151
hard metric, 343
Hardy metric, 168
harmonic mean similarity, 184
Hausdorff dimension, 10
Hausdorff distance up to G, 271
Hausdorff metric, 26
Hausdorff space, 32
healing length, 303
hedgehog metric, 242
Heisenberg metric, 140
helical surface metric, 119
Hellinger distance, 222
Hellinger metric, 183
Hermitian elliptic metric, 74
Hermitian G-metric, 37
Hermitian hyperbolic metric, 75
Hermitian metric, 100
hexagonal Hausdorff metric, 252
hexagonal metric, 248
Hilbert cube metric, 48
Hilbert metric, 57
Hilbert projective metric, 68
Hilbert projective semi-metric, 128
Hilbert space, 56
Hilbert–Schmidt norm metric, 173
Hilditch–Rutovitz metric, 266
Hirst–St-Onge similarity, 282
histogram intersection quasi-distance, 264
histogram quadratic distance, 265
historical distance, 355
Hodge metric, 102
Hofer metric, 109
Hölder metric, 170
homeomorphic metric spaces, 17
homogeneous metric space, 16
hopping distance, 305
horizon distance, 309
horizontal distance, 310
horse-racing distances, 369
hour angle, 314
Hubble distance, 320
hydrodynamic radius, 307
hyper-Kähler metric, 103
hyperbolic Hausdorff distance, 272
hyperbolic metric, 74
hyperboloid metric, 118
hyperconvexity, 9
hypercube metric, 191
hypermetric, 5
hypermetric inequality, 5
hyperspace, 35
hypocentral distance, 312

image compression L_p-metric, 265
immunological distance, 289
imperial length measures, 343
indefinite metric, 37
indel distance, 214
indel metric, 147
indicator metric, 177
indiscrete semi-metric, 21
indiscrete space, 35
induced intrinsic metric, 45
induced metric, 21
induced norm metric, 160
infinite distance, 358
information metric, 187
injective hull, 20
injective metric space, 19
inner product space, 56
instrument distances, 362

integral metric, 165
interaction distance, 305
interatomic distance, 306
interchange distance, 214
interionic distance, 307
interior metric, 45
internal metric, 45
International Metric System, 342
Internet AS metric, 284
Internet IP metric, 284
intersection distance, 219
interval distance, 214
interval norm metric, 136
intrinsic metric, 63
inverse triangle inequality, 325
inverse-square distance laws, 304
inversion distance, 121
inversive distance, 247
ISO paper sizes, 344
isolation-by-distance, 299
isometric subgraph, 195
isometry, 16
Itakura–Saito quasi-distance, 276
Ivanov–Petrova metric, 87

J-metric, 38
Jaccard similarity of community, 218
Janis–Newman–Wincour metric, 330
Jaro similarity, 148
Jaro–Winkler similarity, 149
Jeffrey divergence, 185
Jensen–Shannon divergence, 185
Jiang–Conrath distance, 282
Jin–Nei gamma distance, 293
Johnson distance, 25
Joyner–Boore distance, 312
Jukes–Cantor nucleotide distance, 293
Jukes–Cantor protein distance, 295

k-oriented distance, 129
k-th minimum distance, 210
Kähler metric, 101
Kähler supermetric, 109
Kähler–Einstein metric, 101
Kalmanson semi-metric, 194
Kaluza–Klein metric, 338
Kantorovich–Mallows–Monge–Wasserstein metric, 188
Karlsruhe metric, 242
Kaufman semi-metric, 164
Kawaguchi metric, 93
Kemeny distance, 141
Kendall τ distance, 152
Kendall shape distance, 269
Kerr metric, 329
Kerr–Newman metric, 330
Kerr–Schild metric, 329
Kimura 2-parameter distance, 293
Kimura protein distance, 296
king-move metric, 248
kinship distance, 291
Klatt slope metric, 277
knight metric, 251
Kobayashi metric, 105
Kobayashi semi-distance, 105
Kobayashi–Busemann metric, 106
Kobayashi–Royden metric, 105
Kolmogorov–Smirnov metric, 179
Kolmogorov–Tihomirov dimension, 10
Korányi metric, 140
Kottler metric, 329
Koutras–McIntosh metric, 334
Kristeva non-metric space, 355
Kropina metric, 92
Kruglov distance, 182
Kruskal–Szekeres metric, 328
Kuiper distance, 179
Kulczynski similarity 1, 220
Kulczynski similarity 2, 220
Kullback–Leibler distance, 185
Kurepa–Fréchet distance, 43
Ky-Fan metric K, 177
Ky-Fan metric K^*, 178
Ky-Fan norm metric, 162

L_p-Hausdorff distance, 27
l_p-metric, 54
L_p-metric, 55
L_p-metric between densities, 178
L_p-metric between variables, 177
L_p-space, 55
L_p-Wasserstein distance, 188
Λ-metric, 114
Lasker distance, 299
latitude, 313
lattice metric, 248
lattice valuation metric, 143
Le Cam distance, 181
Leacock–Chodorow similarity, 281
Lebesgue covering dimension, 10
Lebesgue space, 55
Lee metric, 25
left-invariant metric, 135
length of metric space, 11
length space, 63
Lerman semi-metric, 164
Lesk similarities, 282

Levenstein metric, 147
Levi-Civita metric, 332
Levy metric, 179
Lewis metric, 331
lexicographic metric, 153
lift metric, 242
light-travel distance, 322
Lin similarity, 282
Lindelöf space, 33
line-line distance, 48
line-of sight comoving distance, 320
line-of-sight distance, 362
linguistic distance, 278
link distance, 244
linkage metric, 217
Lipschitz distance, 28
Lipschitz distance between measures, 29
Lipschitz mapping, 17
localization metric, 234
locally compact space, 34
locally convex space, 34
log likelihood ratio quasi-distance, 276
logarithmic distance, 254
long range dependence, 366
long range order, 305
longitude, 313
lookback distance, 320
Lorentz distance, 327
Lorentz metric, 326
Lorentz–Minkowski distance, 327
Lorentzian distance, 221
Lorentzian metric, 89
lower Minkowski dimension, 10
loxodromic distance, 309
luminosity distance, 322
lunar distance, 315
Lund–Regge supermetric, 94

(m, s)-simplex inequality, 36
(m, s)-super-metric, 36
m-hemi-metric, 36
M-relative metric, 241
m-simplex inequality, 36
Ma–Deng–Manjunath distance, 264
Macbeath metric, 125
magic constant, 12
Mahalanobis distance, 223
Mahalanobis semi-metric, 178
Malecot's distance model, 299
Manhattan metric, 240
Mannheim distance, 212
map distance, 297
map's distance, 310
marital distance, 299
marking metric, 148
Martin cepstrum distance, 277
Martin distance, 163
Martin metric, 150
mass-distance function, 366
matrix norm metric, 159
matrix p-norm metric, 160
Matusita semi-metric, 183
maximum absolute column metric, 160
maximum polygon distance, 127
McClure–Vitale metric, 124
mean censored Euclidean distance, 222
mean molecular radius, 308
mean width metric, 123
measure metric, 25
mechanic distance, 301
medial axis and skeleton, 268
Meehl distance, 222
Menger convexity, 9
metabolic distance, 297
metallic distance, 306
meter-related terms, 343
metric, 3
metric ball, 7
metric basis, 14
metric between angles, 247
metric between directions, 247
metric between intervals, 140
metric bornology, 40
metric center, 13
metric compactum, 23
metric cone, 15
metric connection, 100
metric convexity, 9
metric coordinates, 14
metric curve, 8
metric diameter, 13
metric dimension, 9
metric entropy, 9
metric extension, 21
metric interval, 8
metric length measures, 343
metric median, 13
metric of bounded curvature, 114
metric of motions, 139
metric of negative curvature, 113
metric of non-negative curvature, 113
metric of non-positive curvature, 113
metric of positive curvature, 113
metric projection, 14
metric property, 16
metric quadrangle, 8
metric radius, 12

metric Ramsey number, 18
metric recursion, 216
metric recursion of a MAP decoding, 216
metric scheme, 3
metric space, 3
metric sphere, 7
metric subspace, 21
metric tensor, 83
metric topology, 7
metric transform, 15, 44
metric tree, 8
metric triangle, 8
metric triple, 23
metric with alternating curvature, 114
metrically homogeneous metric space, 16
metrication, 342
metrics between partitions, 142
metrics on determinant lines, 109
metrics on natural numbers, 154
metrics on Riesz space, 144
metrizable space, 35
metrization theorems, 7
midset, 14
millibot metrics, 233
Milnor metric, 109
minimum distance, 208
Minkowski difference, 126
Minkowski metric, 326
Minkowski–Bouligand dimension, 9
Minkowski–Hölder distance, 69
Minkowskian metric, 69
Mirkin–Tcherny distance, 141
Misner metric, 337
miss distance, 361
modified Hausdorff distance, 271
modified Minkowskian distance, 232
modular distance, 138
modular metric space, 8
modulus metric, 79
Moho distance, 311
molecular rms radius, 308
Monge–Kantorovich metric, 188
monkey saddle metric, 120
monomorphism norm metric, 137
monotone metric, 96
Montanari metric, 266
Moore space, 32
moral distance, 354
Morisita–Horn similarity, 228
Morris–Thorne metric, 336
Moscow metric, 242
motion, 16
motion planning metric, 230
Motyka similarity, 219
multi-cut semi-metric, 194
multiplicatively weighted distance, 254
multiply-sure distance, 193
multiset metric, 147

natural metric, 155
natural norm metric, 160
nautical distance, 309
nautical length units, 344
nD-neighborhood sequence metric, 250
near-metric, 3
nearest neighbor interchange metric, 204
nearness principle, 357
Needleman–Wunsch–Sellers metric, 149
negative ♮-reflection distance, 121
negative reflection distance, 121
negative type inequality, 4
Nei minimum genetic distance, 291
Nei standard genetic distance, 291
Nei–Tajima–Tateno distance, 290
neighborhood sequence metric, 249
network distances, 256
network metric, 244
network tomography metrics, 286
Neyman χ^2-distance, 184
nice metric, 245
Nikodym metric, 124
No. of differences, 292
No. of protein differences, 295
non-Archimedean metric, 6
non-commutative metric space, 30
non-degenerate metric, 84
non-linear elastic matching distances, 267
non-linear Hausdorff metric, 272
norm metric, 23, 51
normal space, 32
normalized compression distance, 148
normalized information distance, 147
normalized information metric, 188
normalized l_p-distance, 222
nuclear norm metric, 173

obstacle avoidance metric, 233
Ochiai–Otsuka similarity, 227
octagonal metric, 249
odd cut semi-metric, 193
offset distance, 361
Oliva et al. perception distance, 298
operating distance, 364
operator norm metric, 172
optical distance, 303
order norm metric, 137
order of congruence, 14

oriented cut quasi-semi-metric, 195
oriented multi-cut quasi-semi-metric, 195
oriented triangle inequality, 4
Orlicz metric, 169
Orlicz–Lorentz metric, 169
Orloci distance, 227
Ornstein $\overline{d}$-metric, 188
orthodromic distance, 309
Osserman Lorentzian metric, 90
Osserman metric, 87
Osterreicher semi-metric, 183
overall non-dominated vector ratio, 237

(p, q)-relative metric, 241
p-adic metric, 155
p-distance, 292
p-relative metric, 156
p-th order mean Hausdorff distance, 270
packing radius, 12
PAM distance, 295
parabolic distance, 80
paracompact space, 34
parallax distance, 322
parameterized curves distance, 267
Paris metric, 242
part metric, 168
partial Hausdorff quasi-distance, 271
partial metric space, 38
partially ordered distance, 42
path metric, 23, 191
path quasi-metric in digraphs, 191
path similarity, 281
path-generated metric, 250
pattern difference, 226
Pearson χ^2-distance, 184
Pearson ϕ similarity, 226
peeper distance, 258
penetration depth distance, 126
Penrose shape distance, 221
Penrose size distance, 221
perfect matching distance, 205
periapsis distance, 315
perigee, 316
perihelion, 316
perimeter deviation, 122
perimeter distance, 244
periodic metric, 245
permutation editing metric, 152
permutation norm metric, 139
phone distances, 277
phonetic word distance, 278
phylogenetic distance, 289
physical distance, 321
piano movers distance, 243
plane wave metric, 334
Poincaré metric, 76
Poincaré–Reidemeister metric, 109
point-line distance, 49
point-plane distance, 49
point-set distance, 26
polar distance, 314
Polish space, 23
polychromatic number, 12
polyhedral chain metric, 130
polyhedral metric, 130
polynomial bar metric, 158
polynomial bracket metric, 158
polynomial norm metric, 158
Pompeiu–Eggleston metric, 123
Pompeiu–Hausdorff–Blaschke metric, 123
pose distance, 233
poset distance, 212
positive ♮-reflection distance, 121
positive reflection distance, 121
positively homogeneous metric, 135
Pospichal–Kvasnička chemical distance, 308
post-office metric, 242
power distance, 254
power (p, r)-distance, 221
power series metric, 59
power transform metric, 45
precise distance measurement, 361
prefix-Hamming metric, 149
Prevosti–Ocana–Alonso distance, 290
primary-satellite distances, 316
prime distance, 248, 266
prime number distance, 49
probabilistic metric space, 40
probability distance, 178
probability-distance hypothesis, 298
Procrustes distance, 269
product metric, 24, 47
product norm metric, 137
projective determination of a metric, 70
projective metric, 24, 67
projectively flat space, 66
Prokhorov metric, 180
proper distance, 321
proper metric space, 23
proper motion distance, 321
propinquity effect, 351
proportional transport semi-metric, 267
protective action distance, 363
proximity fuse, 361
proximity graph of metric space, 9
proximity sensors, 361
proximity space, 38

pseudo-distance, 5
pseudo-elliptic distance, 74
pseudo-hyperbolic distance, 77
pseudo-metric, 2
pseudo-Riemannian metric, 88
pseudo-sphere metric, 119
Ptolemaic graph, 196
Ptolemaic inequality, 5
Ptolemaic metric, 5
Pythagorean distance, 54

q-gram similarity, 149
Q-metric, 115
Q_0-difference, 226
quadric metric, 117
quality metrics, 365
quantum metrics, 338
quartet distance, 205
quasi-distance, 4
quasi-hyperbolic metric, 78
quasi-isometry, 18
quasi-metric, 4
quasi-semi-metric, 4
quaternion metric, 157
quefrency-weighted cepstral distance, 277
quickest path metric, 244
Quillen metric, 109
quotes on "near-far" distances, 357
quotient metric, 58
quotient norm metric, 137

radar discrimination distance, 246
radar distance, 323
radar screen metric, 242
radial metric, 242
radio distance measurement, 362
radio distances, 362
Rand similarity, 223
Randers metric, 92
range distances, 364
range of fundamental forces, 305
range of molecular forces, 307
rank distance, 213
Rao metric, 95
raspberry picker metric, 242
Ray–Singler metric, 109
real tree, 20
rectilinear distance with barriers, 244
rectilinear metric, 240
reflection distance, 268
regular G-metric, 37
regular metric, 112
regular space, 32
Reidemeister metric, 109
Reignier distance, 142
Reissner–Nordström metric, 329
relative metric, 156
relaxed four-point inequality, 6
relaxed four-point inequality metric, 6
relaxed tree-like metric, 192
rendez-vous number, 12
Rényi distance, 186
resemblance, 3
resistance metric, 192
resistor-average distance, 186
Resnik similarity, 282
restricted edit distance, 203
retract subgraph, 195
reversal metric, 153
Reynolds–Weir–Cockerham distance, 291
Ricci-flat metric, 87
Riemannian distance, 85
Riemannian metric, 24, 84
Riesz norm metric, 57
right ascension, 314
right logarithmic derivative metric, 97
right-angle metric, 240
right-invariant metric, 134
ring semi-metric, 141
RMS log spectral distance, 275
RNA structural distance, 296
road travel distance, 310
Roberts similarity, 218
Robertson–Walker metric, 331
Robinson–Foulds metric, 204
Robinson–Foulds weighted metric, 204
robot displacement metric, 232
Roger distance, 290
Roger–Tanimoto similarity, 224
rook metric, 251
Rosenbloom–Tsfasman distance, 213
Rosenbloom–Tsfasman metric, 162
rotating C-metric, 338
rotation distance, 231
rotation surface metric, 118
routing protocol semi-metric, 286
RTT-distance, 285
Rummel sociocultural distances, 352
rupture distance, 312
Russel–Rao similarity, 224
Ruzicka similarity, 218

Sabbath distance, 369
safe distancing, 364
Sangvi χ^2 distance, 291
Sasakian metric, 110
scalar and vectorial metrics, 144

scale-free network, 279
scaled Euclidean distance, 231
Schatten norm metric, 162
Schatten p-class norm metric, 174
Schoenberg transform metric, 45
Schwartz metric, 171
Schwarzschild metric, 328
Schwarzschild–anti-de Sitter metric, 329
Schwarzschild–de Sitter metric, 329
search-centric change metrics, 286
second-countable space, 33
segmented signal-to-noise ratio, 274
seismogenic depth distance, 312
Seittenranta metric, 79
Selkow distance, 203
semantic proximity, 281
semi-metric, 2
semi-metrics on resemblances, 164
semi-norm semi-metric, 52
semi-pseudo-Riemannian metric, 90
semi-Riemannian metric, 90
separable metric space, 22
separable space, 33
separation distance, 126, 184
set-set distance, 26
Sgarro metric, 144
Shannon distance, 187
Shannon information quantity, 187
Sharma–Kaushik distance, 211
sharp triangle inequality, 38
Shen metric, 92
Shephard metric, 124
short mapping, 19
shortest path distance, 63
shortest path distance with obstacles, 256
Shriver–Boerwinkle stepwise distance, 292
shuttle metric, 242
Sibony metric, 107
Sibony semi-distance, 107
Sierpinski metric, 154
sight distances, 363
signed distance function, 15
signed reversal metric, 153
similarity, 2
similarity ratio, 228
Simone Weil distance, 356
simplicial metric, 130
simplicial supermetric, 94
Simpson similarity, 224
single linkage, 217
size function distance, 268
size-distance paradox, 358
skew distance, 259
skew divergence, 185
skin depth, 306
skip distance, 310
Skorokhod metric, 182
Skorokhod–Billingsley metric, 181
slant distance, 310
slope distance, 310
snowmobile distance, 258
$SO(3)$-invariant metric, 103
Sobolev distance, 124
Sobolev metric, 170
social distance, 351
soft metric, 343
Sokal–Sneath similarity 1, 223
Sokal–Sneath similarity 2, 224
Sokal–Sneath similarity 3, 224
solvent migration distance, 363
Souslin space, 23
space form, 86
space of constant curvature, 86
space of elliptic type, 64
space of geodesics, 63
space over algebra, 42
space-time link distance, 312
spacing, 237
spacing distances, 365
spanning distance, 126
spatial coherence length, 304
Spearman ρ distance, 151
Spearman footrule distance, 152
Spearman rank correlation, 228
spectral magnitude-phase distortion, 275
spectral phase distance, 275
sphere metric, 117
spherical distance, 309
spherical extension distance, 46
spherical metric, 73
spheroid metric, 118
spheroidal distance, 309
split semi-metric, 193
splitting-merging distance, 204
spray distance, 363
standoff distance, 361
static isotropic metric, 330
statistical metric, 97
Steiner distance of a set, 197
Steiner ratio, 13
Steinhaus distance, 25
Steinhaus metric, 125
Stenzel metric, 103
Stepanov distance, 167
Stephens et al. shared allele distance, 289
stop-loss metric of order m, 178
stopping sight distance, 363

Stover distance, 278
straight space, 64
straight spinning string metric, 333
straight-line distance, 310
strip projective metric, 68
strong triangle inequality, 2
subgraph-supergraph distances, 197
subgraphs distances, 201
subjective distance, 359
submetry, 19
subordinate norm metric, 160
subtree prune and regraft distance, 205
subway metric, 245
sup metric, 166
super-knight metric, 251
surname distance model, 299
suspension metric, 129
Swedenborg heaven distances, 356
swept volume, 230
symbolic distance effect, 359
symmetric, 2
symmetric χ^2-distance, 223
symmetric χ^2-measure, 222
symmetric difference metric, 25
symmetric metric space, 16
symmetric surface area deviation, 125
syntenic distance, 297
Szulga metric, 180

t-bottleneck distance, 271
t-spanner, 197
t-truncated metric, 44
T_0-space, 32
T_1-space, 32
T_2-space, 32
T_3-space, 32
T_4-space, 33
T_5-space, 33
Tajima–Nei distance, 293
Tamura 3-parameter distance, 294
Tamura–Nei distance, 294
tangent distance, 269
Tanimoto distance, 26
Taub–NUT de Sitter metric, 335
Taub–NUT metric, 104
taxicab metric, 240
taxonomic distance, 298
technology distances, 353
technology-related distancing, 354
Teichmüller metric, 107
template metric, 122
tensegrity, 190
tensor norm metric, 58
terminal distance, 301
tetrahedron inequality, 36
Thompson part metric, 128
tight extension, 20
tight span, 20
time series video distances, 272
Titius–Bode law, 316
Tits metric, 66
token-based similarities, 149
tolerance distance, 310
Tomimatsu–Sato metric, 334
top-down edit distance, 203
topological dimension, 10
Topsøe distance, 186
toroidal metric, 246
torus metric, 119
total convexity, 9
total variation metric, 178
totally bounded metric space, 22
totally bounded space, 34
trace norm metric, 162
trace-class norm metric, 174
transactional distance, 355
transform metric, 44
translation discrete metric, 135
translation distance, 231
translation invariant metric, 51
translation proper metric, 135
transmission distance, 362
transverse comoving distance, 321
traveling salesman tours distances, 200
tree bisection-reconnection metric, 205
tree edit distance, 202
tree rotation distance, 199
tree-like metric, 192
triangle inequality, 2
triangle operation, 40
triathlon race distances, 369
triples distance, 205
trivial metric, 21
true anomaly, 316
truncated metric, 193
turning function distance, 267
Tversky similarity, 225
typographical length units, 345

UC metric space, 23
Ulam metric, 152
ultrametric, 6
unicity distance, 211
uniform metric, 166
uniform space, 39
unit cost edit distance, 203
unit quaternions metric, 232

unitary metric, 55
universal metric space, 30
Urysohn space, 30

Vajda–Kus semi-metric, 183
valuation metric, 58
Van Stockum metric, 332
Van-der-Waals contact distance, 307
variable exponent space metrics, 170
vehicle distances, 364
vertical distance between lines, 48
vertical separation distance, 363
Verwer metric, 266
very small length units, 345
Victor–Purpura distance, 298
video quality metrics, 272
Vidyasagar metric, 235
Vinnicombe metric, 236
visibility shortest path distance, 256
vision distances, 357
Voronoi distance for arcs, 259
Voronoi distance for areas, 260
Voronoi distance for circles, 260
Voronoi distance for line segments, 259
Voronoi distances of order m, 261
Voronoi generation distance, 253
Vuorinen metric, 79

Ward linkage, 217
warped product metric, 48
Wasserstein metric, 188
weak quasi-metric, 4
weak ultrametric, 5
Web hyperlink quasi-metric, 285
Web similarity metrics, 285
Weierstrass metric, 77
weighted cut metric, 252
weighted Euclidean distance, 231
weighted Hamming metric, 149
weighted likelihood ratio distance, 276
weighted Manhattan distance, 232
weighted Minkowskian distance, 231
weighted path metric, 191
weighted tree metric, 192
weighted word metric, 136
Weyl distance, 167
Weyl metric, 333
Weyl semi-metric, 234
Weyl–Papapetrou metric, 332
Weyl–Petersson metric, 108
wheel distance, 310
Wigner–Yanase metric, 98
Wigner–Yanase–Dyson metrics, 98
Wils metric, 334
word metric, 135
word page distance, 286
workspace metric, 231
Wu metric, 107
Wu–Palmer similarity, 282

Yule Q similarity, 225
Yule Y similarity of colligation, 225

Zelinka distance, 197
Zelinka tree distance, 198
zenith angle, 315
zero bias metric, 155
Zipoy–Voorhees metric, 333
Zoll metric, 87
Zolotarev semi-metric, 180

The authors welcome distance definitions that may be suitable to include in future editions (Michel.Deza@er.s.fr).